Other volumes in this series:

Volume 1 **Power Circuit Breaker Theory and Design** C.H. Flurscheim (Editor)
Volume 4 **Industrial Microwave Heating** A.C. Metaxas and R.J. Meredith
Volume 7 **Insulators for High Voltages** J.S.T. Looms
Volume 8 **Variable Frequency AC Motor Drive Systems** D. Finney
Volume 10 **SF_6 Switchgear** H.M. Ryan and G.R. Jones
Volume 11 **Conduction and Induction Heating** E.J. Davies
Volume 13 **Statistical Techniques for High Voltage Engineering** W. Hauschild and W. Mosch
Volume 14 **Uninterruptible Power Supplies** J. Platts and J.D. St Aubyn (Editors)
Volume 15 **Digital Protection for Power Systems** A.T. Johns and S.K. Salman
Volume 16 **Electricity Economics and Planning** T.W. Berrie
Volume 18 **Vacuum Switchgear** A. Greenwood
Volume 19 **Electrical Safety: a guide to causes and prevention of hazards** J. Maxwell Adams
Volume 21 **Electricity Distribution Network Design, 2nd Edition** E. Lakervi and E.J. Holmes
Volume 22 **Artificial Intelligence Techniques in Power Systems** K. Warwick, A.O. Ekwue and R. Aggarwal (Editors)
Volume 24 **Power System Commissioning and Maintenance Practice** K. Harker
Volume 25 **Engineers' Handbook of Industrial Microwave Heating** R.J. Meredith
Volume 26 **Small Electric Motors** H. Moczala *et al.*
Volume 27 **AC-DC Power System Analysis** J. Arrillaga and B.C. Smith
Volume 29 **High Voltage Direct Current Transmission, 2nd Edition** J. Arrillaga
Volume 30 **Flexible AC Transmission Systems (FACTS)** Y-H. Song (Editor)
Volume 31 **Embedded generation** N. Jenkins *et al.*
Volume 32 **High Voltage Engineering and Testing, 2nd Edition** H.M. Ryan (Editor)
Volume 33 **Overvoltage Protection of Low-Voltage Systems, Revised Edition** P. Hasse
Volume 36 **Voltage Quality in Electrical Power Systems** J. Schlabbach *et al.*
Volume 37 **Electrical Steels for Rotating Machines** P. Beckley
Volume 38 **The Electric Car: Development and future of battery, hybrid and fuel-cell cars** M. Westbrook
Volume 39 **Power Systems Electromagnetic Transients Simulation** J. Arrillaga and N. Watson
Volume 40 **Advances in High Voltage Engineering** M. Haddad and D. Warne
Volume 41 **Electrical Operation of Electrostatic Precipitators** K. Parker
Volume 43 **Thermal Power Plant Simulation and Control** D. Flynn
Volume 44 **Economic Evaluation of Projects in the Electricity Supply Industry** H. Khatib
Volume 45 **Propulsion Systems for Hybrid Vehicles** J. Miller
Volume 46 **Distribution Switchgear** S. Stewart
Volume 47 **Protection of Electricity Distribution Networks, 2nd Edition** J. Gers and E. Holmes
Volume 48 **Wood Pole Overhead Lines** B. Wareing
Volume 49 **Electric Fuses, 3rd Edition** A. Wright and G. Newbery
Volume 50 **Wind Power Integration: Connection and system operational aspects** B. Fox *et al.*
Volume 51 **Short Circuit Currents** J. Schlabbach
Volume 52 **Nuclear Power** J. Wood
Volume 53 **Condition Assessment of High Voltage Insulation in Power System Equipment** R.E. James and Q. Su
Volume 55 **Local Energy: Distributed generation of heat and power** J. Wood
Volume 56 **Condition Monitoring of Rotating Electrical Machines** P. Tavner, L. Ran, J. Penman and H. Sedding
Volume 57 **The Control Techniques Drives and Controls Handbook, 2nd Edition** B. Drury
Volume 58 **Lightning Protection** V. Cooray (Editor)
Volume 59 **Ultracapacitor Applications** J.M. Miller
Volume 62 **Lightning Electromagnetics** V. Cooray
Volume 63 **Energy Storage for Power Systems, 2nd Edition** A. Ter-Gazarian
Volume 65 **Protection of Electricity Distribution Networks, 3rd Edition** J. Gers
Volume 66 **High Voltage Engineering Testing, 3rd Edition** H. Ryan (Editor)
Volume 67 **Multicore Simulation of Power System Transients** F.M. Uriate
Volume 68 **Distribution System Analysis and Automation** J. Gers
Volume 69 **The Lightening Flash, 2nd Edition** V. Cooray (Editor)
Volume 70 **Economic Evaluation of Projects in the Electricity Supply Industry, 3rd Edition** H. Khatib
Volume 72 **Control Circuits in Power Electronics: Practical issues in design and implementation** M. Castilla (Editor)
Volume 73 **Wide Area Monitoring, Protection and Control Systems: The enabler for Smarter Grids** A. Vaccaro and A. Zobaa (Editors)
Volume 74 **Power Electronic Converters and Systems: Frontiers and applications** A. M. Trzynadlowski (Editor)
Volume 75 **Power Distribution Automation** B. Das (Editor)
Volume 76 **Power System Stability: Modelling, analysis and control** A.A. Sallam and B. Om P. Malik
Volume 78 **Numerical Analysis of Power System Transients and Dynamics** A. Ametani (Editor)
Volume 79 **Vehicle-to-Grid: Linking electric vehicles to the smart grid** J. Lu and J. Hossain (Editors)
Volume 81 **Cyber-Physical-Social Systems and Constructs in Electric Power Engineering** S. Suryanarayanan, R. Roche and T.M. Hansen (Editors)
Volume 82 **Periodic Control of Power Electronic Converters** F. Blaabjerg, K.Zhou, D. Wang and Y. Yang

IET ENERGY ENGINEERING SERIES 238

AI for Status Monitoring of Utility Scale Batteries

Volume 86 **Advances in Power System Modelling, Control and Stability Analysis** F. Milano (Editor)
Volume 87 **Cogeneration: Technologies, Optimisation and Implentation** C. A. Frangopoulos (Editor)
Volume 88 **Smarter Energy: from Smart Metering to the Smart Grid** H. Sun, N. Hatziargyriou, H. V. Poor, L. Carpanini and M. A. Sánchez Fornié (Editors)
Volume 89 **Hydrogen Production, Separation and Purification for Energy** A. Basile, F. Dalena, J. Tong and T.N.Veziroğlu (Editors)
Volume 90 **Clean Energy Microgrids** S. Obara and J. Morel (Editors)
Volume 91 **Fuzzy Logic Control in Energy Systems with Design Applications in Matlab/Simulink®** İ. H. Altaş
Volume 92 **Power Quality in Future Electrical Power Systems** A. F. Zobaa and S. H. E. A. Aleem (Editors)
Volume 93 **Cogeneration and District Energy Systems: Modelling, Analysis and Optimization** M. A. Rosen and S. Koohi-Fayegh
Volume 94 **Introduction to the Smart Grid: Concepts, technologies and evolution** S.K. Salman
Volume 95 **Communication, Control and Security Challenges for the Smart Grid** S.M. Muyeen and S. Rahman (Editors)
Volume 96 **Industrial Power Systems with Distributed and Embedded Generation** R Belu
Volume 97 **Synchronized Phasor Measurements for Smart Grids** M.J.B. Reddy and D.K. Mohanta (Editors)
Volume 98 **Large Scale Grid Integration of Renewable Energy Sources** A. Moreno-Munoz (Editor)
Volume 100 **Modeling and Dynamic Behaviour of Hydropower Plants** N. Kishor and J. Fraile-Ardanuy (Editors)
Volume 101 **Methane and Hydrogen for Energy Storage** R. Carriveau and D. S-K. Ting
Volume 104 **Power Transformer Condition Monitoring and Diagnosis** A. Abu-Siada (Editor)
Volume 106 **Surface Passivation of Industrial Crystalline Silicon Solar Cells** J. John (Editor)
Volume 107 **Bifacial Photovoltaics: Technology, applications and economics** J. Libal and R. Kopecek (Editors)
Volume 108 **Fault Diagnosis of Induction Motors** J. Faiz, V. Ghorbanian and G. Joksimović
Volume 109 **Cooling of Rotating Electrical Machines: Fundamentals, modelling, testing and design** D. Staton, E. Chong, S. Pickering and A. Boglietti
Volume 110 **High Voltage Power Network Construction** K. Harker
Volume 111 **Energy Storage at Different Voltage Levels: Technology, integration, and market aspects** A.F. Zobaa, P.F. Ribeiro, S.H.A. Aleem and S.N. Afifi (Editors)
Volume 112 **Wireless Power Transfer: Theory, Technology and Application** N.Shinohara
Volume 114 **Lightning-Induced Effects in Electrical and Telecommunication Systems** Y. Baba and V. A. Rakov
Volume 115 **DC Distribution Systems and Microgrids** T. Dragičević, F.Blaabjerg and P. Wheeler
Volume 116 **Modelling and Simulation of HVDC Transmission** M. Han (Editor)
Volume 117 **Structural Control and Fault Detection of Wind Turbine Systems** H.R. Karimi
Volume 119 **Thermal Power Plant Control and Instrumentation: The control of boilers and HRSGs, 2nd Edition** D. Lindsley, J. Grist and D. Parker
Volume 120 **Fault Diagnosis for Robust Inverter Power Drives** A. Ginart (Editor)
Volume 121 **Monitoring and Control using Synchrophasors in Power Systems with Renewables** I. Kamwa and C. Lu (Editors)
Volume 123 **Power Systems Electromagnetic Transients Simulation, 2nd Edition** N. Watson and J. Arrillaga
Volume 124 **Power Market Transformation** B. Murray
Volume 125 **Wind Energy Modeling and Simulation Volume 1: Atmosphere and plant** P.Veers (Editor)
Volume 126 **Diagnosis and Fault Tolerance of Electrical Machines, Power Electronics and Drives** A.J. M. Cardoso
Volume 128 **Characterization of Wide Bandgap Power Semiconductor Devices** F. Wang, Z. Zhang and E.A. Jones
Volume 129 **Renewable Energy from the Oceans: From wave, tidal and gradient systems to offshore wind and solar** D. Coiro and T. Sant (Editors)
Volume 130 **Wind and Solar Based Energy Systems for Communities** R. Carriveau and D. S-K. Ting (Editors)
Volume 131 **Metaheuristic Optimization in Power Engineering** J. Radosavljević
Volume 132 **Power Line Communication Systems for Smart Grids** I.R.S Casella and A. Anpalagan
Volume 134 **Hydrogen Passivation and Laser Doping for Silicon Solar Cells** B. Hallam and C. Chan (Editors)
Volume 139 **Variability, Scalability and Stability of Microgrids** S. M. Muyeen, S. M. Islam and F. Blaabjerg (Editors)
Volume 143 **Medium Voltage DC System Architectures** B. Grainger and R. D. Doncker (Editors)
Volume 145 **Condition Monitoring of Rotating Electrical Machines** P. Tavner, L. Ran, C. Crabtree
Volume 146 **Energy Storage for Power Systems, 3rd Edition** A.G. Ter-Gazarian
Volume 147 **Distribution Systems Analysis and Automation 2nd Edition** J. Gers
Volume 151 **SiC Power Module Design: Performance, robustness and reliability** A. Castellazzi and A. Irace (Editors)
Volume 152 **Power Electronic Devices: Applications, failure mechanisms and reliability** F Iannuzzo (Editor)
Volume 153 **Signal Processing for Fault Detection and Diagnosis in Electric Machines and Systems** M. Benbouzid (Editor)
Volume 155 **Energy Generation and Efficiency Technologies for Green Residential Buildings** D. Ting and R. Carriveau (Editors)
Volume 156 **Lithium-ion Batteries Enabled by Silicon Anodes** C. Ban and K. Xu (Editors)
Volume 157 **Electrical Steels, 2 Volumes** A. Moses, K. Jenkins, Philip Anderson and H. Stanbury
Volume 158 **Advanced Dielectric Materials for Electrostatic Capacitors** Q Li (Editor)
Volume 159 **Transforming the Grid Towards Fully Renewable Energy** O. Probst, S. Castellanos and R. Palacios (Editors)
Volume 160 **Microgrids for Rural Areas: Research and case studies** R.K. Chauhan, K. Chauhan and S.N. Singh (Editors)
Volume 161 **Artificial Intelligence for Smarter Power Systems: Fuzzy Logic and Neural Networks** M. G. Simoes
Volume 165 **Digital Protection for Power Systems 2nd Edition** Salman K Salman

Volume 166 **Advanced Characterization of Thin Film Solar Cells** N. Haegel and M Al-Jassim (Editors)
Volume 167 **Power Grids with Renewable Energy** Storage, integration and digitalization A. A. Sallam and B. OM P. Malik
Volume 169 **Small Wind and Hydrokinetic Turbines** P. Clausen, J. Whale and D. Wood (Editors)
Volume 170 **Reliability of Power Electronics Converters for Solar Photovoltaic Applications** F. Blaabjerg, A.l Haque, H. Wang, Z. Abdin Jaffery and Y. Yang (Editors)
Volume 171 **Utility-scale Wind Turbines and Wind Farms** A. Vasel-Be-Hagh and D. S.-K. Ting
Volume 172 **Lighting interaction with Power Systems, 2 volumes** A. Piantini (Editor)
Volume 174 **Silicon Solar Cell Metallization and Module Technology** T. Dullweber (Editor)
Volume 180 **Protection of Electricity Distribution Networks, 4th Edition** J. Gers and E. Holmes
Volume 811 **Modelling and Simulation of Complex Power Systems** A. Monti and A. Benigni
Volume 182 **Surge Protection for Low Voltage Systems** A. Rousseau (Editor)
Volume 184 **Compressed Air Energy Storage: Types, systems and applications** D. Ting and J. Stagner
Volume 186 **Synchronous Reluctance Machines: Analysis, optimization and applications** N. Bianchi, C. Babetto and G. Bacco
Volume 191 **Electric Fuses: Fundamentals and new applications 4th Edition** N. Nurse, A. Wright and P. G. Newbery
Volume 193 **Overhead Electric Power Lines: Theory and practice** S. Chattopadhyay and A. Das
Volume 194 **Offshore Wind Power Reliability, availability and maintenance, 2nd edition** P. Tavner
Volume 196 **Cyber Security for Microgrids** S. Sahoo, F. Blaajberg and T. Dragicevic
Volume 198 **Battery Management Systems and Inductive Balancing** A. Van den Bossche and A. Farzan Moghaddam
Volume 199 **Model Predictive Control for Microgrids: From power electronic converters to energy management** J. Hu, J. M. Guerrero and S. Islam
Volume 204 **Electromagnetic Transients in Large HV Cable Networks: Modeling and calculations** Ametani, Xue, Ohno and Khalilnezhad
Volume 208 **Nanogrids and Picogrids and their Integration with Electric Vehicles** S. Chattopadhyay
Volume 211 **Blockchain Technology for Smart Grids: Implementation, management and security** Gururaj H L, Ravi K V, F. Flammini, H. Lin, Goutham B, Sunil K. B R and C Sivapragash
Volume 212 **Battery State Estimation: Methods and Models** S. Wang
Volume 215 **Industrial Demand Response: Methods, best practices, case studies, and applications** H. H. Alhelou, A. Moreno-Muñoz and P. Siano (Editors)
Volume 213 **Wide Area Monitoring of Interconnected Power Systems 2nd Edition** A. R. Messina
Volume 217 **Advances in Power System Modelling, Control and Stability Analysis 2nd Edition** F. Milano (Editor)
Volume 225 **Fusion-Fission Hybrid Nuclear Reactors: For enhanced nuclear fuel utilization and radioactive waste reduction** W. M. Stacey
Volume 905 **Power system protection, 4 volumes**

AI for Status Monitoring of Utility Scale Batteries

Shunli Wang, Kailong Liu, Yujie Wang, Daniel-Ioan Stroe, Carlos Fernandez and Josep M. Guerrero

The Institution of Engineering and Technology

Published by The Institution of Engineering and Technology, London, United Kingdom

The Institution of Engineering and Technology is registered as a Charity in England & Wales (no. 211014) and Scotland (no. SC038698).

First published 2022

The Institution of Engineering and Technology
Futures Place
Kings Way, Stevenage
Herts, SG1 2UA., United Kingdom

www.theiet.org

British Library Cataloguing in Publication Data
A catalogue record for this product is available from the British Library

ISBN 978-1-83953-738-7 (hardback)
ISBN 978-1-83953-739-4 (PDF)

Typeset in India by Exeter Premedia Services Private Limited

Cover image: Andriy Onufriyenko via Getty Images

Contents

About the Authors

Shunli Wang is a professor at Southwest University of Science and Technology, Sichuan, China. He is an expert in the field of new energy research. He is the head of NELab, conducting modeling and state estimation strategy research for lithium-ion batteries. He has undertaken over 40 projects and 30 patents, published over 100 research papers, and won 20 awards such as the Young Scholar, and Science & Technology Progress Awards.

Kailong Liu is an assistant professor at the University of Warwick, UK. His research experience lies at the intersection of AI and electrochemical energy storage applications, especially data science in battery management. His current research is focusing on the development of AI strategies for battery applications.

Yujie Wang is an associate professor with the Department of Automation, University of Science and Technology of China. He received his PhD degree in control science and engineering from the University of Science and Technology of China in 2017. He has co-authored over 60 SCI journal papers in battery-related topics. His research interests include energy saving and new energy vehicle technology, complex system modelling, simulation and control, fuel cell system management and optimal control.

Daniel-Ioan Stroe is an associate professor with AAU Energy, Aalborg University, Denmark and the leader of the Batteries research group. He received his PhD degree in lifetime modeling of Lithium-ion batteries from Aalborg University in 2010. He has co-authored one book and over 150 scientific peer-review publications on battery performance, modeling and state estimation. His research interests include energy storage systems for grid and e-mobility, lithium-based batteries' testing, modeling, lifetime estimation, and their diagnostics.

Carlos Fernandez is a is a senior lecturer at Robert Gordon University, Scotland. He received his PhD in electrocatalytic reactions from The University of Hull, and then worked as a consultant technologist in Hull and in a post-doctoral position in Manchester. His research interests include analytical chemistry, sensors and materials and renewable energy.

Josep M. Guerrero is a full professor with AAU Energy, Aalborg University, Denmark. He is the director for the Center for Research on Microgrids (CROM). He has published more than 800 journal articles in the fields of microgrids and renewable energy systems, which have been cited more than 80,000 times. His research interests focus on different microgrid aspects, including hierarchical and cooperative control, and energy management systems.

Foreword

The development of battery system modeling is a strategic measure to alleviate the world energy crisis and achieve the goal of carbon neutrality. Focusing on artificial intelligence (AI) for battery status monitoring of large-capacity lithium-ion batteries, this book systematically describes the key AI-based technologies in battery state estimation together with modeling and parameter identification. It mainly includes various AI methods used in equivalent circuit modeling, parameter identification, state of charge estimation, state of health evaluation, state of power determination, state of energy calculation, and remaining useful life prediction. The AI methods are introduced for the core state estimation and prediction processes, including machine learning, artificial neural networks, deep learning, etc. It provides practical references for the design and application of large-scale lithium-ion battery systems, which are based on the technical battery application requirements compiled from the status monitoring perspective of lithium-ion batteries. Various AI-based methods and techniques are introduced for state estimation and have rich examples that can serve as a guide and reference material. It can be used as a textbook for control science and engineering, automation, electrical engineering, and other related majors in colleges and universities. These methods can also be used as a reference for related researchers in the field of new energy measurement and control in large-scale energy storage for electric vehicles, micro and urban grids, hospitals, universities, etc.

Preface

At present, the economy is in the stage of rapid growth, and the demand for fossil energy is higher and higher. Therefore, the threat of energy depletion is earlier and more serious than before. In recent years, new energy has developed very fast in the automotive field. Under the dual problems of economic energy shortage and deteriorating environment, many countries have spent a lot of financial and human resources on the research and popularization of new energy. The withdrawal of traditional fuel vehicles has become an inevitable trend, and the development of electric vehicles will be unstoppable.

Power battery is one of the key technologies of electric vehicle research and development. Through the relevant analysis and research of lithium-ion batteries, it is found that compared with other material batteries, the lithium-ion battery has obvious advantages: monomer has relatively high working voltage, long charge, and discharge life, high specific energy, low self-discharge rate, no memory, clean and environmental protection, etc. The lithium-ion battery has gradually become the first choice of electric vehicle power battery. The use and protection of a power battery cannot be separated from the battery management system (BMS). A fully functional BMS can ensure the safe and efficient operation of the power battery. The BMS mainly includes the following main functions: battery status monitoring, battery status analysis, battery safety protection, energy control management, and battery information management.

With the continuous progress of relevant algorithms and research in the field of AI, its relevant algorithms are more and more used in battery state estimation. The continuous improvement of data accumulation and computing power will greatly promote the practical application of AI algorithms in the field of power battery state estimation. In this book, Chapter 1 mainly introduces the reasons and advantages of AI technology in the field of batteries. Chapter 2 mainly introduces the classification, basic characteristics, and application of large-capacity lithium-ion batteries. Chapter 3 mainly introduces the battery equivalent modeling and parameter identification based on AI technology. Chapter 4 mainly introduces the selection of AI technology and the monitoring of basic battery parameters. Chapter 5 mainly introduces the state of charge (SOC) estimation of lithium-ion batteries based on a variety of AI-related algorithms in detail. In Chapter 6, the state of health (SOH), state of power (SOP), state of energy (SOE), and remaining useful life (RUL) estimation of lithium-ion battery based on various AI-related algorithms are introduced in detail, and the estimation results are analyzed and summarized in detail. In Chapter 7, the examples and case studies for utility-scale battery operation are introduced. In Chapter 8, the summary and prospect of battery condition monitoring based on AI technology are described in detail.

List of contributors

Bowen Li Southwest University of Science and Technology, Mianyang 621010, China

Carlos Fernandez Robert Gordon University, Aberdeen AB10-7GJ, United Kingdom

Chao Wang Southwest University of Science and Technology, Mianyang 621010, China

Chenyu Zhu Southwest University of Science and Technology, Mianyang 621010, China

Chuangshi Qi Southwest University of Science and Technology, Mianyang 621010, China

Chunmei Yu Southwest University of Science and Technology, Mianyang 621010, China

Chuyan Zhang Southwest University of Science and Technology, Mianyang 621010, China

Dan Deng Southwest University of Science and Technology, Mianyang 621010, China

Donglei Liu Southwest University of Science and Technology, Mianyang 621010, China

Daniel-Ioan Stroe Aalborg University, Pontoppidanstraede 111 9220 Aalborg East, Denmark

Fan Wu Southwest University of Science and Technology, Mianyang 621010, China

Haiying Gao Southwest University of Science and Technology, Mianyang 621010, China

Haotian Shi Southwest University of Science and Technology, Mianyang 621010, China

Heng Zhou Southwest University of Science and Technology, Mianyang 621010, China

Hong Xu Southwest University of Science and Technology, Mianyang 621010, China

Huan Li Southwest University of Science and Technology, Mianyang 621010, China

Ji Wu	Hefei University of Technology, Hefei 230027, China
Jialu Qiao	Southwest University of Science and Technology, Mianyang 621010, China
Jian Wang	Southwest University of Science and Technology, Mianyang 621010, China
Jiani Zhou	Southwest University of Science and Technology, Mianyang 621010, China
Jiawei Zeng	Southwest University of Science and Technology, Mianyang 621010, China
Jie Cao	Southwest University of Science and Technology, Mianyang 621010, China
Jin Li	Southwest University of Science and Technology, Mianyang 621010, China
Jingsong Qiu	Southwest University of Science and Technology, Mianyang 621010, China
Jinhao Meng	Sichuan University, Chengdu 610065, China
Junjie Yang	Southwest University of Science and Technology, Mianyang 621010, China
Kailong Liu	WMG, University of Warwick, Coventry, CV4 7AL, United Kingdom
Ke Liu	Southwest University of Science and Technology, Mianyang 621010, China
Lei Chen	Southwest University of Science and Technology, Mianyang 621010, China
Lili Xia	Southwest University of Science and Technology, Mianyang 621010, China
Long Zhou	University of Shanghai for Science and Technology, Shanghai 200093, China
Mengyun Zhang	Southwest University of Science and Technology, Mianyang 621010, China
Mingfang He	Southwest University of Science and Technology, Mianyang 621010, China
Nan Hai	Southwest University of Science and Technology, Mianyang 621010, China
Paul	Southwest University of Science and Technology, Mianyang 621010, China
Peng Yu	Southwest University of Science and Technology, Mianyang 621010, China

Pu Ren	Southwest University of Science and Technology, Mianyang 621010, China
Qingyun Ma	Southwest University of Science and Technology, Mianyang 621010, China
Ran Xiong	Southwest University of Science and Technology, Mianyang 621010, China
Renjun Feng	Southwest University of Science and Technology, Mianyang 621010, China
Shunli Wang	Sichuan University, Chengdu 610065, China
Shunli Wang	Southwest University of Science and Technology, Mianyang 621010, China
Siyu Jin	Aalborg University, Pontoppidanstraede 111 9220 Aalborg East, Denmark
Tao Long	Southwest University of Science and Technology, Mianyang 621010, China
Wanting Peng	Southwest University of Science and Technology, Mianyang 621010, China
Weihao Shi	Southwest University of Science and Technology, Mianyang 621010, China
Weijia Xiao	Southwest University of Science and Technology, Mianyang 621010, China
Weikang Ji	Southwest University of Science and Technology, Mianyang 621010, China
Wenhua Xu	Southwest University of Science and Technology, Mianyang 621010, China
Xi He	Southwest University of Science and Technology, Mianyang 621010, China
Xianyi Jia	Southwest University of Science and Technology, Mianyang 621010, China
Xianfeng Shen	Southwest University of Science and Technology, Mianyang 621010, China
Xianpei Chen	Southwest University of Science and Technology, Mianyang 621010, China
Xianyong Xiao	Sichuan University, Chengdu 610065, China
Xiaoyong Yang	Southwest University of Science and Technology, Mianyang 621010, China
Xiao Yang	Southwest University of Science and Technology, Mianyang 621010, China

Xiaoxia Li	Southwest University of Science and Technology, Mianyang 621010, China
Xueyi Hao	Southwest University of Science and Technology, Mianyang 621010, China
Yang Li	Southwest University of Science and Technology, Mianyang 621010, China
Yangtao Wang	Southwest University of Science and Technology, Mianyang 621010, China
Yanxin Xie	Southwest University of Science and Technology, Mianyang 621010, China
Yawen Liang	Southwest University of Science and Technology, Mianyang 621010, China
Yixing Zhang	Southwest University of Science and Technology, Mianyang 621010, China
Yongcun Fan	Southwest University of Science and Technology, Mianyang 621010, China
Yunlong Shang	Shandong University, Jinan 250100, China
Yujie Wang	University of Science and Technology of China, Hefei 230027, China
Yuyang Liu	Southwest University of Science and Technology, Mianyang 621010, China
Zhi Wang	Southwest University of Science and Technology, Mianyang 621010, China

Chapter 1
Introduction

1.1 Motivation for utility-scale battery deployment

Some researchers estimate that the fossil fuel reserves can support the world's demand for power for 50 years, whereas others say it will be 100–120 years [1]. Though there is no accurate number about how many years the fuel reserves will deplete, at least it shows that the current energy source is shortly facing a serious shortage [2]. The infrastructure for the current power grid is rapidly aging [3]. The radical changes in the environment show the great negative effects brought by the consumption of fossil resources. Based on these factors, the concept of "smart grids" has been advanced, requiring the modernization of the distribution and the transmission system [4]. One of the priorities of the smart grid is to deploy and integrate distributed resources, including renewable energy resources. One way in which utility-scale battery storage is particularly helpful is in being paired with renewable resources, such as solar or wind farms [5]. These resources provide a forecastable, predictable amount of generation every hour but, due to the nature of sun and wind, do not generate at their maximum capacity every hour. By pairing utility-scale batteries with solar and wind, resource developers can smoothen the output from these resources and ensure that renewable energy is injected onto the grid at the times when it is most needed [6]. Renewable resources are substantial for the long term and can be used to generate electricity with low or zero CO_2 emissions. In some European countries, more than 30% of electricity comes from renewable energy including photovoltaic (PV) power, wind, and other resources.

Meanwhile, renewable resources become more and more obtainable and affordable due to the development of technologies and the enactment of government policies [7]. In the renewable Portfolio Standard, California will generate 33% of the total energy from renewable energy resources before 2020. Industry experts estimate that about 5,000 MW of PV will be connected to the power grid for the resource mix in only 10 years [8]. The goal of New Mexico is to generate 20% of the total energy from renewable energy resources by 2020 [9]. Considering this aggressive goal, a large amount of PV power needs to be installed in the next 10 years. The PV power produces no CO_2 pollution to the environment. The output from the solar power aligns reasonably well with the daytime consumption on the electricity grid, which reduces the need for fossil fuel power stations that cause serious pollution. Renewable energy is penetrating the power system more and more [10]. A technical issue for the PV power is its great fluctuation due to the variation in weather. The output of a PV panel on a sunny day is relatively smooth. But on a cloudy day, a PV panel produces a variable power output due to the motion of clouds. The movement and the size of clouds can be random and irregular [11]. Therefore, the output of a PV panel is also irregular. Such variations in a power system can bring issues on reliability in a power grid and high risks to electrical equipment. Another key characteristic of renewable energies (solar or wind) is their dependence on weather conditions [12]. They are not controlled by people but by natural factors. These two problems need to be addressed. One is called smoothing [13]. It means how to make PV output smooth to bring less impact on the power system. Another is shifting; it is about how to shift the consumer daily load profile to maximize the use of PV output and minimize the utility's peak load [14]. Around these problems, energy storage showed great usage in these two aspects.

The most important enabling technology for the use of renewable energy on the utility scale is energy storage to match the power demand [15]. This is important because of the intermittent nature of renewable energy sources and the fairly predictable behavior of electrical demand [16]. The use of energy storage would enable power plants to run at a higher percentage of capacity and ensure that electrical demand was met at all times, thus reducing the need for peaking power plants that generally have the lowest efficiency, greatest harmful emissions, and highest operating cost [17]. The entire concept of the peaking power plant could be dismissed if adequate and efficient electrical storage could be utilized.

Utility-scale storage, also commonly referred to as large-scale or grid-scale storage, has historically been provided by resources such as pumped hydro system [18]. In a pumped hydro system, a facility will pump water uphill into a reservoir at times when the cost of electricity is inexpensive (in the middle of the night, for instance) and then run that water back downhill through a turbine when electricity costs are higher and extra energy is needed in the grid [19]. With the combination of declining battery energy storage costs and the increased introduction of renewable energy, batteries are beginning to play a different role at the grid scale.

The size and functionality of utility-scale battery storage depend upon a couple of primary factors, including the location of the battery on the grid and the mechanism or chemistry used to store electricity [20]. The most common grid-scale battery solutions today are rated to provide either 2, 4, or 6 h of electricity at their rated capacity; though, it is not unrealistic to anticipate that longer-duration batteries will be available someday shortly [21]. Generally, grid-scale batteries are either paired with a generating resource, such as a wind farm, or placed in the transmission and distribution system, such as at substations, to help balance the local electric supply and demand, as shown in Figure 1.1.

With the declining cost of energy-storage technology, solar batteries are becoming an increasingly popular addition to solar installations [22]. However, it is not just residential and commercial solar shoppers who benefit from installing energy storage [23]. Utility-scale battery storage is increasingly playing a major role in the operation of the electric grid, providing cost savings, environmental benefits, and new flexibility for the grid.

(a) (b)

Figure 1.1 *Utility-scale storage for the application of solar and wind energy. (a) Utility-scale storage for solar energy (b) Utility-scale storage for wind energy.*

As opposed to residential energy storage systems (ESSs), solar batteries' technical specifications are expressed in kilowatts, utility-scale battery storage is spoken about in megawatts (1 megawatt = 1,000 kilowatts) [24]. A typical residential solar battery will be rated to provide around 5 kilowatts of power, and to be able to store between 10 and 15 kilowatt-hours of usable energy, as is the case for the Tesla Powerwall 2 and LG Chem RESU 10H [25]. A typical utility-scale battery storage system, on the other hand, is rated in megawatts, and hours of duration, such as Tesla's Mira Loma Battery Storage Facility, which has a rated capacity of 20 megawatts and a 4-h duration (meaning it can store 80 megawatt-hours of usable electricity).

There are many ways in which battery energy storage can support the grid [26]. The analogy used most frequently by utilities and grid operators to describe energy storage is that it is a "new tool in the toolbox." There are lots of benefits of utility-scale battery storage [27]. Storage can inherently act like load (charging from the grid when electricity prices and demand are both low) or like a generator (pushing electricity back onto the grid when demand and prices are both high) [28]. What is more, whereas power plants may take minutes or even hours to turn on, battery storage can start injecting electricity onto the grid in milliseconds. This level of flexibility from a resource is unprecedented, and the possibilities for how to harness this capability are endless.

Energy-storage technologies also can help utilities provide the power quality and reliability required by increasingly complex and sensitive equipment, maximize the use of electrical capacity, and make possible the transition to intermittent renewable energy sources. Energy storage devices improve system responsiveness, reliability, and flexibility while reducing capital and operating costs for both suppliers and customers. Suppliers can use energy storage for transmission line stabilization, spinning reserve, and voltage control, which means customers would receive improved power quality and reliability [29]. Technologies such as ultracapacitors, flywheels, batteries, and superconducting magnetic energy storage can be used for quality and reliability purposes. These applications require a large power output over very short timescales, typically from tenths of a second to a few minutes. Large amounts of energy need not be stored in these applications because the power is delivered over such a short timescale.

For load leveling and peak shaving, the opposite is necessary. The power is delivered over longer timescales from minutes to hours [30]. These systems need to store large amounts of energy but do not necessarily need to deliver power as high as for UPS or power quality applications. This chapter will focus on AI for utility-scale energy storage use in making intermittent electricity a more usable commodity [31]. Hydrogen has been proposed as the preferred energy-storage medium of the future. It could be used for storage at the utility level via electrolyzers and then redistributed along electrical transmission lines during peak periods. It could be stored and piped out to the city or county level, from which it could be transmitted along electrical lines [32].

It could be distributed to the house or building level and consumed in a fuel cell to provide electricity and space heaters, or consumed in fuel cell vehicles. Or it could be a combination of these scenarios. Hydrogen has higher specific energy (energy/mass) than conventional fuels like gasoline, natural gas, and coal. Unfortunately, it also has a very low energy density (energy/volume) compared to those fuels. Therefore, to store it in any useful form, even in the stationary context where the size of the storage system is of smaller importance, it must be altered to achieve a higher energy density. A few different methods have emerged to accomplish this [33]. One method is to mechanically change its pressure or temperature by compressing it and storing it in large tanks or underground caverns or at varying pressure in a pipeline. Other methods under development are to store it chemically by bonding hydrogen to different atoms or molecules (e.g., metal hydrides) or to store it in nanoscale structures. The latter two will probably prove to be of larger importance to the portable fuel cell industry because energy density is a much more important consideration in that area than at the utility scale. Only if these other methods obtain extremely high efficiencies they will become realistic candidates for utility-scale energy storage [34]. Even then they are likely to be much more expensive than methods using compression.

As an alternative to developing and using hydrogen storage technologies, a combination of conventional rechargeable batteries, flow batteries, pumped hydro, and/or compressed air could be used instead. Depending on local conditions, one or more of these could meet the needs of particular utilities [35]. At times of low demand, the excess can be stored by one of these methods for redistribution during times of high demand. Since the energy is transformed back into electricity, it can be efficiently sent down to high-voltage transmission lines for use by consumers.

There are several factors to evaluate when looking at electrical storage on the utility scale:

1. Energy efficiency
2. Environmental impact
3. Location dependence
4. Lifetime
5. Economics
6. Space and weight requirements

This part is meant to be a forward-looking inquiry into future storage possibilities and how they can be compared to each other. Therefore, on an extended time scale, we can minimize the economic considerations. The point is to find a direction for a long-term policy that itself can directly affect the economics of these technologies in the future. The focus of economics for the long term can be viewed as the amount of energy and mass of materials required to build and operate these various storage systems. Additionally, on the utility scale, the importance of size and weight requirements can be minimized [36]. If a largely renewable electrical generation is assumed, then the size of the generator will be significantly larger than the size of the storage system. Though this will not be a deciding criterion for choosing a preferred electric storage technology, the energy densities will be compared to justify a reasonable scope.

Energy efficiency refers to the percentage of the input energy that is retrieved from the storage system. This is important because more energy-efficient systems require less generating capacity to charge, thus minimizing the size of the generating facility. Lifetime of the systems is important, as it would not do any good having to replace expensive systems every few years. They need to last under heavy cycling conditions for upward of 20 years to be useful [37]. Local geography plays a large role because not all of the systems are well suited for all places. For example, the Netherlands has no mountains for conventional pumped storage or caverns that could be used for compressed air or hydrogen. Ultimately, all batteries will reach their end of life and must be recycled or disposed of in some manner. Therefore, the system must ideally use a minimal number of hazardous materials or have sufficient engineering to eliminate the impact of those materials. The ability to recycle the majority of the constituents making up the battery will be most desirable to eliminate both financial and energy costs of disposal and a new production of the material.

The conventional energy storage technologies include three ways:

(1) Secondary batteries

The use of rechargeable batteries in utility applications is not new. Lead-acid varieties have been used for several different utility applications in the past. Currently, there are large installations of both lead-acid and nickel-cadmium batteries for the use of load leveling and other purposes. These, however, are on a much smaller scale than would be needed for large-scale renewable projects. The Golden Valley Electrical Association recently commissioned a 40 MW/(80 MWh) nickel-cadmium battery to improve reliability and supply power for essentials during outages. Both the lead-acid and nickel-cadmium batteries are made of fairly toxic materials and their use in much larger-scale projects might cause disposal and recycling concerns in the future. There are better options to look forward to that will eventually have a longer life, higher energy density, and be much less toxic.

One of the top candidates being considered for utility-scale is the sodium-sulfur battery (NAS). A NAS battery consists of liquid (molten) sulfur at the positive electrode and liquid (molten) sodium at the negative electrode as active materials separated by a solid beta alumina ceramic electrolyte. Positive sodium ions pass through the electrolyte and combine with the sulfur to form sodium polysulfides. During discharge, positive Na-ions flow through the electrolyte, and electrons flow

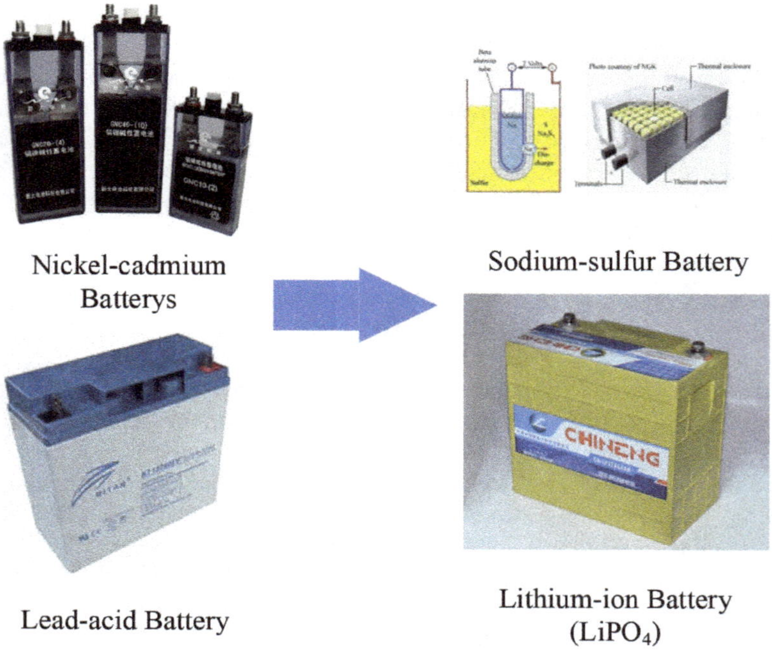

Figure 1.2 Secondary batteries

in the external circuit of the battery. This process is reversible as charging causes sodium polysulfides to release the positive sodium ions back through the electrolyte to recombine as elemental sodium. The battery operates at about 300°C. NAS batteries boast a high energy density of around 650 MJ/m^3. The energy efficiency of this battery is around 75% and so would be suitable for bulk storage applications, as shown in Figure 1.2.

Lithium-ion and lithium polymer batteries, while primarily used in the portable electronics market, could have some application in future utility-scale energy storage. The cathode in these batteries is a lithiated metal oxide ($Lisa_2$, $LiMO_2$, etc.) and the anode is made of graphitic carbon with a layer structure. The electrolyte is made up of lithium salts (such as $LiPF_6$) dissolved in organic carbonates. When the battery is charged, lithium atoms in the cathode become ions and migrate through the electrolyte toward the carbon anode where they combine with external electrons and are deposited between carbon layers as lithium atoms. This process is reversed during discharge. The lithium-polymer variation has a plastic film that does not conduct electricity but allows ions to pass through it. The 60°C operating temperature requires a heater, reducing efficiency. The lithium-ion batteries (LIBs) have a high energy density of about 720 MJ/m3 and have low internal resistance and so will achieve efficiencies in the 85% range and above [38]. Their high energy efficiencies make LIBs excellent candidates for storage on the utility scale. This book focuses on the application of LIBs in utility-scale battery deployment.

(2) Flow batteries

Another possibility for large-scale electrical storage is the flow battery. Most secondary batteries use electrodes both as part of the electron transfer process and to store the products or reactants via electrode solid-state reactions. Consequently, both energy and power density are tied to the size and shape of the electrodes. Flow batteries store and release electrical energy using a reversible electrochemical reaction between two liquid electrolytes. An electrochemical cell has two compartments, one for each electrolyte, physically separated by an ion-exchange membrane. Electrolytes flow into and out of the cell through separate manifolds and are transformed electrochemically inside the cell. The chemical energy in the electrolytes is turned into electrical energy and vice versa for charging. They all work in the same general way but vary in the chemistry of electrolytes.

There are some advantages to using the flow battery over a conventional secondary battery. The capacity of the system is scalable by simply increasing the number of solutions. This leads to cheaper installation costs as the systems get larger. The battery can be fully discharged with no ill effects and has little loss of electrolyte over time. There are three types of flow batteries that are closing in on commercialization: vanadium redox, polysulfide bromide, and zinc bromide. The vanadium redox flow battery (VRB) was pioneered at the University of New South Wales, Australia, and has shown the potential for long life cycle and energy efficiencies of over 80% in large installations. The VRB uses compounds of the element vanadium in both electrolyte tanks. This eliminates the possible problem of cross-contamination of the electrolytes and makes recycling easier.

Currently, there are only demonstration-scale operations in existence. The polysulfide bromide battery (PSB) utilizes two salt solution electrolytes, sodium bromide, and sodium polysulfide. PSB electrolytes are separated in the battery cell by a polymer membrane that only allows positive sodium ions to go through. The PSB battery is being developed by Regenesys Technologies, which is currently constructing two 120 MWh commercial plants. These plants should attain energy efficiencies of around 75%. Though the salt solutions themselves are only mildly toxic, a catastrophic failure by one of the tanks could release bromine gas, which is an extremely hazardous substance. However, the Tennessee Valley Authority released a finding of no significant impact for one of the 120 MWh facilities and deemed it safe, as shown in Figure 1.3.

In each cell of the zinc bromide (ZnBr) battery, two different electrolytes flow past carbon–plastic composite electrodes in two compartments separated by a microporous membrane. During discharge, Zn and Br combine into zinc bromide. During charge, metallic zinc is deposited as a thin film on one side of the composite electrode. Meanwhile, bromine evolves as a dilute solution on the other side of the membrane, reacting with other agents to make thick bromine oil that sinks at the bottom of the electrolytic tank. It is allowed to mix with the rest of the electrolyte during discharge. The zinc bromide battery has an energy efficiency of nearly 80%.

Exxon developed the ZnBr battery in the early 1970s. Over the years, many multi-kWh ZnBr batteries have been built and tested. Media demonstrated a 1 MW/4 MWh ZnBr battery in 1991 at Kyushu Electric Power Company. Some

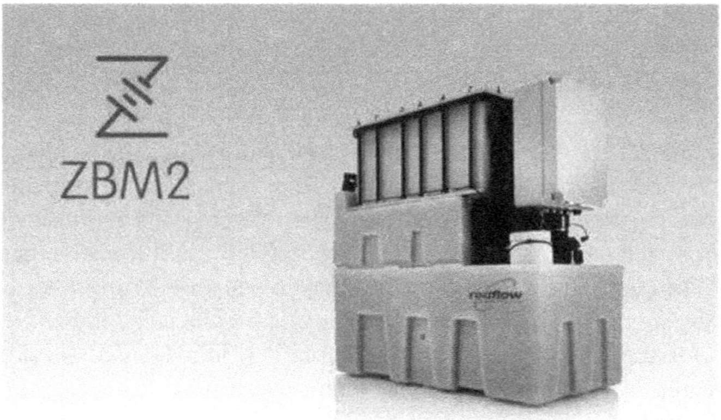

Figure 1.3 ZBM2 zinc-bromide flow battery—Redflow

multi-kWh units are now available preassembled, complete with plumbing and power electronics.

Another viable energy storage technology is compressed air energy storage (CAES). CAES depends on coupling with a gas turbine power plant. Conventional gas turbines consume about two-thirds of their input fuel to compress air at the time of generation. CAES precompresses air using electricity from the power grid at off-peak times and utilizes that energy later along with about 40% of the usual quantity of gaseous fuel to generate electricity as needed. The overall efficiency of energy storage is about 75% [13]. The compressed air is stored in appropriate underground mines, and caverns are created inside salt rocks or possibly in aquifers. The first commercial CAES was a 290 MW unit built in Hundorf, Germany, in 1978. The second commercial CAES was a 110 MW unit built in McIntosh, Alabama, in 1991. The third commercial CAES is a 2,700 MW plant that is under construction in Norton, Ohio. This 9-unit plant will compress air to about 100 bars in an existing limestone mine 2,200 feet underground. To be compatible with a renewable future, the plant would have to be able to run on renewable fuel. One could imagine a bio-fuel of some kind to be the fuel source for the gas turbine engines. There would still be other emission issues but the system would be fully carbon neutral.

(3) Pumped hydro storage

Pumped hydro is the oldest and largest of all the commercially available energy storage technologies, with facilities up to 1,000 MW. Conventional pumped hydro uses two water reservoirs, separated vertically. During off-peak hours, water is pumped from the lower reservoir to the upper reservoir. When required, the water flow is reversed to generate electricity. Some high-dam hydro plants have a storage capability and can be dispatched as pumped hydro storage. Underground pumped storage, using flooded mine shafts or other cavities, is also technically possible but probably prohibitively expensive. The open sea can also be used as the lower reservoir if a suitable upper reservoir can be built in proximity. A 30-MW seawater pumped hydro plant was first built in Yanbaru, Japan, in 1999.

Pumped hydro is most practical at a large scale with discharge times ranging from several hours to a few days. There is over 90 GW of pumped storage in operation worldwide, which is about 3% of the global generation capacity. Pumped storage plants are characterized by long construction times and high capital expenditure. Pumped storage is the most widespread ESS in use on power networks. Its main applications are energy management, frequency control, and provision of the reserve. Pumped hydro storage has the limitation of needing to be a very large capacity to be cost-effective but can also be used as storage for several different generation sites. The efficiency of these plants has greatly increased in the last 40 years. Pumped storage in the 1960s had efficiencies of 60% compared with 80% for new facilities. Innovations in variable-speed motors have helped these plants operate at partial capacity and greatly reduced equipment vibrations, increasing plant life.

1.2 Definition of AI in the context of battery management

Urgent and massive deployment of emission energy sources, such as nuclear and renewable, is required. Renewable energy sources are fluctuating, and hence, their deployment has to be accompanied by efficient energy storage, where rechargeable batteries are at the forefront for short-term-to-medium-term storage, due to operational efficiency and flexibility [39]. Among them, LIBs constitute one of the most influential technologies in modern society, which has enabled the wide emergence of portable electronic devices and which is triggering the growth of the electric vehicle (EV) market. Even if LIBs have been very significantly improved, by more than 200% in energy density since the first LIB cells were successfully commercialized by Sony in 1991, their massive deployment for EV or stationary applications requires them to be even further optimized in terms of performance, durability, safety, cost, as well as reducing their CO_2 footprint and increasing their reusability and recyclability. This is true for both current LIBs and any next-generation batteries currently being developed or produced.

Several international initiatives have been created to develop novel tools and protocols for reducing the number of experiments in battery research by a factor of 3 and, more generally, for boosting the pace of material discovery for energy applications by a factor of 10. Artificial intelligence (AI), and particularly its fruitful branch known as machine learning (ML), stands out as a promising approach that could lead to a paradigm shift in the way we do battery R&D, hopefully enabling us to overcome the major challenges dealing with a vast number of variables and a large quantity of data.

(1) Battery R&D is a complex multivariable problem, where very different properties, such as performance, life-cycle analyses, safety, cost, environmental effects, and resource issues, are contained. Furthermore, the overall battery circular economy should eventually be included from the mining, production, and assembly stage via the long usage phase to the final reuse and recycling processes [40]. The present research workflow, however, relies heavily on a forward trial-and-error approach and is largely materials-centered: synthesizing materials, manufacturing electrolytes

and electrodes, assembling cells, and finally assessing performance. Even considering only these aspects, there are > 10^{100} possibilities to synthesize active materials and prepare electrolytes, almost an infinite number of possibilities for choosing the electrode manufacturing parameters, and dozens of possible cell formats, which is far greater than what a human brain can handle. This makes difficult the emergence of inverse design tools enabling the prediction of the battery component properties needed for a given performance target and cell format.

(2) The size of battery R&D data grows exponentially, following the world data-sphere trend. For example, BASF, the second-largest chemical producer in the world, recently announced that they produce >70 million battery characterization data points per day, and in an academic context, an example, the French Network on Electrochemical Energy Storage (RS2E) with its academic partners generates about 1 petabyte of battery data per year [41]. These enormous data sets are currently not accessible to the scientific community as a whole, but actions have been taken toward establishing open and FAIR14 battery databases. Furthermore, there is already a massive amount of data spread out in scientific publications: almost 30,000 LIB publications already exist, and this number is growing rapidly [42]. A researcher reading 200 papers per year will need nearly 150 years to read all of the LIB publications available today.

AI and ML thus need to assist researchers in efficiently solving the parameters and data challenges of LIBs as well as assist the R&D of battery technologies beyond LIBs such as Na-ion, all-solid-state, and Li–S batteries and electrochemical capacitors (supercapacitors). For this to become true, several challenges need to be tackled, for instance, defining widely accepted standards in battery R&D combined with systematic data disclosure, the identification of the most suited descriptor(s) for a certain ML model, or the determination of the associated error, among others. In addition, different battery technologies bring different challenges, and AI- and ML-based approaches can be already helpful in many aspects, such as ML-assisted operando imaging techniques aiming to study Li dendrite formation and growth for all-solid-state batteries (ASSBs). Other examples could be the increase in time and length scales of current physics-based simulations or the development of innovative multiscale approaches.

AI is ubiquitous in our modern world, equipping many modern digital devices [43]. AI equips Internet search engines like Google to learn from our search habits and suggest the most relevant results to us. It is implemented in social networks, like Facebook or Twitter, and Amazon for personalizing news feeds, recognizing people or objects in photos, offering machine translations, or detecting inappropriate content, among other uses. Online video-on-demand services, like Netflix, use AI to personalize movie offerings, and our cell phones use AI as personal assistants (e.g., Siri, Google Now, Bixby) [44]. Other widely adopted AI applications have capabilities spanning from sorting spam and performing speech recognition to making personalized sales offers in e-commerce, among others. Another widely known example of application is gaming, whose breakthroughs are the chess-playing computer Deep Blue, AlphaGo, and Watson. AI is also at the heart of the development of modern robotics, autonomous driving, and smart power grids. Chemistry fully

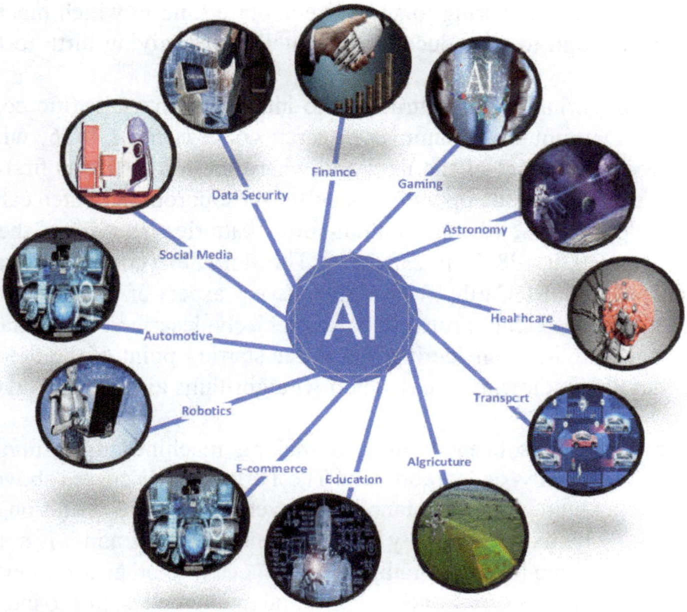

Figure 1.4 AI applications

follows this trend. Aiming to decrease the cost and increase the quality of their products, chemical industries are investing in AI and digitalization to accelerate their R&D, while academics intend to use AI and ML to accelerate research on materials, pharmaceuticals, catalysts, and more, as shown in Figure 1.4.

Despite this, for the vast majority of its history, AI was not as widely accepted as it is today. Even if typically associated with the fields of informatics and computer science, the concept of AI also belongs to fields like philosophy and psychology, interrogating the relationships between human beings and machines. From the beginning of human history, the development of new machines and tools guaranteed the survival of humankind, resulting in a strong relationship between humans and machines since early times [45]. An example of this can be found in the Egyptian society, in which the announcement of the next pharaoh was indicated to the population by the god Amon statue, through a mechanically moveable arm.

However, the emergence of the AI concept and the development of computers, both originating from the English mathematician Alan Turing, triggered a revolution in this relationship. For the first time in human history, the question about the capability to develop machines able to reason like humans was raised from the ground, as formalized in the philosophical question "Can machines think?" by Alan Turing himself. In his publication of 1950, he proposed a test, today known as the Turing test, whose aim is to verify if a human being, who is asked to interact with either a human or a machine through a few questions and without knowing her/his/its identity, is capable of distinguishing machines from humans [45]. Despite the limitations of such a test, it enabled Turing to speculate about a time in which machines will become smart enough to reproduce human intelligence, giving birth to the era of modern AI.

The idea of Turing rapidly attracted the interest of the scientific community, leading to the Dartmouth AI summer research conference in 1956, widely considered the founding event of the field and where the term AI was first proposed [46Ando, T., et al., High-rate operation of sulfur/mesoporous activated carbon composite electrode for all-solid-state lithium-sulfur batteries. Journal of the Ceramic Society of Japan, 2020. 128(5): p. 233-237.]. The aim of this conference was defined by its organizer, John McCarthy, who stated "Every aspect of learning or any other feature of intelligence can in principle be so precisely described that a machine can be made to simulate it." Considering this as the starting point of the field, it is not surprising that the first attempts to develop AI algorithms aimed to simulate human brain behavior.

Historically, AI has been defined as making machines think humanly, act humanly, think rationally, or act rationally. The Turing test discussed above required the machine to act humanly, for instance. However, clearly, the definition of what is acting or thinking humanly/rationally is in constant evolution, and it is not linked to computer science alone [47]. It is rather interconnected to other disciplines such as philosophy, psychology, neurobiology, logic, and mathematics, just to cite a few. AI could also be defined as "the science and engineering of making computers behave in ways that, until recently, required human intelligence." However, similarly to the previous case, which behaviors we classify as requiring human intelligence or not are time- and society-dependent.

Some decades ago, it would have been believed by many that playing games or interpreting human behaviors to send personalized feeds would require human intelligence, while today these are tasks that we recognize machines can do [48]. All the above makes AI a moving target, whose exact definition is not trivial [49]. However, the majority of AI systems in current use have in common the capability of learning from experience. The most widely adopted approach to making machines do so is through algorithm architectures known as ML, which are the ones employed nowadays in battery R&D and will be the main subject of this book. These algorithms have tremendous capabilities to assess multidimensional data sets (i.e., data sets containing multiple variables), discover patterns in data, and unlock applications that are difficult to exploit by using other approaches.

This is of high relevance for the fields of battery material discoveries or battery manufacturing optimization, in which a multitude of parameters should be considered simultaneously. The discovery capabilities of modern ML algorithms rely on the quantity, quality, and veracity of data. Therefore, the first step for any ML-based approach is to build a suitable and complete enough data set. Afterward, the ML model should be trained and, when possible, evaluated. In the most common case (supervised models), this is achieved by using a part of the data set to train the algorithm (training step), whose predictive capability is assessed by comparing values predicted by the model and data that were not used for the training step. This is generally referred to as a test step [50]. If the so-obtained model proves to be trustable along this step, the supervised ML algorithm is ready to be used.

ML algorithms can be classified as supervised, unsupervised, or semisupervised methods. Supervised approaches employ data sets that are pretreated to define certain variables as inputs and others as outputs. This prior information is missing in the case of unsupervised ML algorithms, whose goal is to find patterns in the data set. Within supervised ML, it is possible to distinguish between regression and classification, where the latter indicates an ML approach analyzing the data set in terms of classes, while the former analyzes it in terms of continuous values. The classes used for a supervised ML can come from the operator or an unsupervised ML. Semisupervised approaches are somewhere in between the two and utilize data sets containing both labeled and unlabeled data. Besides the type used, classical ML algorithms rely on data and are rather agnostic to physics, meaning that they could aim, for instance, to determine the relationship between different variables interpolating the training data, rather than offering any physical interpretation of such a relationship. However, physical-informed ML approaches exist, e.g., when using ML algorithms to solve or discover partial differential equations, among others.

As for the importance of data and good practices, all ML algorithms rely on data, which is vital to developing accurate ML models. As widely known, the size of data is critical, and it is typically believed that higher amounts of data lead to more accurate ML models. Even though this is generally true, the data quality is not a negligible factor and it should be considered, as well. Data sets containing too little data or containing poor quality data (e.g., data difficult to reproduce or affected by significant errors) can lead to incorrect ML predictions, biasing the associated result interpretation. In this context, the first step to developing a reliable ML model is building a data set representative of the problem under analysis. Good experimental practices in terms of both designs of experiments and experimental procedures are needed to ensure the reliability of data sets and ML results.

Defining effective experimental strategies is even more critical when applying ML-driven methods to rare failure scenarios, as recently highlighted by Finegan *et al.* In terms of variables to be considered, it is a good practice to consider as many variables as possible to have a global perspective on the problem under study. However, to simplify the ML model development for the case of multivariable problems, unsupervised techniques, such as principal component analysis (PCA), can be used. In particular, PCA can project the original data onto a low-dimensional subspace identified by convenient axes, also known as principal components, arising from linear combinations of the original variables, ordered by the variance they represent in the data set. Once the principal components accounting for the vast majority of the variance are identified, the ML model can be trained using these instead of the original variables, leading to a dimensionality reduction.

Similar to experimental measurements, AI algorithms themselves should be subjected to good standards and protocols to ensure that no bias is made during data processing and predictions. For instance, the data set and how the quality of the trained model was evaluated (if this was possible) should be systematically disclosed. The latter is relatively easy for the case of supervised models (the most commonly employed ones in the battery field), but evaluating the quality of unsupervised ones can be rather challenging, as it will be shortly discussed in section 1.3.

Taking the case of supervised methods, the model accuracy can be assessed by comparing the outputs predicted by the ML model when considering inputs not used during the training step and the real outputs, which were previously measured. The data set used to carry out this procedure is typically known as a test set. If the results predicted by the ML algorithm are equal, or close enough, to the real outputs, the model can be considered correct and can be used for predictions. A simple way to quantify this is through regression plots, which are obtained by plotting the data in the test set (true results) and the predictions coming from the trained ML model (predicted results). The predictive accuracy of the model can be quantified as the R-square of the points obtained when compared to the first bisector (true results = predicted results) of the regression plot. As is intuitive, an R-square ranging from 0 to 1 stands for a predictive accuracy ranging from 0% to 100%. However, the use of only the R-square as a metric to evaluate the predictive accuracy could lead to errors in certain cases. An example of this is when using an ML algorithm to predict the energy of a molecule while the energy of the system is shifted for instability (or other) reasons, which could lead to high R-squared and high mean-squared error (that, on the contrary, should be minimized).

Therefore, other metrics, such as the root-mean-square error or the mean absolute error, can be substituted for or can be associated with the R-square to better verify the predictive accuracy of the model. In addition, it should be stressed here that each model has a limit of validity that should be taken in mind. Indeed, if the model is used to predict results associated with inputs that are significantly different from the ones used for the training and test steps, it is likely that the predictions will not be as accurate as desired.

In the battery management system (BMS), software design to perform a state of charge (SOC) and state of health (SOH) estimation accurately is significant. Most of the practical SOH methods for capacity estimation are based on amp-hour (Ah) counting between precise reference SOC points. SOH methods for resistance estimation are more varied and range from simple averaging of delta voltage divided by delta current to recursive algorithms such as recursive least squares or advanced Kalman filter (KF) algorithms. In the case of SOC estimation, one of the simplest methods is based on open-circuit voltage and coulomb counting. However, more robust and sophisticated methods are preferred to handle sensor errors and uncertain model knowledge. Many approaches employ an equivalent circuit model combined with KF variants for SOC estimation. To make these approaches work well, significant battery testing is needed to model and parameterize the algorithms.

ML data-driven approaches to battery state estimation have been driven by recent advances in AI in fields such as computer vision and autonomous vehicles. The Venn diagram shows how the field of AI is subdivided, including ML and its subsequent divisions of representation learning and deep learning, as shown in Figure 1.5.

Figure 1.5 also shows the state-of-the-art ML SOC and SOH estimation methods for electrified vehicles (EVs), which is bounded within the AI field while permeating

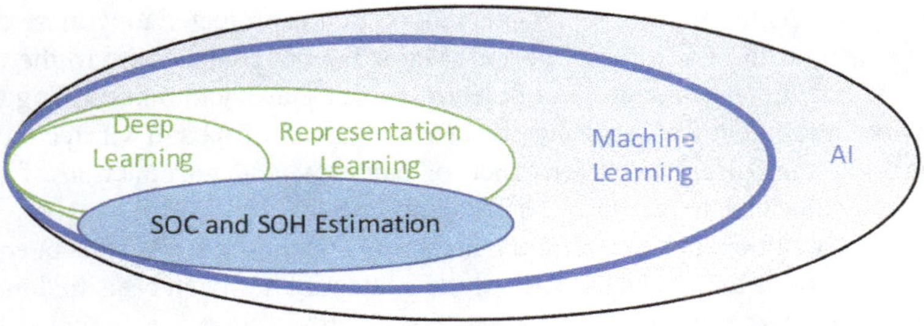

Figure 1.5 *SOC and SOH estimation based on AI and ML*

all subfields of ML. The structured summary of the SOC and SOH estimation methods is considered and analyzed, as shown in Figure 1.6.

As battery technology grows and matures, a significant amount of data is being collected and analyzed in a partially or fully automated fashion to improve battery design and usage. This plethora of data has made it possible to improve BMS performance via big data, the Internet of things (IoT), cloud computing, and the ML methods investigated here. In the case of SOC and SOH estimation based on ML methods, the main computational load demanded by these approaches happens during its off-line training phase, making it feasible for implementation on typical BMS hardware.

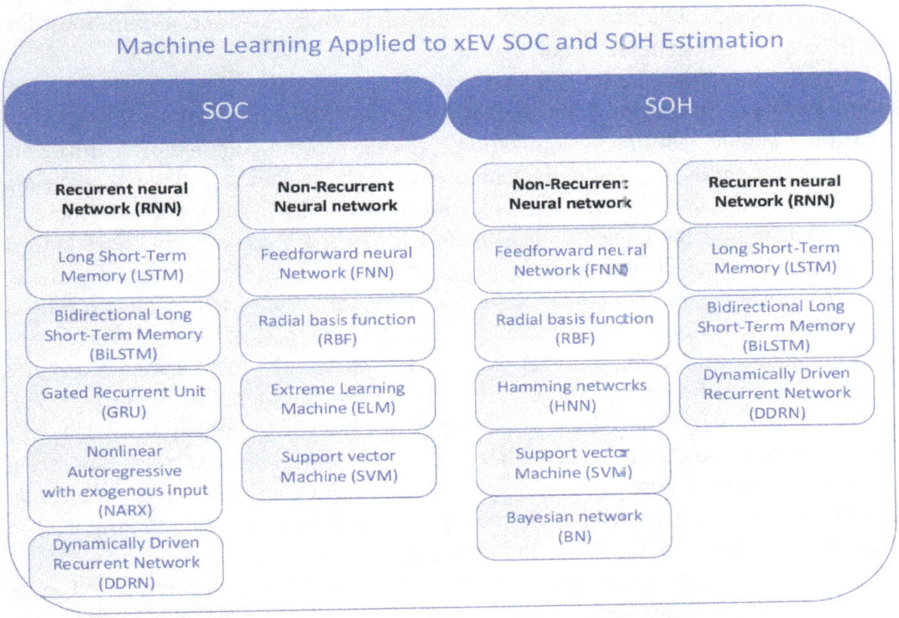

Figure 1.6 *Structured summary of the ML methods*

The global AI in the battery management market is projected to witness considerable growth during the forecast period. Major factors contributing to the market growth include increased adoption of electronic rickshaws and bikes, rising awareness toward intelligent BMSs, rising need for utilization of AI and ML technologies in enhancing shelf-life and performance of batteries, and introduction of battery intelligence solutions, particularly for EVs and ESSs.

Based on components, the AI in the battery management market has been bifurcated into solutions and services. The solution category has been classified into lifecycle management, analytics, and others. Battery lifecycle management solutions assist in the management of all stages of battery life cycle, such as monitoring and diagnostics, control of charge cycle, and predicting the replacement of batteries.

Based on industry, the AI in the battery management market has been classified into consumer electronics, automotive, medical, military, and others, wherein others include ESSs and security systems. Among these, the automotive category is expected to witness significant growth during the forecast period. This is attributed to the increasing requirement for automated battery management operations, optimized battery performance, reduced cell aging, and increasing demand for hybrid vehicles. In addition, the consumer electronics category is also projected to exhibit rapid growth in the coming years, due to the surging number of consumer applications and growing use of portable electronic devices.

Globally, Asia Pacific Accreditation Cooperation (APAC) is projected to be one of the most lucrative markets for intelligent BMSs, owing to factors such as increasing demand for battery operated EVs, awareness among the populace toward harmful emissions, and rising focus of governments on the introduction of laws to reduce emissions, which would increase the demand for EVs and in turn boost the AI in battery management market growth.

The introduction of intelligent battery management platforms is one of the key drivers of AI in the battery management market. AI-integrated battery management platforms enable real-time visualization, provide recommendations related to the battery's current usage, and deliver predictive alerts related to excessive charging, unused batteries, and high temperature. Players operating in the market are increasingly focusing on the launch of intelligent BMSs for EV makers and electric fleet managers to improve battery performance and achieve a better return on investment.

For instance, in September 2019, ION Energy Pvt. Ltd. launched Edison Analytics, a cloud-based battery management platform. It uses ML and AI that support enhancing the performance of batteries and extend the battery life up to 40%. The platform would help automotive original equipment manufacturers by providing automated suggestions through the implementation of technologies, such as ML and AI, which would result in the collection of real-time insights, thereby supporting in reducing the total cost of ownership and better user experience. Thus, an increasing number of AI-integrated battery management platforms is likely to propel AI in battery management market growth in the coming years.

Players in the AI in the battery management market are focusing on business strategies, including product launches, mergers and acquisitions, client wins, partnerships, and other developments, to increase their presence in the market and enhance their product offerings to capture larger customer base. For instance, in April 2019, Advanced Charging Technologies (ACT) launched first, an industrial battery room management system. It utilizes the Best-One-First-Out (BOFO) algorithm, which selects the best battery to use based on temperature, cool downtime, and SOC at end of the charge cycle. Some of the major players in the global AI in battery management market include Robert Bosch GmbH, ION Energy Pvt. Ltd., Advanced Charging Technologies (ACT), Texas Instruments Incorporated, Moixa Energy Holdings Ltd., Voltaiq, Nuvation Energy, Headsun Technology Co. Ltd., Energsoft Inc., and Schneider Electric SE.

1.3 Advantages of using AI for battery management

LIBs are one of the most influential technologies in modern society, enabling the widespread emergence of portable electronic devices and triggering the growth of the EV market [52]. Since Sony successfully commercialized the first LIBs in 1991, their energy density has increased by more than 200%. Although lithium batteries have been significantly improved, their large-scale deployment in EVs or stationary applications requires them to be further optimized in terms of performance, durability, safety, cost, etc., to improve their reusability and recyclability sex. This is a current requirement for LIBs and any next-generation batteries under development or production, as battery R&D is a complex multivariate problem with very different characteristics such as performance, life cycle analysis, safety, cost, environmental impact, and resource issues [53]. In addition, the overall circular economy of batteries should ultimately include the process from mining, production, and assembly, through the lengthy use phase, and finally, reuse and recycling.

However, the current research pipeline relies heavily on forwarding trial-and-error approaches and is primarily material-centric: synthesizing materials, fabricating electrolytes and electrodes, assembling batteries, and finally evaluating performance. Even considering only these aspects, there are thousands of possibilities for synthesizing active substances and preparing electrolytes, an almost infinite number of possibilities for choosing electrode fabrication parameters, and dozens of possible battery forms, which are far beyond what the human brain can hold, the range that can be handled. This makes designing tools difficult. Therefore, to improve the efficiency of battery research, many new tools have been developed to meet this requirement [54]. Among them, AI is a promising method that promises to help people overcome dealing with large quantities of variables and large amounts of data.

AI, and ML in particular, is the promise to lead in overcoming the major limitations of battery optimization. The most widely adopted approach to making machines do so is through algorithm architectures known as ML, which are the ones employed nowadays in battery R&D. These algorithms have tremendous capabilities to assess multidimensional data sets (i.e., data sets containing multiple variables), discover patterns in data, and unlock applications that are difficult to exploit by using other approaches. This is of high relevance for the fields of battery material discoveries

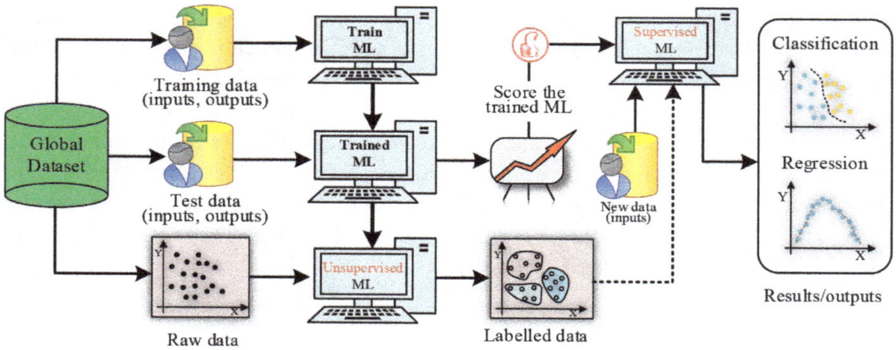

Figure 1.7 Principle of ML approach for classification/regression methods

or battery manufacturing optimization, in which a multitude of parameters should be considered simultaneously [55]. The discovery capabilities of modern ML algorithms rely on the quantity, quality, and veracity of data.

Therefore, the first step for any ML-based approach is to build a suitable and complete enough data set. Afterward, the ML model should be trained and, when possible, evaluated. In the most common case (supervised models), this is achieved by using a part of the data set to train the algorithm (training step), whose predictive capability is assessed by comparing values predicted by the model and data that were not used for the training step. This is generally referred to as a test step. If the so-obtained model proves to be trustable along this step, the supervised ML algorithm is ready to be used. The overall working principle of an ML approach for supervised/unsupervised and classification/regression methods is shown in Figure 1.7.

ML algorithms can be classified as supervised, unsupervised, or semisupervised methods [56]. Supervised approaches employ data sets that are pretreated to define certain variables as inputs and others as outputs. This prior information is missing in the case of unsupervised ML algorithms, whose goal is to find patterns in the data set. Within supervised ML, it is possible to distinguish between regression and classification, where the latter indicates an ML approach analyzing the data set in terms of classes, while the former analyzes it in terms of continuous values. The classes used for a supervised ML can come from the operator or an unsupervised ML. Semisupervised approaches are somewhere in between the two and utilize data sets containing both labeled and unlabeled data. Besides the type used, classical ML algorithms rely on data and are rather agnostic to physics, meaning that they could aim, for instance, to determine the relationship between different variables interpolating the training data, rather than offering any physical interpretation of such a relationship.

ML-based approaches may allow for navigating such chemical, formulation, and operating condition spaces in a selective manner, and hopefully, reduce the number of experiments and/or computations required [57]. From a theoretical point of view, ML can support the development of efficient force fields, creating new opportunities for material simulation and reliable alternative models. This can also facilitate the development of multiscale modeling frameworks with reasonable computational costs. Furthermore, ML has the potential to be a powerful experimental enhancement tool in terms of identifying reaction mechanisms directly from electrochemical results such as cycling.

On the one hand, some ML applications in the battery field have been extensively studied in the scientific literature, such as online and offline estimation of SOH, SOC, and RUL for batteries. On the other hand, in the battery field, several promising applications of AI/ML are surprisingly understudied. Among them, battery fabrication and battery material characterization are obvious examples. The use of data-driven methods will profoundly impact the industrial facilities of modern society, leading to the Industry 4.0 revolution. Battery manufacturing is no exception and will require dedicated data warehouses shortly.

Despite this apparent trend, academic research on this topic is still rare in the literature, and greater efforts are needed in this direction. Academia should provide the industry with new data-driven approaches to help them overcome this revolution. Similar comments can be made in the case of battery material characterization, for which the scientific literature is still scarce. One of the first and more important applications of ML technology is image analysis, which makes ML particularly suitable for tomographic image segmentation, and ML algorithms are expected to play a leading role in this field in the future. Another area where AI is expected to play a key role in battery research in terms of data retrieval and analysis is text mining.

This provides access to massive data sets and "just" restores information already available in the scientific literature, which will greatly simplify the analysis of the chemistry and electrode/battery fabrication spaces discussed at the beginning of this section. However, a serious concern for its applicability is the systematic data-missing properties of key electrodes and batteries, such as electrode porosity, electrolyte volume, electrochemical testing protocols, etc., which may often be overlooked in scientific reporting [58]. Furthermore, despite the challenges, as ML begins to emerge in other chemistry-related fields, it is expected to drive the emergence of self-driving battery labs to automate experiments and data collection. The workflow of ML methods commonly used in battery research and development, including neural networks, decision trees, support vector machines, and k-nearest neighbors (k-NN), is shown in Figure 1.8.

The automobile industry is an important force to promote a new round of scientific and technological revolution and industrial transformation, important support for building a strong manufacturing country, and an important pillar of the national economy. The healthy and sustainable development of the automobile industry is related to the daily travel of the people, the smooth circulation of social resources, and the overall leap of ecological civilization. At present, the integration of new generation information communication, new energy, new materials, and other technologies with the automobile industry is accelerating, the industrial ecology has undergone profound changes, and the competitive landscape has been reshaped in an all-around way. The development of a traditional BMS has problems such as weak software modularity, short life cycle, and frequent replacement of hardware processors, requiring software redesign. The high-safety intelligent BMS, as a key

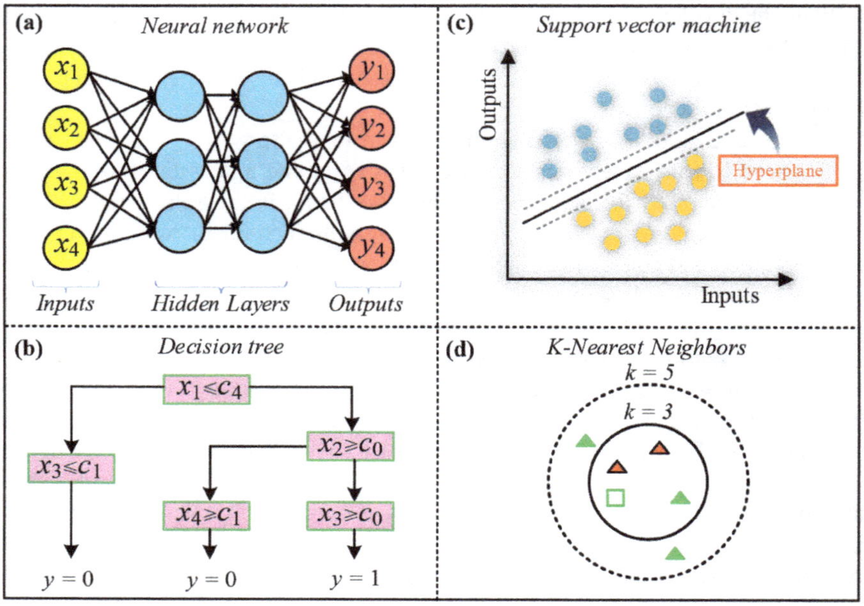

Figure 1.8 *Workflows of some of the most common ML techniques: (a) neural network, (b) decision tree, (c) support vector machine, and (d) k-nearest neighbors (k-NN)*

component of EVs, can improve the utilization rate of the battery, prevent the battery from overcharging and overdischarging, prolong the service life of the battery, and monitor the status of the battery.

The transition to a fully electrified future reportedly hinges on lower-cost, higher-performing, and safer batteries. The use of next-generation battery chemistries, such as solid-state methods, to optimize the energy density and power of batteries has achieved varying degrees of success [59]. However, no one has yet reached the commercialization stage to meet the exploding demand for advanced technologies such as electric vehicles, medical devices, drones, and energy storage solutions. Smart cars are not only limited to the automotive field but also require the Internet, AI, big data, information, and communication, as well as multicollar collaborative innovation, including semiconductor chips, to complete their manufacturing from planning and to mass production. It can be called a model of cross-discipline, cross-field, and cross-industry integration. At present, the development of smart cars has become a commanding height in the strategic competition of the world's major auto powers, and "software-defined cars" have also become a consensus in the development of the industry. Whoever can take the lead in grabbing the strategic high ground can control the wealth code of the automobile industry and then build a new model for future competition advantage.

Smart cars are an important part of the smart transportation system, an important carrier for realizing smart transportation, and an important application field of smart transportation technology, and smart transportation is an important part of smart city construction [60]. The innovative development of the smart car industry can improve road traffic efficiency, reduce urban traffic congestion, reduce the probability of traffic accidents, and protect people's safety and property safety; the promotion of smart cars can also reduce air pollution and carbon emissions, and slow down the global warming trend. Also, benefit the ecological environment, realize the perfect integration of people, vehicles, and cities, give full play to the mission of automobiles to benefit the society, and greatly improve the people's life happiness index.

The new generation of batteries must be able to charge quickly without fail. These batteries also need to exceed the current performance standards, keep weight low, and be constructed from materials that are easy to mass-produce. Researchers have spent decades exploring solutions, and progress has been slow due to slow experiments, long turnaround times, and a difficult discovery process. AI can help address these long-term challenges and shorten the process of evaluating battery materials, cell architectures, and chemistries from years to months.

(1) Solve the problem of a long evaluation period

The traditional method of generating battery performance data is to continuously inject energy into the battery cells until the battery is depleted. The researchers had to spend years charging and discharging the battery thousands of times to get the desired results. Predicting battery degradation in this way is critical for developing safer, less flammable batteries. However, given that some relatively new applications are booming, such as EVs and home solar combined with energy storage, there is not much time left to waste. Battery scientists take a systems-level approach through AI to more efficiently test and understand battery packs, their integration, and expected performance. Such AI applications also include various cell types, their different chemistries, and expected performance, and help determine the best way to distribute energy among multiple cells or battery packs.

(2) Find materials faster and more efficiently

Previously, researchers faced the daunting task of narrowing down the range of alternative materials needed for next-generation battery applications. This process requires evaluators to analyze large amounts of data collected from the testing process. Researchers can only operate as fast as machines that compute information, and this often takes years to improve. With AI, some useful material combinations can be discovered that would not otherwise be considered. The application of AI to the material discovery process has produced interesting results in many fields such as superconductors and will have promising applications in the field of batteries.

(3) Use AI to optimize battery structure

Over the past few decades, much effort has been devoted to improving batteries through battery chemistry. However, changing the physical properties of batteries has been shown to improve key battery performance indicators such as density, capacity, and safety. Battery scientists can use AI to better understand structure–property relationships at the electrode level to design optimal battery structures for any given application. Based on how the battery is being used and other technical specifications, AI can make valuable recommendations on possible structural designs to optimize battery performance. AI algorithms can even be tweaked to suggest possibilities based on emerging technologies and chemistries that have not yet been applied. It is like having a fast-build battery prototype factory. From a time- and cost-saving point of view, there are benefits to the entire value chain. For example, the performance of an EV depends heavily on the battery cells. Combining this with AI to better understand how to improve cell performance, not just BMSs, has important implications. This will help lay the foundation for the development of next-generation batteries for EV applications.

In the field of battery science, although AI is still an emerging application, there are already many examples showing its great potential. For example, researchers at Stanford University, MIT, and the Toyota Research Institute used AI to determine the best way to charge an EV battery in 10 min. Where traditional methods require a 500-day evaluation process, the team used a highly targeted AI algorithm to identify the optimal charging method from 224 options in just 16 days. Not only researchers but also many large companies are taking this approach. Volkswagen is working with Google to use AI and quantum computing to simulate and optimize the structure of high-performance batteries. Panasonic claims that thanks to AI, it can drastically reduce the number of times the battery needs to be charged and discharged when testing new designs. These are just a few examples, and as ML technology develops, the associated applications and benefits will explode. Despite all the hopes for AI and ML, there is still a long way to go before data-driven approaches are widely used in the battery world. The challenges to be addressed can be summarized as (1) descriptors, (2) data scarcity and indetermination, (3) lack of standard and immature representation, (4) user-friendly tools, and (5) bridging scales. How to solve these questions and challenges is the direction of researchers in the future.

It is clear that the capabilities and potential of AI in general and ML in particular are attracting a growing interest to attain new insights on batteries at all scales: from material design and synthesis to manufacturing, and from material to electrochemical characterization. Many hopes are pinned on data-driven approaches applied to batteries, yet significant progress in the field is needed before leading to the revolution that AI promises. Creative and innovative AI solutions are emerging in other fields, such as in music creation where a composer can request an AI to compose music by mixing predefined music styles. In the not-so-distant future, these solutions may become adapted for the battery field as support of researchers' creativity; for example, we can imagine AI algorithms building multiscale models by picking up automatically from the literature a collection of models with the degree of fidelity desired. Another example is Alpha fold, an AI network that recently demonstrated being able to determine the 3D structure of proteins starting from their amino-acid sequence. Similar approaches can be envisioned for batteries, as computational frameworks are able to indicate how cell/electrode components organize in the space as a function of their chemical nature, ratio, and manufacturing conditions, as well as how the arising 3D mesostructure affects the cell lifetime, electrochemical, and mechanical properties.

All the AI programs discussed above are also categorized as "weak AI" in computer science, i.e., a set of informatics programs mimicking human intelligence. Since Alan Turing's times, strong debate persists about the possibility or not of designing AIs which can think and have a genuine understanding and conscious thoughts, i.e., the so-called "strong AI" [61]. If strong AI emerges one day, it will revolutionize even the battery field. Still, ethical aspects should be carefully considered in their design for both "weak" and "strong" AIs to ensure strong synergies between humans and machines.

Overall, to become an unavoidable driving force to foster innovation, AI experts should mandatorily and strongly collaborate with battery experts from both an experimental and computational point of view. If the battery community will be able to successfully integrate data-driven approaches into their routine work, data could open the door to new fascinating discoveries in the field at an unseen rate, boosting the development of the new generation of batteries.

Chapter 2
Utility-scale lithium-ion battery system characteristics

2.1 Overview of lithium-ion batteries

2.1.1 *Battery working principle*

(1) Brief introduction of lithium-ion battery

Lithium-ion battery system is a complex system integrating chemical, electrical, and mechanical characteristics [62]. So the requirements of the various characteristics must be considered for the design of a lithium-ion battery. In particular, the safety and life attenuation characteristics contained in the chemical characteristics of the battery cell cannot be measured directly by the equipment, and it is not easy to predict in a short time. Therefore, when designing the battery system, battery technology, group technology, and BMS (battery management system) technology should be adopted to ensure the safety, reliability, and durability of the battery.

Lithium-ion batteries usually come in cylindrical and rectangular shapes [63]. Spiral wound structure is adopted inside the cylindrical battery, which is made of a very fine and highly permeable film isolation material separated between the positive and cathode, mainly including polyethylene, and polypropylene composite materials [64]. The rectangular lithium-ion battery is formed by laminating, placing a diaphragm on the cathode, then placing the anode, etc. The cathode includes a lithium-ion collector composed of lithium-containing materials (such as one or more mixed materials of lithium cobalt oxide, lithium manganate, and nickel-cobalt lithium manganate) and a current collector composed of the aluminum film. The anode is composed of a lithium-ion collector and a current collector. The battery is filled with organic electrolyte solution and equipped with a safety valve and positive temperature coefficient resistor (PTC) elements. It has the advantages of low thermal resistance, high heat exchange efficiency, not easy to burn, safety and reliability, and can effectively prevent the battery from being damaged in an abnormal state or output short circuit.

Under the influence of abnormal factors such as short circuits, high heat, and overcharge, high-pressure gas is easy to be generated in the battery, which will cause the deformation of the battery shell and even the risk of explosion [64, 65]. To use it safely, the battery must be equipped with a safety valve and abnormal discharge of gas to avoid an explosion. When the pressure inside the battery container rises to an

Table 2.1 The conditions that an ideal cathode material needs to meet

Serial number	Characteristic
1	High delithiation/intercalation lithium potential and stable charge–discharge platform
2	Small chemical equivalent
3	High stability
4	High diffusion coefficient
5	During battery operation, the cathode material should be electrochemically inert to the electrolyte
6	Environmentally friendly, cheap

abnormal state, the safety valve can quickly open and expel the gas and providing safety protection in case of an abnormal situation [66]. Since the PTC element is in the low-resistance state under normal temperature, the danger of battery overheating can be prevented by adjusting the current in case of abnormality. For example, in case of overheating due to an abnormally large current caused by a short circuit or overcharge, the PTC element is converted to a very high-resistance state to reduce the current in the circuit [67]. Therefore, the PTC element is usually used to prevent battery overcurrent and overheating.

(2) Composition of lithium-ion battery

The lithium-ion battery is mainly composed of four parts, i.e., cathode material, anode material, diaphragm, and electrolyte. Common anode materials include lithium manganate, lithium cobalt oxide, and nickel-cobalt manganate, which mainly provide lithium ions for batteries. The cathode material is one of the important components of the lithium-ion battery, and the requirements to be met are shown in Table 2.1.

The anode material is mainly graphite. Its main function is to store lithium ions and realize the intercalation/deintercalation of lithium ions in the process of charge and discharge. The characteristics of an ideal anode material are shown in Table 2.2.

Table 2.2 The conditions that an ideal anode material needs to meet

Serial number	Characteristic
1	Low electrochemical reaction potential
2	Strong lithium-ion storage capacity
3	Small change in voltage platform
4	High ionic and electronic conductivity
5	A stable SEI (Solid Electrolyte Interphase) film can be formed on the anode surface
6	Environmentally friendly, cheap

Table 2.3 The conditions that an ideal electrolyte needs to meet

Serial number	Characteristic
1	Low viscosity, high ionic conductivity, and low activation energy for solvation and desolvation of lithium ions
2	Wide chemical reaction window
3	Wide working temperature range, high boiling point, and low melting point
4	Chemically inert to battery components such as current collectors and separators
5	Environmentally friendly, cheap

Diaphragm is a special composite mode to prevent electrons from shuttling freely between the positive and anode, allowing only lithium ions in the electrolyte to pass freely. The diaphragm is essentially an insulator and cannot contain free electrons, so it is not conductive [68]. In a battery, the elements exist in the form of ions, which can easily pass through the diaphragm, while the electrons separate from their elements and go to the new carrier (anode material or cathode material) [69]. When in contact with the diaphragm, the diaphragm cannot absorb the free electrons on the electrode, thus blocking the passage of electrons. The common materials of the diaphragm are single-layer PP (Polypropylene) film, PE (Polyethylene) film, and three-layer composite film composed of PP and PE. In future, the development trend of diaphragm performance is lighter, thinner, and safer, which can meet the requirements of large capacity, long life, high power, and high safety of lithium batteries.

The electrolyte is a bridge for lithium ions to transfer between positive and negative electrodes. It is one of the four key materials of lithium-ion batteries and has an important impact on the performance of lithium-ion batteries. The function of the electrolyte is to realize the conduction of lithium ions between the positive and negative electrodes of the battery. It is mainly composed of lithium salts, solvents, and additives. The three together determine the performance of the electrolyte. The characteristics of an ideal electrolyte are shown in Table 2.3.

There are lithium ions, metal ions, oxygen ions, and carbon layers inside the battery. The reaction of the battery is completed by the movement of these ions. The diaphragm of the battery acts as a barrier to separate the two poles of the battery. The internal structure of the battery is shown in Figure 2.1.

The lithium-ion battery is an indispensable portable electrical energy storage element at present. There are many external parameters to characterize its performance of it, such as voltage, current, and internal resistance. Lithium-ion batteries have a series of advantages, first, the combustion heat of them is very high, i.e., the heat released per unit volume is very high [70]. Second, lithium-ion battery is relatively environmentally friendly and meets the requirements of the development of a green society. Third, under normal conditions, lithium-ion batteries can be charged/discharged hundreds of times, so they can be used for a long time. Fourth, there is no memory effect. During the working process of an ordinary battery, its capacity

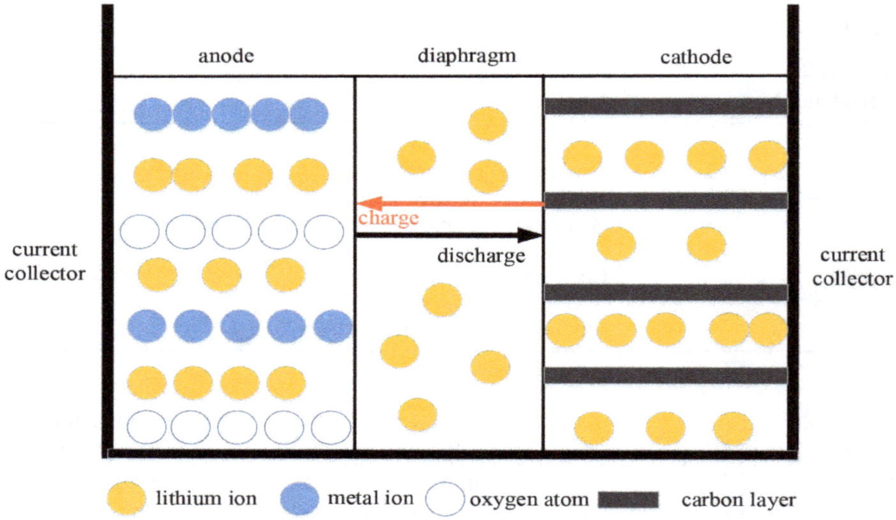

Figure 2.1 Schematic representation of the lithium-ion battery structure

will be lost, resulting in less capacity, and this problem does not exist in lithium-ion batteries. Of course, it has many other advantages, such as good safety performance, low self-discharge, fast charging, and a wide working temperature range. With the above advantages, lithium-ion batteries are widely used in various industries.

(3) How lithium-ion batteries work

The internal chemical reaction of lithium-ion battery is a basic redox reaction, which is also its working principle in the actual use process. That is, the electric energy is converted into heat energy through a chemical reaction. According to the chemical reaction equation, the charge and discharge process of lithium-ion batteries is the process of lithium-ion embedding and de-embedding. When the battery is charged, the lithium atom of the cathode will undergo an oxidation reaction, losing electrons and becoming lithium ions. The lithium ions in the anode are embedded into the micropores of the carbon layer. The more lithium ions are embedded, the higher the charging capacity [71]. During discharge, the anode will undergo an oxidation reaction, and the lithium ions embedded in the carbon layer of the anode will escape and move back to the cathode. The more lithium ions return to the cathode, the higher the discharge capacity [72]. During charging, lithium ions are generated at the cathode of the battery. The generated lithium ions move to the anode through the electrolyte. The internal chemical reaction process of lithium batteries is shown in Figure 2.2.

The anode, cathode, and total reaction equation are described as follows.

The cathode reaction is shown in (2.1).

$$LiM_xO_y = Li_{(1-x)}M_xO_y + xLi^+ + xe^- \tag{2.1}$$

The anode reaction is represented in (2.2).

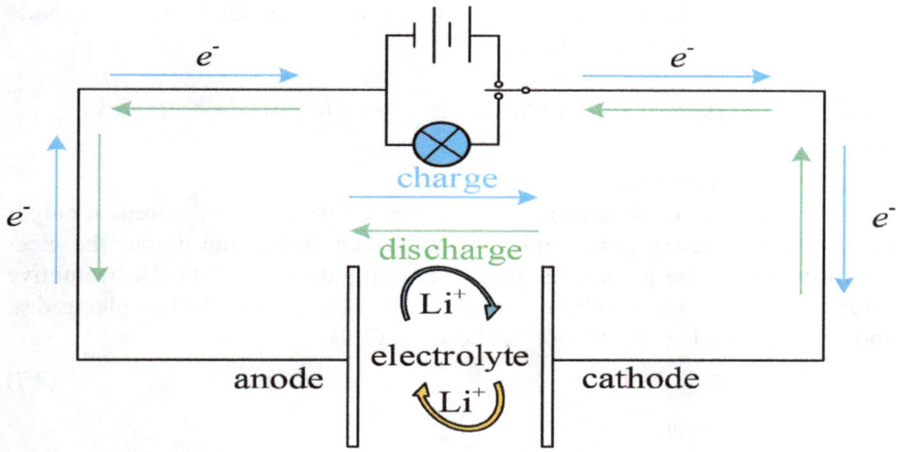

Figure 2.2 *The working process of lithium-ion battery*

$$nC + xLi^{+} + xe^{-} = Li_{x}C_{n} \tag{2.2}$$

The total battery response is shown in (2.3).

$$LiM_{x}O_{y} + nC = Li_{(1-x)}M_{x}O_{y} + Li_{x}C_{n} \tag{2.3}$$

In the three equations, M can be Co, Mn, Fe, and Ni, respectively, representing the lithium cobalt oxide, lithium manganese oxide, lithium iron phosphate, and lithium nickel oxide battery [73]. The chemical reaction process of lithium-ion battery discharge and charging is opposite to each other. If there is an external electric field, the Li^{+} in the battery cathode material can be detached and embedded from the lattice under the action of the electric field.

Taking the $LiCoO_2$ as an example, the chemical reaction expression of the cathode is shown in (2.4).

$$LiCoO_{2} \rightarrow xLi^{+} + Li_{1-x}CoO_{2} + xe^{-} \tag{2.4}$$

The chemical reaction expression of the anode is shown in (2.5).

$$xe^{-} + xLi^{+} + 6C \rightarrow Li_{x}C_{6} \tag{2.5}$$

The total equation of the battery reaction is shown in (2.6).

$$LiCoO_{2} + 6C \Leftrightarrow Li_{1-x}CoO_{2} + Li_{x}C_{6} \tag{2.6}$$

Lithium-ion batteries have three important parts of the diaphragm and the cathode and anode, its work relies mainly on the back and forth movement of ions between the anode and the cathode. During charging, the Li^{+} is removed from the cathode and embedded into the anode through the corresponding electrolyte. After a series of chemical reactions, the cathode is in the less lithium state, and the anode is in the more lithium state. At the same time, its compensation charge is supplied to the cathode from the external circuit. During discharge, the Li^+ is removed from the anode and embedded into the cathode again through the action of electrolyte.

2.1.2 Principles of status monitoring of utility-scale batteries

(1) Voltage

1) Electromotive force

Electromotive force is one of the indexes to measure the theoretical output energy of the battery [74]. Under the same conditions, the higher the electromotive force, the greater the theoretical output energy. The electromotive force of a battery is the difference between the potentials of the two electrodes, and its mathematical expression is shown in (2.7).

$$E = \varphi_+ - \varphi_- \tag{2.7}$$

In this equation:

E, Battery electromotive force; φ_+, Positive balance potential; φ_-, Negative balance potential.

2) Open circuit voltage

Open circuit voltage refers to the potential difference between the two poles of the battery in an open circuit state (when almost no current flows). It depends on the activity of the battery's positive and negative electrodes, electrolyte and temperature conditions, etc. and has nothing to do with the battery's geometry and size. In general, the open-circuit voltage of the battery is less than its electromotive force.

3) Rated voltage

Rated voltage is also called the nominal voltage, which refers to the standard voltage at which the battery works under specified conditions. The chemical system of the battery can be distinguished by the rated voltage.

4) Working voltage

Working voltage refers to the voltage displayed during the discharging process after the battery is connected to the load, also known as the load voltage or discharge voltage [75]. Its mathematical expression is shown in (2.8). After the battery is connected to the load, due to the existence of ohmic internal resistance and polarization internal resistance, the operating voltage of the battery is lower than the open-circuit voltage, and of course, it must be lower than the electromotive force.

$$V = E - IR_i = E - I\left(R_\Omega + R_f\right) \tag{2.8}$$

In this equation: I, Battery working current; R_i, Polarized internal resistance; R_Ω, Ohm resistance; R_f, Polarized internal resistance.

5) The end-of-discharge voltage

The end-of-discharge voltage is also called the discharge cutoff voltage, which refers to the lowest operating voltage value at which the battery discharges, the voltage drops to the point where it is no longer suitable for discharging [76]. According to different types of batteries and different discharge conditions, the requirements for the capacity and life of the battery are also different, and the discharge cutoff voltage is also different. Generally, when discharging at a low temperature or high current, the final voltage is regulated lower. When discharging with the small current for a long time or intermittently, the final voltage is regulated higher. Internal resistance is one of the important parameters of lithium batteries

6) The end-of-charge voltage

The end-of-charge voltage refers to the voltage at which the battery reaches full charge during the specified constant current charging period. If continue to charge after reaching the end-of-charge voltage, it is overcharged, which generally damages battery performance and life.

7) The average voltage

The average voltage refers to the value obtained by dividing the number of watt-hours by the number of ampere-hours during the specified charging and discharging process. In the case of constant current, it is the average voltage over a certain period.

(2) Capacity

The amount of electricity that the battery can release under certain discharge conditions is called battery capacity, which is represented by the symbol C. The unit is usually expressed in Ah or mAh.

1) Theoretical capacity

The theoretical capacity is the amount of electricity that can be provided assuming that all active materials participate in the flow reaction of the battery. It can be accurately calculated according to the amount of electrode active material in the battery reaction formula and the electrochemical equivalent of the active material calculated according to Faraday's law. The calculation process is shown in (2.9).

$$Q = \frac{zmF}{M} \tag{2.9}$$

Faraday's law states that when current passes through the electrolyte solution, the amount of chemically reacted substance on the electrode is proportional to the amount of electricity passed,

where,

Q, The amount of electricity passed in the electrode reaction (Ah)

z, The electronic measurement coefficient in the electrode reaction formula

m, The mass of the reactive material (g)

M, The molar mass of the active substance (g/mol)

F, The Faraday constant is about 965,000 C/mol or 26.8 Ah/mol

It can also be understood that the active material with mass m can release electricity Q after a complete reaction. The electric quantity Q is the theoretical capacity (C_0) of the electrode active material, so the above equation can also be written in the form of (2.10).

$$C_0 = 26.8z\frac{m}{M} = \frac{1}{K}m \tag{2.10}$$

where K represents the electrochemical equivalent of the active material, $K = \frac{M}{26.8z}$ (g/Ah) is the mass of active material required to obtain 1 Ah of electricity. The above equation is the calculation formula for the theoretical capacity of the electrode active material.

2) The rated capacity (Cg)

The rated capacity is the minimum capacity that the battery should discharge under certain discharge conditions (such as temperature, discharge rate, termination voltage, etc.) according to the standards stipulated by the country or relevant departments.

3) The actual capacity (C)

The actual capacity refers to the amount of electricity that the battery discharges when discharged under actual application working conditions. It is equal to the integral of the discharge current and the discharge time. The actual discharge capacity is greatly affected by the discharge rate. It is often in the lower right corner of the letter C. The discharge rate is indicated by Arabic numerals, such as $C_5 = 50$ Ah, which means that the discharge rate is completed in 5 hours. The obtained capacity is 50 Ah. The actual capacity calculation method is as follows.

When discharging with constant current, the actual capacity calculation method is shown in (2.11): ion movement toward

$$C = It \tag{2.11}$$

When discharging with variable current, the actual capacity calculation method is shown in (2.12):

$$C = \int_0^T I(t)\, \mathrm{d}t \tag{2.12}$$

In this equation:

I: The discharge current, which is a function of the discharge time t

T: The time to discharge to the end voltage

Due to the existence of internal resistance and various other reasons, the active material cannot be fully utilized, i.e., the utilization rate of the active material is always less than 1 [77]. Therefore, the actual capacity and the rated capacity of the chemical power supply are always lower than the theoretical capacity. The actual capacity of the battery is closely related to the discharge current. During high-current discharge, the polarization of the electrode increases, the internal resistance increases, the discharge voltage drops quickly, and the energy efficiency of the battery decreases, so the actual discharge capacity is lower [78]. Correspondingly, under low-rate discharge conditions, the discharge voltage drops slowly, and the actual capacity of the battery is often higher than the rated capacity.

4) The remaining capacity

The remaining capacity refers to the remaining usable capacity of the battery after being discharged at a certain discharge rate. The estimation and calculation of the remaining capacity are affected by factors such as the discharge rate and discharge time of the battery, the degree of battery aging, and the application environment. Therefore, there are certain difficulties in accurate estimation of it.

(3) Internal resistance

Internal resistance is one of the important parameters of lithium batteries. The size of the internal resistance directly affects the battery's operating voltage, operating current, and battery capacity. The internal resistance of a lithium-ion battery is not static, in the process of battery charging and discharging, it is changed with the influence of changes in temperature, charge–discharge rate, charge–discharge time, electrolyte concentration, and active material quality.

The internal resistance of the battery includes the ohmic internal resistance R_0 and polarization internal resistance R_p. The ohmic internal resistance comes from the resistance of the electrolyte, the diaphragm, and the contact resistance of various parts. An electric field is generated during the internal oxidation–reduction reaction of the battery. Under the action of this electric field, the dielectric will generate a polarization charge due to the polarization effect [79]. The polarization resistance is the resistance of the polarization charge to the current.

During charging and discharging process, due to the internal resistance generated by the polarization reaction, the polarization reaction is mainly electrochemical polarization and concentration polarization. The nature of the active material, the structure of the electrode, the manufacturing process of the battery, and the operating conditions of the battery will all lead to the difference in the polarization internal resistance. The main influencing factor of the polarization internal resistance is the operating conditions of the battery. The calculation formula of internal resistance is shown in (2.13).

$$R = R_0 + R_p \tag{2.13}$$

The internal resistance is most sensitive to temperature, and the value of internal resistance can vary greatly at different temperatures. An important reason for the degradation of battery performance at low temperatures is that the internal resistance of the battery is too large at low temperatures. Under normal circumstances, lithium batteries are used as power sources. From the external analysis, the internal resistance is usually as small as possible. Especially in the case of high-power applications, small internal resistance is a necessary condition.

(4) Energy density

The energy density of a battery refers to the energy that a battery can output per unit mass or unit volume, correspondingly called mass–energy density (Wh/kg) or volume–energy density (Wh/L), also known as mass-specific energy or volume ratio energy [80]. Specific energy is an important indicator for evaluating whether power batteries can meet the application needs of new energy vehicles, and it is also an important indicator for comparing the performance of different types and types of batteries.

In the application process of new energy vehicles, the installation of battery packs requires corresponding components such as battery boxes, connecting wires, and current and voltage protection devices [81]. Therefore, the actual specific

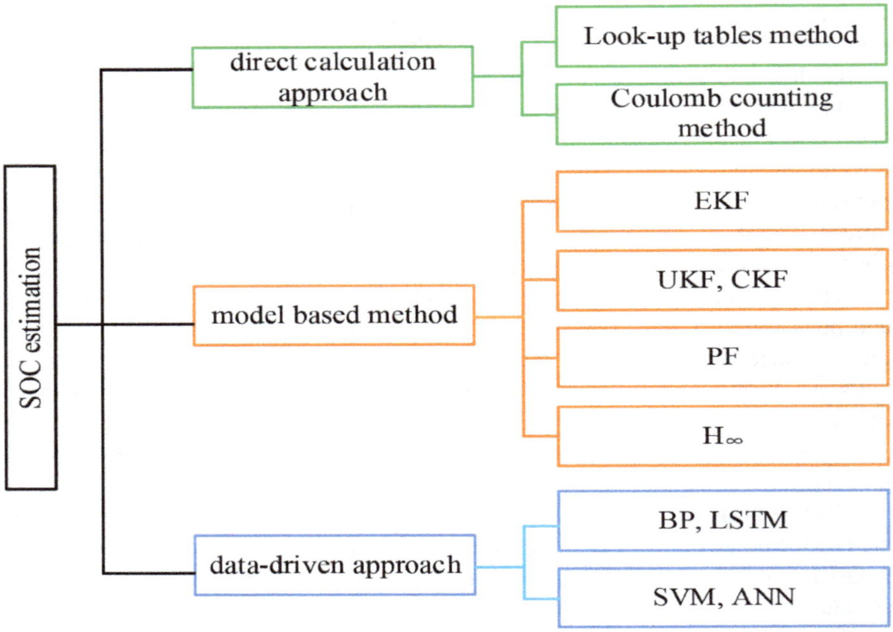

Figure 2.3 The SOC estimation methods

energy of the battery pack is less than the specific energy of the battery, and the specific energy of the battery pack is one of the important parameters of new energy vehicles. The smaller the gap between the specific energy of the battery and the specific energy of the battery pack, the higher the design level of the battery pack, and the higher the integration degree of the battery pack. Therefore, the specific energy of the battery pack often becomes an important measure of the performance of the battery pack.

(5) Power density

The power output per unit mass or unit volume of the battery is called power density, also known as specific power, and the unit is W/kg or W/L. The specific power represents the size of the working current that the battery can withstand. The battery's specific power means that it can withstand high current discharge. Specific power is an important indicator for evaluating whether batteries and battery packs meet the acceleration and climbing capabilities of new energy vehicles.

(6) The state of charge

The state of charge (SOC) describes the remaining power of the battery and is an important parameter of lithium-ion batteries. This parameter is related to the battery's charge and discharge history and the size of the charge and discharge current. The SOC value is a relative quantity, generally expressed as a percentage, and the value of SOC is $0\% \leq SOC \leq 100\%$. The commonly used SOC estimation methods are shown in Figure 2.3.

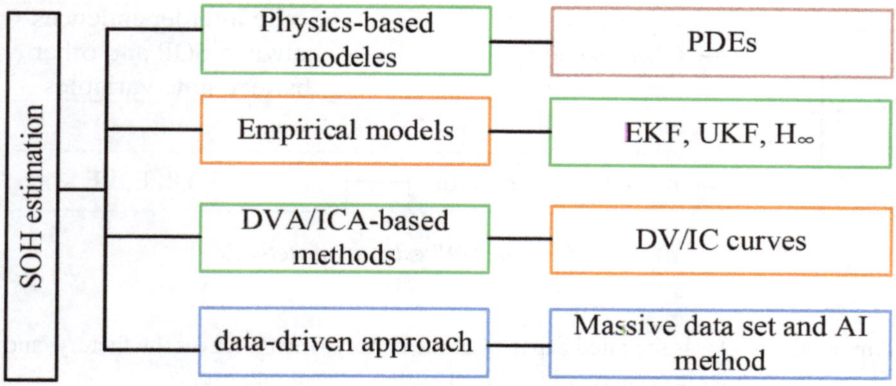

Figure 2.4 The SOH estimation methods

The remaining power of the battery is affected by the basic characteristic parameters of the power battery (terminal voltage, working current, temperature, capacity, internal pressure, internal resistance, and the number of charge and discharge cycles) and the characteristics of the power battery, making it difficult to measure the battery pack's SOC. At present, the simpler method for the research on the power of the battery pack is to equate the battery pack as a battery cell and find out the relationship between the SOC and these parameters by measuring the current, voltage, internal resistance, and other external parameters of the battery pack, then indirectly ground test the SOC value of the battery [82].

In the application process, to ensure the safety and service life of the battery pack, the SOC of the cells with poor energy in the battery pack is often used to define the SOC of the battery pack. Currently, commonly used SOC estimation methods include the open-circuit voltage method, ampere-hour accumulation method, electrochemical test method, battery model method, neural network method, impedance frequency advance method, Kalman filter method, etc.

(7) The state of health

The state of health (SOH) characterizes the ability of current lithium-ion batteries to store electrical energy relative to new lithium-ion batteries. With a form of percentage, SOH represents the battery's state from the beginning to the end of its life and is used to quantitatively describe the lithium-ion battery's performance. The commonly used SOH estimation methods are shown in Figure 2.4.

China's automotive industry standard "QC/T743-2006" issued in 2006 stipulates the end-of-life conditions for lithium-ion batteries: the available capacity decays to 80% of the standard capacity. At present, the definition of SOH is mainly embodied in several aspects such as capacity, power, internal resistance, number of cycles, and peak power. The SOH expression defined by capacity is shown in (2.14).

$$SOE(t) = \frac{E(t)}{E} \times 100\% \tag{2.14}$$

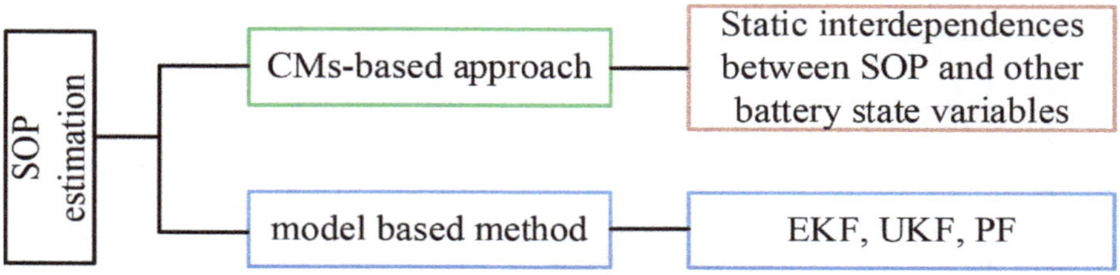

Figure 2.5 The SOP estimation methods

Among them, Q_{uite} is the rated capacity of the battery when it leaves the factory, and Q_{aged} is the available capacity of the battery after it is put into use.

(8) The state of power

The state of power (SOP) represents the power output capability of the battery during charging and discharging, and peak power is often used as its constant index. The peak power of the power battery refers to the maximum power that can be continuously discharged for *t* time based on the current state of the power battery without violating the preset use restrictions of the battery and is one of the measurement standards for the power performance of electric vehicles. The commonly used SOP estimation methods are shown in Figure 2.5.

The peak power capability of the power battery can evaluate the charge and discharge power capabilities of the power battery in different SOH and SOC, optimize the matching relationship between the power battery and the vehicle dynamics, and maximize the regenerative braking energy recovery performance of electric vehicles. For vehicle battery pack power design, if a conservative battery power design is adopted to meet the power requirements, it is easy to cause excessive battery cost and weight, which will affect the performance of the vehicle. If the battery power design is relatively small, it may cause overdischarge damage to the battery during use. The peak power of the power battery will change with changes in the discharge temperature, battery SOC, and battery health status. Especially, the peak power drops significantly under low temperature, low SOC, and low SOH conditions. If a constant peak power is used as a limiting condition, it may cause harm to the vehicle and the driver. Therefore, real-time high-precision estimation of peak power is essential.

(9) The state of energy

The state of energy (SOE) of lithium-ion batteries is a direct reflection of the remaining energy of the battery and a key indicator for energy optimization and management. The definition of SOE is shown in (2.15):

$$SOE(t) = E_c - E_d(t) \tag{2.15}$$

In (2.15), *SOE*(*t*) is the remaining energy of the battery at time point *t*, which is the total energy of the battery, *E* is the maximum energy that a battery can release out, and *E*(*t*) is the remaining energy of the battery at time *t*. Normally, the SOE reaches 1 when the battery is fully charged, and the SOE is 0 when the battery is discharged to the cutoff voltage. The commonly used SOE estimation methods are shown in Figure 2.6.

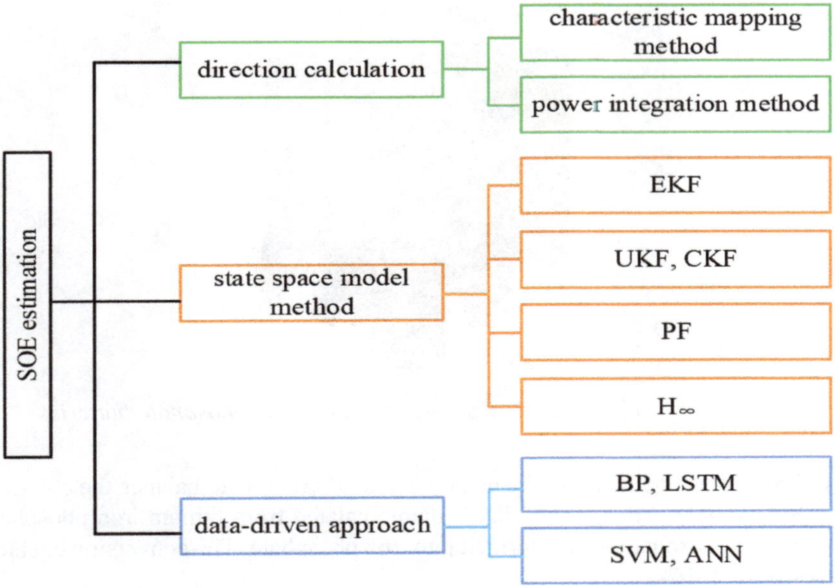

Figure 2.6 The SOE estimation methods

SOE refers to the remaining available energy of the lithium battery, which directly reflects the energy that the battery can release and is a key parameter for the management and optimization of the power battery pack.

2.2 Lithium-ion batteries

2.2.1 Lithium iron phosphate batteries

A lithium iron phosphate battery is a lithium-ion battery with lithium iron phosphate ($LiFePO_4$) as the cathode material and carbon as the anode material [83]. The rated voltage of the unit is 3.2 V, and the charging cutoff voltage is 3.6 –3.65 V. The cylindrical shape of lithium iron phosphate batteries is shown in the Figure 2.7.

The charge–discharge reaction of the $LiFePO_4$ battery is carried out between $LiFePO_4$ and $FePO_4$. During charging, $LiFePO_4$ is gradually separated from lithium ions to form $FePO_4$. During discharge, lithium ions are embedded into $FePO_4$ to form $LiFePO_4$. When the battery is charged, lithium ions migrate from the lithium iron phosphate crystal to the crystal surface, enter the electrolyte under the action of electric field force, then pass through the diaphragm, and then migrate to the graphite crystal surface through the electrolyte, which is then embedded in the graphite lattice [84]. At the same time, electrons flow through the conductor to the aluminum foil collector of acid batteries because of their positive pole, through the pole lug, the positive pole of the battery, the external circuit, the negative pole, and the negative ear to the copper foil collector of the negative pole of the battery, and

Figure 2.7 The cylindrical shape of lithium iron phosphate batteries

then through the conductor to the negative pole of graphite to balance the charge of the negative pole. After lithium ion is deintercalated from lithium iron phosphate, lithium iron phosphate is transformed into iron phosphate. The conversion equation is shown in (2–16).

$$LiFePO_4 - xLi^+ - xe^- \rightarrow xFePO_4 + (1-x)\,LiFePO_4 \tag{2.16}$$

When the battery is discharged, lithium ions are de-embedded from the graphite crystal into the electrolyte, then migrate through the diaphragm, through the electrolyte to the surface of the lithium iron phosphate crystal, and then re-embed themselves within the lithium iron phosphate lattice [85]. At the same time, electrons flow through the conductive body to the copper foil collector of the negative electrode, through the pole lug, the negative cell post, the external circuit, the positive pole post, the positive pole lug to the aluminum foil collector of the positive electrode of the cell, and then through the conductive body to the positive electrode of lithium iron phosphate, so that the charge of the positive electrode reaches equilibrium. After the lithium ions are embedded into the iron phosphate crystal, the iron phosphate is converted into lithium iron phosphate. The conversion equation is shown in (2.17).

$$FePO_4 + xLi^+ + xe^- xLiFePO_4 + (1-x)\,FePO_4 \tag{2.17}$$

In conclusion, the composition of lithium-ion batteries is some compounds. The reaction is done by the ion movement towards taking the diaphragm acts as a barrier, which separates the battery into two sides, as shown in Figure 2.8.

Lithium iron phosphate batteries have a high energy density. It is reported that the energy density of the square aluminum shell lithium iron phosphate battery produced in batch 2018 is about 160 Wh/kg. In 2019, some excellent battery manufacturers may be able to achieve the level of 175–180 Wh/kg. Some powerful manufacturers use stacking technology to achieve more production capacity of 185 Wh/kg. The electrochemical performance of lithium iron phosphate cathode material with good safety performance is relatively stable, which determines that it has

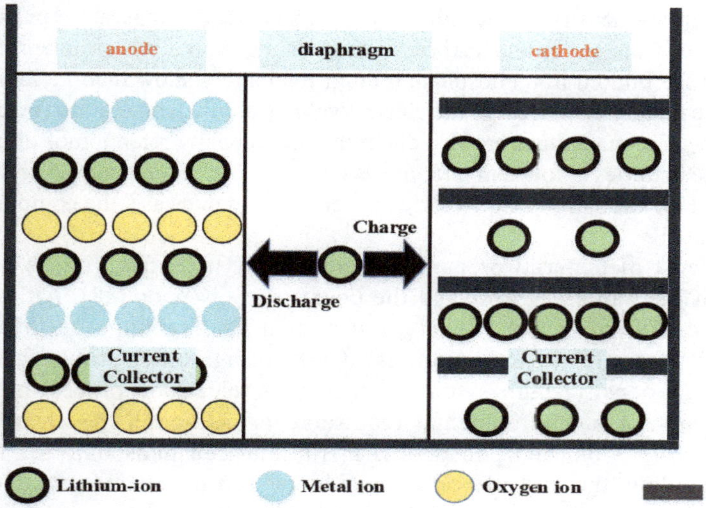

Figure 2.8 Schematic representation of the lithium-ion battery structure

a smooth charging and discharging platform [86]. Therefore, the battery structure will not change during charging and discharging, will not burn and explode, and is still very safe even in short circuits, overcharge, extrusion, pinhole, and other special cases.

2.2.1.1 Long cycle life

The cycle life of lithium iron phosphate battery 1C can generally reach 2,000 times or even 3,500 times, while the requirements for the energy storage market should reach more than 4,000–5,000 times to ensure that the life is 8–10 years, which is more than 1,000 times higher than that of ternary battery, while the cycle life of a long-life lead–acid battery is about 300 times. Rechargeable batteries without memory effect often work under the condition of full ignition, and the capacity will quickly fall below the rated capacity [87]. This phenomenon is called the memory effect. Like nickel–metal hydride batteries, nickel–cadmium batteries also have a memory function, while lithium iron phosphate batteries do not (lithium-ion batteries usually do not produce a memory effect). No matter what state the battery is in, it can be used during charging, during non discharging and before charging.

However, lithium iron phosphate batteries also have some disadvantages. In the sintering process of preparing lithium iron phosphate, iron oxide may be reduced to singlet iron in a high temperature reducing atmosphere. Monatomic iron will cause a microshort circuit in the battery, which is the tabooest material in the battery. This is the main reason why this material is not used as the cathode material of power lithium-ion batteries in Japan. Lithium iron phosphate has some performance defects, such as very low vibration density and compaction density, resulting in the low energy density of the lithium-ion battery. Low-temperature performance is poor, even nanosize, and carbon coating cannot solve this problem. The test results of the lithium iron phosphate lithium-ion battery show that the lithium iron phosphate battery cannot drive the electric vehicle at low temperature (below 0°C). Although some manufacturers also claim that the capacity retention of lithium iron phosphate batteries at low temperatures is not poor, i.e., when the discharge current is low and the discharge cutoff voltage is very low, in this case, the equipment cannot start at all.

The cost of material preparation and battery manufacturing is high, the output of the battery is low, and the consistency is poor [88]. Although the nanosized lithium iron phosphate and carbon coating improve the electrochemical properties of the materials, it also brings other problems, such as low energy density, high synthesis cost, poor electrode processing performance, and harsh environmental requirements. Although the chemical elements lithium, iron, and phosphorus in lithium iron phosphate are rich and the cost is low, the cost of preparing lithium iron phosphate products is not lowlithium manganese battery is. Even excluding the preliminary research and development cost, the process cost of materials and the high cost of preparing batteries will also make the final cost of unit energy storage power higher, and the consistency of products is poor. Whether it is material preparation or manufacturing, product consistency is difficult to guarantee, and the voltage platform of lithium iron phosphate is narrow, which increases the difficulty of battery observation.

Lithium iron phosphate battery is widely used in passenger cars, buses, logistics vehicles, low-speed electric vehicles, and other fields with its advantages of safety and low cost [89]. Although in the current field of new energy passenger vehicles, the national subsidy policy for new energy vehicles once dominated by ternary batteries with the advantage of energy density, lithium iron phosphate still occupies irreplaceable advantages in buses, logistics vehicles, and other fields. The latest data show that lithium iron phosphate batteries account for half of the total battery shipments.

Starting lithium iron phosphate battery not only has the characteristics of a powerful lithium battery but also has instantaneous high-power output capacity. The traditional lead–acid battery with energy lower than that of a powerful lithium battery is used, and the BSG (Belt-Driven Starter Generator) motor is used to replace the traditional starting motor and generator. It not only has the function of idle start and stop but also has the functions of engine stop sliding, sliding and braking energy recovery, acceleration assistance, and electronic cruise.

Lithium iron phosphate battery has a series of unique advantages such as high working voltage, high energy density, long cycle life, low self-discharge rate, no memory effect, green environmental protection, and supports stepless expansion. It is suitable for large-scale electric energy storage [90]. It has a good application prospect in the fields of safe grid-connected power generation of renewable energy power stations, grid peak shaving, distributed power plants, UPS power supply, emergency power supply system, etc.

With the rise of the energy storage market, in recent years, some power battery companies have arranged energy storage business to develop new applications for the lithium iron phosphate market [91]. On the one hand, lithium iron phosphate can be transferred to the field of energy storage due to its ultralong service life, safety, high capacity, and green characteristics, which will extend the value chain and promote the establishment of a new business model. On the other hand, lithium iron phosphate auxiliary energy storage system has become the mainstream choice in the market. According to the report, lithium iron phosphate batteries have been tried for frequency regulation on electric buses, electric trucks, user side, and grid side.

(1) The safety of wind power generation, photovoltaic power generation, and other renewable energy power generation to the power grid [92]. The inherent randomness, intermittence, and volatility of wind power generation determine that its large-scale development will have a significant impact on the safe operation of the power system. Photovoltaic power generation is affected by ambient temperature, sunlight intensity, and weather conditions. Photovoltaic power generation is characterized by random fluctuation. China presents the development trend of "decentralized development, low-voltage local access" and "large-scale development, medium- and high-voltage access," which puts forward higher requirements for power grid peak shaving and safe operation of the power system.

Therefore, high-energy storage products have become the key factor to solve the contradiction between the power grid and the renewable energy. Lithium iron phosphate energy storage system has the characteristics of fast working condition conversion, flexible operation mode, high efficiency, safety, environmental protection, and strong scalability. It has been applied in the national scenic energy storage and transmission demonstration project, which will effectively improve equipment efficiency, solve local voltage control problems, improve the reliability of renewable energy power generation, improve power quality, and make renewable energy sustainable and stable power supply. With the continuous expansion of capacity and scale and the continuous maturity of integration technology, the cost of energy storage systems will be further reduced. After long-term safety and reliability test, the lithium iron phosphate energy storage system is expected to be widely used in wind power generation, photovoltaic power generation, and other renewable energy power generation, safe grid connection, and improve power quality.

(2) Peak grid. The main means of power grid peak regulation is a pumped storage power station. As the pumped storage power station needs to build two reservoirs, the upper and lower reservoirs are not easy to build on the plain due to geographical constraints, covering a large area, and high maintenance costs. The lithium iron phosphate energy storage system is used to replace the pumped storage power station to deal with the peak load of the power grid. It is not limited by region. It can be selected freely, with less investment, less land occupation, and low maintenance cost. It will play an important role in the peak load regulation of the power grid.

(3) Distributed power plant. A large power grid has its defects, which is difficult to ensure the quality, efficiency, safety, and reliability of the power supply [93]. Important units and enterprises often need dual power supplies or even multiple power supplies as backup and security. Lithium iron phosphate energy storage

system can reduce or avoid power outages caused by power grid failure and various emergencies, and plays an important role in ensuring a safe and reliable power supply for hospitals, banks, command, and control centers, data processing centers, chemical materials industry, and precision manufacturing industry.

2.2.2 Lithium cobaltate oxide batteries

The lithium cobaltate battery with the chemical formula $LiCoO_2$ is an inorganic compound that is generally used as the cathode material for lithium-ion batteries. $LiCoO_2$ is used in the current commercial lithium-ion batteries for the liquid-phase synthesis process of the cathode material (lithium drill) for lithium-ion secondary batteries with a layered structure, and it uses an aqueous solution of polyvinyl alcohol (PVA) or polyethylene glycol (PEG) as the solvent. It uses PVA or PEG aqueous solution as the solvent, dissolves lithium salt and drills salt in PVA or PEG aqueous solution, heats the mixed solution, concentrates it and forms a gel, then heats the resulting gel to decompose, then calcines it at high temperature, grinds and sieves the burnt powder to obtain lithium drill powder [94]. The charging and discharging phase understand a chemical reaction and the essential response and working process of lithium ion in Figure 2.9.

The reaction of lithium diphosphate battery during charging and discharging is described. The response during charging is shown in (2.18).

$$\begin{cases} LiCo_2 \rightarrow Li^{1-x}CoO_2 + xLi^+ + xe^- \\ \quad 6C + xLi^+ + xe^- = LixC_6 \end{cases} \tag{2.18}$$

The reaction during discharge is shown in (2.19).

$$\begin{cases} Li^{1-x}CoO_2 + xLi^+ + xe^- \rightarrow LiCo_2 \\ \quad LixC_6 = 6C + xLi^+ + xe^- \end{cases} \tag{2.19}$$

Advantages of $LiCoO_2$ batteries: $LiCoO_2$ battery is a lithium battery with excellent electrochemical performance. The capacity attenuation rate is less than 0.05%, and the specific capacity of the first discharge is greater than 135 mAh/g. The battery has stable performance, good consistency, easy synthesis in the process, and good safety performance. The operating temperature of the lithium cobalt acid battery is −20°C to 55°C.

Deficiencies of $LiCoO_2$ batteries:

(1) Cobalt has a high price and is only produced in some parts of Africa, with the risk of regional disputes and price changes. In $LiCoO_2$ with rock salt structure, the proportion of movable lithium is only about 50%, i.e., when overcharged, the basic structure will be damaged and the reversible charge–discharge cycle will be lost, which make the overcharge of lithium cobalt acid batteries have potential safety hazards, and additional circuit protection boards are required. The thermal stability and toxicity indexes are not ideal, and the countermeasures are more complex.

(2) Lithium cobalt acid battery preparation, main technical performance in lithium powder manufacture: Lithium cobalt acid battery adopts liquid-phase synthesis

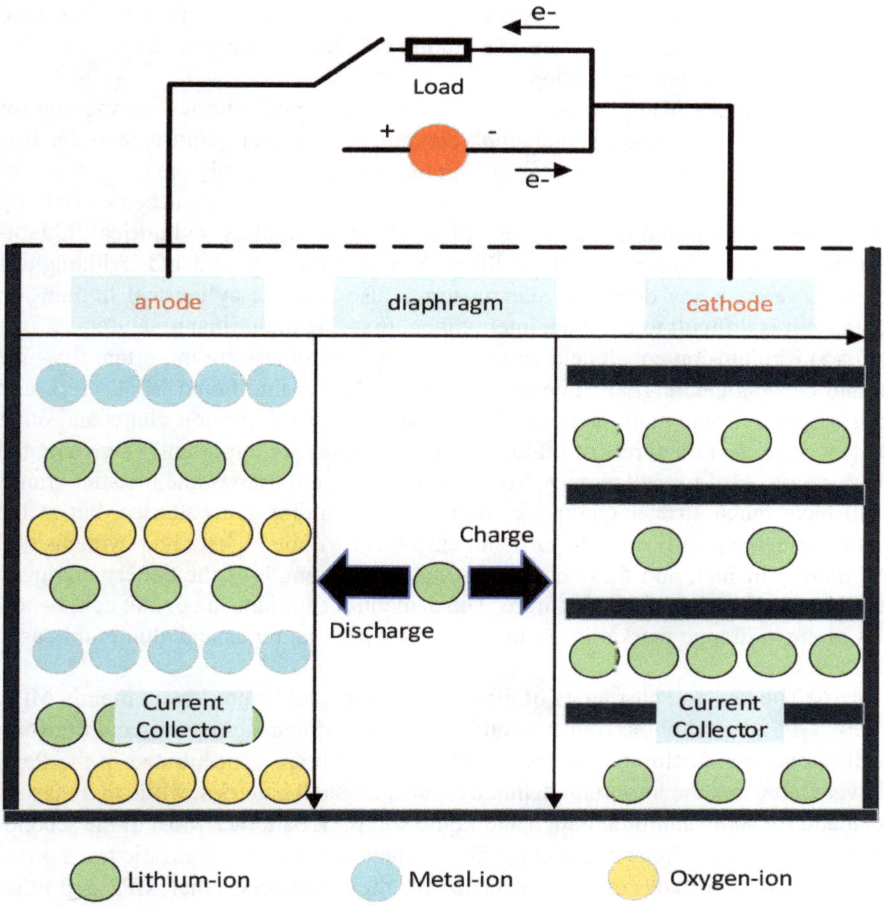

Figure 2.9 Schematic structure of lithium-ion batteries

process, dissolves lithium and cobalt salt in PVA and PEG solution, heats the mixed solution, and then condenses it into the gel, and the gel is calcined at high temperature to form a powder, grinding and screening to obtain lithium cobalt acid powder.

(3) Application of lithium cobalt acid batteries: Lithium cobalt acid batteries have become the mainstream of lithium-ion batteries because of their easy synthesis, high platform voltage, moderate specific energy, and especially excellent cycle performance [95]. However, due to the insufficient reserves of cobalt, the toxicity and overcharge of its preparation overcome the cost of lithium cobaltate battery. Therefore, the lithium cobaltate market is generally located in portable devices and is not suitable for high-power use.

Lithium cobaltate oxide battery is a lithium-ion battery with superior electrochemical performance, the capacity decay rate is less than 0.05%, the first discharge specific capacity is more than 135 mAh/g, the battery performance is stable and consistent, in addition, the process is easy to synthesize, and the safety performance is good. The working temperature of a lithium cobalt-ion battery is –20°C –to 55°C. So it also has relevant applications.

(4) The high-end application of lithium-ion cobalt acid batteries is important for notebook computers because notebook computers have high requirements for battery capacity and safety, so the current application can meet only the consistency of cylindrical lithium-ion cobalt acid batteries and currently can be accepted only by the international brand manufacturers of notebook computers' cylindrical lithium-ion battery manufacturers such as Sony, Sanyo, Samsung, and LG. Although in these 2 years, many domestic manufacturers also produce cylindrical lithium-ion battery, it is difficult to enter the international first-line major manufacturers.

(5) Lithium-ion cobalt acid battery midrange customers are important first-line brand cell phones, digital cameras, mobile DVD, first-line brand MP4, GPS, and digital photo frame manufacturers. The domestic four lithium-ion giants and other larger-scale manufacturers of lithium-ion battery sales are importantly concentrated in this area. Also for cell phones, MP4, and digital photo frames, the first-tier brands will have much stricter quality requirements for lithium-ion batteries, but at the same time, the profits of first-tier brand customers will be higher and payments will be more punctual, and these customers value the strength of the battery company and the quality of the batteries more. The demand for lithium-ion cobalt acid batteries in the midrange market is important for soft pack batteries and aluminum-cased batteries.

(6) The low-end customers of lithium-ion cobalt acid batteries are mainly MP3, MP4, GPS, digital photo frame manufacturers, and domestic second- and third-tier cell phone manufacturers and cottage phones, which are concentrated in the Pearl River Delta, and the important demand is for steel-shell batteries, while there is also demand for some aluminum-shell and liquid soft pack batteries. Most of the second- and third-tier cell phone and cottage phone manufacturers are domestic trade, price is the most important competitive factor for their products; other MP3 and other manufacturers are mainly foreign trade-type business, because in the lower end of their industry chain, the price is often a factor that they attach great importance to. This customer group with price as an important indicator is the target customers of 200 or 300 small- and medium-sized lithium-ion battery cell factories or pack factories scattered in the Pearl River Delta. This part of the low-end customers tends to focus on price rather than quality, and many companies have a bad business reputation, which leads to low-end market battery suppliers not only low profits but also high risk and fierce competition.

2.2.3 Lithium manganese oxide batteries

Lithium manganese battery is a kind of lithium-ion battery. Lithium-ion batteries are classified because their cathode is made of carbon, so the main classification of the industry is based on the positive electrode material and the negative electrode material of lithium-ion battery respectively [96]. The so-called lithium manganese battery refers to the positive battery made of lithium

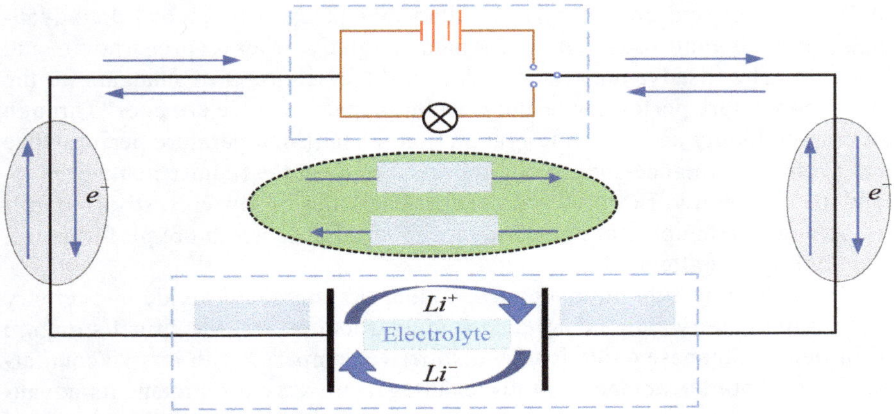

Figure 2.10 *Lithium-ion battery working principle*

manganese material. It is made of EMD (Electric Motor Driven) (raw material, special material for mercury-free alkaline manganese battery) and lithium carbonate (raw material) and corresponding additives through mixing, sintering, and other steps. The more lithium ions are embedded, the higher the charging capacity is. The internal chemical reaction process of the lithium-ion battery is described as shown in Figure 2.10.

Lithium manganese spinel structure refers to the crystal shape, which is suitable for lithium-ion batteries. When lithium manganese is not suitable for lithium-ion batteries, it is a laminated structure. Spinel structures are relatively more stable than laminated structures (although it seems that the stability of different geometries can also be considered according to chemical properties), so spinel structures are currently used in practical applications. Lithium manganese ion batteries can generally reach more than 800 cycles. Compared with traditional lead–acid batteries, the service life is not short. The reaction equation is

$$Li_{1-x}MnO_4 + xLi^+ + xe^- = LiMnO_4 \tag{2.20}$$

In the current three power batteries, the energy density of lithium iron phosphate batteries is lower than that of ternary batteries and lithium manganese batteries, but the price of lithium iron phosphate cathode material is higher than that of lithium. The actual capacity manganese batteries. The requirements are not too high, so most electric bicycle companies do not consider carrying lithium iron phosphate batteries. Compared with ternary batteries, lithium manganese oxide batteries have good safety, and the price is much lower than ternary batteries [97]. Therefore, it has become the first choice for many companies to produce electric bicycle batteries.

A lithium manganese battery is a battery made using lithium manganese material as the positive electrode. The nominal voltage of a lithium manganese battery is 2.5–4.2 V. Lithium manganese battery is widely used because of its low cost and good safety. Lithium manganese battery has the advantages of high multiplication performance, relatively easy preparation, and low cost. The disadvantage is that due to the dissolution of manganese, the high-temperature performance and circulation performance are poor. Through aluminum doping and sintering granulation, its high-temperature performance and cycle performance are greatly improved to meet the requirements of practical use. Generally, $LiMn_2O_4$ has the characteristics of low cost, strong stability, strong low-temperature performance, poor high-temperature performance, and slightly fast attenuation.

In addition to lithium manganese oxide, lithium cobalt oxide and ternary lithium-ion battery cathode are also spinel structures, but the spinel structure of lithium manganese oxide is very distinctive compared with its two counterparts, i.e., both advantages and disadvantages are very prominent. Its advantages are low-temperature resistance, good multiplicative properties, and ease of preparation. The disadvantages are instability of the material itself, the need to mix it with other materials, poor high-temperature performance, poor cycling performance, and fast decay. These disadvantages of $LiMn_2O_4$ come from the properties of manganese [98]. However, due to the widespread availability of manganese, it has significant cost advantages. Lithium manganese batteries are those that use lithium manganese as the positive electrode. Lithium manganese batteries have a nominal voltage of 2.5–4.2 V. Lithium manganese batteries are widely used because of their low cost and good safety.

At present, the demand for lithium batteries is mainly concentrated in lithium manganese oxide batteries, ternary batteries, or mixtures of these two batteries. Some companies that are optimistic about lithium manganese batteries said that it is the general trend that lithium batteries will replace lead–acid batteries and nickel–metal hydride batteries in the future electric bicycle market. Lithium manganese batteries can easily occupy a leading position in terms of performance and price. Lithium batteries also have good expectations in other fields. When the new energy subsidy policy completely withdraws from the market, the cost advantage of lithium manganese batteries will be greatly reflected. It may also make breakthroughs in the field of electric tools and electric vehicles, which is shown in Figure 2.11.

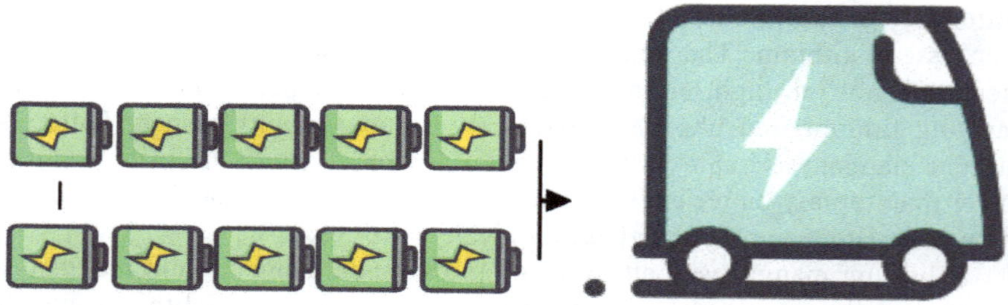

Figure 2.11 Automotive applications of lithium manganese batteries

The so-called lithium battery life refers to that after some time, the battery capacity decays to 70% of the nominal capacity (the battery capacity under room temperature 25°C, standard atmospheric pressure, and 0.2C discharge), which can be regarded as the end of the life. In industry, the cycle life is usually calculated according to the cycle times of fully charged and discharged lithium batteries. During use, irreversible electrochemical reactions will occur inside the lithium battery, resulting in a reduced capacity, such as electrolyte decomposition, inactivation of active substances, collapse, and reduction of positive and negative structures. Many lithium ions are inserted and extracted. Experiments show that a higher discharge rate will lead to faster capacity attenuation. If the discharge current is low, the battery voltage will be close to the equilibrium voltage, which can release more energy.

Due to the huge market share of electric bicycles, the demand for lithium-ion batteries for electric bicycles will increase year by year after the formal implementation of the new national standards and is expected to reach 20 GWh in 2021. At the same time, the application proportion of lithium manganese batteries in electric bicycles will increase year by year, and the application rate of lithium manganese batteries will reach 80%, by 2021.

Lithium manganese materials are used as negative electrode materials for batteries. Due to their characteristics, in the charge–discharge cycle reaction, the material and electrolyte often interact and produce gas precipitation. Therefore, an ordinary lithium manganese battery is easy to expand, resulting in battery expansion and greatly reduced electrical performance, which greatly shortens the theoretical cycle life of a lithium manganese battery. The test data show that the ordinary lithium titanate battery will expand after about 1,500 to 2,000 cycles, resulting in abnormal use, which is also an important reason for limiting the large-scale application of lithium manganese batteries.

The improvement of lithium manganese battery performance is the comprehensive embodiment of the improvement of single material performance and the organic combination of key materials [99]. According to the requirements of fast charging and long service life, in addition to the cathode material, other key raw materials of lithium-ion battery (including cathode material, diaphragm, and electrolyte) should also be combined with special engineering process experience to finally form gas-free lithium titanate laptop battery products, which will be applied in batch on electric passenger cars for the first time.

The test data show that under the conditions of 6C charge, 6C discharge, and 100% discharge, the cycle life of lithium manganese titanate to a single battery exceeds 25,000 cycles and the residual capacity exceeds 80%. At the same time, the flatulence produced by the battery is not obvious, which does not affect its service life. The practical application of fast charging pure electric bus also shows that the electrical performance after battery assembly is also quite excellent, which can ensure the daily commercial operation of pure electric bus.

From the physical point of view, the meter energy level of the material cost determines the voltage of its working platform. When graphite is used as the negative electrode, it differs from lithium metal by 0.2 V, so the working voltage of lithium iron phosphate battery is 3.2 V, lithium cobaltate battery is 3.6 V, and lithium manganese is 3.8 V. Lithium nickel manganese replaces a quarter of the manganese, and the price state of nickel can be changed. How effective? The elevated voltage plateau is advantageous compared with lithium iron phosphate and ternary batteries. It is well known that in today's post-subsidy era for new energy vehicles and energy storage applications, lithium iron phosphate batteries become more valuable and ultimately wait until the market begins to say rather than subsidies say. As we all know, lithium iron phosphate batteries have high value. At present, the highest specific energy of lithium iron phosphate monomer is 180 Wh/kg, the battery pack reaches 150 Wh/kg. Recent bus battery project and then efforts, the monomer can reach 200 Wh/kg, and the system can reach 170 Wh/kg.

2.3 Large capacity lithium-ion batteries

2.3.1 Application areas of utility-scale batteries

There are many applications of batteries, including household appliances, electronic products, aerospace, and military fields [100]. Household appliances, such as radios, clocks, toys, remote controls, etc., mostly use primary batteries. Portable electronics, such as laptop computers, cell phones, and electric cars, all use secondary batteries. In practice, the battery pack is usually made of a battery or several batteries connected in series and parallel. Electric vehicles such as electric bicycles and electric motorcycles power lithium batteries include four main types: valve-regulated lead–acid batteries, colloidal lead–acid batteries, nickel–hydrogen batteries, and lithium-ion batteries. The vast majority of electric vehicles are equipped with lead–acid batteries because of their low- and high-cost performance.

The first batteries used in laptop batteries were nickel–cadmium (NiCd) batteries, but they had a memory effect and had to be discharged before each charge, making them inconvenient to use. They were soon replaced by NiMH batteries, which not only have the advantage of no memory effect but also add 10% more power per unit weight. Cell phone batteries generally use lithium-ion and NiMH batteries. Milliampere (mA) is the unit of battery capacity.

There are five main types of new energy electric vehicle batteries: lead–acid batteries, nickel–cadmium, nickel–hydrogen batteries, lithium-ion batteries, lithium iron phosphate ion batteries, and fuel-powered batteries. However, in larger devices, energy storage batteries need to be employed. With the development of society, energy storage battery occupies an increasingly important position in society. The energy storage battery industry to achieve large-scale development, the formation of a more complete industrial system, especially lithium batteries and lithium polymer batteries class, becomes a new growth point of the energy sector economy.

The application of renewable energy generation in the power system is increasing, and the power industry is moving towards low-carbon and environmental protection. In the process of grid connection, the power volatility of

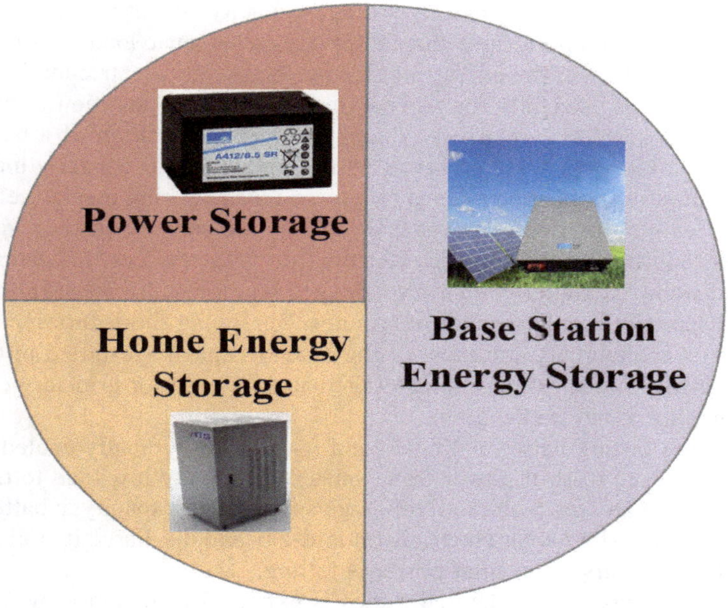

Figure 2.12 Energy storage battery applications

renewable energy generation is large, and the power system will become unstable when the power supply is large. Energy storage technology can be used as a buffer for the power system, which plays an important role in improving the quality of power generation and enhancing the stability of the power grid. Energy storage technology can be applied to many aspects of power generation, distribution, transmission, and consumption in the power system and can also be widely used in electronic products, electric vehicles, and other fields. As a mature and efficient technology, large- and medium-sized battery energy storage have become one of the priority development directions for energy storage.

Currently, in the energy storage battery market, there are three main applications: power storage, home energy storage, and base station energy storage, of which the base station backup power field currently accounts for a larger proportion, accounting for nearly half of the entire energy storage battery market share. Home energy storage by Tesla set off a wave of energy home, there is greater room for further development and expansion, but currently mainly concentrated in the United States, Germany, Australia, and Japan and other overseas regions. In addition, home energy storage systems are used as an emergency power source in remote areas, as well as in areas with a high incidence of natural disasters such as earthquakes and hurricanes, to eliminate the inconvenience of frequent power outages due to disasters or other reasons, which are shown in Figure 2.12.

More than 40% of the battery storage capacity operating in the United States in 2020 could perform both grid service and electric load shifting applications. About 40% of the energy storage performs only electric load shifting, and about 20% is used only for grid service. The average duration of batteries used for grid service is relatively short (the average duration of a battery is the time it takes to provide power at its nameplate power capacity until it is depleted duration) of batteries used for electric load shifting is relatively long. Batteries with a duration of fewer than 2 hours are considered short-duration batteries, and almost all batteries can provide grid services that help maintain grid stability. Batteries that provide grid service will discharge for short periods, sometimes only seconds or minutes. Deploying short-duration storage batteries is economical, and most of the battery capacity installed in the late 2010s consisted of short-duration storage batteries used for grid services. But over time, this trend is changing.

Batteries lasting between 4 hours and 8 hours are typically cycled once a day and are used to shift power from times of relatively low load to times of higher load. In an area with relatively high solar power capacity, a battery that cycles daily can store solar electricity at mid-day and discharge it at night during peak load hours when solar power is falling.

Battery storage in the United States is expected to increase by 10 GWh by the end of 2023, with more than 60% of battery capacity used in conjunction with solar power plants. As of 2020, the majority of battery storage units installed in solar facilities are used to shift electrical loads for an average duration of more than 4 hours. The following is an introduction to different areas of energy storage methods.

Domestic energy storage: Domestic energy storage lithium-ion battery systems are mainly energy storage systems installed on residential properties. Its operation mode includes independent operation, operation with small wind turbines, rooftop photovoltaic power generation, and other renewable energy generation equipment, as well as home energy storage thermal equipment and other supports. Applications of home energy storage systems include electricity bill management, electricity control (low charge and high discharge) power supply reliability, distributed renewable energy acquisition, application of electric vehicle energy storage batteries, etc.

PV energy storage: PV (Photovoltaic) energy storage is of great importance to smooth the power curve, improve the accuracy of power prediction, enhance the overdistribution capability of the PV power system, and improve the friendliness of the grid. Suitable for photovoltaic, solar energy storage lithium-ion BMSs, the product performance is stable and highly secured. As the energy storage device of the PV system, the energy storage efficiency of a lithium battery can be increased to 95%, which can greatly reduce the cost of solar power generation.

Distributed energy storage system: Distributed lithium-ion battery pack energy storage system can optimize demand-side distribution, regulate load regulation of microgrids, improve the economic efficiency of microgrid systems, and solve the mismatch between peak power consumption and peak power generation. The flexibility and rapid response capability are fully utilized to achieve the optimal allocation of power resources. With the escalation of supply–demand contradiction in the grid system and related policies, the domestic distributed energy storage market has taken off, and the market is gradually heating up.

Microgrid energy storage system: Due to the influence of the environment and weather, renewable energy sources such as photovoltaic and wind power have strong intermittency and volatility, which will bring huge potential problems to the power quality and stability of the grid and may lead to accidents. Therefore, to better coordinate the grid with renewable energy sources, make full use of their value and obtain higher economic benefits; microgrid technology has emerged as a more a transition metal ions plicable, efficient, and stable technology and will soon become a more mature grid structure, which can integrate DC power sources and loads, such as photovoltaic arrays, fuel cell modules, battery storage, motor-driven loads, and converter–generator loads. Among them, DC microgrid technology has been used in power distribution systems and data centers as well as islanding applications (e.g., space stations, aircraft, ships, remote areas). More importantly, microgrid technology provides a stable source of energy for infrastructure using renewable energy sources to operate with unity.

A microgrid is a collection of load and microenergy, a small power generation and distribution system composed of distributed power supply, energy storage device, energy conversion device, load, monitoring, protection device, etc. It operates in the networked mode under normal conditions and can operate independently in emergencies. Lithium-ion battery pack energy storage is an important part of the microgrid energy storage power plant. A high-performance lithium iron phosphate lithium-ion battery pack energy storage system can store excess energy and send it to the load when needed to make full use of various energy sources. In small hydropower areas, areas with declining power supply capacity, power shortage areas such as islands without electricity, energy storage systems independently or in conjunction with other energy sources can provide continuous power supply for important loads when the microgrid loses power, importantly solving the problem of power supply stability and security. It plays a crucial role in the whole microgrid.

New energy vehicles: The development of new energy vehicles cannot be separated from the construction of charging infrastructure, but there are bottlenecks in the construction of charging stations in cities. Due to the shortage of grid capacity, the construction of charging stations will cause high pressure on the local grid peaking, and large-scale access to charging stations and PV will have an impact on the grid. The construction of supporting energy storage facilities is conducive to improving the quality of local grid electricity, and adding new charging station sites can be selected.

Diesel-powered areas: Energy storage technology can replace diesel generators, reduce the cost of power generation, and at the same time reduce the air and noise pollution caused by diesel generators.

Renewable energy power station self-built: Auxiliary renewable energy to the grid, effectively solves the abandoned light problem, improves economic efficiency, reduces transient power changes, and reduces the impact on the grid. Tracking plan scheduling improves grid controllability, improves power generation forecast accuracy, and improves grid friendliness. Due to the high cost of energy storage, even after considering the investment of energy storage can reduce the amount of abandoned light, its return on investment is still low, power station party initiative to invest in supporting energy storage is not strong.

Trading with renewable energy power station: Energy storage system participates in the paid peaking auxiliary service of electricity, thus helping to make up for the shortcomings of power supply peaking capacity. At noon, the energy storage system will buy the waste photoelectric at a low price, and sell the photovoltaic on grid price during the power shortage at night, so as to achieve a win-win situation between the energy storage system and the power station. Regarding the old power station with high electricity prices, such as model has strong feasibility.

FM auxiliary services: Energy storage systems can play a supporting purpose for the grid by participating in FM auxiliary services. Some countries open their FM market and adopt competitive bidding mechanisms, and the participation of battery energy storage is relatively high, such as the United States, but the volume of the FM market is limited.

In recent years, a new frequency modulation mode of new energy wind power participating in power grid frequency modulation has gradually emerged in the world. The main reason is that the traditional frequency modulation mode has its limitations and shortcomings and has been unable to meet the frequency quality requirements of the modern power grid. For example, because the response time and climbing rate of the thermal power unit are affected by the factors of the unit itself, it is not suitable to adjust the network frequency in a short time, and the actual value of the frequency modulation capacity of the unit participating in the primary and secondary frequency modulation may not reach the theoretical value of the frequency modulation capacity. And automatic generation control (AGC) instructions cannot be accurately tracked. The units serving for the frequency adjustment of the grid not only increase the loss of equipment but also increase the consumption of fuel, thus greatly increasing the operating cost. Hydropower units have many advantages, such as short FM response time, but they have great limitations, such as geographical and seasonal constraints, so the FM capacity they can provide is extremely limited. Therefore, in the rapid growth of new energy power generation, wind power in power grid frequency modulation way has become the present developing and research object, but after a study found that although involved in wind power grid frequency control can effectively reduce the frequency deviation, but due to the volatility of its generation and the uncertainty of wind speed, leading to the grid frequency the second fall, and then affect the economy of the wind farm operation.

Therefore, the way wind turbines participate in frequency modulation is still not enough to meet the frequency modulation demand of smart grid in the future, and battery energy storage technology has become one of the research hotspots of power grid-assisted frequency modulation means in the current era due to its excellent power response ability. The study found that the battery energy storage technique can quickly, accurately, and effectively solve the disadvantages of the current wind power frequency modulation way, put energy storage battery into the wind power system, can effectively alleviate the secondary frequency drop problem, to restore the system frequency phase, energy storage battery system to provide the part of the power output, leave the speed of the wind turbines recovery time allowance. Thus, the second drop of power grid frequency is effectively restrained and the stability of power grid frequency is greatly improved. And energy storage battery power output in the short term and better absorption ability are wind FM units, make its frequency modulation for higher efficiency than wind power frequency modulation way and can minimize the spare energy capacity utilization in power system. When the power system operation failure occurs, this section can be used to save the backup active capacity for load power supply. Of course, it can also be used for power network peak regulation, so it can be used as an effective means of auxiliary frequency regulation.

As an important way of energy storage, battery energy storage is not bound by factors such as geographical conditions, which can realize flexible capacity configuration, FM the advantages of fast response, and the other in terms of energy storage technology is more suitable for mass production and large-scale applications, in the stability of power grid has unique advantages in the application of auxiliary means. The efficiency of energy storage batteries varies with different types, generally ranging from 60% to 90%. At present, lithium iron phosphate batteries, vanadium liquid flow batteries, and lithium-ion batteries are widely used in practical grid application research.

In recent years, as the battery energy storage system has a stable power output power response ability, quickly and accurately make it to the power grid load frequency control in areas such as a powerful auxiliary means, it can be either according to work-like synchro frequency regulating properties to control their complete power to absorb and emit, achieving regulation of power grid frequency. It can also use the area control error signal determined in AGC to control the charge and discharge of the energy storage battery, to guide its participation in the second frequency modulation of the power system. By the end of 2019, energy storage battery technology is developing rapidly in the application of power grid frequency modulation. Energy storage battery technology is developing rapidly in the field of power grid frequency modulation. In terms of region, it is mainly represented by the United States, China, Chile, France, and Israel. Therefore, in terms of the advantages of the energy storage battery in the field of frequency modulation, it is necessary to conduct an in-depth exploration of its frequency modulation technology, which not only provide theoretical support for its application in actual working conditions but also break through the technical bottleneck that has not been solved before, laying a solid foundation for the development of smart grid in the future.

2.3.2 Characteristics of utility-scale battery systems

Lithium-ion battery energy storage power plants embody greater potential in promoting the integration of renewable resources and improving the flexibility of system operation due to their advantages such as flexible resource allocation, fast response time, and independence from the weather as well as geographical location [101]. This is mainly because of the following features: lithium-ion battery packs are easily combined in multiple ways to meet higher operating voltages and larger operating currents. Testable and diagnosable battery capacity and performance enable the control system to realize the scheduling control of power plant load with predicted battery capacity and performance. It has high security and reliability. Under normal use, the normal service life of the battery is not less than 15 years. Within the limit, even if a failure occurs, it is within the controlled range and should not explode, burn, and other failures that endanger the safe operation of the power station.

Having a good fast ring response, large multiplier charge and discharge capacity generally requires 5–10 times the charge and discharge capacity. Lithium-ion battery packs almost do not maintain. According to the current situation of domestic and foreign energy storage power plant use and battery characteristics, the proposed energy storage power plant battery selection is closely for lithium iron phosphate ion batteries.

With high efficiency, high energy density, and long cycle life, lithium battery energy storage is a suitable technical route for engineering applications at this stage. According to the incomplete statistics of the global energy storage project library of Zhongguancun Energy Storage Industry Technology Alliance, by the end of 2019, the cumulative installed capacity of energy storage projects commissioned worldwide reached 184.6 GWh, and the cumulative installed capacity of electrochemical energy storage was 9520.5 MW, among which, the cumulative installed capacity of lithium-ion battery energy storage was the largest, reaching 8453.9 MW. If the battery system is not equipped with the advanced BMS, it may lead to poor performance and premature aging of the battery monomer or battery pack.

The energy storage battery used for new energy power station is an energy type lithium battery, the charge and discharge rate of this type of lithium battery are low, generally at 1C and below, in some off-grid new energy power station projects, the charge and discharge rate of lithium battery will even be as low as 0.1C (energy storage length of 10 hours), such lithium batteries focus on continuous discharge, safety, and cycle life and other characteristics.

The type of lithium battery used in energy storage batteries in new energy power plants is not the same, China's energy storage batteries mainly use lithium iron phosphate batteries. While in South Korea and the United States, energy storage batteries are mainly used in the ternary lithium battery lithium nickel cobalt manganate (NCM) batteries. Lithium iron phosphate ($LiFePO_4$(LEP)) battery is a lithium battery with LFP material as the positive electrode of the battery, the price of such lithium batteries is low, stable structure, in the process of repeated charging and discharging can maintain the stability of the structure, high cycle reversibility.

LFP is an ordered olivine structure, which belongs to the Pnma space group, and its high-temperature stability is significantly better than other materials. It also has thermal stability and low reactivity with organic electrolytes. But on the other hand, the PO_4 tetrahedra in the crystal structure of LFP restrict the free movement of Li^+, so that Li^+ only has a one-dimensional transport channel, which reduces the diffusion migration rate of Li^+ and therefore also affects the discharge multiplicity performance of the LFP battery.

During the charging and discharging process of the LFP battery, the positive material LFP will be converted into iron phosphate ($FePO_4$), but the volume after conversion will only increase by about 6.9%. Therefore, the volume change rate of positive material is small during the charging and discharging process, so the LFP battery has good cycle performance.

According to the types of cathode materials, ternary lithium batteries can be divided into lithium NCM batteries and lithium nickel cobalt aluminate batteries. Because ternary lithium battery has good overall performance and high specific energy, it is widely used in passenger car power batteries. In the energy storage batteries of new energy power stations, NCM batteries are mainly used. According to the relative content of transition metal ions, the types of NCM batteries can be divided into NCM111 ($LiNi1/3Co1/3Mn1/3O_2$), NCM424, NCM523, etc., of which NCM111 is the most widely used.

The cathode material of the NCM battery is the α-$NaFeO_2$ lamellar structure, which belongs to the R-3m space group. The role of each transition metal ion in $Li(Ni, Co, Mn)O_2$ is different. The cost of Mn is lower, and Mn^{4+} helps to improve the safety and structural stability of the cathode material in NCM battery; however, the research shows that the high content of Mn will destroy the lamellar structure of the material and make the specific capacity of the material decrease. Co^{3+} can stabilize the laminar structure of the material and help to improve the cycling and charge–discharge multiplier performance of the material while Ni^{2+} helps to increase the bulk energy density of the material, and the higher the Ni content, the higher its theoretical specific capacity, but it is more difficult to take off the Li^+ into the Ni^{2+} position, i.e., the increase of Ni content means the cycle performance of lithium batteries is reduced. Meanwhile, in ternary lithium batteries with high Ni content, Li^+/Ni^{2+} mixing will occur in the cathode material, thus causing Li^+ precipitation. The electrochemical reaction of ternary lithium batteries during charging is complex, and different reaction processes exist at different potentials, which is shown in Table 2.4

Due to the difference between positive and negative materials and electrolyte materials, as well as the difference in process, there will be some differences in the performance of lithium batteries [102]. The key performance of lithium batteries includes energy density, charge–discharge ratio, cycle life, safety, and working temperature. The Table 2.4 shows the performance of lithium batteries.

(1) Energy density

The energy density of the battery refers to the electric energy released by the unit volume or unit mass of the battery, which is largely determined by the performance of the positive and negative materials of the battery. Theoretically, the mass energy density of NCM batteries is about 1.5–1.8 times that of LFP batteries.

Table 2.4 The lithium battery comparison

	NCM	LFP
Energy density	The mass energy density of NCM batteries is about 1.5–1.8 times that of LFP batteries.	
Charge–discharge ratio	**1C–2C**	**<1C**
Cycle life	Under the same working conditions, the cycle life of an NCM battery is only about 1/2 of that of an LFP battery.	
Security	The LFP safety is better than that of an NCM battery.	
Operating temperature	The low-temperature applicability of the LFP battery is not as good as the NCM battery.	

The battery used in the power station requires high energy density, which can save land, to reduce the transportation, installation, and other costs of the energy storage battery.

(2) Charge discharge ratio

At present, the charge–discharge ratio of mainstream LFP batteries in the market is 1C or below, which is mainly due to electricity. The performance and cycle life of the pool are considered. The charge–discharge ratio of the NCM battery is mostly 1–2C. The common charge–discharge rates of lithium batteries for energy storage batteries are 0.25C (4 hours energy storage time), 0.5C (2 hours energy storage time), and 1C (1 hour energy storage time). With the decrease in the cost of lithium batteries, a 1C lithium battery is usually used to meet the power requirements of the energy storage system with a charge–discharge ratio of 2C in early projects. The method is safer and the service life of lithium batteries is longer. At the same time, because the energy storage system with a charge–discharge ratio of 1C can have a variety of profit models such as frequency modulation and peak valley arbitrage, it is more economical than the energy storage system with a charge–discharge ratio of 2C.

(3) Cycle life

When the ambient temperature is 25°C, the charge–discharge ratio is 1C and the remaining battery power (EOL) is 80%, the cycle life of an LFP battery with a cell capacity of 260 Ah is 6,000 times. The above other conditions remain unchanged. Under the condition of EOL of 60%, the cycle life of an LFP battery with a cell capacity of 260 Ah can reach more than 7,000 times. Under the same working conditions, the cycle life of an NCM battery is only about 1/2 of that of an LFP battery. This is mainly because the discharge voltage platform of the LFP battery is stable, and there is no phase change during charging and discharging, while the discharge voltage platform of the NCM battery operates unevenly and there is phase change during charging and discharging, which is easy to cause the attenuation of its cell capacity and shorten the cycle life. At a low charge–discharge ratio (below 0.5C), the cycle life of the NCM battery can be significantly improved to 5,000–6,000 times. At present, new energy power stations at home and abroad require that the lithium battery used for its energy storage battery has a cycle life of 5,500 times or a service life of 15 years (under the condition of one cycle per day).

(4) Security

Thermal runaway in lithium battery is a chain accumulation of heat in the process of battery heat release, followed by explosion and fire when the heat production intensity and quantity completely suppress the heat dissipation intensity with the increase of battery temperature. When the temperature of oxygen reaches the limit of 300°C, the battery is easily overheated and collides, and a large number of batteries will be easily ignited under the condition of a high temperature of NCM ~300°C. The thermal stability of the LFP battery is good. In the case of thermal runaway, the oxygen in its crystal structure exists in the structure of phosphorus oxygen tetrahedron and will not release oxygen. Therefore, its safety is better than that of an NCM battery. Extreme working conditions such as extrusion and collision are unlikely to occur in the energy storage battery for a new energy power station, but the cell capacity of the energy storage battery for a new energy power station is much higher than that of the power battery for passenger cars, and it is more difficult in temperature control. In addition, the safety requirements of new energy power stations are also higher.

(5) Operating temperature

Taking 25°C as the reference temperature, the discharge capacity of the NCM battery and LFP battery measured at 55°C has no obvious attenuation compared with that measured at 25°C. However, at low temperatures, especially below −20°C, the discharge capacity of the NCM battery is significantly higher than that of the LFP battery. Therefore, for the energy storage system built-in high latitudes, due to the large temperature difference between day and night and the low temperature at night, the low-temperature applicability of the LFP battery is not as good as the NCM battery. It is not suitable to adopt the direct arrangement of outdoor containers. By adopting the station building arrangement, the LFP battery can be placed indoors, which can improve the operating environment of the energy storage system. This problem can be solved.

2.3.3 Operational challenges of utility-scale battery systems

The connection of battery storage power station to power system involves many security and stability problems. For example, the optimization strategy of the battery control system, the emergency strategy of battery system communication failure, lightning protection, and fire protection technology. And one of the most important parts is the effectiveness of the BMS. Considering the safety and reliability of energy storage power station operation, the optimization strategy of the battery control system is one of the operational challenges. The control system of utility-scale battery systems includes BMS, power conversion system (PCS) protection system, background monitoring system, relay protection, and safety automatic device, as shown in Figure 2.13.

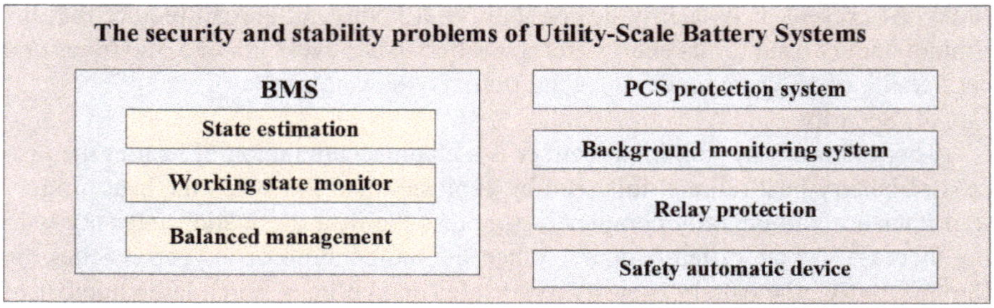

Figure 2.13 The security and stability problems of utility-scale battery systems

BMS can realize battery status monitoring, operation control, insulation monitoring, balance management, protection alarm, and communication functions and ensure the normal, stable, and safe operation of the system through real-time monitoring of battery status. BMS is divided into three levels: master control unit, and slave control unit [103]. The master control unit centrally manages the energy storage battery stack and is responsible for the strategy implementation, data collection, and external communication of the battery system. The main control unit is responsible for the management of battery clusters, including total voltage detection, current detection, insulation detection, charge and discharge process management, fault alarm processing, etc. The slave control unit has the function of monitoring the voltage and temperature of the single battery in the battery pack and can safely manage the charging and discharging process of the battery pack.

The PCS protection control system monitors the operating conditions of the energy storage PCS and can trigger the protection action shutdown under fault conditions such as overvoltage, overcurrent, and BMS protection signal. It has the function of fault recording. The PCS controller receives the command from the background monitoring system and adjusts the PC working mode according to the command, such as charge and discharge mode and active and reactive power.

The background monitoring system monitors and controls all electrical operation equipment and energy storage equipment in the station. In addition to the electrical monitoring system contained in the conventional substation, it also integrates the energy management system (EMS) to receive dispatching instructions and realize AGC, AVC, and other functions.

Relay protection and safety automatic devices include public measurement and control device, 10-kV line protection measurement and control device, station transformer protection measurement and control device, anti-islanding protection device, frequency, and voltage emergency control device, and source network load interactive terminal.

Second, the emergency strategy for battery system communication failure is also an operational challenge. The communication system of utility-scale battery systems can be divided into station control layer, compartment layer, and energy storage unit layer. Station control layer equipment includes monitoring host, historical data server, etc. Bay level equipment includes a bay level switch, public measurement, control device, etc. Energy storage unit layer equipment includes an energy storage unit layer switch, PCS secondary system, BMS, and local monitoring device. In case of communication failure, an effective emergency strategy is to improve the safe operation level of utility-scale battery systems.

In case of a lightning stroke, utility-scale battery systems may be affected, such as equipment damage, abnormal operation, and even fire and explosion. Therefore, lightning protection design is particularly important. The lightning protection of primary equipment includes the lightning protection of the whole station and each primary equipment. To protect utility-scale battery systems from lightning, metal lightning receiving pins can be installed on the top of the station to introduce lightning into the ground to realize the protection of the whole station. The high-voltage distribution equipment and battery energy storage unit in utility-scale battery systems may also be damaged by lightning attacks. Lightning arresters and surge protectors can be installed at the ports of such equipment. In addition, thunderstorm weather is often accompanied by lightning induction, which may interfere with or even damage the secondary equipment. Therefore, the lightning protection of secondary equipment is also very important, mainly following the following principles: secondary circuit grounding, equipotential connection of secondary equipment, cable shielding layer grounding, and add signal surge protector. In the secondary system design, the overall layout of the line should be deeply studied to reduce the probability of being struck by lightning.

Utility-scale battery systems take lithium-ion batteries as the carrier. During operation, overheating, short circuits, overcharge, and discharge may occur and even cause a fire. Therefore, the fire protection design of utility-scale battery systems is essential. In recent years, a variety of energy storage battery fire-fighting technologies have been proposed, such as fire-fighting technology based on the fully closed fire-fighting pipeline, fire-fighting technology based on physical isolation, etc. The emergence of these technologies is expected to improve the safety of energy storage battery systems.

In addition to the parameter characteristics of the battery module, the service performance of the battery for energy storage is closely related to the function of the BMS. The BMS is of great significance to protect battery safety, improve battery utilization, and prolong battery life. The BMS of utility-scale battery systems is the brain of the energy storage battery system. It is mainly used for real-time monitoring, fault diagnosis, SOC estimation, short-circuit protection, leakage detection, and display alarm of the energy storage battery, to ensure the safe and reliable operation of the battery system. It is an important part of the whole battery system.

BMS can monitor and collect the state parameters of the battery module in real time, calculate and process the relevant state parameters, get more system parameters, and effectively control the battery system according to the specific control strategy. At the same time, BMS can exchange information with other external equipment (converter, energy management unit, fire fighting, etc.) through its communication interface and analog/digital input and output interface to form the linkage of the whole energy storage system. All system components can be used to monitor the whole energy storage system efficiently and reliably through reliable physical and logical connections.

As the core component of the battery system, the BMS is the bridge between the battery pack and external equipment, which determines the utilization rate of the battery. Its performance is very important to the use cost and safety performance of the battery system. With the development of technology, many functions have been gradually added to BMS. For utility-scale battery systems, the effective control of battery pack charge and discharge by BMS can achieve the purpose of prolonging service life, reducing operation cost, and ensuring the safety and reliability of battery system application.

At present, lithium-ion battery packs are generally formed in series and parallel to meet the use requirements of energy storage batteries. In the process of use, due to the performance differences of single batteries, changes in ambient temperature, overcharge and discharge, and other factors, the performance of the battery pack depends on the single cell with poor energy. Therefore, it is necessary to effectively manage the energy of the lithium-ion battery pack through BMS, to improve the service efficiency of the lithium-ion battery, prolong the service life of the battery pack, reduce the operation cost, and improve the reliability of the battery pack.

In the whole life cycle of the battery pack, the monitoring and regulation of the core parameter SOC by BMS will affect the effect and safety of energy storage. Therefore, it is very necessary to monitor the change of this parameter in real time to ensure the working performance of lithium utility-scale battery systems. Due to the immature BMS technology, the inaccurate prediction of energy storage caused by the imperfect SOC estimation in the process of use, as well as the potential safety hazards such as energy imbalance and thermal runaway, has seriously restricted the development of utility-scale battery systems.

The main functions of BMS are as follows:

(1) Accurately estimate the state of charge of the power battery pack

Accurately estimate the SOC of the power battery pack, i.e., the remaining power of the battery, ensure that the SOC is maintained within a reasonable range, and prevent the damage to the battery due to overcharge or overdischarge, to predict how much energy remains in the utility-scale battery systems energy storage battery or the SOC of the energy storage battery at any time.

(2) Dynamically monitor the working state of the battery system

In the process of battery charging and discharging, the terminal voltage and temperature of each battery in the power battery pack, charging and discharging current, and the total voltage of the battery pack are collected in real time to prevent overcharge or overdischarge of the battery.

(3) Balance between single cells and battery packs

That is the balance between single batteries and battery packs to make each battery in the battery pack reach a balanced and consistent state. Battery equalization is generally divided into active equalization and passive equalization. At present, BMS mostly adopts passive equilibrium. Equalization technology is a key technology of battery EMS that are being studied and developed around the world.

Chapter 3
AI-based equivalent modeling and parameter identification

3.1 Overview of battery equivalent circuit modeling

Battery management systems (BMSs) are widely used in utility-scale batteries as an effective way of managing energy. A key function of the BMS is to assess and monitor the performance of the traction battery through accurate characterization of various battery states [104]. These states include the state of charge (SOC – quantity of deliverable Ampere-hour charge at any time), state of health (SOH – ability of a battery to provide its nominal capacity over its service lifetime), state of power (SOP – a quantity describing the battery's power capability), state of function (SOF – a binary yes/no parameter indicating the battery's ability to complete a task), and so on.

Whilst direct measurement techniques such as coulomb-counting (integration of battery current over the charge or discharge period) are easy to implement for SOC estimation, they suffer largely from erroneous initialization of SOC, and drifts caused by current sensor noise and battery capacity variations due to temperature and SOH. Moreover, the direct measurement of the other battery states of interest (i.e., SOH, SOP, and SOF) for real-time applications is somewhat impossible. Hence, battery models are often utilized within the BMS to indirectly infer and monitor the battery's operation through the measurement of its terminal voltage, current, and surface temperature. In addition to accurate characterization of the battery status, a candidate model is also desired to be computationally efficient. In other words, there should be a balance between model accuracy and complexity so that it can easily be embedded on a simple and inexpensive microprocessor unit, similar to those found in electric vehicle (EV) BMS hardware.

The equivalent circuit models (ECMs), parameter identification, and state-space description play a very important role in the SOC estimation process, which involves multidisciplinary integration. The establishment of the battery model mainly includes the following aspects:

1. Selecting the appropriate model. The technical route of battery modeling is determined according to the performance characteristics, application, and model accuracy of the batteries. The appropriate model can be selected from the

appropriate model set, including the model structure with unknown parameters obtained according to the mechanism, an electrochemical model established based on the internal electrochemical reaction of the battery, or a black-box model.

2. Designing experimental schemes and identifying model parameters. The identification of battery model parameters should be based on the actual measurement data. The identification of model parameters aims to minimize the error between the fitted curve and the measured curve. The commonly used identification methods of model parameters include the least-square (LS) method, genetic algorithm, neural network (NN) method, Kalman filtering algorithm, etc.
3. Verification of model accuracy. There are mainly four ways of verification: (1) Use prior knowledge verification to judge whether the model is practical or not according to the existing knowledge of the system. (2) After a model is identified by one set of data, the applicability of the model is verified by the another set of data that is not involved in the identification. (3) Analyzing the actual response and impulsion response by comparison. (4) The autocorrelation function of the excitation signal is used for verification.

Three types of the most commonly used models of lithium-ion batteries are: electrochemical models based on chemical reactions within the battery, NN models that simulate the work of the human brain, and ECMs built using electronic components [105]. Electrochemical models, including the pseudo-two-dimensions (P2D) model and single-particle model (SPM), are rarely used in engineering applications because of their complex structure and high computational cost [106]. The ECM is currently the most widely used battery model, and its estimation accuracy and robustness are better than the other two types of models. Since the battery is a highly nonlinear system, which is affected by temperature, working conditions, aging degree, and other conditions. It will increase the complexity of calculation of these factors taken into full consideration in battery modeling, and the general controller cannot meet the requirements. The NN model has a large estimation error in the case of insufficient sample data. The convergence and stability of the algorithm are temporal, which cannot guarantee the timeliness of the model. However, with the increase of data samples, it is more and more able to accurately reproduce the nonlinear characteristics of the battery. The key points of the model establishment are described as follows.

The research of the power battery model must fully consider the characteristics of the application environment. How to develop an accurate, maintenance-free, and reliable power battery model in a complex and changeable vehicle environment is the focus and difficulty of EV BMS research, and how to objectively and systematically evaluate the credibility of the battery model also needs serious consideration.

Although the electrochemical model can better evaluate and simulate the key performance indicators of the power battery than other models, there are many parameters involving the internal materials of the battery, and it is difficult to update and calibrate regularly when the real car is applied, and the model operation is large, and it is generally only used for battery performance analysis. Although the simplified electrochemical model can reduce the amount of calculation of the model, the dynamic simulation accuracy of the model is poor because it ignores the nonlinear characteristics of the battery and the characteristic parameters of the SOC. Therefore, from the perspective of fine management of EV power batteries, this type of model has certain limitations [108]. The black-box model and analysis model represented by the NN model can better simulate the nonlinear characteristics of the power battery. However, in model training, the choice and number of input variables affect the accuracy and amount of computation of the model, and the error of the model is affected by the training data and training method. Therefore, NNs are more suitable for mass production and mature products with good consistency of power batteries. At the current battery level, this type of model is constrained by at least three challenges: the size of the training set, the suitability of the model, and the fidelity of the model. Therefore, this model is generally used in laboratory environments and BMS predevelopment work.

Compared with the above two types of models, the ECM of the centralized parameter type has the advantages of simple structure, few parameters, and ease of identification, so this type of model is widely used [109]. In recent years, researchers have developed ECMs for a variety of structures, including Rint models, Thevenin models, partnership for a new generation of vehicles (PNGV) models, Dual polarization (DP) models, general nonlinear (GNL) models, etc. And it seems that the more complex the model, the better the accuracy, so the number of resistor–capacitor (RC) networks seems to determine the prediction accuracy of the model. However, for a given power battery system, the relationship between the number of RC networks and the accuracy of the model is still not resolved. Not the more RC networks, the better, too many RC networks increase the complexity and computation of the model. On the other hand, the characteristics of the power battery are directly related to its aging state, working temperature, etc., and a perfect battery model needs to achieve accurate prediction under uncertain application conditions. Therefore, how determining the balance between the structural complexity of the power battery and its prediction accuracy has become the focus and difficulty of the development of the power battery model.

In addition, the uncertainty of the use environment and application conditions poses the following challenges to the accurate modeling of the battery:

1. The structure of the ECM is too complex, and the model parameters are too fixed, which makes the promotion ability poor in different application environments.
2. The electrochemical model ignores the dynamic characteristics of the power battery, making the dynamic voltage prediction accuracy poor.
3. The model of using a "large monomer" to establish a power battery pack ignores the inconsistencies between the monomers and will greatly reduce the reliability of the system.
4. The use of screening into groups, and then the use of power battery pack monomers to characterize the entire battery pack do not ensure the consistency of all

monomers in their service cycle, when the consistency becomes worse after the reliability of the battery pack model will be significantly reduced.

3.2 Modeling types and concepts

The techniques of battery modeling can be classified into mechanism modeling, experimental modeling, and hybrid modeling methods. Based on scientific equations such as physical equations and concepts of chemical reactions, the mathematical modeling framework is constructed. The managed object is regarded as a black box, and the model is constructed by experimenting with the variant law on the dynamic characteristic parameters of the target object [110]. The battery is a nonlinear electrochemical energy storage system, and its internal interactions and reactions are difficult to be explained by precise equations. Experimental data models, such as NNs, involve a great deal of feedback and learning, so hybrid modeling is more prevalent. The electrical characteristic model is usually divided into three parts: the black-box model, electrochemical model, and an ECM.

Black-box models based on NNs or machine learning methods usually require a large amount of test data as a training set to improve the prediction accuracy and adaptability of the model [111]. Power battery complex internal chemical reaction process and uncertain temperature, aging conditions, dynamic conditions and other application environments, many factors, complex relationship, can also be simplified to the black-box model to study, its essence is through the nonlinear relationship of the input amount to characterize its external characteristics, this type of model can accurately reproduce the nonlinear characteristics of the power battery, which is shown in Table 3.1.

The ECM uses electrical components, such as inductors, resistors, capacitors, voltage sources, current sources, etc., to describe the charging and discharging characteristics of lithium-ion batteries through different combinations, which belong to the semiempirical model ECM by numerically expressing electrical components and combining parameter identification algorithms, such as KF, EK F, UK F, and other algorithms, to simulate the charge and discharge characteristics of lithium-ion batteries, and combine the state diagnostic equation to achieve the diagnosis and estimation of the relevant states of lithium-ion batteries. To obtain the battery state conveniently and accurately, it is necessary to select the correct battery model to get a suitable scheme [112]. The commonly used ECMs are given as follows.

1. The Rint model, also known as the internal resistance model, equates the lithium-ion battery as a structure formed by a series of ideal voltage sources and ohmic internal resistance, the voltage U_{OC} is the open-circuit voltage (OCV) of the battery, R_0 is the internal resistance of the battery, which are all functions of the battery temperature, SOC and health state.

Table 3.1 Model types

Classes	Black-box model	Electrochemical model	Equivalent circuit model
Structure	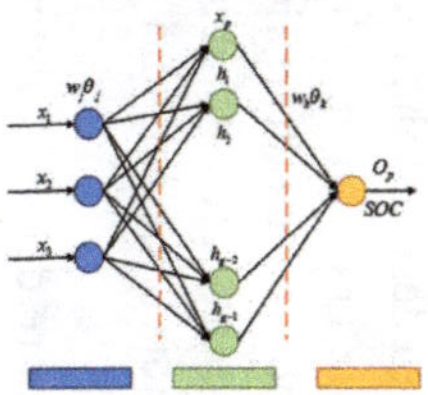	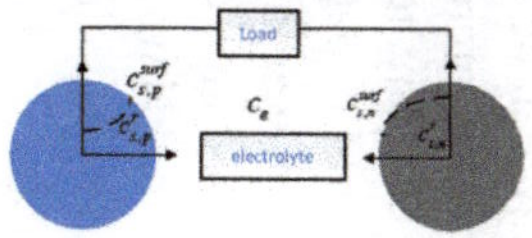	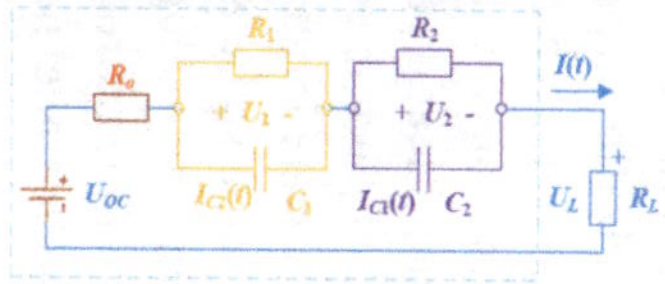
Advantages	Accurately reproduce the nonlinear characteristics of the power battery	Very precise description of the characteristics of the battery	Simple to implement, easy to apply
Disadvantages	Usually require a large amount of test data	The calculation is complex and the application is difficult	The accuracy is poor and the accuracy needs to be further improved

The Rint model has the simplest structure and the fewest number of parameters, but the scope of application of the Rint model is small because the transfer polarization and diffusion polarization are not considered. When the current through the lithium-ion battery is large, the deviation between the simulation results of the Rint model and the measured values is too large [113]. Therefore, this model uses more parameter identification algorithms (such as Kalman filtering) and realizes a rough estimate of parameters such as the SOC of lithium-ion batteries based on the OCV charge state lookup method.

2. The Thevenin equivalent model can also be accomplished by the combination of the internal resistance model with the analog circuit of the resistance capacity model, considering the effects of temperature and polarization.

   The Thevenin model adds an RC parallel network consisting of load-transfer impedance and capacitor based on the Rint model, and the RC parallel network is mainly used to describe the electrochemical polarization characteristics of lithium-ion batteries during charge and discharge [114]. Thevenin model will be lithium-ion battery ohmic polarization and electrochemical polarization at the same time, the model structure is relatively simple, the amount of calculation is small, has a good practical value. It can be in the case of lithium-ion battery health state does not change significantly, more accurately simulate the lithium-ion battery in the constant current constant temperature condition charge and discharge behavior, to achieve lithium-ion battery charge state, health state, or power state diagnosis. However, with the aging of lithium-ion batteries or large temperature changes, the internal impedance characteristics of lithium-ion batteries will gradually change from a single impedance arc to a double impedance arc or the capacitance characteristics will gradually change to the characteristics of normal phase components, resulting in a decrease in the simulation accuracy of the model. For this long-timescale, a combination of lookup tables or adaptive algorithms is a viable way to implement Thevenin model engineering applications.

3. The battery test manual proposes an equivalent model of the PNGV. The circuit is connected to the series with the capacitor C_b based on the Thevenin model. Where internal resistance is R_0, polarization resistance is R_p, polarization capacitance is C_p, load current is $I(t)$, and terminal voltage is U_L. The PNGV model uses the series capacitor C_Q to describe the change of the OCV of a lithium-ion battery with the current-time integration, which reflects both the capacity of the battery and the DC response characteristics of the battery. Therefore, it is possible to simultaneously realize the estimation of the SOC, power status, and usable capacity of lithium-ion batteries, and realize the estimation of battery health status.

4. The GNL model makes a more detailed distinction between the internal characteristics of lithium-ion batteries, especially the introduction of the influence of self-discharge factors integrates the advantages of the above four ECMs, and considers the influence of ohmic polarization, electrochemical polarization, concentration polarization, and self-discharge factors on the charge and discharge behavior of lithium-ion batteries, because the structure of the model

is closer to the internal characteristics of lithium-ion batteries, so it has a better simulation effect on the magnification response characteristics of lithium-ion batteries. And it is more applicable and more accurate [115]. At present, it is mainly used to realize the estimation of the SOC and power of lithium-ion batteries by simulating the charge and discharge characteristics of lithium-ion batteries.

As shown in Table 3.2, it shows the comparison between and among the commonly used battery models. Among the four standard models, the internal resistance model does not consider the transient characteristics of the battery in the electrochemical response stage, and its conversion process cannot be defined with due convenience and efficiency in calculation. The resistance-capacity equivalent model can better simulate the dynamical properties of the battery and increase the definition of the surface impact, temperature, and cell polarization effects that are ignored. Because considering the current accumulation effect, the PNGV model has high accuracy, but the introduction of the series-connected condenser C_b makes the system vulnerable to cumulative errors in long-term simulation. The error in the polarization effect can be resolved by the Thevenin model. The steps are short and the theory is established. It is very appropriate for the battery charging and discharge study of transient energy. The Thevenin model is a simple structure compared to the PNGV model, the GNL model, and other models that are a nonlinear lower-order model and requires fewer parameters, so it can fulfill engineering application requirements in terms of precision.

Based on the porous electrode theory and the concentrated solution theory, the electrochemical model describes the charge and discharge behavior of lithium-ion batteries from the electrochemical mechanism level by Americanizing the microscopic reaction processes such as electrochemical reaction kinetics, mass transfer, and heat transfer inside the lithium-ion battery. Therefore, the electrochemical mechanism model has been widely used in the optimal design of lithium-ion batteries, simulation of charge–discharge behavior, SOC, SOH, and diagnosis of thermal state. At present, the electrochemical model of lithium-ion batteries mainly includes an SPM, a P2D model, and a simplified quasi-two-dimensional model.

1. The P2D model is an electrochemical model suitable for constant current and insulation systems, which was first proposed by Doyle *et al.* The model equates a lithium-ion battery with a sandwich structure consisting of an infinite number of spherical solid-phase particles (positive and negative), a diaphragm, and electrolyte. The model argues that within the same plane of the polar ear, the differences between the various properties of the battery are negligible, taking into account only the chemical reaction kinetics in the direction (x-direction) of the vertical polar ear. Therefore, the dimension of the lithium-ion battery includes the x-direction and the radial r-direction of the spherical particles, so it is called a P2D model. The model uses Fick's law of diffusion to describe the concentration distribution of lithium ions in the solid phase particles of the electrode, calculates the lithium-ion concentration in the electrolyte and the

Table 3.2 Comparison between and among the commonly used battery models

Models	Structure	Advantages	Disadvantages
Rint	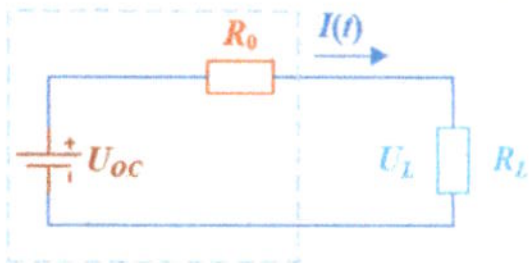	The structure is relatively simple, the calculation is simple, and it is easy to implement	The application range is small, and the simulation results are largely offset from the measured values when the current is too large
Thevenin	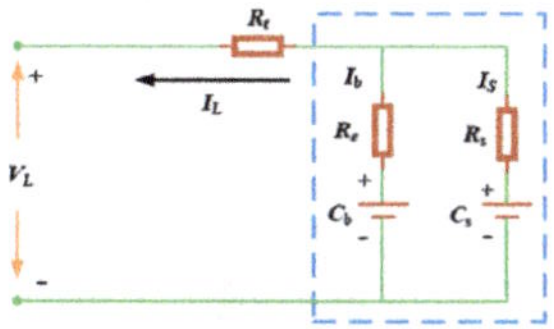	Simple structure, clear physical significance, easy parameter identification, predictable SOC response to load under constant temperature conditions, combined with lookup table method or adaptive algorithm can achieve engineering applications	The model parameters are constant, but due to battery aging, temperature changes have a greater impact and low accuracy
PNGV	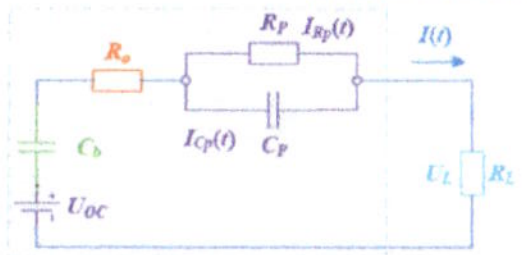	It can describe the open-circuit voltage, capacity variations, and the reaction process inside the battery	Series capacitors cause cumulative errors and are not highly accurate
GNL	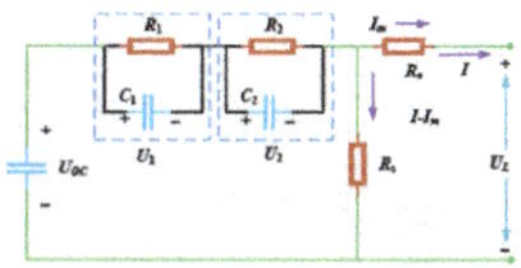	The simulation accuracy is high, the applicability is wide, and the effects of self-discharge and overcharge are additionally considered	The model is complex, the noise has a large impact, the parameters are difficult to adjust, and the calculation is more complicated

diaphragm based on the law of conservation of charge and conservation of matter, calculates the potential in the solid phase electrode based on Ohm's law, calculates the liquid phase potential in the electrolyte and the diaphragm based on Kirchhoff's law and Ohm's law, and calculates the electrode reaction kinetics using the Butler–Volmer formula.

The P2D model of lithium-ion batteries is too complex, computationally large, and its analytical solution cannot be obtained, so the P2D model is more suitable for laboratory research, which is used to assist in analyzing the attenuation aging mechanism of lithium-ion batteries and diagnosing their state, and to provide theoretical support for the optimal design of lithium-ion batteries (such as material particle design, diffusion coefficient adjustment direction, etc.) through simulation.

The P2D model of lithium-ion batteries is too complex, computationally large, and its analytical solution cannot be obtained, so the P2D model is more suitable for laboratory research, which is used to assist in analyzing the attenuation aging mechanism of lithium-ion batteries and diagnosing their state, and to provide theoretical support for the optimal design of lithium-ion batteries (such as material particle design, diffusion coefficient adjustment direction, etc.) through simulation.

2. The SPM is the simplest electrochemical model of a lithium-ion battery, which is a simplified P2D model. As shown in Figure 3.1, the SPM uses two spherical particles to represent the positive and negative electrodes of the lithium-ion battery, respectively, assuming that the embedding and expulsion of lithium ions occur on the spherical particles and that the concentration of the electrolyte and its internal potential is considered to be constant. The SPM has a simple structure and a small amount of calculation, which is easy to implement in online applications. At present, the SPM is mainly used in the diagnosis of the SOC of lithium-ion batteries. However, at the same time, the lithium-ion battery SPM has some unavoidable disadvantages, that is, under the condition of large-rate charge and discharge, the assumption of the model is unreasonable, resulting in simulation bias.

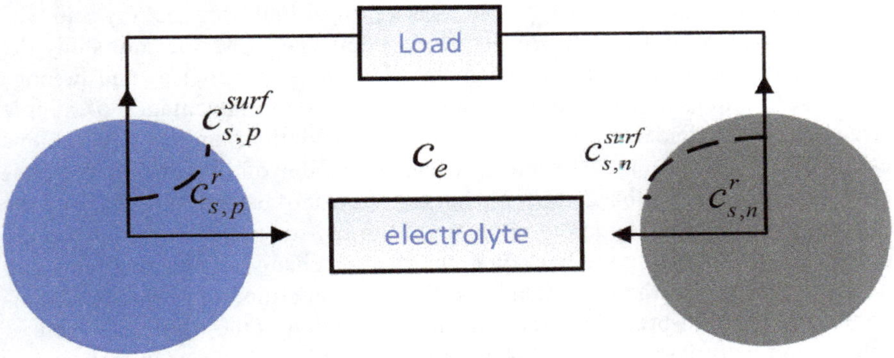

Figure 3.1 The single-particle model of the lithium-ion battery

3. Because the control equation of the lithium-ion battery P2D model is too complex, resulting in the P2D model cannot achieve real-time online application, and the applicability of the SPM is relatively poor, so many scholars are committed to the reasonable simplification of the P2D model, for different application scenarios, using different simplification methods to obtain a simplified quasi-two-dimensional model that meets the corresponding accuracy requirements and timeliness. The existing simplification methods mainly include geometric structure simplification, simplification of solid–liquid phase diffusion process, and transformation through mathematical algorithms to achieve the purpose of simplifying calculations, so the polynomial approximation method, two-parameter, and three-parameter parabolic approximation method, penetration depth method, green function transformation method based on feature root and other approximate simplification methods of SPM are proposed.

The model structure of the simplified P2D model has been simplified, which greatly reduces the amount of computation, and at the same time, compared with the simple SPM, the distribution and diffusion of lithium ions inside the lithium-ion battery are taken into account, so the simulation of large-rate charge–discharge behavior is more applicable. At present, the simplified quasi-two-dimensional model has good application for the diagnosis of the charge state, health state, and thermal state of lithium-ion batteries under laboratory conditions. However, due to the existence of the electrochemical mechanism model itself, the lithium-ion battery is one of the new generations of clean and efficient devices in energy storage equipment, and its sluggish management technology research and development have seriously limited its application in mobile energy storage and stationary energy storage equipment. The lithium-ion battery is a very complex system, given the intricate interaction relationship between its internal parts and the limited experimental conditions in spacecraft applications, it is of great significance to study the charge and discharge behavior of lithium-ion batteries by establishing a lithium-ion battery model and optimizing the battery management.

This section reviews two broad classes of lithium-ion battery models, namely, the ECM and the electrochemical model. The ECM does not study the internal microscopic reaction mechanism of lithium-ion batteries and belongs to the semiempirical simulation model, which has the advantages of simple structure, few parameters, a small amount of calculation, and easy to achieve engineering applications. The electrochemical model of lithium-ion batteries numerically describes the charge–discharge behavior of batteries by studying the internal microscopic reaction mechanism of lithium-ion batteries, which has the advantages of reflecting the battery attenuation mechanism, high accuracy, and good versatility. On the other hand, the fixed model structure of the ECM cannot realize the high-precision modeling and simulation of the whole life cycle of lithium-ion batteries. The complex control equations of electrochemical models and the high load of computational tasks limit the application of electrochemical models in practical engineering. Therefore, given the above deficiencies, an ECM can be dynamically adjusted according to the state of the lithium-ion battery by establishing an ECM; establishing an electrochemical simplification model by developing a collaborative coupling technology that can adaptively adjust between different simplification methods is one of the research directions of high-precision and high-applicability simulation models for lithium-ion batteries in the future.

The model structure of the simplified P2D model has been simplified, which greatly reduces the amount of computation, and at the same time, compared with the simple SPM, the distribution and diffusion of lithium ions inside the lithium-ion battery are taken into account, so the simulation of large-rate charge–discharge behavior is more applicable. At present, the simplified quasi-two-dimensional model has good application for the diagnosis of the charge state, health state, and thermal state of lithium-ion batteries under laboratory conditions. However, due to the existence of the electrochemical mechanism model itself, the lithium-ion battery is one of the new generations of clean and efficient devices in energy storage equipment, and its sluggish management technology research and development have seriously limited its application in mobile energy storage and stationary energy storage equipment. The lithium-ion battery is a very complex system, given the intricate interaction relationship between its internal parts and the limited experimental conditions in spacecraft applications, it is of great significance to study the charge and discharge behavior of lithium-ion batteries by establishing a lithium-ion battery model and optimizing the battery management.

This section reviews two broad classes of lithium-ion battery models, namely, the ECM and the electrochemical model. The ECM does not study the internal microscopic reaction mechanism of lithium-ion batteries and belongs to the semiempirical simulation model, which has the advantages of simple structure, few parameters, a small amount of calculation, and easy to achieve engineering applications. The electrochemical model of lithium-ion batteries numerically describes the charge–discharge behavior of batteries by studying the internal microscopic reaction mechanism of lithium-ion batteries, which has the advantages of reflecting the battery attenuation mechanism, high accuracy, and good versatility. On the other hand, the fixed model structure of the ECM cannot realize the high-precision modeling and simulation of the whole life cycle of lithium-ion batteries. The complex control equations of electrochemical models and the high load of computational tasks limit the application of electrochemical models in practical engineering. Therefore, given the above deficiencies, an ECM can be dynamically adjusted according to the state of the lithium-ion battery by establishing an ECM; establishing an electrochemical simplification model by developing a collaborative coupling technology that can adaptively adjust between different simplification methods is one of the research directions of high-precision and high-applicability simulation models for lithium-ion batteries in the future.

3.3 Equivalent circuit modeling methods

3.3.1 Basic equivalent circuit modeling methods

The establishment of the lithium battery equivalent model has two methods: theoretical analysis and experimental analysis: theoretical analysis is to derive the dynamic equation of the object change law based on understanding the internal law of the research object; experimental analysis needs to collect the input and output signals of the object and establish an equivalent model according to the collected signal. The established model is also processed for parameter identification. Lithium batteries' internal physical and chemical changes are very complex, and the battery dynamic equations derived based on electrochemical theory are difficult to apply in practice. Therefore, in most cases, the method of experimental analysis is often used.

Many different models have been built to predict the behavior of batteries, but none of them can fully and accurately simulate the dynamic behavior of batteries under any one condition. The equivalent model of the lithium battery can be divided into the empirical model and an electrochemical model according to modeling methods. The empirical model is a battery dynamics alternative model obtained based on sample data and system identification methods, which has the advantages of controllable complexity and low modeling cost [116]. The disadvantage is that the physical and chemical significance of the model parameters is not obvious the robustness is poor. The electrochemical model obtains the mathematical expression of the battery kinetic properties through the mechanism analysis method, which has the advantage of high accuracy of the model and can reveal the evolution mechanism of the battery characteristics at a deep level, and the disadvantage is that the complexity of the model is high, which is not conducive to its engineering application.

Depending on the mechanism of battery model building, it can be divided into simple electrochemical models, intelligent mathematical models, and ECMs. Electrochemical models are more complex and difficult to apply to actual products, mainly to assist in the design and manufacture of batteries; intelligent mathematical models are mainly NN models (Albertus, 2009 #3,109), which can theoretically complete battery modeling, but due to the need for a large amount of actual data for training, high technical thresholds, and long processing time, its practical application is limited; the ECM is widely used because of its clear physical significance and simple mathematical expressions [117]. Due to the nonlinear characteristics of the battery such as current, power, SOC, and temperature, battery modeling to fully consider these factors, will increase the complexity of the calculation, the general controller cannot meet the requirements, and model establishment points are as follows.

1. Logic structure

   The logic structure of the circuit model is shown in Figure 3.2.

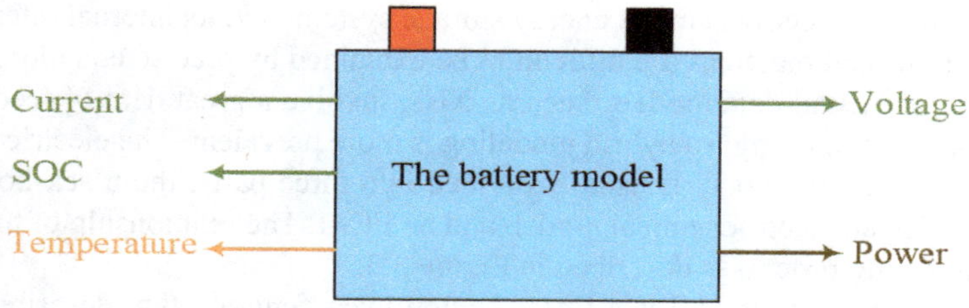

Figure 3.2 *Logic structure diagram of the circuit model*

2. Principal description
   The battery simulation model is to verify the accuracy of parameters set in the model, so the input is current, and the output is terminal voltage. The current and terminal voltage are both inputs in the actual BMS.

3. Modeling steps
   (1) Selecting the appropriate ECM; (2) determining the input, output, and state variables; (3) writing the state-space equation; (4) simulation software modeling; (5) parameter identification by experimental data; (6) modifying model parameters according to variables (temperature, SOC, etc.); (7) simulate the circuit and verify its characterization effect.

4. Model selection
   According to the needs, the selection of different battery equivalent models greatly influences the accuracy and accuracy of the battery model [118]. At the same time, the more complex the structure, the higher the accuracy of the model, the more complex the operation, and the hardware requirements are increased accordingly. In the equivalent model, the accuracy of the equivalent circuit containing the RC loop is higher. Generally, the higher the RC order, the higher the accuracy.

5. Parameter identification
   According to the experimental results of the hybrid pulse power characterization (HPPC) test, curve fitting can be performed by combining with the LS method to identify relevant parameters.
   The techniques of battery modeling can be classified into mechanism modeling, experimental modeling, and hybrid modeling methods [119]. The mathematical modeling framework is constructed based on scientific equations such as physical equations and concepts of chemical reactions. The managed object is regarded as a black box, and the model is constructed by experimenting with the variant law on the dynamic characteristic parameters of the target object. The battery is a

nonlinear electrochemical energy storage system, and its internal interactions and reactions are difficult to be explained by precise equations. Experimental data models, such as NNs, involve a great deal of feedback and learning, so hybrid modeling is more prevalent. The electrical characteristic model is usually divided into three parts: the black-box model, an electrochemical model, and an ECM. The relationship of the modeling process is described in Figure 3.3.

The black box model is a linear or nonlinear function that describes the characteristics of battery voltage response, it shifts the focus from the internal mechanism of the battery to the data itself, which determines that it can be very flexible to establish the model structure and parameterization, but lacks essential physical significance, the difficulty of black-box modeling is that the model performance is sensitive to the quantity and quality of training data, and usually needs to take data-driven algorithms as a support, such as NNs, support vector regression, etc.

The most common are electrochemical models and ECMs. The original intention of the research of the electrochemical model is for the design of the battery structure, through the macroscopic data outside the battery and the analysis of the particle activity of the internal microscopy, combined with the basic mechanism of the battery to generate energy, to obtain the internal state of the battery, the model mainly reflects the chemical reaction situation inside the battery. Electrochemical models can analyze the internal properties of a battery from a microscopic perspective, but to represent these properties of a battery, an electrochemical model usually requires the creation of multiple sets of complexes, and time-varying partial differential equations, which usually take a lot of time to solve.

The ECM is to use basic circuit components to form a specific circuit network to characterize the working characteristics of the circuit, the model can establish the relationship between the external characteristics exhibited during the operation of the battery and the amount of internal state of the battery itself. The ECM is intuitive, easy to

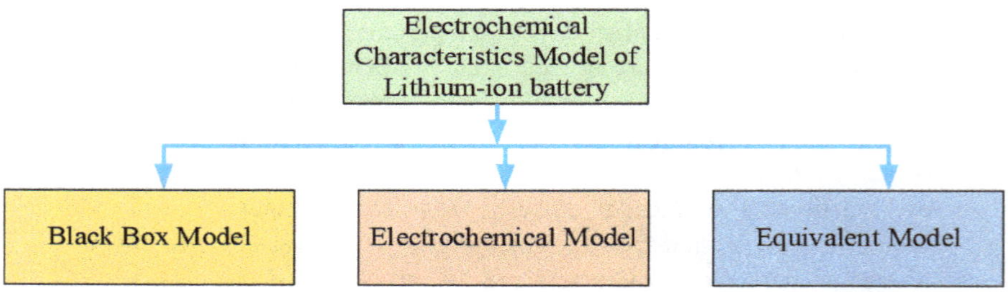

Figure 3.3 Connection process of the modeling concept

process, moderate in computing, and the parameters of the model are easy to identify, which is suitable for simulation experiments with circuits. Therefore, ECMs are more widely used in practical engineering applications.

The ECM is easy to analyze using circuit and mathematical methods and has become the most widely used battery model, such as the Rint model, the RC model, the Thevenin model, and the PNGV model. The following is the analysis of several commonly used ECMs.

1. Rint model

The Rint model is also called the internal resistance model. In the Rint model, the battery is equivalent to the series connection between the ideal voltage source U_{OC} and the resistance R^0. It is a relatively rough battery model, which consists of only a constant voltage source and a resistor in series. Due to the polarization, characteristics of the battery are not considered in this model, the accuracy of the model is low, and can only approximately represent the ohmic resistance of lithium-ion batteries, as shown in Figure 3.4.

In Tables 3.6, the Rint model regards the lithium-ion battery pack as an ideal power source. U_{OC} is a constant value that represents the OCV of the battery, R_0 represents the ohmic resistance of the battery, and U_L is the terminal voltage between

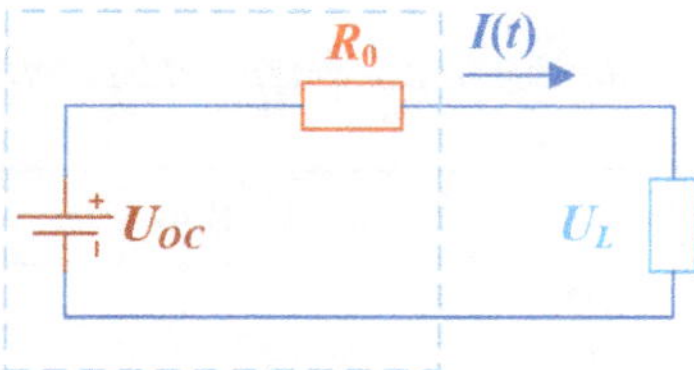

Figure 3.4 Internal resistance equivalent structure diagram

Table 3.3 Equivalent model parameter identification result

SOC/100%	U_{oc}/V	R_0/mΩ	R_{p1}/mΩ	C_{p1}/F	R_{p2}/mΩ	C_{p2}/F
1	4.1838	1.3095	0.0410	25,058	0.5043	32,260
0.9	4.0581	1.3716	0.0541	19,712	0.6114	30,896
0.8	3.9434	1.3759	0.0432	20,625	0.5859	25,449
0.7	3.8380	1.3846	0.0586	19,176	0.7824	28,540
0.6	3.7354	1.3911	0.0615	19,140	0.8830	29,978
0.5	3.6527	1.4524	0.0299	27,093	0.3942	34,396
0.4	3.6162	1.4740	0.0402	29,987	0.5798	43,135
0.3	3.5904	1.5058	0.0488	29,257	0.7104	43,919
0.2	3.5408	1.5411	0.0389	15,665	0.4792	29,026
0.1	3.4641	1.6169	0.0992	3476	0.5655	18,549

Table 3.4 Parameter fitting polynomial coefficients

	R_o	R_{p1}	R_{p2}	C_{p1}	C_{p2}	U_{oc}
a_0	0.001006	0.001006	0.009941	1.567e+04	1.702e+05	3.626
a_1	0.001583	–0.01929	–0.2116	–3.468e+05	–3.495e+06	–5.006
a_2	–0.02599	0.1496	1.75	2.744e+06	2.888e+07	51.47
a_3	0.1317	–0.5873	–7.111	–4.517e+06	–1.116e+08	–219.1
a_4	–0.3275	1.272	15.69	–7.005e+06	2.318e+08	481
a_5	0.4266	–1.538	–19.15	2.757e+07	–2.68e+08	–566.6
a_6	–0.2782	0.9704	12.13	–2.784e+07	1.627e+08	341.9
a_7	0.07154	–0.2488	–3.113	9.402e+06	–4.041e+07	–83.16

Table 3.5 SOC–OCV curve parameter

Parameter	P_1	P_2	P_3	P_4	P_5	P_6	P_7
Value	37.69	–121.5	148.3	–84.49	22.54	–1.904	3.47

Table 3.6 BBDST working condition parameter description table

Step	**P/kW**	**t/s**	**All time/s**	**Condition**
1	15	21	21	Starting
2	80	12	33	Speedup
3	12	16	49	Glissade
4	–15	6	55	Braking
5	45	21	76	Speedup
6	12	16	92	Glissade
7	–15	6	98	Braking
8	80	9	107	Speedup
9	100	6	113	Rapidly speedup
10	45	21	134	Constant speed
11	12	16	150	Glissade
12	–15	6	156	Braking
13	80	9	165	Speedup
14	100	6	171	Rapidly speedup
15	45	21	192	Constant speed
16	12	16	208	Glissade
17	–35	9	217	Braking
18	–15	12	229	Braking
19	12	71	300	Stop

the positive and negative ends of the batteries. In the application process, the battery operating characteristics are characterized by equating the battery as a series connection of a voltage source and a resistor. Due to the existence of the internal resistance, during the charging and discharging process, some of the energy converted into electricity is consumed by the internal resistance in the form of heat generation.

Then, the charge–discharge efficiency is analyzed and carried out to establish the mathematical expression of the working characteristics. To simplify the calculation, U_{OC} and R_0 are set as constant values in the model, ignoring the characteristic that the internal resistance of the battery varies with electrolyte concentration and SOC value, which is inconsistent with the actual parameters of the batteries. The Rint model has low accuracy and is unsuitable for an application used to estimate the actual SOC estimation of lithium-ion batteries. The state-space description of the Rint model is expressed in (3.1).

$$U_L = U_{OC} - I(t) * R_0 \tag{3.1}$$

From the analysis of the state-space shown in (3.1), it can be observed that in the equivalent simulation process of the internal resistance model of the lithium-ion battery pack, the actual output voltage of the battery pack is quite different from the output voltage of the model. Also, the transient characteristics of the electrochemical reaction process are not considered in the model, which cannot accurately characterize the changes in the working process of the batteries.

2. RC model

The RC model has two capacitors and three resistors where the high-capacity C_b demonstrates its energy storage potential. The C_s capacitance defines the battery's surface potential and diffusion effect. R_t is the battery terminal resistance, R_s is the surface resistance, and R_e is the end resistance. The RC model structure is shown in Figure 3.5.

When the model is used in an operating environment to simulate the battery behavior at a specific SOC value, the parameters of the model can then be considered constants. However, if the influence of these parameters on the change of temperature needs to be considered, a large SOC working range must be simulated,

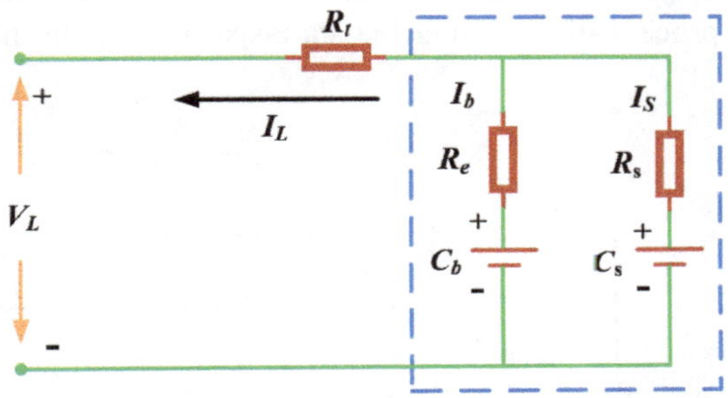

Figure 3.5 Schematic diagram of the RC model

which makes the process of arranging and processing model parameterization complicated. The battery terminal voltage is shown in (3.2).

$$\begin{cases} \begin{bmatrix} V_{cb}^{*} \\ V_{cs}^{*} \end{bmatrix} = \begin{bmatrix} -\frac{1}{(Re+Rs)C_b} & \frac{1}{(Re+Rs)C_b} \\ \frac{1}{(Re+Rs)C_s} & -\frac{1}{(Re+Rs)C_s} \end{bmatrix} \begin{bmatrix} V_{C_b} \\ V_{C_s} \end{bmatrix} + \begin{bmatrix} \frac{R_s}{(Re+Rs)C_b} \\ \frac{Re}{(Re+Rs)C_s} \end{bmatrix} I_L \\ V_L = \begin{bmatrix} \frac{R_s}{Re+Rs} & \frac{Re}{Re+Rs} \end{bmatrix} \begin{bmatrix} V_{C_b} \\ V_{C_s} \end{bmatrix} + \begin{bmatrix} \frac{R_s+Re}{Re+Rs} + R_t \end{bmatrix} I_L \end{cases} \tag{3.2}$$

3. Thevenin model

Based on the internal resistance model, the Thevenin model is constructed by considering the internal polarization effect during the charging and discharging process of the lithium-ion battery pack. Thevenin model considers that the characteristics of batteries are similar to those of capacitors. In the model, the ideal voltage source U_{oc} describes the OCV of batteries, the resistance R_0 is the internal resistance of batteries, and the capacitor C_p and resistor R_p are in parallel to describe the overpotential of batteries. Compared with the Rint model, this model has the advantage of adding an RC circuit, which can fully reflect the battery voltage lag phenomenon during charging and discharging.

The model enhances the nonlinear characterization effect of the model by characterizing the obstruction of polarization effect to current. It realizes the effective characterization of electrochemical polarization and concentration difference polarization effect of the battery through the parallel loop of R_p and C_p. Since the Thevenin model is a first-order model, there are some errors in the dynamic output voltage curve of the model compared with the actual battery output in practical applications. The structure of the Thevenin model is shown in Figure 3.6.

In Figure 3.6, U_{OC} is the OCV, R_0 is the ohmic resistance, R_p is the polarization resistance, and C_1 is the polarization capacitance. The parallel circuit of R_p and C_1 describes the polarization process. U_L is the closed-circuit voltage after the battery is connected to the external circuit. The model takes into account the polarization process of the battery, and the first-order model constructed is relatively simple to calculate. In the process of obtaining the state-space equation, first, according to

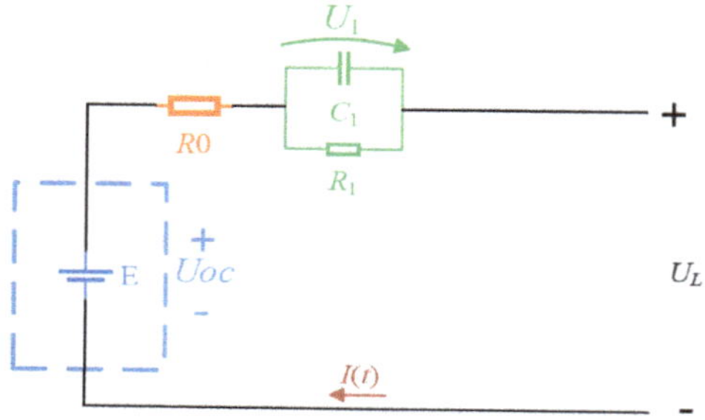

Figure 3.6 Thevenin equivalent model structure diagram

the basic operating characteristics of the capacitor components. The relationship between the current flowing through the battery polarization capacitor C_p and its two closed-circuit voltage is shown in (3.3).

$$I_{C_p}(t) = C_p \frac{dU_{Cp}(t)}{dt} \tag{3.3}$$

In (3.3), based on the analysis of the equivalent circuit composition, through the application of the knowledge of circuit science and Kirchhoff's voltage laws (KVL), the voltage relationship in the equivalent circuit of can be obtained, as shown in (3.4).

$$R_o * I_{C_p}(t) + R_0 * \frac{U_{Cp}(t)}{Rp} + U_{Cp}(t) = U_{OC} - U_L \tag{3.4}$$

In (3.4), taking $U_{Cp}(t)$ as the state variable, and the above two expressions are combined to describe the comprehensive working state. The voltage $U_{Cp}(t)$ at both ends of the equivalent, the capacitor is set as the state variable, and the calculation process is analyzed to obtain the state-space of the equivalent model, as shown in (3.5).

$$R_0 * C_P \frac{dU_{Cp}(t)}{dt} + \left(1 + \frac{R_0}{R_p}\right) U_{Cp}(t) = U_{OC} - U_L \tag{3.5}$$

In (3.5), U_{OC} is the OCV, R_0 is the ohmic resistance, R_p is the polarization resistance, C_p is the polarization capacitance, and U_L is the closed-circuit voltage after being connected to the external circuit. The parameters in the equivalent model are fixed values and cannot be modified and adjusted. After the internal resistance is equivalently processed, the equivalent model takes into account the influence of the polarization effect. The RC parallel circuit of the capacitive device C_p is added with the dynamic characteristics, which have better dynamic simulation characteristics than the battery model.

4. PNGV model

PNGV model is the standard battery model in the "PNGV Battery Test Manual". The PNGV circuit model for lithium-ion batteries contains two resistors and two capacitors, one less than the RC circuit model. The characteristic of this model is that it takes into account the error caused by the current accumulation during the charging and discharging process of the battery OCV characteristics. Compared with the Thevenin model, the PNGV model adds a large capacitance C_b to describe the OCV change. Because the circuit model has the characteristics of battery polarization and ohmic resistance, the circuit model is more accurate. The model structure is shown in Figure 3.7.

PNGV model is a typical nonlinear lumped parameter circuit, which can be used to predict the terminal voltage variation of battery under the condition of hybrid power pulse capability characteristic pulse load. In Figure 3.6, U_{OC} is the ideal voltage source, R_0 is the ohmic resistance, R_p is the polarization resistance, and C_p is the polarization capacitance. The parallel circuit of R_p and C_p is used to reflect the production of the battery polarization process to greatly reduce the differences in the working process and the impact on the working environment. C_b is the self-discharge capacitance, which is used to describe the influence of the current

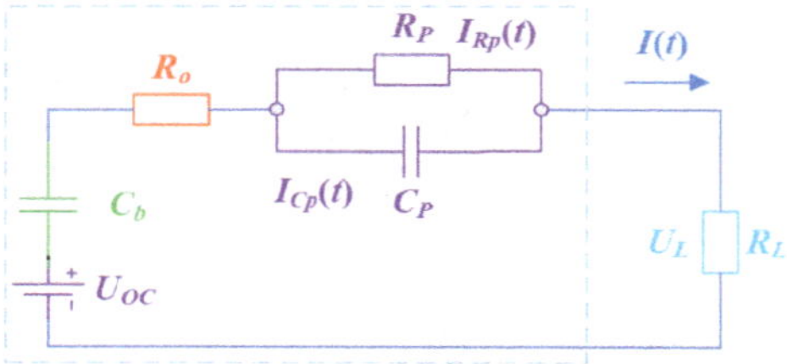

Figure 3.7 Lithium-ion battery PNGV model equivalent

accumulation effect on battery OCV. U_{OC} and C_b together indicate the differences in the OCV, such as voltage, temperature, and capacity, between the cells, and the differences will increase with the aging of the battery. The model does not consider the effect of self-discharge and temperature. By analyzing the illustrated ECM, the state-space equation of each parameter can be obtained, as shown in (3.6).

$$U_L = U_{OC} - C_b\left(\int i(t)\,dt\right) - R_0 I_L - R_p I_p \tag{3.6}$$

In (3.6), parameter identification of the PNGV model is combined with the HPPC test. In the HPPC experiment for parameter identification, four different sample times of a–d are selected for the mathematical description of the state-space. After the matrix description is performed, the relationship between different sample times can be obtained, as shown in (3.7).

$$\begin{bmatrix} U_{La} \\ U_{Lb} \\ U_{Lc} \\ U_{Ld} \end{bmatrix} = \begin{bmatrix} 1 & \int_0^a I_a(t)\,dt & I_a(t) & I_a(t) \\ 1 & \int_0^b I_b(t)\,dt & I_b(t) & I_b(t) \\ 1 & \int_0^c I_c(t)\,dt & I_c(t) & I_c(t) \\ 1 & \int_0^d I_d(t)\,dt & I_d(t) & I_d(t) \end{bmatrix} \begin{bmatrix} U_{OC} \\ \frac{1}{C_b} \\ R_0 \\ R_p \end{bmatrix} \tag{3.7}$$

In (3.7), U_{OC} represents OCV, R_0 represents ohmic resistance, R_p represents polarization resistance, C_p represents polarization capacitance, I_L represents current, I_p represents the current through polarization internal resistance, and U_L represents external load closed-circuit voltage. The subscript parameters a–d, respectively, represent four different sample times, which are used to characterize the state values of the parameters at different times.

3.3.2 Second-order RC ECM

At present, there are many types of lithium-ion battery ECMs. The second-order RC model is one of the commonly used ones, which can well reflect the dynamic and static characteristics of the battery, with moderate calculations and fewer identification parameters [120]. Therefore, the second-order RC model is used as the battery ECM, as shown in Figure 3.8.

In Figure 3.8, U_{OC} represents the OCV of the circuit; U_L represents the terminal voltage. R_0 is the ohmic internal resistance, U_R is the ohmic voltage, which characterizes the battery voltage drop effect at the time point the battery is discharged and

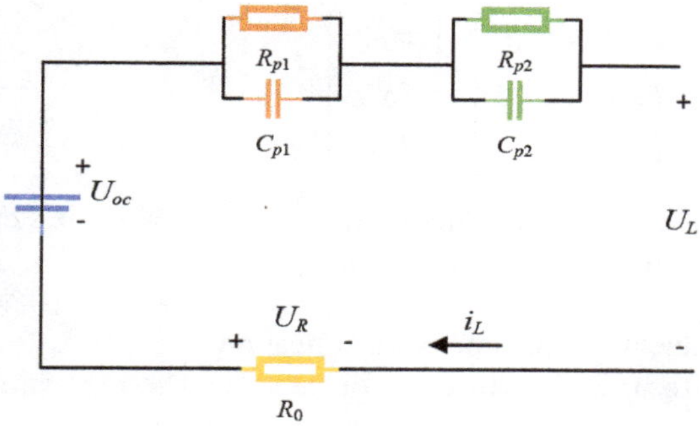

Figure 3.8 *Second-order Thevenin model*

the end of the discharge, i_L is the battery charge–discharge current, and the discharge direction is set to positive. The RC loop composed of R_{p1} and C_{p2} characterizes the internal chemical reaction of the battery. The stage where the voltage changes rapidly during the process. According to Kirchhoff's circuit law, the equivalent model circuit can be listed, as shown in (3.8).

$$U = Uoc - UR - Up1 - up2 \tag{3.8}$$

The RC loop composed of R_{p2} and C_{p2} represents the stage where the voltage changes slowly during the chemical reaction within the batteries, as shown in (3.9).

$$\begin{cases} \frac{dUp1}{dt} = -\frac{Up1}{Rp1Cp1} + \frac{i}{Cp1} \\ \frac{dUp2}{dt} = -\frac{Up2}{Rp2Cp2} + \frac{i}{Cp2} \end{cases} \tag{3.9}$$

The equation for the SOC definition is obtained, as shown in (3.10).

$$SOC_t = SOC_{t_0} - \frac{\int_{t_0}^{t} I(t)\,\eta dt}{Q_0} \tag{3.10}$$

Combined with the SOC definition, select state-space variables, input variables, and output variables, as shown in (3.11).

$$x_k = \left[SOC_k, U_{p1,k}, U_{p2,k}\right]^T \tag{3.11}$$

The discrete state-space expression can be obtained, as shown in (3.12) and (3.13).

$$\begin{cases} \begin{bmatrix} SOC_{k+1} \\ U_{p1,k+1} \\ U_{p2,k+1} \end{bmatrix} = \begin{bmatrix} 1 & 0 & 0 \\ 0 & e^{-\frac{\Delta t}{\tau_1}} & 0 \\ 0 & 0 & e^{-\frac{\Delta t}{\tau_2}} \end{bmatrix} \begin{bmatrix} SOC_k \\ U_{p1,k} \\ U_{p2,k} \end{bmatrix} + \begin{bmatrix} -\frac{\Delta t}{Q_N} \\ R_1(1 - e^{-\frac{T}{\tau_1}} \\ R_2(1 - e^{-\frac{T}{\tau_2}} \end{bmatrix} I_k + \begin{bmatrix} w_{1,k} \\ w_{2,k} \\ w_{3,k} \end{bmatrix} \\ \tau_1 = R_1 C_1,\ \tau_2 = R_2 C_2 \end{cases} \tag{3.12}$$

$$U_{L,k} = U_{oc,k} - R_{0,k}I_k + \begin{bmatrix} 0 \\ -1 \\ -1 \end{bmatrix}^T \begin{bmatrix} SOC_k \\ U_{p1,k} \\ U_{p2,k} \end{bmatrix} + v_k \tag{3.13}$$

In (3.12) and (3.13), Δt is the sampling time interval, w is the state error, and v is the measurement error.

1. HPPC experiment and parameter identification

The test platform is described as follows: (1) The experimental lithium-ion battery is a ternary lithium iron phosphate battery with a nominal capacity of 72 Ah and an actual calibrated capacity of 69.23 Ah; (2) the battery test system (NEWAREBTS-4000) is used to test the charge–discharge of the tested lithium-ion batteries. The equipment can record the voltage, current, capacity, temperature, and other data during the experiment, and the sampling interval is 0.1 seconds; (3) a constant temperature (25°C) environment incubator is provided for the test batteries.

HPPC is performed on the battery as an identification experiment. The battery model parameters are identified by analyzing the working characteristics of the battery during the experiment. The parameters that need to be identified are U_{OC}, R_0, R_{p1}, C_{p1}, R_{p2}, C_{p2}. The steps of the HPPC experiment are as follows. (1) Charge the battery with constant voltage and constant current to full capacity and let it stand fully; (2) Pulse discharge with 1 C (69.23 A) current for 10 seconds; after discharge, let it stand for 40 seconds; (3) Use 1 C current, pulse charging lasts for 10 seconds, and rests for 40 seconds after the end of the charge; (4) Discharge at a constant current of 1 C for 6 minutes, so that the SOC value drops by 0.1, and stand for 30 minutes after the discharge ends; (5) Cycle the Test Steps (2)–(4) until the SOC value drops to 0.1. The HPPC experiment current curve and voltage response curve are shown in Figure 3.9.

Single HPPC voltage and current change curve can be obtained, as shown in Figure 3.10.

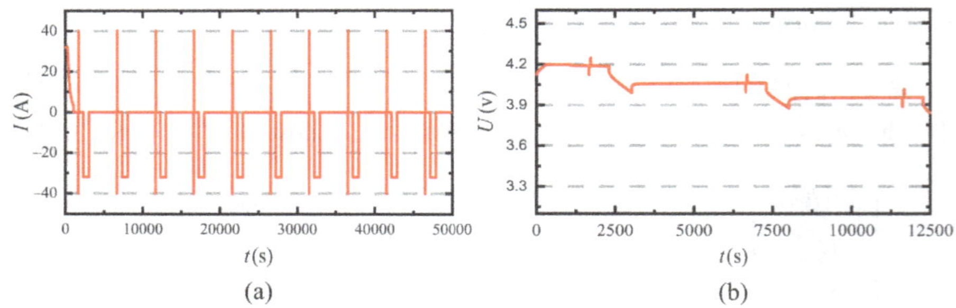

Figure 3.9 The HPPC experiment current curve and voltage response curve (a) Current curve and (b) Voltage curve

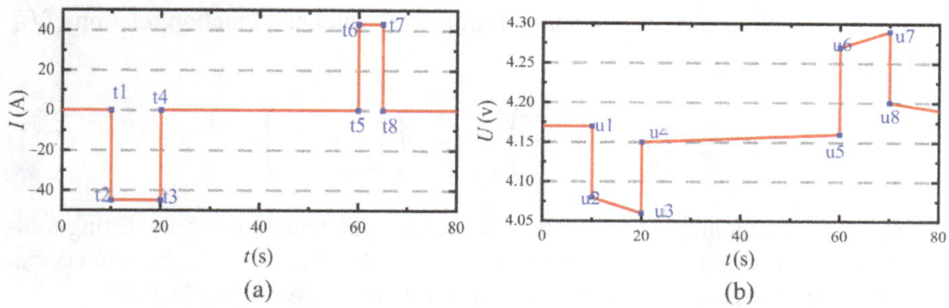

Figure 3.10 *Single HPPC voltage and current change curve: (a) single HPPC current change curve and (b) single HPPC voltage change curve*

The graph of the HPPC experimental voltage variation curve is shown, and the variation characteristics can be extracted to obtain the model parameters. Analyzing, the following four stage characteristics can be obtained.

1. Discharge starts at t_1. The sudden drop of the voltage at the lithium-ion battery terminal from U_1 to U_2 is mainly due to the voltage change caused by the ohmic internal resistance of the lithium-ion batteries.
2. During the period from t_2 to t_3, the voltage of the lithium-ion battery terminal slowly drops from U_2 to U_3. It is due to the existence of the battery polarization effect. The process of charging the polarization capacitor by the discharge current is the zero-state response of the double RC series circuit.
3. During the period from t_3 to t_4, the voltage of the lithium-ion battery terminal rises suddenly from U_3 to U_4, which is also due to the voltage change caused by the ohmic internal resistance of the lithium-ion batteries.
4. During the period from t_4 to t_5, the voltage of the lithium-ion battery terminal slowly rises from U_4 to U_5, which is the process of discharging the polarizing capacitor to the polarizing resistance, which is the zero-input response of the dual RC circuit.

The terminal voltage of a lithium-ion battery suddenly changes at the time point of discharging and at the time point of stopping, both of which are caused by the ohmic internal resistance [121]. Therefore, the value of the ohmic internal resistance can be calculated in (3.14).

$$R_0 = \frac{(U_2 - U_1) + (U_4 - U_3)}{2I} \tag{3.14}$$

During the period from t_2 to t_3, the voltage of the lithium-ion battery terminal slowly drops from U_2 to U_3. This is due to the existence of the battery polarization effect. The process of charging the polarized capacitor by the discharge current is the zero-state response of the double RC series circuit [122]. Carry on the time domain, analysis to the circuit, choose the data from t_2 to t_3, can get a function relation of U_L and time t, as shown in (3.15).

$$U_L\left(t\right) = U_2 - i_L R_{p1}\left(1 - e^{-\frac{t}{\tau_1}}\right) - i_L R_{p2}\left(1 - e^{-\frac{t}{\tau_2}}\right) \tag{3.15}$$

Curve fitting is conducted for the HPPC experimental data by using the fitting toolbox to fit as the target formula, and then the U_{OC}, R_{p1}, C_{p1}, R_{p2}, C_{p2} parameters can be identified. The parameter identification results are shown in Table 3.3.

The SOC value is a variable x, and $f(x)$ is a polynomial function of the parameter and SOC value fitting. Considering the error and correlation between the fitting curve and the discrete data, the fitting equation is shown in (3.16).

$$f\left(x\right) = a_0 + a_1 x + a_2 x^2 + a_3 x^3 + a_4 x^4 + a_5 x^5 + a_6 x^6 + a_7 x^7 \tag{3.16}$$

The identification results are shown in Table 3.4.

U_2 is the model voltage simulation curve, and (b) is the error curve. It can be seen that the overall tracking of the model is relatively accurate, the maximum error is 1.5%, and the overall error is kept within 1%, indicating that the selected second-order RC ECM can more accurately reflect the operating characteristics of the lithium-ion batteries.

3.3.3 Compound equivalent circuit modeling

The BMS is an electronic control unit that is mounted onboard an EV to overcome the limitations of the battery. A well-designed BMS is required to ensure the battery's optimal performance and operating safety. State estimation of batteries is one of the main functions in the BMS of EVs, which includes the SOC, SOH, SOP, state of energy, etc. estimation, which plays a crucial role in EVs, smart grids, etc. The BMS can estimate and monitor the time-varying evolution of the physical–chemical and degradation characteristics inside the lithium-ion battery (LIB) by analyzing the transitions in the values of ECM parameters. BMS devices are also used to improve the reliability of the battery pack. It is also in charge of monitoring the transient dynamic, behavior, managing every cell inside the battery system, and ensuring that its operations are bound within the safe interval to improve the operations performance of the battery system. Furthermore, the excessive charging and discharging currents and power transients are monitored by the BMS, avoiding possible damage and deterioration of the battery cell.

The need to design compact, lightweight, and fast devices requires finding a suitable trade-off between effectiveness and efficiency. Three main battery modeling approaches are electrochemical models, ECMs, and data-driven models. In various references, it is well emphasized the application of electrochemical-based methods. Electrochemical battery models often provide better accuracy, but they require a great amount of computing power, which makes them unsuitable for many real-time battery applications, such as the P2D model is computationally challenging and requires significant simplifications. The electrochemical model is based on the physical characterization of the processes occurring inside the battery during functioning. Despite that, the major challenges of this technique are solution deduction complexity, high computation load, and its usage in real-time battery state estimation.

By far, the most used modeling method is the ECM. The ECM is a battery model that is commonly used to monitor and regulate lithium-ion batteries in BMS. ECMs are electrical components that characterize the dynamic characteristics of batteries simply and straightforwardly with a limited number of parameters and states. The ECM is used in many batteries real-time applications due to its instantaneous implementation time, simplicity, and relatively high precision. Generally, an ECM consists of three components, which include an OCV source, an ohmic resistor, and RC circuit networks. Conventional ECMs, such as the Rint, GNL, PNGV, Thevenin, RC, etc., are too simple and unable to represent the dynamics of the cell, as presented in Figure 3.11.

Figure 3.11 shows the construction of the different types of ECMs that are commonly used to control and monitor the dynamic characteristics of the battery in the estimation of state parameters. Generally, I_L is the load current flowing through the circuit, R_0 is the internal ohmic resistance, U_{oc} is the OCV, and U_L is the terminal *voltage* of the battery. R_{p1} and C_{p1} are the polarization electrochemical resistance and capacitance. In Figure 3.11 (c), U_d and U_{pn} are the voltage across $1/U_{oc}$ and C_{pn}. Also, R_{pn}, C_{pn}, and U_{pn} are the *n*th-order polarization resistance and capacitance, and the voltage drop across the *n*th-order circuit, respectively. The Thevenin and RC ECMs usually consist of three main components: a component describing the thermodynamic effects of the battery's chemistry as the OCV, which is a function of the SOC. Another component represents the dynamic characteristics of the cell's internal impedance and acts as a source or load to complete the circuit for the charge or discharge.

An accurate and reliable ECM is necessary for the optimal design and performance of the BMS to control and monitor the operational electrical behavior to guarantee the efficient, safe, and long life of lithium-ion batteries. It has been demonstrated to function in the lead-acid, nickel–metal hydride, lithium-ion, sodium-ion, and zinc-ion batteries. The simplest model only takes the internal ohmic resistance into *account*, such as the Rint, which is not adequate to represent the battery dynamics during operation. Even though this model is linear and has a basic structure, it cannot be used in real time because it only represents the battery's static features. The GNL model is difficult to apply in practical engineering owing to its complex topology and difficult parameter identification.

The PNGV model can accurately simulate various operational statuses of the battery. Nevertheless, because of the large accumulated error of the current, the model is not conducive to long-term simulation. As a result, the Thevenin ECM, which features an extra RC pair to combine with the internal ohmic resistance, is commonly utilized because it strikes a fair balance between precision and simplicity. Also, the RC ECM is composed of a resistor and a capacitor in parallel to form a network structure that represents the transient response characteristics. It also has a resistor connected in series to represent the ohmic resistance of the battery. Although

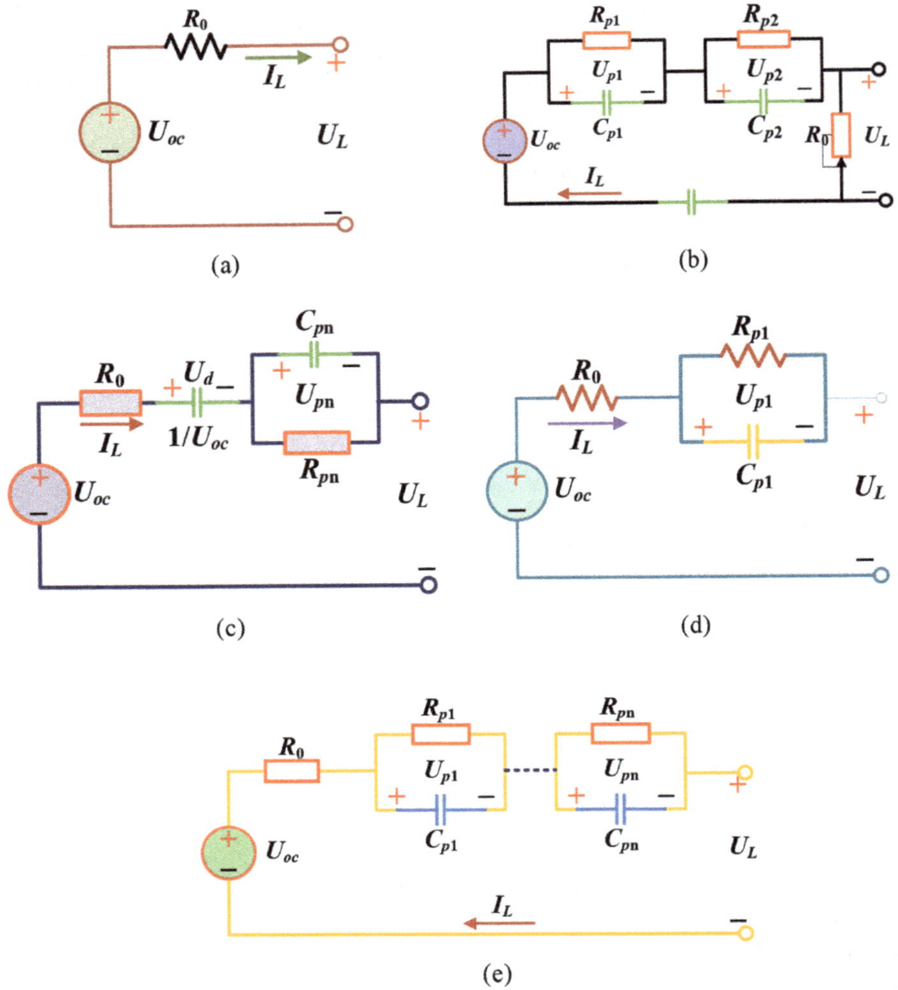

Figure 3.11 *Types of conventional equivalent circuit models: (a) Rint equivalent circuit model, (b) GNL equivalent circuit model, (c) PNGV equivalent circuit model, (d) Thevenin equivalent circuit model and (e) An* nth-*order RC equivalent circuit model*

increasing the number of RC networks ensures a better characterization of the battery behavior, it makes the model parameter identification and SOC estimation more demanding for the BMS, resulting in increased mathematical complexity. Also, they lack the physical interpretation and can only describe the high-level physical parameters. However, if the battery is pushed to its limits, these conventional ECMs are unable to complete the model extrapolation under a wide range of operating situations. As a result, it is not a battery model that is commonly utilized for applications that require high current rates or operate at extremely low temperatures.

Due to its low computational cost, the ECM is frequently favored for onboard implementation in BMS for parameter and state estimation and control due to its low computational cost, ease of parameterization, and desirable accuracy. Besides, the ECM does not require a 100% charged battery, making it more suitable for onboard applications. Also, due to limitations in data storage and processing capability among the various battery models used in the BMS, the ECM remains the most preferred and has been widely used for batteries with other chemistries, such as lead-acid, nickel–metal hydride, lithium-ion, and zinc-ion. For most lithium-ion batteries, these models are computationally efficient and have reasonable accuracy. Most models, on the other hand, use an OCV vs. SOC curve to construct a functional link between the hidden state (i.e., SOC) and a directly observable variable like battery current and terminal voltage.

The voltage response dynamics were directly described using the data-driven model, which is a black-box modeling method. However, the data-driven model's performance is highly dependent on sample quality, and its generalization performance cannot be guaranteed for unknown operating conditions. ECMs can be built in either the frequency domain or the time domain. NNs are used more often in modeling lithium-ion batteries. The application of AI-based ECMs has emerged as a promising solution able to adequately characterize the dynamics of the battery with high computational efficiency with a simple model structure using NNs. For estimating state parameters of batteries, machine learning models have lately gained prominence, such as Gaussian processes, feedforward NNs, time-delayed NNs, recurrent NNs, and long short-term memory NNs. The fact that NN models are black boxes is one of their fundamental drawbacks. With random weight initialization, convergence training outcomes at distinct local optimums could lead to varying battery performance under different operating situations. Furthermore, determining the core cause and explaining the effects are challenging. Nonetheless, utilizing several machine learning methods estimates accurate results for state parameters.

Other data-driven methods are pretty famous as they require less pretesting work than other methods. Evolutionary and swarm optimization techniques, such as fuzzy inference systems, support vector machines, particle swarm optimization, white-box ECM, grey-box ECM, and other improved variants, including an artificial neural network (ANN)-based thermal-coupled ECM method to establish the empirical model between these three *parameters* and SOC and temperature. These methods have effectively proved successful by accurately determining the characteristic parameters of batteries. The model structures of the ANN and SVM networks commonly used to represent the ECM are presented in Figure 3.12.

Neural network equivalent circuit model (NN ECMs) are designed to model the battery's characteristics based on test and training data to give accurate voltage traction without being directly programmed to do so. It keeps a physical description of the battery cell while approximating the dynamics of each component. The NN describes the linear or nonlinear mapping functions of the battery by measuring

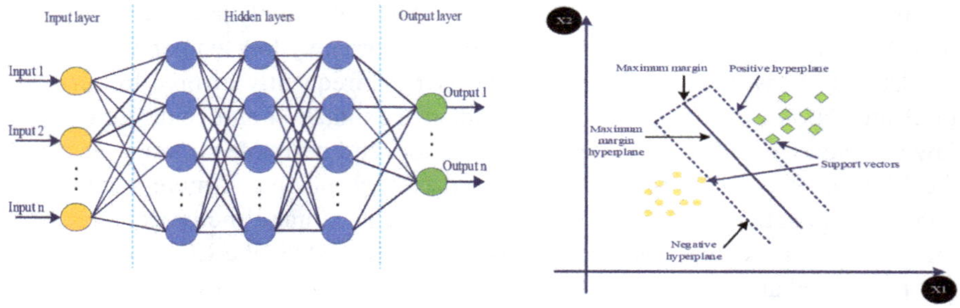

(a) (b)

Figure 3.12 *Model structure for artificial neural network and support vector machine;(a) Artificial neural network and (b) Support vector machine*

the voltage response characteristics. Indeed, with the support of a suitable data set, an NNECM can be configured to represent a given ECM, modeling each lumped parameter through the proposed NN model.

Specifically, the NNECM approach can preserve the interpretability aspect characterized by the ECM and simultaneously solve the main issues regarding the inability to take into account the relevant input parameters. The data-driven model has the advantage of not requiring a thorough understanding of electrochemistry or time-consuming parameter adjustment. However, for model training to be effective, a vast amount of data is usually required. The inputs to train the network are voltage, current, and temperature. Also, the NN can eliminate the internal battery's complexity, but it requires a large amount of data to train the model. The training data set and training procedure have a significant impact on the model's measurement errors.

In general, transfer learning for NN models can be accomplished in two ways: one is to use the original pretrained network as a fixed feature extractor for the new task, except for the last fully connected layer, and the other is to fine-tune the parameters of the pretrained network using data for the new task. The size and resemblance of the target data set to the original data set used in pretraining are the most important factors in selecting an appropriate transfer learning technique. It is usually assumed that the first few layers of a network are used to extract low-level features, whereas the next few levels are utilized to extract high-level features. To learn nonlinear combination of these features, the fully connected layers are used.

There have been several NN methods developed using ECMs for extracting internal parameters of the battery based on input data. A physical inspired-neural network equivalent circuit model (PI-NNECM) has been proposed that has an ensemble architecture that relies on a fractional-order extended single-particle of the lithium-ion cell formulation. The PI-NNECM is designed to approximate the ESP transfer functions referred to as the ohmic effects, electrolyte diffusion, and the nonuniform charge distribution in the cell. This proposed model is tested using three publicly available data sets, investigating the model's behavior according to two different training strategies and with different input configurations.

In the ANN-based thermo-ECM, the complete model is obtained to predict the coupled electrical and thermal dynamics during battery operation. The ECM first calculates SOC and terminal voltage using the applied current and circuit parameters OCV, internal ohmic resistance, polarization resistance, and polarization capacitance. The OCV model calculates the circuit parameter's OCV, whereas the ANN calculates internal ohmic resistance, polarization resistance, and polarization capacitance based on the SOC and temperature. The irreversible heat generation rate is then calculated using OCV, current, and terminal voltage, while the reversible heat generation rate is calculated using entropic heat coefficients OCV, temperature, and current. The thermal model is then used to calculate the temperature using the sum of both irreversible and reversible heat generation rates.

3.3.4 Voltage matching equivalent circuit modeling

1. Voltage characteristic

End voltage refers to the potential difference between the positive and negative electrodes of the lithium battery, which can be divided into OCV and working voltage according to the circuit operation [123]. The OCV is the end voltage of the lithium battery without load or other power supply. Generally, because the battery has an internal resistance, the operating voltage in the discharge state is lower than the OCV, and the working voltage when charging is higher than the OCV. The highest voltage allowed to reach the end of the battery is called the charge cutoff voltage, and the lowest voltage allowed to reach after the battery discharge is called the discharge cutoff voltage. Beyond this limit, the battery will suffer some irreversible damage, and the cutoff voltage is an important safety indicator.

Operating voltage refers to the potential difference between the positive electrode and the negative electrode when the battery is operating. The operating voltage will be below the OCV due to internal resistance inside the battery. Similar to the rated voltage classification method, according to the working state, the lithium battery working voltage is also divided into rated voltage, theoretical voltage, OCV, and load working voltage. The rated voltage is directly calibrated by the battery manufacturer before the battery production; the theoretical voltage is the theoretical value of the positive and negative electrode potential difference after the internal chemical reaction balance, which is represented by E. The calculation process is as shown in (3.17).

$$E = \varphi_{+} - \varphi_{-} \tag{3.17}$$

In the above equation, φ_{+} is the battery positive electrode potential, and φ_{-} is the battery negative electrode potential. The OCV can be measured by connecting the multimeter or voltmeter directly to both the positive and negative electrodes of the lithium battery [124]. The OCV is the voltage used in the equivalent model simulation in this book. Because lithium battery is not an ideal power supply, electrolyte and electrode materials have internal resistance, so the OCV value is slightly less than the electromotive potential value, in the internal resistance pole hour, the battery OCV can be treated as the electromotive potential value.

Under actual working conditions, the working instantaneous voltage of lithium battery is also dynamically changed, and overcharging and overdischarge are unavoidable, so the upper and lower limit voltage in the actual use; if the ceiling voltage is too high, the safety hazard; if too low, the battery charging may occur. Similarly, if the lower limit cutoff voltage is set too low, the battery safety hazard may be too low, but if the setting is too high, the battery power adoption efficiency is too low. After research, in the case of high discharge rate C value and low ambient temperature, the discharge of lithium battery can set the lower discharge cut voltage a little smaller.

The OCV of the lithium battery is the voltage value of both ends of the lithium battery after full rest. When the battery has a current through, the potential deviates from the equilibrium potential phenomenon, called electrode polarization. Electrode polarization can be divided into concentration polarization and chemical polarization. Affected by the ohmic effect and polarization effect, the battery voltage fluctuates during the use process, so the acquisition of OCV usually needs to hold the battery for a long enough time. Because OCV corresponds one-to-one to SOC, if you know the current OCV value of the battery, you will know the corresponding SOC value. This is known as the OCV method. To study the battery's OCV characteristics, that is, to obtain the OCV and SOC relationship curve, the battery cycle discharge is a shelved experiment at 23°C at room temperature. The specific experimental steps are as follows:

1. Charge the battery with constant current and constant voltage. Charge at 1 C times to an upper stop voltage of 4.2 V, and then charge at constant voltage with a charging cutoff current of 0.05 C. After charging, the battery is shelved to stabilize the battery voltage. Because the capacity of the selected lithium battery is small, so the selected stand time is 30 minutes.
2. The SOC sampling points were selected as 1, 0.95, 0.9, 0.85, 0.8, 0.7, 0.6, 0.5, 0.4, 0.3, 0.2, 0.15, 0.1, and 0.05. Because the voltage changes significantly in the early stage and the end stage of lithium battery discharge, so the SOC sampling point interval is smaller, to obtain a more accurate OCV–SOC characteristic curve.
3. Continuous current discharge of the lithium battery at 1 C times, and the discharge cutoff voltage is 2.75 V. When each discharge reaches the SOC sampling point, stand the battery for 30 minutes, think that the battery end voltage is the OCV, and then continue to discharge until the battery SOC is 0, ending the experiment.

Experimental data were fitted by origin, OCV–SOC relationship is shown in Figure 3.13.

As shown from the figure, with the increase of the battery SOC, the OCV of the battery also increases. In the middle stage of the battery discharge, the voltage change tends to flatten out, which is called the discharge platform effect of the

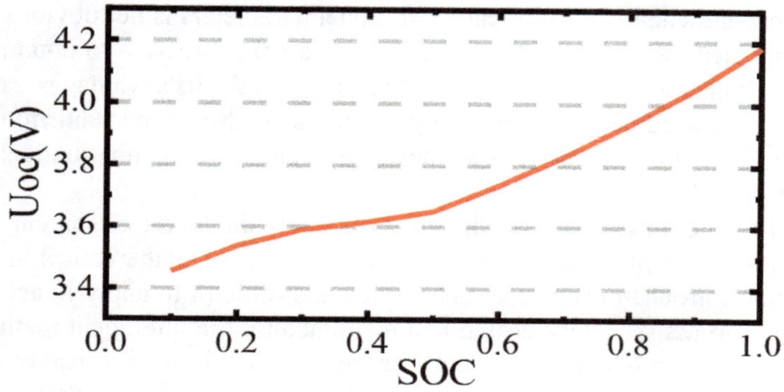

Figure 3.13 OCV–SOC relationship curve

battery. At the early and late stages of the battery discharge, the OCV of the battery changes significantly. At this point, the small battery OCV change error will cause drastic changes in the battery SOC. Therefore, the method of using the OCV to obtain the SOC is not suitable for the online estimation of the SOC values of lithium batteries. The MATLAB® was used, and the experimental data were processed. The seventh-order polynomial of OCV–SOC is shown in (3.18).

SOC_K is the battery SOC value at time k, and $U_{OC,K}$ is the corresponding OCV values, which is shown in (3.18).

$$U_{OC,K} = f(SOC_K) = 75.232 * SOC_K^7 - 270.75 * SOC_K^6 + 389.23 * SOC_K^5 - 287.21 * SOC_k^4 + 117.89 * SOC_k^3 - 27.427 * SOC_k^2 + 3.9258 * SOC_k + 3.2377 \tag{3.18}$$

2. Overview of battery modeling

The establishment of a lithium battery equivalent model includes theoretical analysis and experimental analysis: theoretical analysis is the dynamic equation of understanding the internal law of the research object; experimental analysis requires the input and output signals of the collected object, establishes the equivalent model according to the collected signal, and performs the parameter identification processing of the established model. For lithium batteries, the internal physical and chemical changes are very complex, and the battery dynamic equations derived based on the electrochemical theory are difficult to be applied in practice. Therefore, in most cases, the method of experimental analysis is often used.

Many different models have been built to predict the battery behavior, but no one model can completely and accurately simulate the dynamic behavior of the battery under anyone's working condition. The equivalent model of a lithium battery can be divided into the empirical model and an electrochemical model according to the modeling method. An empirical model is an alternative model of battery dynamics obtained based on sample data and system identification method. Its advantages are controllable complexity and low modeling cost, and its disadvantages are that the physical and chemical significance of model parameters is not obvious and less robust. The electrochemical model is a mathematical expression to obtain the battery dynamic characteristics through mechanism analysis. Its advantages are that the model has high accuracy and can deeply reveal the evolution mechanism of battery characteristics, while the model complexity is high, which is not conducive to its engineering application.

According to the different mechanisms of battery model establishment, it can be divided into the simple electrochemical model, intelligent mathematical model, and ECM. The electrochemical model is complex and difficult to apply in actual products, mainly to assist battery design and manufacture; the intelligent mathematical model is mainly an NN model, which can be completed, but because of practical data training, high technical threshold, and long processing time; ECM is widely used due to clear physical meaning and simple mathematical expression. Due to the nonlinear characteristics of the battery such as current, power, SOC, and temperature, battery modeling to fully consider these factors, will increase the complexity of calculation, the general controller cannot meet the requirements, and the key points of model building are described as follows.

1. Logical structure of the circuit model
   Logical structure of the circuit model is shown in Figure 3.14.

2. Principal description
   The battery simulation model is designed to verify the accuracy of the parameters set in the model, so its input is current and its output is end voltage; in the actual BMS, both current and end voltage are input quantities.

3. Modeling steps
   (1) Selects appropriate ECM; (2) determines input, output, and state variables; (3) column writes state equation; (4) uses MATLAB modeling; (5) makes parameter identification through experimental data; (6) corrects model parameters in real time according to variables (temperature, SOC, etc.); and (7) simulation.

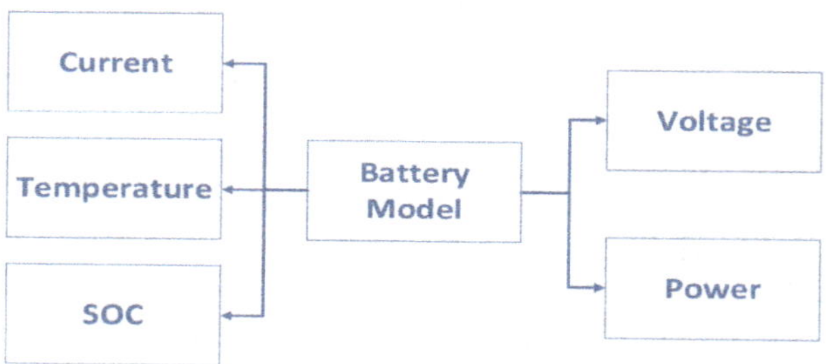

Figure 3.14 Logic structure diagram of the circuit model

4. Model selection
   According to the need, choosing different battery equivalent models has a great impact on the accuracy and accuracy of the battery model. At the same time, the more complex the structure and the higher the accuracy of the model, the more complex the operation, and the hardware requirements. In the equivalent model, the equivalent circuit containing the RC circuit has higher accuracy, and the higher the RC order, the higher the accuracy.

5. Parameter identification
   According to the experimental results of the HPPC test and the LS method to identify the relevant parameters.

3. Battery modeling type and concepts

The modeling of lithium batteries is divided into mechanism modeling, experimental modeling, and hybrid modeling. Mechanism modeling: the mathematical model is established according to the physical equation and the chemical reaction principle. Experimental modeling: the accused object is regarded as a "black box," and the model is constructed by trial recording the change rules of the characteristic parameters of the target object. Hybrid modeling: both mechanism modeling and experimental modeling are combined to build a model. Battery belongs to the highly complex nonlinear electrochemical energy storage device, on the one hand, through the precise equation to describe the internal process in the control interaction and reaction is difficult, on the other hand, completely using the experimental data modeling (such as the NN) requires a lot of data input and learning, so mixed modeling method is more common.

In the selection of mechanism modeling, electrochemical model, thermodynamic model, coupling model, and ECM can be adopted. In contrast, the ECM does not need to conduct an in-depth analysis of the electrochemical reaction inside the battery. It describes the OCV, DC internal resistance, and polarization internal resistance of the battery through the circuit relations, to realize the characterization of the external characteristics of the battery. The model constructed in this design is within the range of the electrical characteristic model. Electrical characteristic models are generally divided into three categories: black-box model, electrochemical mechanism model, and ECM, as shown in Figure 3.15.

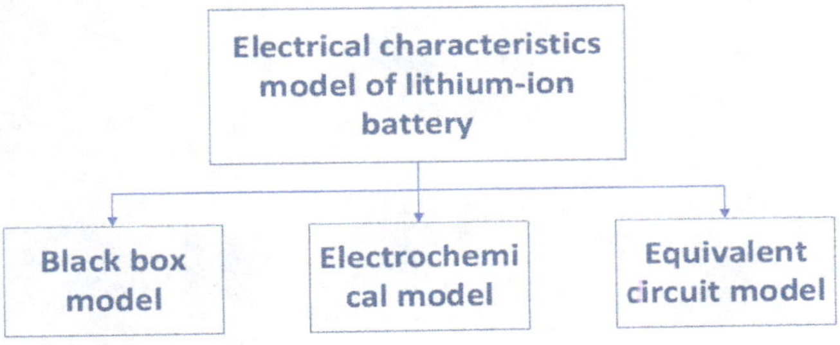

Figure 3.15 Equivalent model classification

A black-box model is a description of a linear or nonlinear function, it will focus on the battery's internal mechanism to the data itself, which determines that it can flexibly establish model structure and parameterization, but lacks essential physical significance, the difficulty of black-box modeling is that the model performance is sensitive to the quantity and quality of training data, usually need to take data-driven algorithm for support, such as NN, support vector regression, etc.

The electrochemical mechanism model is to establish the partial differential equation of electrode and electrolyte dynamics by analyzing the internal reaction mechanism of the battery, it can accurately predict the macroscopic physical parameters, also can simulate its distribution, so very suitable for lithium battery optimization design and safety analysis, but electrochemical model complexity is higher, the actual system computing power may not meet the theoretical requirements, so the current research mainly focuses on how to simplify the model, to apply to lithium battery management technology.

An ECM uses an RC circuit network and a constant voltage source to simulate the dynamic voltage response characteristics of the battery. The relationship between the parameters in the model is direct and obvious. Generally, contains a relatively small number, which makes the mathematical description of state-space description easy. So, it is widely used in system simulation and practical management. Battery ECM contains a variety of model frameworks. The advantage of constructing a simple combination structure is to achieve fast and less computational models, but the simulation accuracy is low. While the more comprehensive combination structure improves the accuracy results in greater complexity, which increases the difficulty of identification parameters and calculation process. Therefore, in a real-time application, the advantages and disadvantages of the existing ECM are weighed, and the appropriate circuit model is selected after a comprehensive analysis, which is shown in Figure 3.16.

The following several more used ECMs are analyzed and compared. Here is a brief description of several related parameters. Internal ohmic resistance: it is composed of electrode material, electrolyte, diaphragm internal resistance, and contact resistance of each part.The schematic diagram of the experimental current for the transient pulse is shown in Figure 3.17.

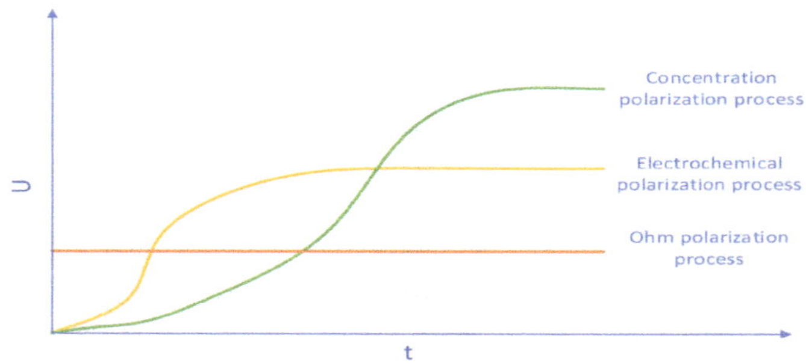

Figure 3.16 Internal curve of Po in the constant current phase

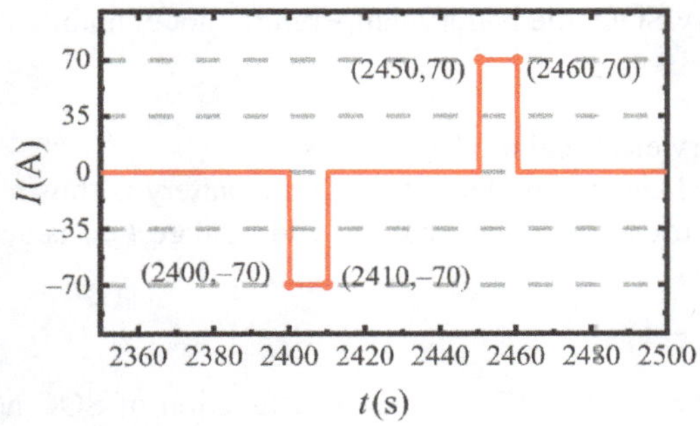

Figure 3.17 *Schematic diagram of the experimental current for the transient pulse*

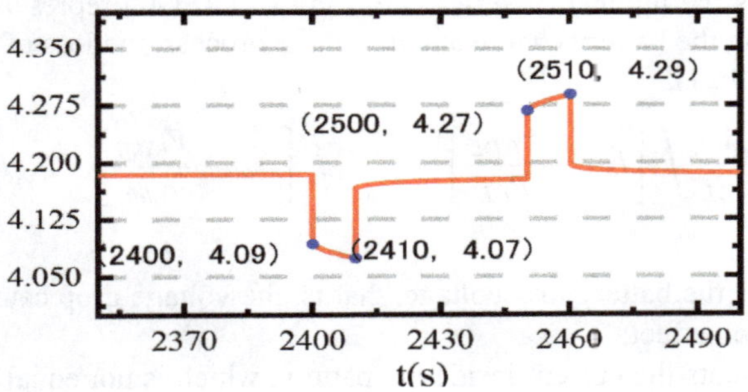

Figure 3.18 *Schematic diagram of the experimental voltage changes of the composite pulse*

Internal polarization resistance: the resistance caused by the polarization reaction phenomenon during the electrochemical reaction, including the resistance caused by the electrochemical polarization and the concentration polarization. It forms the resistance circuit in parallel with the polarized capacitance, which is used to simulate the dynamic characteristics of the battery polarization and greatly reduce the process, which is shown in Figure 3.18.

There are several sets of experimental data plots in the constant current phase, which are shown in Figure 3.6(a). The composite pulse experimental current is shown in Figure 3.6(b). The experimental voltage change of the composite pulse is shown in Figure 3.6(c).

4. Battery model establishment

The battery model includes an electrical model of the battery and a battery temperature model. The electrical model (also known as the ECM) and the temperature model together constitute the body model of the battery. In the battery body model, the electrical model mainly calculates the battery voltage, battery self-discharge current, and battery SOC; the battery temperature model mainly calculates the battery temperature.

1. Lithium battery electrical model analysis
   The load-based electric model of the lithium battery is shown in Figure 3.19.
   From the electrical model, the battery end voltage V_{bat} is accurately given in (3.19).

$$V_{bat} = V_{bat,0} - V_{loss} \tag{3.19}$$

$V_{bat,0}$ represents the battery OCV, which is a function of SOC and temperature if the battery is composed of multiple cells in series. The total voltage is obtained as shown in (3.20).

$$V_{bat,0} = n_{cell} V_{cell} (SOC, T_{bat}) \tag{3.20}$$

n_{cell} represents the number of series cells, and V_{cell} (SOC) represents the OCV of the cell. During the battery charge and discharge process, the loss of voltage V_{loss} is calculated in (3.21).

$$V_{loss} = \frac{1}{C_{DL}} \int \left[I_{bat} - \frac{V_{CDL}}{R_{DL}} \right] + \frac{1}{C_{diff}} \int \left[I_{bat} - \frac{V_{Cdiff}}{R_{diff}} \right] + I_{bat} R_{bat} (SOC, T_{bat}) \tag{3.21}$$

V_{loss} represents the battery loss voltage, that is, the voltage drop caused by the battery polarization effect;

I_{bat} represents the current inside the battery, which is not equal to the external end current of the battery;

C_{DL} represents the current inside the battery, which is not equal to the external end current of the battery (double-layer capacitance);

R_{DL} represents the battery double-layer resistance value (double-layer resistance);

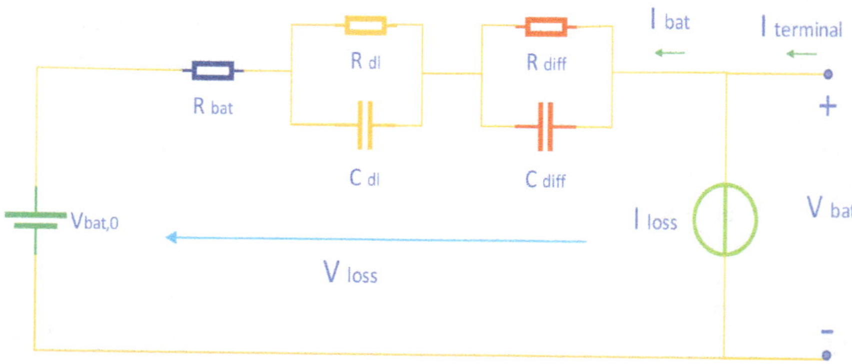

Figure 3.19 Electrical model of lithium battery

C_{diff} represents the battery diffusion capacitance capacity (diffusion capacitance);
R_{diff} represents the battery diffusion resistance value (diffusion resistance);
V_{CDL} represents the loss voltage of the double capacitor;
V_{Cdiff} represents the loss voltage of the diffusion capacitor;
R_{bat} represents the battery resistance;
T_{bat} represents the battery temperature;
SOC represents the state of the battery charge.

The self-discharge current k of lithium batteries at different temperatures can be approximated in (3.22).

$$I_{loss} = I_0 n_{cell} V_{cell} \left(SOC, T_{bat}\right) \tag{3.22}$$

I_0 indicates the battery self-discharge current at standard temperature conditions; T_{bat} represents battery temperature.

The battery SOC is an important parameter to characterize its available remaining energy, and the accurate estimation of the SOC is one of the core tasks of the BMS. Using Ampere integral method to calculate battery SOC is the best feasible among various existing battery SOC calculation methods. However, this algorithm has the biggest drawback, it needs to know the initial SOC_0 value of SOC, so Kalman filter can be used to calculate the initial SOC value of the battery SOC_0. For batteries, the battery end current is defined as $I_{terminal}$, the battery self-discharge current as I_{loss}, charging if $I_{terminal} - I_{loss} > 0$, and discharging if $I_{terminal} - I_{loss} < 0$. The battery SOC can be expressed in the form of (3.23).

$$SOC = SOC_0 + \frac{1}{K_N}\int \left(I_{terminal} - I_{loss}\right) dt \tag{3.23}$$

SOC_0 indicates the initial charge state of the battery;
K_N represents battery nominal capacity is indicated;
$I_{terminal}$ represents the current outside of the battery;
I_{loss} represents the self-discharge current of the battery.

2. Lithium battery temperature model analysis
   The main task of the temperature model established for the lithium battery is to calculate the change in the battery temperature during the heat dissipation process. Theoretically, the electrochemical model of the battery is the best way to calculate the battery temperature because the chemical reactions occurring inside the battery are the direct cause of the battery temperature change. Electrochemical models can accurately describe the chemical reaction mechanism inside the battery. So, they can also accurately calculate the heat released by the battery during the chemical reaction. In theory, the cell temperature calculated by the electrochemical models is the closest to the actual cell temperature. However, the establishment of electrochemical models is greatly influenced by the battery manufacturing process. Due to the enterprise secrets and equipment precision involving battery manufacturers, electrochemical models are only suitable for theoretical research and cannot be applied to engineering control.

The battery temperature model in this section is based on the analysis of the battery electrical model and adds the standard reaction thermal composition of the main chemical reactions inside the battery. The accuracy of the model will be reduced, but as an engineering research object with low-temperature requirements, this method is still feasible. In the cell temperature model, the battery temperature T_{bat} is determined in (3.24).

$$T_{bat} = T_{bat,0} + \frac{1}{C_{bat,0} m_{bat}} \int P_w dt \tag{3.24}$$

$T_{bat,0}$ represents the initial temperature of the battery, generally taking the ambient temperature.

C_{bat} represents the specific heat capacity of the battery, and the specific heat capacity of the lithium battery is generally 800–1000 J/kg*K. The test data returned from the battery manufacturer indicate that the specific heat capacity of the battery sample is 854 J/kg*K.

m_{bat} represents the battery quality, the single battery mass is 1004 g, so the total mass of the battery sample is 10.04 kg.

P_w represents the total heat absorbed by the battery. In this model, the heat absorbed by the battery is mainly composed of the heat generated by the internal loss voltage, the heat generated by the internal chemical reaction, the heat generated by the battery self-discharge current, and the heat exchange between the battery and the environment. So, the total heat PW absorbed by the battery can be expressed in (3.25).

$$P_W = P_{Joule} + P_{MainReaction} + P_{LossReaction} - P_{Radiation} \tag{3.25}$$

Therein, the heat P_{Joule} generated by the internal loss voltage V_{loss} is shown in (3.26).

$$P_{Joule} = \mathrm{I_{bat} V_{loss}} \tag{3.26}$$

The heat $P_{MainReaction}$ generated from the internal chemical reaction is shown in (3.27).

$$P_{MainReaction} = \mathrm{C_{MR} I_{bat}} \tag{3.27}$$

The heat due to the self-discharge current of the battery, $P_{LossReaction}$ is shown in (3.28).

$$P_{LossReaction} = V_{bat} I_{loss} \tag{3.28}$$

The heat $P_{Radiation}$ generated by the battery and ambient heat exchange is shown in (3.29).

$$P_{Radiation} = \varepsilon \mathrm{S_{bat}} \sigma \left(\mathrm{T^4_{bat}} - \mathrm{T^4_{ambient}}\right) \tag{3.29}$$

I_{bat} represents the in-cell current, $I_{bat} = I_{terminal} - I_{loss}$;
V_{loss} represents the battery loss voltage;
I_{loss} represents the self-discharge current of the battery;
V_{bat} represents the battery end voltage;
C_{MR} represents the battery constantly, for lithium manganese oxide battery, is 27 mV;

$T_{ambient}$ represents the ambient temperature;
S_{bat} represents the contact area of the battery and the environment;
σ represents the Stefan–Boltzmann constant, 5.67 * 10 – 8 W/(m^{2}K^4);
ε represents the scattering rate, the emission ratio constant, is generally 0.7–0.9 for general building materials. This model temporarily takes 0.8, because it cannot be determined.

The ultimate goal of the battery model is to simulate the various parameters of the battery and eventually need to run in the embedded processor, so the model variable continuous integral form, changed to step dispersion, set the model running the size of 100 ms. So, it can ensure the accuracy of the model and can ensure that the embedded processor resource occupancy is not too high.

3.3.5 Improved second-order equivalent circuit modeling

By comprehensively comparing the advantages and disadvantages of several lithium-ion battery models, the ECM with clear physical meaning and easy identification of model parameters is selected in this book. Compared with the Rint model, Thevenin model, and PNGV model, the RC model can more accurately describe the dynamic characteristics of the batteries.

The RC loop needs to be reduced, but several order RC models can not only better reflect the dynamic characteristics of the battery but also reduce the amount of calculation [125]. Xu Qiaoqiao of Chongqing University analyzed and compared the first-order, second-order, third-order, and fourth-order RC models, respectively. It is found that the accuracy of the second-order RC model is greatly improved compared with the first-order RC model, while the accuracy of the third-order and fourth-order RC models is limited compared with the second-order RC model, and the amount of calculation is greatly increased.

After comprehensively comparing the advantages and disadvantages of each battery model, the second-order RC model is selected as the research object, as shown in Figure 3.20.

In Figure 3.20, the OCV is represented by U_{OC}, the terminal voltage is represented by U_L, and the ohmic internal resistance is represented by R_0. The first RC

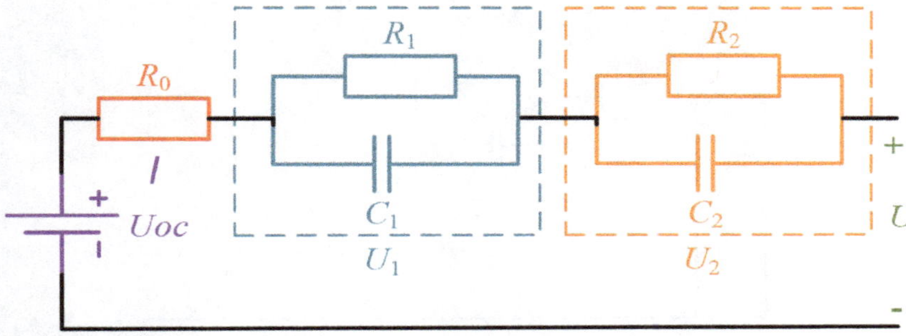

Figure 3.20 Second-order RC equivalent circuit model

parallel circuit is composed of electrochemical polarization resistance R_1 and electrochemical polarization capacitance C_1. The second RC parallel circuit is composed of concentration polarization resistance R_2 and concentration polarization capacitance C_2. R_1C_1 loop represents the process of rapid change of circuit voltage, and the R_2C_2 loop represents the process of slow change of circuit voltage. With the discharge direction as positive, the KVL equation of the circuit is listed as shown in (3.30).

$$\begin{cases} U_L = U_{OC} - I(t) R_0 - U_1 - U_2 \\ \dfrac{dU_1}{dt} = -\dfrac{U_1}{R_1C_1} + \dfrac{I}{C_1} \\ \dfrac{dU_2}{dt} = -\dfrac{U_2}{R_2C_2} + \dfrac{I}{C_2} \end{cases} \tag{3.30}$$

where $[\mathrm{SOC}\ U_1\ U_2]^T$ is selected as the state variable. By discretizing the equivalent circuit, the state space expression of the model can be obtained as shown in (3.31).

$$\begin{cases} \begin{bmatrix} SOC_{k+1} \\ U_{1,k+1} \\ U_{2,k+1} \end{bmatrix} = \begin{bmatrix} 1 & 0 & 0 \\ 0 & e^{-\Delta t/\tau_1} & 0 \\ 0 & 0 & e^{-\Delta t/\tau_2} \end{bmatrix} \begin{bmatrix} SOC_k \\ U_{1,k} \\ U_{2,k} \end{bmatrix} + \begin{bmatrix} -\frac{\Delta t}{Q_N} \\ R_1(1 - e^{-T/\tau_1}) \\ R_2(1 - e^{-T/\tau_2} \end{bmatrix} I_k + w_k \\ U_{L,k+1} = U_{OC,k+1} - U_{1,k+1} - U_{2,k+1} - IR_0 + v_k \end{cases} \tag{3.31}$$

Due to the existence of self-discharge, the performance of the lithium-ion batteries in identity is not excellent, and the second-order RC battery model determined above does not fully describe the unity of the battery. Therefore, an improved second-order RC model based on the second-order RC model and considering the factors of battery self-discharge is designed, as shown in Figure 3.21.

Comparing Figures 3.20 and 3.21, it can be found that the improved second-order RC model is based on the second-order RC model, and a self-discharge resistance R_d is connected in parallel to the voltage source. Considering the self-discharge effect of battery is very important for a BMS, so it is very necessary to take the improved second-order RC model as the research object of battery model, and carry out mathematical formula derivation and parameter identification.

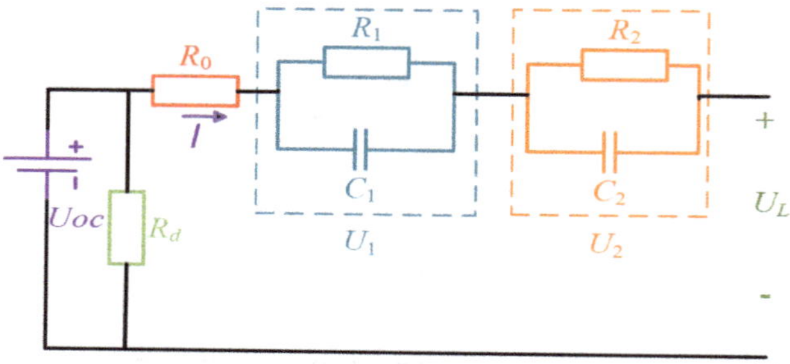

Figure 3.21 Improved second-order RC equivalent circuit model

For parameter identification of lithium-ion battery model, there are two kinds: online identification and offline identification. Offline identification refers to the method of obtaining battery model parameters after certain processing and calculation of battery-related data such as temperature, terminal voltage, charge, and discharge current. Online parameter identification refers to the online recursive estimation of model parameters by common methods such as the LS method. For the improved second-order RC model adopted, the model parameters to be identified include battery OCV U_{OC}, battery ohmic resistance R_0, self-discharge resistance R_d, battery electrochemical polarization internal resistance R_1, electrochemical polarization capacitance C_1, battery concentration polarization resistance R_2, and concentration polarization capacitance C_2.

During the charging and discharging process of the lithium-ion battery used in this book, there is a certain mapping relationship between the battery SOC and the electromotive force, and this mapping relationship is relatively stable. This stable mapping relationship is the theoretical basis for the offline identification of the battery's OCV U_{OC}. When the current in the circuit is zero, the voltage on the RC circuit is approximately zero. At this time, the OCV of the battery and the electromotive force of the battery is approximately equal. The electromotive force of the battery is difficult to be measured directly, so the OCV after standing for a long time is often equivalent to the electromotive force of the battery in the experimental process, to obtain the SOC–OCV curve and provide a reference for the SOC estimation results. The specific experimental method is to conduct a pulse charge and discharge experiment on the battery, and obtain the OCV value of the battery after standing for a long time, to obtain the SOC–OCV curve of the battery.

The experimental data of HPPC working conditions are used for the extraction of the SOC–OCV relation curve and online battery parameter identification. The whole process of the HPPC experiment is shown in Figure 3.22.

After the HPPC experiment is completed, the obtained experimental data are saved, and then the terminal voltage data are extracted. The terminal voltage data obtained under HPPC working condition are shown in Figure 3.23.

As can be seen from Figure 3.23, the voltage decreases continuously with continuous discharge, and the battery SOC gradually decreases from 1 to 0.1 by 0.1 each time. The current data obtained under HPPC working condition are shown in Figure 3.24.

According to the above analysis, the OCV of the battery at this time can be equivalent to the battery electromotive force. The SOC–OCV relationship curve extracted after the experiment is shown in Figure 3.25.

The OCV corresponding to SOC 0–1 from the voltage curve is extracted, and the corresponding points are found, respectively. The SOC–OCV curve of lithium-ion battery is fitted with MATLAB, and the functional relationship is shown in (3.32).

$$OCV = P_1 SOC^6 + P_2 SOC^5 + P_3 SOC^4 + P_4 SOC^3 + P_5 SOC^2 + P_6 SOC + P_7 \tag{3.32}$$

As shown in (8.19), according to different attempts and adjustments, it is found that the fitting effect of six-order polynomial function on SOC–OCV is the best. The specific values of the identified parameters are shown in Table 3.5.

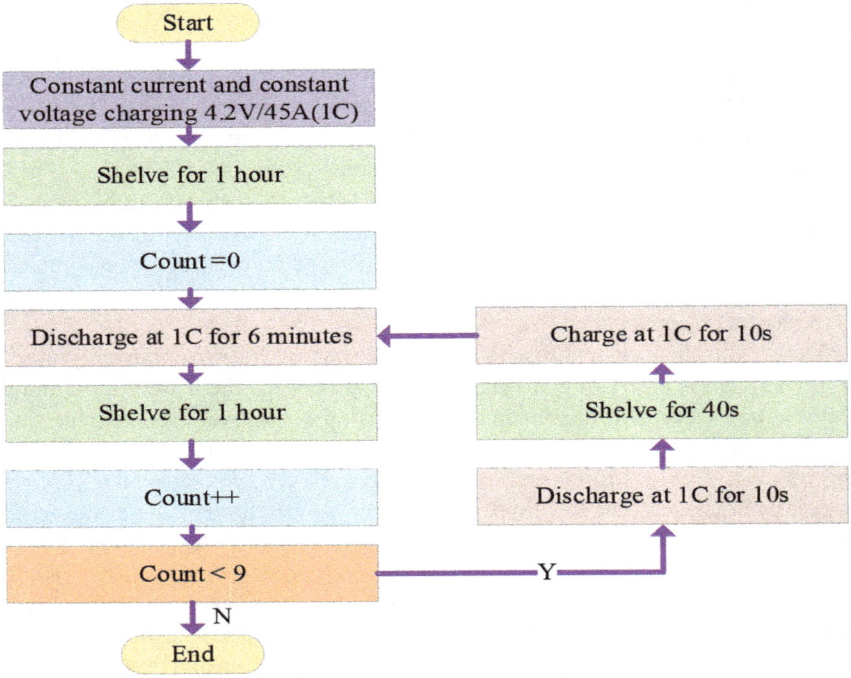

Figure 3.22 Whole process of the HPPC experiment

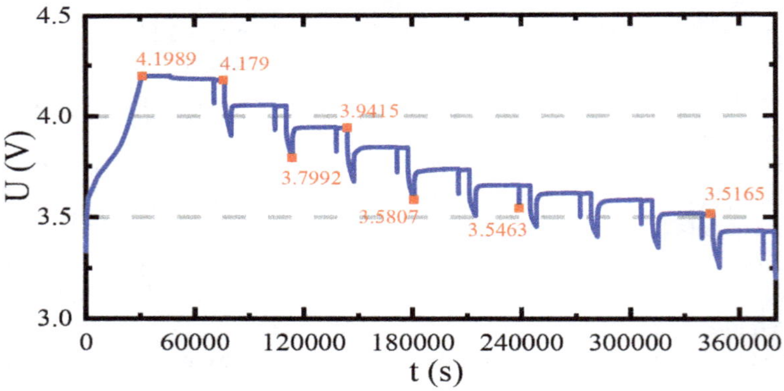

Figure 3.23 Voltage data under HPPC working condition

Ohmic resistance R_0 represents the ohmic polarization effect of the battery. For the lithium-ion battery used in the experiment, the ohmic polarization effect of the battery will disappear when the current stops. Therefore, the specific method to identify ohmic resistance R_0 is: the pulse discharge of the battery.

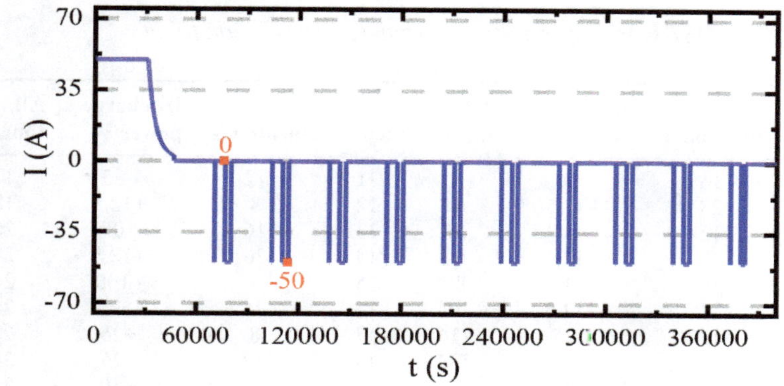

Figure 3.24 *Current data under HPPC working condition*

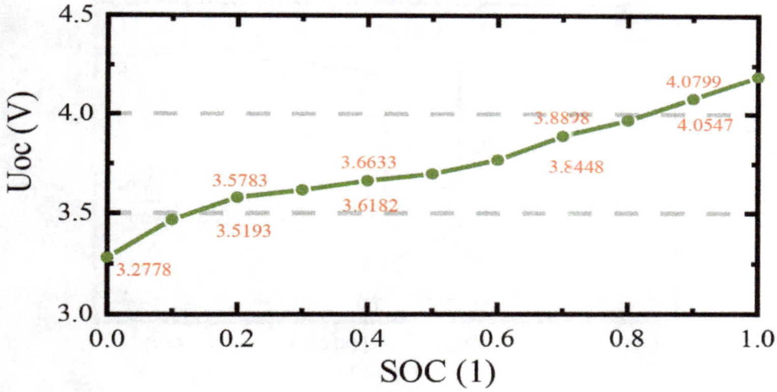

Figure 3.25 *SOC–OCV relation curve*

The voltage difference at the beginning and end of pulse discharge is mainly caused by ohmic resistance, and the pulse time is very short. Therefore, the voltage difference $(U_A - U_B)$ and $(U_D - U_C)$ generated at the beginning and end of pulse discharge is almost equal. In the calculation, to reduce the error, the two voltage differences can be averaged. The calculation formula of ohmic internal resistance is shown in (3.33).

$$R = \frac{(U_A - U_B) + (U_D - U_C)}{2I} \tag{3.33}$$

In (8.19), I represents the discharge current at the current time. The terminal voltage and current response curve of a unidirectional pulse discharge sequence are shown in Figure 3.26.

Table 3.7 DST working condition parameter description table

Step	Duration/s	Discharge power/%	All time/s	Step	Duration/s	Discharge power/%	All time/s
1	16	0	16	11	12	–25	176
2	28	–12.5	44	12	8	+12.5	184
3	12	–25	56	13	16	0	200
4	8	+12.5	64	14	36	–12.5	236
5	16	0	80	15	8	–100	244
6	24	–12.5	104	16	24	–62.5	268
7	12	–25	116	17	8	+25	276
8	8	+12.5	124	18	32	–25	308
9	16	0	140	19	8	+50	316
10	24	–12.5	164	20	44	0	360

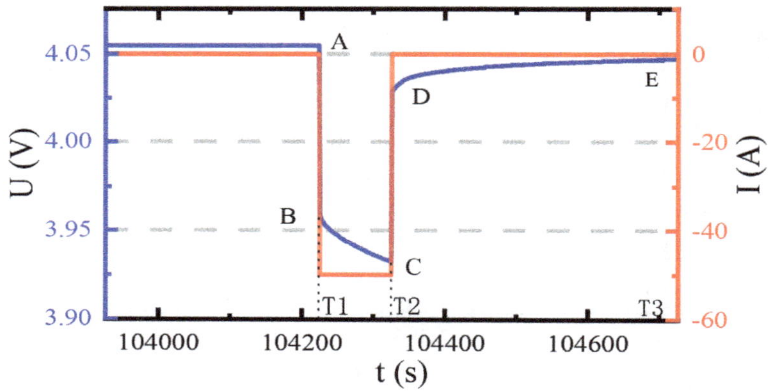

Figure 3.26 Terminal voltage and current response curve

During the static period of the battery after discharge, that is, from Points C to E in Figure 3.7, the current is 0, the second-order RC circuit has 0 input voltage response, and the expression of battery terminal voltage response is shown in (3.34).

$$U = U_{OC}\left(1 + \frac{R_0}{R_d}\right) - U_1(0)\, e^{\frac{-\Delta T}{R_1C_1}} - U_2(0)\, e^{\frac{-\Delta T}{R_2C_2}} \tag{3.34}$$

In (3.34), ΔT is the sampling time, and the value is 0.1 seconds. U_1 (0) and U_2 (0) represent the initial values of polarization voltages U_1 and U_2, respectively. Through the user-defined function in the MATLAB toolbox cftool, the exponential fitting is carried out for the standing time after the battery pulse discharge according to (3.35).

$$y = a(1 + g) - be^{-c\Delta T} - de^{-f\Delta T} \tag{3.35}$$

In (3.35), *a*– *g* represent the abstract parameters, respectively. The solution of model parameters can be obtained by simultaneous (3.34) and (3.35), as shown in (3.36).

$$\begin{cases} U_{OC} = a \\ U_1(0) = b, U_2(0) = d \\ 1/R_1C_1 = c, 1/R_2C_2 = f \\ R_0/R_d = g \end{cases} \tag{3.36}$$

By solving the above equation, the values of polarization resistance–capacitance parameters R_1, C_1, R_2, C_2, and R_d of the improved second-order RC model of the lithium-ion battery can be obtained.

3.4 Introduction of complex testing conditions

3.4.1 HPPC test

HPPC test is the most commonly used test condition experiment for lithium-ion batteries. HPPC experiment can directly and accurately reflect the performance of lithium-ion battery pulse charge and discharge. At the same time, the internal ohmic resistance of the lithium-ion batteries can be calculated directly by obtaining detailed experimental data in the experimental process [126]. By analyzing the causes of voltage changes and designing the calculation equation or fitting equation, we can further obtain the relevant simulation parameters such as polarization capacitance, polarization resistance, and terminal voltage used to characterize the working state model of lithium-ion battery.

The process of a single HPPC experiment is to discharge the battery with a current of 1 C for 10 seconds, then put the battery aside for the 40 seconds, and then charge for 10 seconds, that is, complete the charge and discharge of an HPPC experiment. The charge and discharge current of a single HPPC experiment is 1 C. Taking 70 Ah ternary lithium-ion battery as an example, the specific experiment shows that the current of the corresponding battery 1 C is 70 A, the current design is shown in Figure 3.27(a), and the voltage change of the corresponding terminal is shown in Figure 3.27.

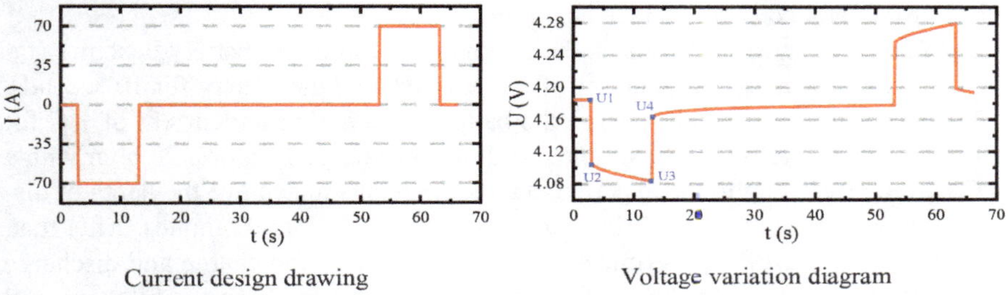

Figure 3.27 Design drawing of a single HPPC experiment

As shown in Figure 3.27, when the current of the HPPC test experiment changes, the voltage will change abruptly, such as U1-U2 and U3-U4 stages. When the current is constant, the battery voltage will also change slightly, such as U2-U3 and U4-U5 stages. With further analysis of its voltage change principle, the following arguments can be drawn:

1. The voltage changes in the U1-U2 stage occur at the moment when the current is generated. From this, it can be known that the voltage change in this stage is caused by the ohmic resistance existing in the battery, so the voltage change can be used as the calculation of the internal ohmic resistance of the battery.
2. When the voltage in the U2-U3 stage changes slowly, the current size remains unchanged. It can be known that the voltage change in this stage can be regarded as the polarization charging effect inside the battery during the discharge process. Therefore, the voltage change can be calculated as the basis for simulating the polarization parameters in the battery state model.
3. The voltage changes in the U3-U4 stage occur at the moment when the current returns to zero. From this, it can be known that the voltage change in this stage is due to the change of the voltage applied to the internal ohmic resistance of the battery, so the voltage change can be regarded as a calculation. The basis of the internal ohmic resistance of the battery.
4. When the voltage in the U4-U5 stage changes slowly, the current flowing through the battery is zero. From this, it can be known that the voltage change in this stage is caused by the polarization discharge effect inside the battery, so the voltage change in this stage is the same as that of the battery. The functional relationship between time can be used as the basis for calculating the polarization parameters in the simulated battery state model.

According to the above demonstration, the internal parameter values of the battery at the depth of discharge in a single HPPC experiment can be specifically calculated, and different depths of discharge can be obtained by charging and discharging, and the HPPC experiment at this depth of discharge can be calculated. The battery parameter value corresponds to the depth of discharge [127]. The experiment combining a single HPPC experiment with a deep discharge experiment is a full-cycle HPPC experiment. The specific steps are: first, use the specified charging voltage and current of the battery to charge the battery with a specified capacity using constant current and constant voltage charging.

After standing for 40 minutes, the internal reaction of the battery becomes stable, and then use a constant current of 1 C to discharge the battery for 10 seconds, set aside for the 40 seconds, charge the battery with a constant current of 1 C for 10 seconds, set it aside for 40 seconds, and then discharge 1/10 or 1/20 hour with a 1 C current according to the experimental requirements to achieve the depth of discharge required for the next HPPC experiment, set aside for 40 minutes. After that, repeat the single HPPC experiment step, and thus repeat the charge and discharge 10 or 20 times to complete the full-cycle HPPC experiment from the fully charged state to the end of the discharge. Taking a 70 Ah ternary lithium-ion battery as an example to illustrate the specific experiment, the current design size of the full-cycle

HPPC experiment is shown in Figure 3.28(a), and the corresponding experimental voltage change diagram is shown in Figure 3.28.

Taking a 70 Ah ternary lithium-ion battery as the test object, the full cycle HPPC test is conducted every 10% of the discharge depth. Each cycle includes pulse charge and discharge, deep discharge, and shelving. Finally, after ten cycles of the cycle test, the battery voltage drops to the cutoff discharge voltage of the battery, which marks the completion of the full cycle HPPC test.

3.4.2 Beijing bus dynamic stress test

According to the relevant requirements of China's QC/T 743–2006 lithium-ion battery for EVs, to accurately characterize the variation law of battery pack working voltage and current with vehicle speed during the operation of EVs. The researchers conducted a 14-day actual operation tracking test on the electric bus of Beijing line 121 and comprehensively collected the relevant information such as battery pack voltage, current, and speed when the bus operates under the provisions of the line [128]. The data of different periods such as morning peak, evening peak, and smooth road sections are classified and processed. Before the bus leaves the bus charging station in the morning, the same charger is used to charge the energy supply battery pack of the electric bus with the same charging method and current.

The output power of the battery pack is closely related to the form and state of the vehicle, so the change in the output power of the battery pack can be used to characterize the driving condition of the vehicle. Through the analysis of the statistical results of 121 electric bus working conditions, we can sort out and obtain four representative small working conditions to form a driving condition with a cycle of 300 seconds to accurately simulate the actual driving condition of an electric bus.

Referring to the typical driving processes such as acceleration, uniform speed, deceleration, and idle speed in the process of vehicle form, the driving conditions obtained above are disassembled and combined at intervals to obtain the power battery pack test condition (Beijing bus dynamic stress test (BBDST)) under the driving condition of electric bus in Beijing, which can accurately characterize the driving condition of electric bus. The specific parameters of working conditions are described in Table 3.6.

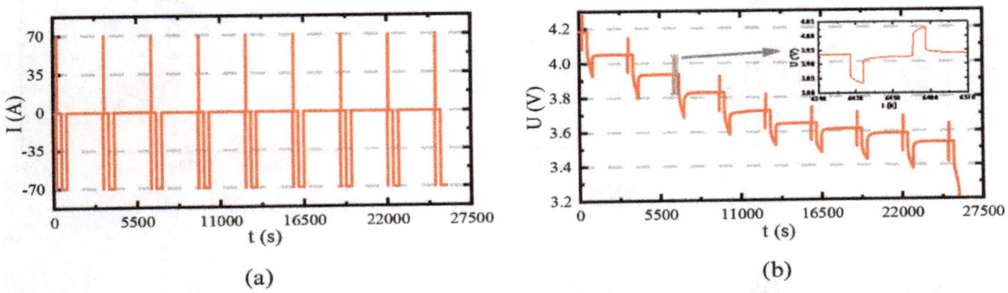

Figure 3.28 Full cycles HPPC experimental design diagram: (a) Current design drawing and (b) Voltage variation diagram

As shown in Tables 3.5, the BBDST working condition describes the operation condition of the electric bus of the Beijing 121 line in detail, and through splitting, interval reorganization, and rational allocation of power distribution, the working condition is more representative. Taking the 70 Ah ternary lithium-ion battery as the test object, the charge and discharge power are proportionally reduced according to the discharge cutoff voltage and the maximum discharge current limit. Finally, a BBDST working condition based on a 70 Ah lithium-ion battery is customized. The power characterization diagram is shown in the figure, the current change diagram is shown in the figure, and the voltage change diagram is shown in Figure 3.29.

As shown in Figure 3.29, since the test condition of BBDST is constant power charge and discharge condition, the charge and discharge current and voltage of the battery will change significantly with the increase of discharge depth. Because the current and voltage will change, BBDST working condition is a typical complex rate charging and discharging working condition commonly used to verify the charging and discharging working state of lithium-ion batteries.

3.4.3 Dynamic stress test

A dynamic stress experiment (DST) is a simplified variable power discharge cycle experiment, which can effectively simulate the dynamic discharge process and can be realized on the equipment of most test laboratories and developers [129]. DST has the same average characteristics as funds state, but the DST test condition has only a 360 seconds power step sequence and only seven discrete power levels, which greatly simplifies the difficulty of the experiment. The DST experiment is scaled according to the percentage of the maximum rated power of the battery to determine the discharge power of each step. The specific step power scaling of each step is shown in Table 3.7.

As shown in Tables 3.6, one cycle of the DST test condition is 360 seconds, and the discharge power of each step is scaled according to the rated power of the test object. DST test conditions can well characterize the charge and discharge characteristics of the battery, and the design of the condition is simple. Therefore, this condition is long and is commonly used in the test laboratory.

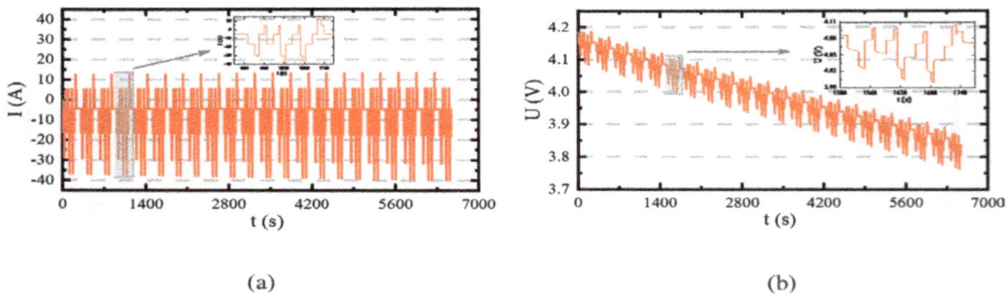

Figure 3.29 BBDST experimental design diagram: (a) Current design drawing and (b) Voltage variation diagram

3.5 Offline parameter identification

3.5.1 Point calculation strategies

The ECM is only a rough description of the characteristics of lithium-ion batteries. It is necessary to identify the parameters of the equivalent model based on the experimental data, which is used to describe the working characteristics of batteries accurately and establish a state-space model. Describing the behavioral characteristics of the system is mainly the identification of system model parameters, which are determined by the functional relationship between the input and output parameters of the system changing with time [130]. The purpose of parameter identification is to make the error between the output data and the actual data within the specified range when the input data are transmitted into the system, after the expression processing of the model, the accuracy of the ECM can be judged, and the subsequent SOC estimation can be provided on reliable basis. The experimental process is shown in Figure 3.30.

The complexity of the working environment and application situation aggravates the difficulty of mathematical modeling of lithium batteries. Considering its aging and the complexity and variability of the environment, the existing research builds an ECM to simulate the voltage response characteristics under different load conditions. The Thevenin ECM is selected, which has a simple structure, involves few parameters, and is easy to identify. The model also takes into account the polarization effect and self-discharge effect of the battery, which well meets the model accuracy requirements, as shown in Figure 3.31.

In Figure 3.31, U_{OC} is the OCV, R_0 is the ohmic resistance, R_p is the polarization resistance, and C_p is the polarization capacitance. The parallel circuit of R_p and

Figure 3.30 Data processing and model validation

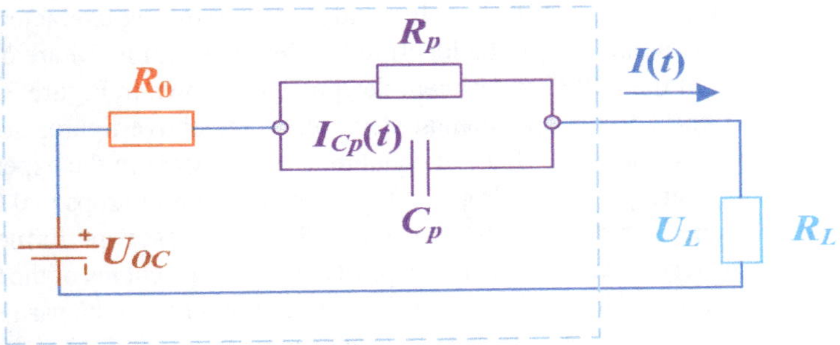

Figure 3.31 Thevenin equivalent model structure diagram

C_p describes the polarization process. U_L is the closed-circuit voltage after the battery is connected to the external circuit. The model takes into account the polarization process of the battery, and the first-order model constructed is relatively simple to calculate. First, the state space equation should be obtained according to the basic operating characteristics of the capacitor components. The relationship between the current flowing through the battery polarization capacitor C_p and its two closed-circuit voltages is shown in (3.37).

$$I_{Cp}(t) = C_P \frac{dU_{Cp}(t)}{dt} \tag{3.37}$$

In (3.37), based on the analysis of the equivalent circuit composition, through the application of the knowledge of circuit science and KVL, the voltage relationship in the equivalent circuit of (3.37) can be obtained, as shown in (3.38).

$$R_0 * I_{Cp}(t) + R_0 * \frac{U_{Cp}(t)}{R_P} + U_{Cp}(t) = U_{OC} - U_L \tag{3.38}$$

In (3.38), taking $U_{Cp}(t)$ as the state variable, the above two expressions are combined to describe the comprehensive working state. The voltage $U_{Cp}(t)$ at both ends of the equivalent circuit and the capacitor is set as the state variable, according to which the calculation process is analyzed to obtain the state-space expression of the equivalent model, as shown in (3.39).

$$R_0 * C_P \frac{dU_{Cp}(t)}{dt} + \left(1 + \frac{R_0}{R_p}\right) U_{Cp}(t) = U_{OC} - U_L \tag{3.39}$$

In (3.39), U_{OC} is the OCV, R_0 is the ohmic resistance, R_p is the polarization resistance, C_p is the polarization capacitance, and U_L is the closed-circuit voltage after being connected to the external circuit. The parameters in the equivalent model are fixed values and cannot be modified and adjusted. After the internal resistance is equivalently processed, the equivalent model takes into account the influence of the polarization effect. The RC parallel circuit of the capacitive device C_p is added with the dynamic characteristics so that the ECM can have better dynamic simulation characteristics of the battery.

At an ambient temperature of 25°C, the performance test of the lithium battery was carried out by using the HPPC test. By studying the working characteristics of lithium batteries, the parameters to be identified in the Thevenin ECM are obtained. The detailed steps of the HPPC experiment designed are shown in Figure 3.32.

According to the HPPC experimental steps set in the above figure, during the cycle test, the SOC is selected as 1-0.1 at equal intervals to perform the experimental operation until the battery capacity drops to 0, or when the current drops to 0.05C, the experiment is stopped. During this process, the HPPC cycle test is performed every time the lithium battery releases 10% of its capacity. The change curves of the terminal voltage and current of the lithium battery during the experiment and the partial amplification curve of one of the HPPC experiments are shown in Figure 3.33.

From the graphical analysis of the experimental results, it can be seen that the voltage changes significantly after one hour of constant current discharge. After

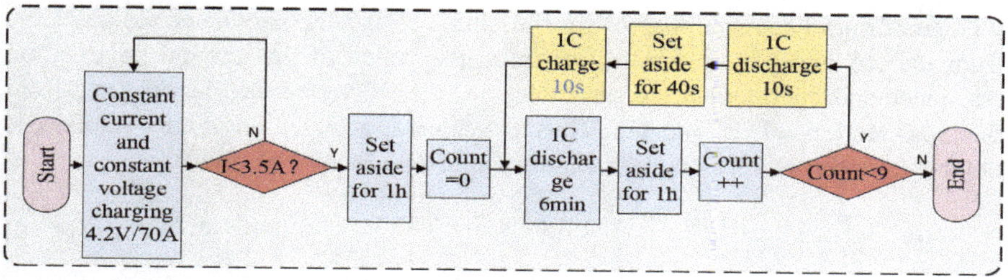

Figure 3.32 Experiment process

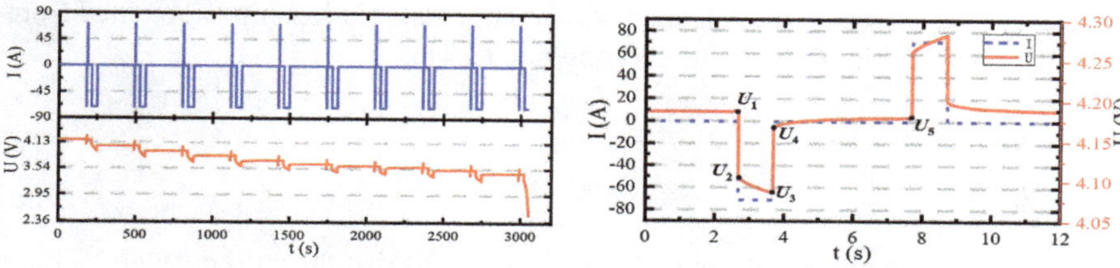

(a)

(b)

Figure 3.33 Lithium battery terminal, voltage and current changes: (a) voltage current diagram and (b) voltage-current curve at SOC=0.1

discharge was completed, the battery voltage gradually stabilizes after prolonged storage, indicating that internal chemical reactions and thermal effects had reached equilibrium. The battery voltage at this time is its OCV. Therefore, the relationship curve between the OCV and SOC is obtained. A comparison of the two curves has shown to have the same fitting. Useful data segments are extracted from the data of the overall process, and then parameter identification is carried out. For the Thevenin model, the parameters that need to be identified are ohmic resistance R_0, polarization internal resistance R_p, and polarization capacitance C_p. These parameters can be identified according to the power pulse test phase of the HPPC test experiment.

The transient and steady-state characteristics of lithium-ion batteries are also reflected. When the pulse discharge starts, the battery voltage drops instantaneously, and then the voltage slowly drops with time during the discharge. When the discharge ends, the battery voltage rebounds immediately. During the resting period, the voltage will gradually rise and stabilize. Meanwhile, the voltage response during charge–discharge is the opposite. There are two methods for parameter identification using pulse test data, one is the calculation method by taking points, and the other is the curve-fitting method.

The identification process of the point-taking calculation method is described as follows. First, five Points *A*, *B*, *C*, *D*, and *E* are obtained. From the Thevenin equivalent model, the rapid change of voltage at the beginning of discharge and charging is due to the instantaneous response caused by the current flowing through R_0. The RC circuit can be used in the Thevenin model to explain the gradual drop in voltage during discharge. Values for the ohmic resistance parameter are available from the *AB* and *CD* sections. In addition, the value of the internal polarization resistance and the polarization capacitance of the RC circuit can be obtained from the cross sections of *BC* and *DE*. Thus, the calculation equation of the parameter is obtained, wherein the calculation expression of R_0 is shown in (3.40).

$$R_0 = \frac{|\Delta U_{AB}| + |\Delta U_{CD}|}{2I} \tag{3.40}$$

In (3.40), ΔU_{AB} is the voltage difference of the *AB* section, ΔU_{CD} is the voltage difference of the *CD* section, and the discharge current is characterized by I. The voltage response of the *DE* segment is analyzed corresponding to the zero-input response in the Thevenin model, and the mathematical relationship is obtained from the KVL relationship of the ECM, as shown in (3.41).

$$\begin{cases} U_D = U_A - U_p e^{-\frac{t_D - t_A}{\tau}} \\ U_E = U_A - U_p e^{-\frac{t_E - t_D}{\tau}} \end{cases} \tag{3.41}$$

In (3.41), U_A is the voltage at Point *A*, which is the OCV at the current stage. Taking Point *D* as the zero time points, the time from Points *D* to *E* is $t_E - t_D$, and U_D and U_E represent Points *D* and *E* voltage, respectively. The calculation equation of the time constant is constructed for τ, as shown in (3.42).

$$\tau = -\frac{t_E - t_D}{lnln\left(\frac{U_A - U_E}{U_A - U_D}\right)} \tag{3.42}$$

After the time constant is obtained, only one parameter of R_p or C_p needs to be calculated, and the other can be derived from (3.43).

$$\tau = R_p C_p \tag{3.43}$$

The voltage response of the *BC* section corresponds to the zero-state response of the Thevenin model. According to the KVL relationship in the model, the set is shown in (3.44).

$$\begin{cases} U_C = U_A - IR_0 - IR_p\left(1 - e^{-\frac{t_C - t_B}{\tau}}\right) \\ U_B = U_A - IR_0 \end{cases} \tag{3.44}$$

In (3.44), U_C is the voltage at Point *C*. Taking Point *B* as the zero-time point, the time from Points *B* to *C* is $t_B - t_C$. Solving it to obtain the calculation equation of polarization internal resistance R_p is shown in (3.35).

$$R_p = \frac{U_B - U_C}{I\left(1 - e^{-\frac{t_C - t_B}{\tau}}\right)} \tag{3.45}$$

In (3.45), using the point calculation method to identify the parameters of each SOC point stage, the ohmic resistance, polarization internal resistance, and polarization capacitance data from SOC from 0.1 to 1 are shown in Table 3.8.

The identified parameters are obtained and drawn in the coordinate system. Taking SOC as the independent variable and the parameters as the dependent variable. A scatter plot of the relationship between U_{OC} R_0, R_p, C_p, and SOC is obtained, and the fitting curve is obtained by the LS polynomial fitting method, as shown in Figure 3.34.

Table 3.8 Parameter fitting result

τ	R_0 **(Ω)**	R_p **(Ω)**	c_p **(F)**
0.36737624	0.004326	0.000527964	6958.360967
0.312561072	0.004835	0.000140234	2228.850513
0.319941747	0.007286	0.000935037	3421.703185
0.369667569	0.005796	0.000265601	1391.814065
0.401429972	0.006542	0.00090463	4437.505023
0.333883416	0.005912	0.000255268	1307.972699
0.348955741	0.005943	0.000235589	1481.206277
0.375727008	0.005905	0.000224778	1671.547113
0.378136003	0.005637	0.000176885	2137.753394
0.332731927	0.006058	0.000279054	1192.355491

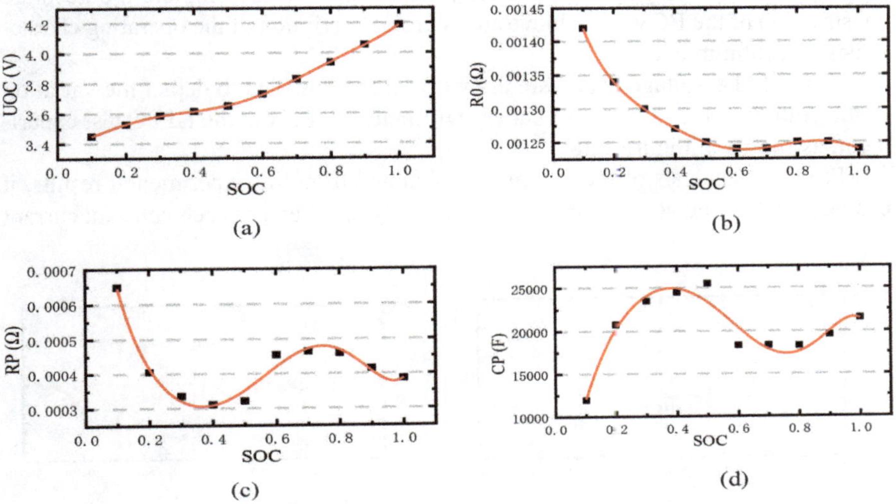

Figure 3.34 Thevenin model parameter identification results: (a) OCV-SOC fitting curve, (b) R_0-SOC fitting curve, (c) RP-SOC fitting curve and (d) CP-SOC fitting curve

According to the identification results, when the SOC takes different values, the parameters to be identified by the model all fluctuate in a small range within a certain range. After the test results of each identification parameter obtained above, check whether the parameter identification using the Thevenin model is accurate and feasible. The verification process is carried out in MATLAB/Simulink. The key to the verification is that the simulation input must be consistent with the above HPPC experimental steps so that the error between the simulation output voltage and the actual voltage reflects the accuracy of the model. The Thevenin model simulation is shown in Figure 3.35.

It can be seen from the error curve shown that the simulation model does not diverge under the entire working condition, and the larger error occurs in the pulse charging and discharging process because of the sudden large current change of the lithium battery at these times will cause the terminal slump/swell voltage occurs. It can be seen from the figure that the average error is within 0.04 V, and the rated voltage of the lithium battery is 4.2 V, so the accuracy of the model can reach more than 98%, which proves that the Thevenin model and the parameter identification method can more accurately characterize the work of the lithium battery process. On the other hand, the existence of errors means that the Thevenin model cannot be fully equivalent to the complex electrochemical reactions inside the lithium battery.

3.5.2 Curve-fitting methods

In the process of data analysis and processing, effective data segments are first extracted from the original experimental data. Then effective processing methods are adopted to obtain the relationship between the internal parameters of the ECM and the SOC by analyzing the extracted data segments to complete the accurate construction of the ECM, to achieve an accurate description of the operating characteristics of lithium batteries.

First, all the voltage data were drawn from the raw data to depict the variation of the voltage at the lithium-ion battery terminals throughout the HPPC test experiment, as shown in Figure 3.36.

From the analysis of the two graphs obtained from the experimental results, it can be seen that the voltage changes significantly at the end of each constant current

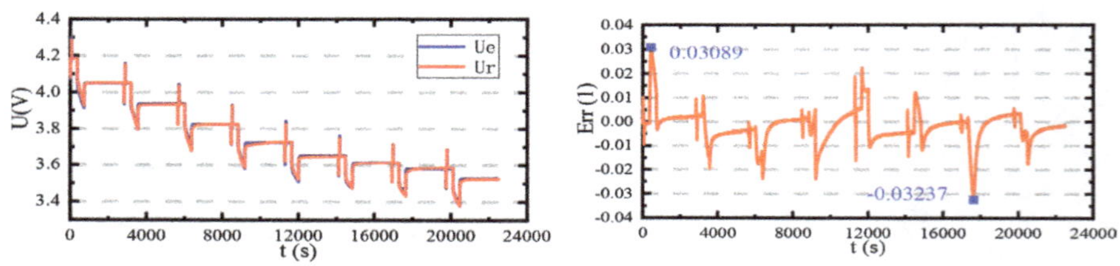

(a) (b)

Figure 3.35 Parameter identification result graph: (a) voltage output comparison (b) voltage error

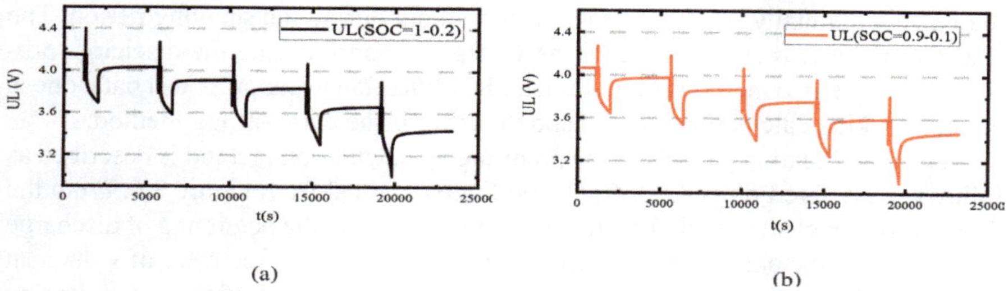

Figure 3.36 HPPC test experimental voltage response curve: (a) SOC from 1 to 0.2 and (b) SOC from 0.9 to 0.1

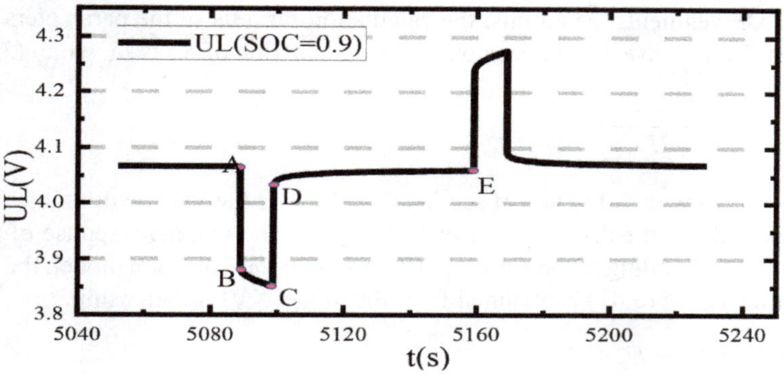

Figure 3.37 Pulse test chart at SOC = 0.9

discharge after one hour of shelving, and the battery voltage will gradually stabilize after a long period of shelving at the end of discharge, which means that its internal chemical reactions and thermal effects have basically reached equilibrium, and the battery voltage at this time is its OCV, so the relationship curve between the OCV and SOC can be obtained from it, and the results are roughly the same as the effect of the OCV fitting curve.

Useful data segments can be extracted from the overall process data before the parameter identification work can be performed [131]. For the Thevenin model, the parameters to be identified are ohmic internal resistance R_o, polarized internal resistance R_p, and polarized capacitance C_p. These parameters can be identified from the power pulse test phase of the HPPC test experiment, so the data segments of the pulse test in each cycle should be extracted first. The impulse response curve of the lithium-ion battery at SOC of 0.9 is shown in Figure 3.37.

The figure reflects both the transient and steady-state characteristics of the lithium battery. At the beginning of the pulse discharge, the battery voltage drops instantly, then the voltage slowly declines with time during the discharge, and when the discharge ends instantly, the battery voltage bounces back immediately, and the voltage gradually rises and tends to smooth out during the shelving period. The charging process is the opposite of the voltage response during the discharge process. There are two methods for parameter identification using pulse test data, one is the point taking calculation method and the other is the curve-fitting method.

The identification process of the point-taking calculation method is described as follows. First, five Points A, B, C, D, and E are obtained from Figure 3.6. From the Thevenin equivalent model, the sudden drop of voltage at the beginning of discharge is the effect of the ohmic internal resistance *Ro*, and the rapid recovery of voltage at the end of discharge is also the effect of *Ro*; the phenomenon of a gradual drop of voltage during discharge can be explained by the RC circuit in the Thevenin model. Therefore, the values of the ohmic internal resistance parameters of the battery can be obtained from the AB and CD segments, and the values of the polarized internal resistance and polarized capacitance of the RC circuit can be obtained from the BC and DE segments. From this, the calculation formula of the parameters can be obtained. In this case, the expression for the calculation of *Ro* is shown in (3.46).

$$R_0 = \frac{|\Delta U_{AB}| + |\Delta U_{CD}}{2I} \tag{3.46}$$

ΔU_{AB} is the voltage difference of section AB, ΔU_{CD} is the voltage difference of section CD, and I is the discharge current. Analyzing the voltage response of the DE section, corresponding to the zero-input response of the Thevenin model, the system of equations as in (3.47) is obtained from the model KVL relationship.

$$\begin{cases} U_D = U_A - U_{p0}e^{-\frac{0}{\tau}} \\ U_E = U_A - U_p e^{-\frac{t_E - t_D}{\tau}} \end{cases} \tag{3.47}$$

In this equation, U_A represents the voltage at Point A, which is the OCV at the current stage, and the time from Points D to E is $t_E - t_D$, with Point D as the zero moments, and UD and UE represent the voltages at Points D and E, respectively. In turn, the system of equations is solved and the time constant is calculated, as shown in (3.48).

$$\tau = \frac{t_E - t_D}{\text{In}\left(\frac{U_A - U_E}{U_A - U_D}\right)} \tag{3.48}$$

After obtaining the time constant τ, only one of the parameters Rp or Cp needs to be calculated, and the other one can be derived from (3.49).

$$\tau = R_p C_p \tag{3.49}$$

The voltage response of the BC section corresponds to the zero-state response of the Thevenin model, and the set of equations is obtained according to the KVL relationship in the model, as shown in (3.50).

$$\begin{cases} U_C = U_A - IR_O - IR_p\left(1 - e^{-\frac{t_C - t_B}{\tau}}\right) \\ U_B = U_A - IR_O \end{cases} \tag{3.50}$$

where U_C denotes the voltage at Point C. Taking Point B as the zero moments, the time from Points B to C is $t_B - t_o$. Solving the system of equations yields the formula for the polarized internal resistance R_p, as shown in (3.51).

$$R_p = \frac{U_B - U_C}{I\left(1 - e^{-\frac{t_C - t_B}{\tau}}\right)} \tag{3.51}$$

Obtaining the key points in each phase of the impulse experiment, the values of *Ro*, *Rp*, and *Cp* parameters in the Thevenin model can be directly calculated using (3.49), (3.50), and (3.51), and finally the model parameters at each SOC point are obtained. However, the point-taking formula method only picks a few marker points from numerous data. It discards the data in the whole impulse response process, which has a low utilization rate of data and easily leads to inaccurate identification results. The formula is also complicated, and the error situation of data banding is easy to occur in the use process, which leads to the error of the final parameter calculation results and brings potential trouble to the identification process.

The curve-fitting method, which uses the entire pulse discharge curve to obtain the model parameters by fitting a formal function with parameters, is a method. It has a much higher data adoption rate than the point-taking formula method. For curve fitting, it is only necessary to give an expression for the fitting function, which contains the symbols of the parameters to be obtained. The curve fitting is mainly for the relatively complicated parameters, such as *Rp* and *Cp* in the Thevenin model, while the ohmic internal resistance *Ro* does not need to use curve fitting and can be obtained directly. The curve segment selected here is the BC segment as in Figure 3.6, the zero-state response curve, and the KVL relationship from the Thevenin model can be obtained, as shown in the circuit expression in (3.52).

$$UL = U_{OC} - IR_O - IR_p\left(1 - e^{-\frac{t}{\tau}}\right) \tag{3.52}$$

Abstraction of this equation yields a parametric expression in the form shown in (3.53).

$$y = a - b\left(1 - e^{-\frac{x}{c}}\right) \tag{3.53}$$

where y denotes the vertical coordinate variable, the terminal voltage *UL*, and x denotes the horizontal coordinate variable. a, b, and c are three of the parameters corresponding to $U_{OC} - IR_O$, IR_p, and τ in (3.53), respectively. According to the known conditions of OCV, current, and ohmic internal resistance, the range of parameters can be set to make the fitted curve better express the real voltage variation, and SOC is 0.9 the fitted curve for the zero-state response of the phase pulse is shown in Figure 3.38.

Parameter identification was performed for each SOC point stage using curve fitting, and the ohmic internal resistance, polarized internal resistance, and polarized capacitance data from SOC from 0.1 to 1 were obtained, as shown in Table 3.9.

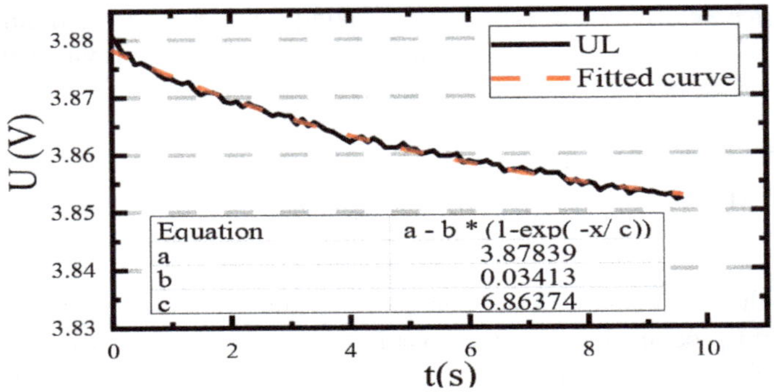

Figure 3.38 *Curve-fitting method to identify parameters*

Table 3.9 *Parameter identification results*

SOC	R_0/Ω	R_p/Ω	$C_p/\mu F$C
1	0.0257	0.01907	188.614
0.9	0.0484	0.02076	187.559
0.8	0.0077	0.02094	181.184
0.7	0.0209	0.02305	177.440
0.6	0.0257	0.01896	230.960
0.5	0.0043	0.01926	263.292
0.4	0.0021	0.01811	243.125
0.3	0.0413	0.02264	217.668
0.2	0.0449	0.02014	215.094
0.1	0.0242	0.01936	203.512

The identified parameters are obtained and plotted in the coordinate system, as shown in Figure 3.39. With SOC as the independent variable and the parameters as the dependent variables, scatter plots of *Ro*, *Rp*, *Cp* versus SOC are obtained, and the fitted curves are obtained by the LS polynomial fitting method.

Where Panels a–c of Figure 3.39 indicate the trends of *Ro*, *Rp*, and *Cp* with SOC, respectively. It can be seen from the figures that the internal parameters of the Thevenin model fluctuate within a certain range with SOC variation, so in cases where the model accuracy is not required, these parameters can be replaced by their average values or the model parameters can be used by using the look-up table method, which has an advantage of being less computationally intensive. If more accurate simulation data are needed, polynomial fitting equations for the curves need to be obtained, and the parameter fitting polynomials corresponding to the figures are shown in (3.54)–(3.56).

$$R_O = -293.452 \times SOC^7 + 1095.861 \times SOC^6 - 1647.308 \times SOC^5 + 1274.801 \times SOC^4 - 538.498 \times SOC^3 + 121.190 \times SOC^2 - 13.11912 \times SOC + 0.551 \tag{3.54}$$

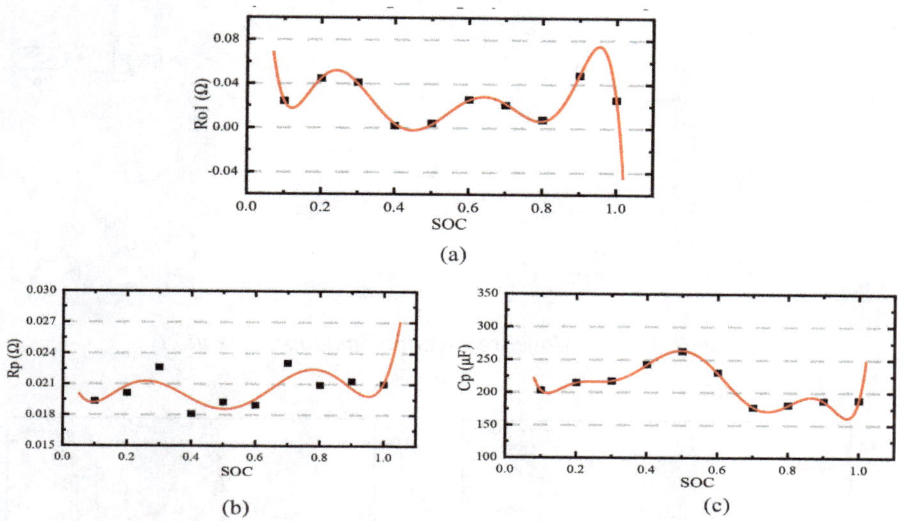

Figure 3.39 *Thevenin model parameter identification results: (a) Ohm internal resistance* Ro, *(b) Internal resistance of polarization Rp and (c) Polarization Capacitor* Cp

$$R_p = 2.944 \times SOC^6 - 9.196 \times SOC^5 + 10.832 \times SOC^4 - 5.963 \times SOC^3 + 1.542 \times SOC^2 - 0.165 \times SOC + 0.0253 \tag{3.55}$$

$$C_p = 860.966 \times 10^{-3} \times SOC^8 - 3.637 \times SOC^7 + 6.336 \times SOC^6 - 5.888 \times SOC^5 + 3.159 \times SOC^4 - 992.688 \times 10^{-3} \times SOC^3 + 177.144 \times 10^{-3} \times SOC^2 - 16.226 \times 10^{-3} \times SOC + 0.0253 \tag{3.56}$$

In the above equation, the unit of ohmic internal resistance *Ro* is Ω, the unit of polarized internal resistance *Rp* is also Ω, and the unit of polarized capacitance *Cp* is F.

3.5.3 Model parameter validation and evaluation

After obtaining the mathematical description of each parameter of the model, it is necessary to verify the accuracy of the parameters. The internal structure of the Thevenin model is shown in Figure 3.40. The identified parameters and current values of the HPPC test are input into the Thevenin ECM. The output voltage response data from the model are compared with the actual voltage data to verify the model and improve the model based on the verified results. The inputs of the Thevenin model are I, R_0, R_p, C_p, and U_{OC} from top to bottom in Figure 3.40. They have a functional relationship with SOC as an independent variable and change along with the SOC and current variation [131]. The inputs are the terminal voltage and the load current. The Ampere-hour integration module obtains the real-time changes

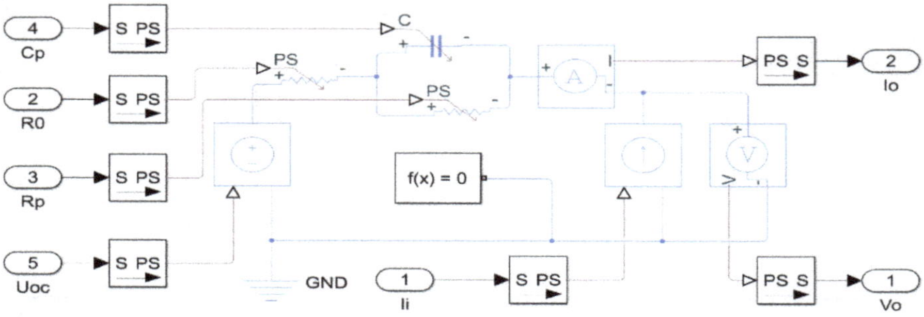

Figure 3.40 Model verification Simulink simulation

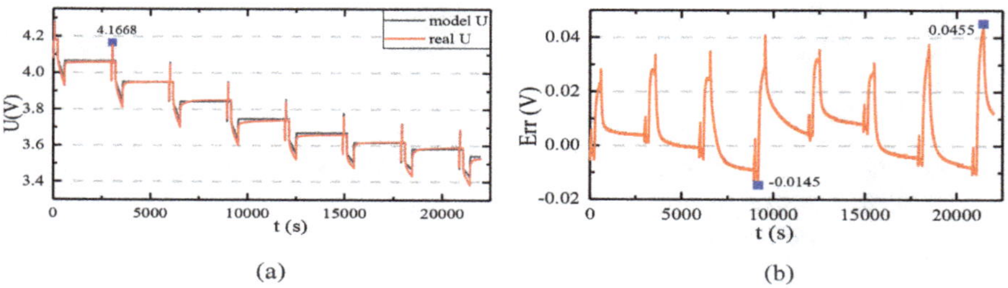

Figure 3.41 Terminal voltage output curve comparison: (a) Output comparison and (b) Output error

of the SOC. Then, it is connected to the input and end of the parameter function, to obtain the model parameters corresponding to the constant changes of the current input. Therefore, the control quantity of the model is the input current value, and the response is the terminal voltage of the model. Based on this model, the working condition of the lithium-ion battery is analyzed. The internal structure of the Thevenin model is shown in Figure 3.40.

In Figure 3.40, the circuit diagram of the Thevenin model is designed, in which each circuit element is a time-varying controllable parameter. The input current I acts on a controllable current source as a load and the output voltage U is obtained by using a voltmeter. As shown on the right side of Figure 3.40, the effectiveness and accuracy of the model parameter identification are analyzed and compared with the actual terminal voltage data of the lithium-ion battery under the same input current. If the deviation is large, the parameter identification is not accurate enough, or the model itself has defects.

As shown in Figure 3.41(a), the real U is the model output voltage, and the model U is the actual output voltage of the battery. It can be observed that the model output voltage is in good agreement with the actual value, indicating the rationality of the Thevenin ECM and the feasibility and reliability of the parameter identification method.

In Figure 3.41(b), the error curve of the simulated and real voltage is described under the periodic current discharge.

1. When the current pulse applies, the battery terminal voltage suddenly changes, causing the error between the experimental and simulated values to increase to 0.3–0.4 V. After several HPPC experimental results analyses under different discharge rates, it is believed that at the time point when the current pulse applies, the large error of the battery terminal voltage is caused by environmental factors and measurement deviations.
2. When the battery discharges within the range of SOC value varying from 10% to 0%, the error between the experimental value and the simulated value of the terminal voltage is slightly larger than that of the SOC value discharges from 100% to 10%, mainly because when the SOC value is less than 10%, the internal electrochemical reaction is changed drastically, and the parameter values in the battery model are changed highly.
3. Except for the above individual time points, when the SOC value of the battery changes from 100% to 10%, the simulated value of the terminal voltage is consistent with the experimental value, and the average error is less than 0.011 V, which accounts for 0.3% of the nominal voltage.

3.5.4 Double exponential fitting strategies

For the parameter identification of the improved PNGV model, the double exponential fitting method is used for parameter identification [132]. The improved PNGV equivalent circuit diagram of the lithium-ion battery is shown in Figure 3.42 .

Take the zero-input response stage of the circuit, where $I_b = 0$, the input equation of battery terminal voltage is shown in (3.57).

$$U_L = U_{OC} - IR_s e^{-\frac{t}{\tau_S}} - IR_L e^{-\frac{t}{\tau_L}} \tag{3.57}$$

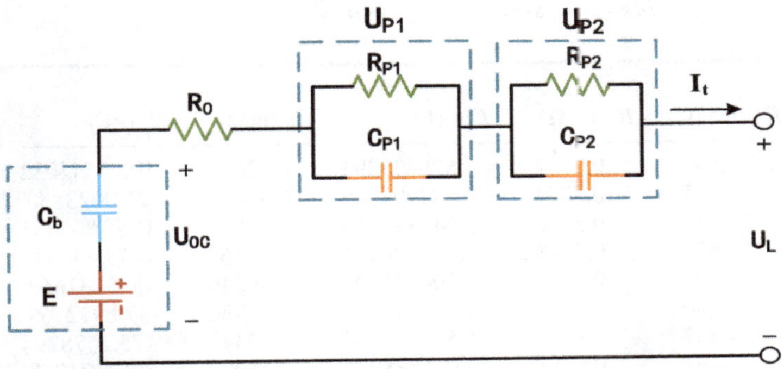

Figure 3.42 Schematic diagram of improved PNGV equivalent circuit

To simplify the parameter identification, replace the coefficient of the equation shown in (3.57), and the result is shown in (3.58).

$$U_L = f - ae^{-ct} - be^{-dt} \tag{3.58}$$

Take f, a, b, c, and d as undetermined parameters, and take (3.58) as the target formula to fit the double exponential curve using the test data. Import the experimental data into MATLAB, and then use the curve-fitting tool to fit the double exponential curve. The fitting interface is shown in Figure 3.43.

Double exponential fitting is conducted, according to which the fitting results are shown in Table 3.10.

Analyzing the parameter identification results, the time constant τ_s of the parallel circuit composed of R_S and C_S is small, which is used to simulate the rapid voltage change process of the battery when the current changes suddenly. The time constant of the parallel circuit is composed of R_L and C_L is larger than τ_S, which is used to simulate the process of slow voltage stabilization. The establishment of the improved PNGV model and the SOC estimation of lithium-ion batteries need to obtain the characteristic curves of R_S, R_L, C_S, C_L, R_0, and U_{OC} for the various SOC

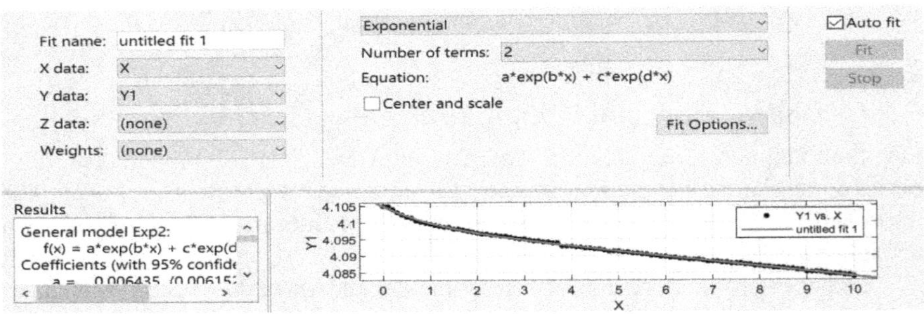

Figure 3.43 Double exponential fitting interface in MATLAB

Table 3.10 Double exponential fitting results

SOC	R_0 (mΩ)	R_S (mΩ)	C_s (kF)	R_L (mΩ)	C_L (kF)	U_{oc} (V)
0.9	2.2867	0.9734	3.879633291	0.22620	108.7544032	3.984
0.8	2.2967	0.5754	2.187991245	0.88600	21.0023850	3.880
0.7	2.4033	0.8038	3.640885484	0.74980	106.7805431	3.786
0.6	2.3833	0.7968	5.548276217	0.19364	147.0450532	3.682
0.5	2.4467	0.7704	5.106321790	0.39240	60.5181662	3.632
0.4	2.4300	0.8178	4.814145114	0.71880	37.8972369	3.599
0.3	2.4467	1.1822	5.598150640	0.37340	175.2678794	3.551
0.2	2.6667	0.8738	3.075588934	0.41000	60.3121758	3.462
0.1	2.9933	1.6028	2.015858355	0.49600	40.9283197	3.417

levels. The U_{OC} varies significantly under different SOC values. A seventh-order polynomial is used as the fitting function with the U_{OC} value for different SOC levels to obtain the characteristic curve of U_{OC} with SOC. The U_{OC} − SOC fitting polynomial is shown in (3.59).

$$f(x) = B_1x^7 + B_2x^6 + B_3x^5 + B_4x^4 + B_5x^3 + B_6x^2 + B_7x + B_8 \tag{3.59}$$

In (3.59), the independent variable x is the SOC value; the dependent variable is the parameter curve to be fitted; B_1–B_8 are the polynomial undetermined coefficients. Using simulation to establish the polynomial curve fitting on the experimental data, the results are obtained, as shown in Table 3.11.

In addition, the characteristic curves of R_S, R_L, C_S, C_L, and R_0 can be obtained by polynomial fitting to the data identified. When using the LS method to fit the data with a polynomial, if the parameter change law is not obvious, the higher the polynomial order (sixth-order and above), the more likely it is to produce oscillations. Therefore, this chapter compares the data. When performing polynomial fitting, it is found that the fitting effect of fourth-order and fifth-order polynomial is similar, so the fourth-order polynomial is selected for fitting, and the fitting polynomial is described, as shown in (3.60).

$$f(x) = P_1x^4 + P_2x^3 + P_3x^2 + P_4x + P_5 \tag{3.60}$$

The meaning of each parameter in (3.60) is similar to that in (3.59), and the fitting results are shown in Table 3.12.

In the same way, the other parameters of the equivalent circuit are fitted with the SOC, and the fitting results obtained are shown in Table 3.12. Then the fitting result function can be brought into MATLAB for parameter identification.

Table 3.11 Open-circuit voltage fitting results

Parameter	**B_1**	**B_2**	**B_3**	**B_4**
U_{oc} (V)	−23.60229	141.34077	−314.9228	345.34531
Parameter	B_5	B_6	B_7	B_8
U_{oc} (V)	−200.1546	60.21383	−7.88447	3.77173

Table 3.12 Polynomial fitting results

Parameter	P_1	P_2	P_3	P_4	P_5
R_0 (Ω)	0.02176	−0.0529	0.04554	−0.01654	0.004568
R_S (Ω)	0.03249	−0.0669	0.04916	−0.0158	0.002803
C_S (F)	1771000	−3139000	1604000	−180400	26,910
R_L (Ω)	−0.06047	0.1192	−0.07955	0.02085	−0.0013
C_L (F)	6417000	−4156000	−6328000	5408000	−14110

3.6 AI-based online parameter identification

3.6.1 Recursive LS (RLS) method

The LS method is a mathematical optimization technique. It finds the best function match for the data by minimizing the sum of the squares of error. The LS method can be used to obtain the unknown data easily and minimize the sum of squares of errors between the obtained data and the actual data [133]. The LS method can also be used for curve fitting, and other optimization problems can also be expressed by minimizing energy or maximizing entropy.

For us, the LS method is the most basic discriminating method, but it has the following two shortcomings:

1. When the model noise is colored noise, the LS parameter estimation is not unbiased and uniform.
2. With the increase of data, the LS method will show what is called "the sum of data." This is because the augmented moment matrix gradually approaches zero with the increase of *K*, and the recursive algorithm slowly loses the positive energy force.

The LS method of bias compensation is to avoid these missing points, one of the LS discriminating methods of class parameters that cannot give unbiased estimation is proposed by using the LS method of plus needle pairs. In addition, to overcome the shortcomings of the LS method, some methods of discrimination based on the LS method have been developed: generalized LS, auxiliary variables (IVA) and augmented matrix (EM), and square methods combining general LS with other square methods. There are correlational analysis – LS method and stochastic approach method.

The LS method is a classic data processing method. The famous scientist Gauss proposed the LS method, which is applied to the calculation of the orbits of planets and comets. When calculating the orbits of planets and comets, six parameter values are estimated to describe the celestial body's motion based on the observations obtained by the telescope. When inferring unknown parameters from observation data, the most appropriate value of the unknown parameter should be such a value

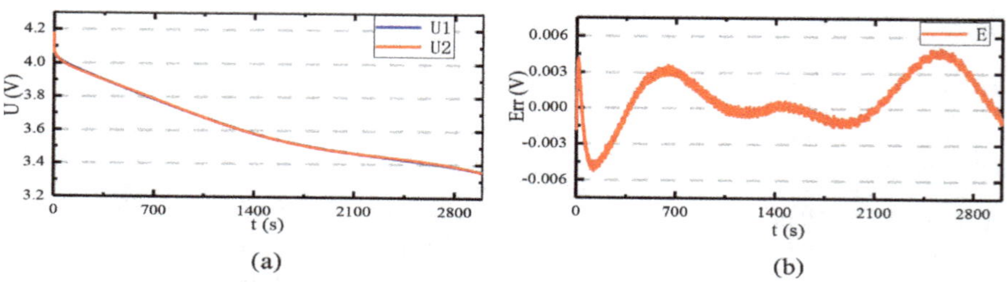

Figure 3.44 The parameter identification and closed-circuit voltage traction results:(a) Simulation voltage and (b) Simulation voltage error

that multiplies the square of the difference between each observation. The calculated value can be obtained by the sum of the measured values making its accuracy to be the smallest. This is the earliest LS idea. Since then, this method has been used to solve many practical problems. For different purposes, it is modified and various corresponding LS algorithms have been proposed.

In the field of parameter identification, the LS algorithm is an estimation method, which can be used for both dynamic and shelved systems. It can be used for both linear and nonlinear systems, which can be introduced into offline and online battery state estimation. In a random environment, the observation data are not required to provide information on its probability and statistics when using the LS method. Its estimation results have quite good statistical characteristics, which are easy to understand master. The identification algorithm developed by the calculation principle is relatively simple to implement, providing a solution to the problem when other parameter identification methods are difficult to be applied. At the same time, the RLS are improved by LS and its improved method can realize the online estimation of the model parameters, which can minimize the influence of temperature, discharge rate, and other conditions on the estimation accuracy of the parameters during the parameter identification process.

The LS method is easy to be understood and has fast convergence speed, which has been widely used in system parameter identification. For a single-input single-output linear system, its mathematical description is described, as shown in (3.61).

$$\begin{cases} A(z^{-1})Y(k) = B(z^{-1})U(k) \\ A(z^{-1}) = 1 + a_1 z^{-1} + a_2 z^{-2} + \cdots + a_n z^{-n} \\ B(z^{-1}) = 1 + b_1 z^{-1} + b_2 z^{-2} + \cdots + b_n z^{-n} \end{cases} \tag{3.61}$$

Wherein, the system input is described by the parameter a and the system output is described by b. And then, the discrete transfer function can be obtained. The input system is described by $U(k)$, and $Y(k)$ is the system output. The discrete transfer function can be obtained, as shown in (3.62).

$$G\left(z^{-1}\right) = \frac{B\left(z^{-1}\right)}{A\left(z^{-1}\right)} = \frac{1 + b_1 z^{-1} + b_2 z^{-2} + \cdots + b_n z^{-n}}{1 + a_1 z^{-1} + a_2 z^{-2} + \cdots + a_n z^{-n}} \tag{3.62}$$

The corresponding difference equation is described, as shown in (3.63).

$$\begin{aligned} y(k) &= -a_1 y(k-1) - a_2 y(k-2) - \ldots - a_n y(k-n) \\ &+ b_1 u(k-1) + b_2 u(k-2) + \ldots + b_n u(k-n) + e(k) \end{aligned} \tag{3.63}$$

Wherein, the included parameters are the parameters to be estimated, and $e(k)$ is the vector equation error.

$$\begin{cases} \phi(k) = \left[-y(k-1), -y(k-2), \ldots, -y(k-n), u(k-1), u(k-2), \ldots, u(k-n)\right] \\ \theta = \left[(a_1, a_2, \ldots, a_n), (b_1, b_2, \ldots, b_n)\right] \end{cases} \tag{3.64}$$

Thus, (3.64) can be rewritten accordingly. The input and output of the system can be expanded upon n dimensions, respectively, where k = 1, 2, …, $N + n$. And then, the matrix expression can be obtained, as shown in (3.65).

$$\begin{cases} Y = \phi\theta^T + e(k) \\ \phi = \begin{bmatrix} -y(n) & -y(n-1) & \dots & -y(1) & u(n) & \dots & u(1) \\ -y(n+1) & -y(n) & \dots & -y(2) & u(n+1) & \dots & u(2) \\ \vdots & \vdots & \vdots & \vdots & \vdots & \vdots & \vdots \\ -y(n+N) & -y(n+N-1) & \dots & -y(N) & u(n+N) & \dots & u(N) \end{bmatrix} \\ Y = [y(n+1)\, y(n+2) \dots y(n+N)]^T, e = [e(n+1) \dots e(n+N)]^T \end{cases} \tag{3.65}$$

The purpose of the LS method is to find a recursive. When $a = b$ exists, the minimum value can be obtained. Taking the objective function $J(\theta)$, it is then described as shown in (3.66).

$$J(\theta) = \sum_{i=1}^{N} (Y - \phi\theta)^2 = \sum_{i=1}^{N} e^2(n+i) = (Y - \phi\theta)^T (Y - \phi\theta) \tag{3.66}$$

Therefore, the first derivative of a can be obtained for (3.66). When $a = b$, the first derivative is zero. The purpose of the LS method is to find the value recursively. When it exists, the minimum value can be obtained. Otherwise, the first derivative is zero. And then, it can be obtained in (3.67).

$$\frac{\partial J}{\partial \theta} = \frac{\partial}{\partial \theta}\left[(Y - \phi\theta)^T (Y - \phi\theta)\right] = 0 \tag{3.67}$$

After that, the LS estimation value of the system can be obtained as shown in (3.68).

$$\hat{\theta} = (\phi^T\phi)^{-1}\phi^T Y \tag{3.68}$$

It is necessary to continuously input and output the latest experimental data to improve the accuracy of parameter estimation in the continuous iterative process until satisfactory accuracy is reached. The specific calculation process is described as shown in (3.69).

$$\begin{cases} \hat{\theta}(k+1) = \hat{\theta}(k) + K(k+1)\left[y(k+1) - \phi^T(k+1)\hat{\theta}(k)\right] \\ K(k+1) = P(k+1)\phi(k+1) \\ P(k+1) = P(k) - \dfrac{P(k+1)\phi(k+1)\phi^T(k+1)}{1 + \phi^T(k+1)P(k)\phi(k+1)} \end{cases} \tag{3.69}$$

Wherein, it is the estimated value of the parameter at the time point of k. The calculated value of the system observation at the time point of k+1. $y(k+1)$ is the observation at the time point of k+1. In each iteration, the algorithm uses the difference between the system observation calculation value and the gain to correct the final estimation value. The initial value can be any value and the larger value is better. And then, I can be initialed as a unit array. Then, the estimated value of the parameter can be obtained at the time point of k. The calculated value of systematic observation is calculated at the time point of k+1. $y(k+1)$ is the observation at the time point of k+1.

As for each iteration, the algorithm uses the difference between the system observation calculation value and the gain to correct the final estimation value. The initial value can be any value. Taking a larger value, I can be initialed as the unit array, and $P(0) = \alpha I$. Wherein, α is a larger value as far as possible and I can be taken as the unit array. Taking the Thevenin model as an example, the model can be expressed as shown in (3.70).

$$\begin{cases} U_a(s) = U_{OC}(s) - U_L(s) = R_0 I(s) + \dfrac{R_P I(s)}{1+\tau s} + e \\ U_a(s) + a \cdot sU_a(s) = b \cdot I(s) + c \cdot s + e \\ a = \tau b = (R_0 + R_P) \quad c = \tau R_0 \end{cases} \tag{3.70}$$

Wherein, e is the measurement noise. Therefore, the discrete system can be obtained with bilinear transformation, which can be expressed as shown in (3.71).

$$y(k) = -ay(k-1) + bu(k) + cu(k-1) + me(k) \tag{3.71}$$

To realize the LS principle for parameter identification, the discrete system formula can be expressed as the LS form as shown in (3.72).

$$\begin{cases} y(k) = x(k)^T \cdot \theta(k) \\ x(k) = [y(k-1) \;\; u(k) \;\; u(k-1) \;\; e(k)\,]^T \\ \theta(k) = [a \; b \; c \; m\,]^T \end{cases} \tag{3.72}$$

According to the above RLS formula, the estimation result can be obtained. The experimental data can be used for the model parameter identification, and the results can be obtained as shown in (3.44).

Wherein, $U1$ is the measured voltages that are obtained by experiment, and $U2$ is the simulation results by the parameters that are identified by using the RLS method. E is the error of simulation voltage. It can be seen from the figure that this method can identify the model parameters well.

3.6.2 Bias compensation RLS method

In the field of parameter identification, the bias compensation RLS algorithm is an estimation method, which can be used for both dynamic and shelved systems. It can be used for both linear and nonlinear systems, which can be introduced into offline and online battery state estimation. In a random environment [134], the observation data are not required to provide information on its probability and statistics when using the bias compensation RLS method. Its estimation results have quite good statistical characteristics, which are easy to understand master. The identification algorithm developed by the calculation principle is relatively simple to implement, providing a solution to the problem when other parameter identification methods are difficult to be applied. At the same time, the bias compensation RLS is improved by LS and its improved method can realize the online estimation of the model parameters, which can minimize the influence of temperature, discharge rate, and other conditions on the estimation accuracy of the parameters during the parameter identification process.

In the system identification method, because the LS method is simple and easy to program, it is the commonly used and the most common. The LS method also has a sudden advantage – strong robustness. However, its deficiency is also obvious, if the model noise is not white noise, then its LS parameter estimation is no longer unbiased. This deficiency of the LS method of needle pairs is very much based on the method that has been put forward and continuous optimization. The LS method of bias compensation is one of them, and in continuous improvement and perfection.

Although the partial compensation LS method has been widely used in the mode discrimination of discrete systems, in the mode discrimination algorithm of continuous systems based on the modulation function method, the minimum square residual estimation of the input and output signals and differentials is more complex than that of the dispersion system model. The calculation of the deviation term is related to the mode truth value and the noise square error, and the recursive calculation procedure of the noise square error is more complicated than that of the dispersion system.

First, a lower correlation expression is used to indicate the input and output of the decoupling linear system in (3.73):

$$y(t) = G(q)\,u(t) + v(t) \tag{3.73}$$

where $u(t)$ is the input signal of the system, and $y(t)$ is the output signal of the system. $v(t)$ is the process disturbance suffered by the system, $g(q)$ is the transmission function number of the system, and the definition of $g(q)$ is as (3.74):

$$G(q) = \frac{B(q^{-1})}{A(q^{-1})} = \frac{b_1 q^{-1} + +b_m q^{-m}}{1 - a_1 q^{-1} - -a_n q^{-n}} \tag{3.74}$$

where q is the backward operator, the output error through the down-line regression in (3.75).

$$\begin{cases} t(t) = \Phi^T \\ \theta + \varepsilon(t)\ \ \theta^T = \left[a^T : b^T\right] = \left[a_1 \cdots a_n : b_1 \cdots b_m\right] \\ \Phi^T = \left[y^T : u^T\right] = \left[y(t-1) \cdots y(t-n) : u(t-1) \cdots u(t-m)\right] \\ \varepsilon(t) = A(q^{-1})\, v(t) \end{cases} \tag{3.75}$$

We first do the following hypothetical identification of the system:

1. $A(q)$ is a Hurwitz multiterm.
2. Input signal $u(t)$ is stable, and continues to exist in.
3. System dry disturbance $v(t)$ is stable.

4. The order of the system (n, m) is given.
5. Input signal $u(t)$ and system dry disturbance $v(t)$ are unrelated.

To establish the common method of machine system identification under colored noise interference, special is false is for the independent identification system; bar on the meaning of dry noise $v(t)$ is colored; item is strongly added because the first item is an unbiased estimation of the system parameter vector θ by using data, in which N represents the number of data points. When the order n of the system has no effect, or the false hypothesis is violated, an order n that exceeds the parameters of the unknown system can be selected. This method is often used in unknown order systems. The formula is shown in (3.76).

$$E\left[u(t)\, v(t)\right] = 0, \forall t, k \tag{3.76}$$

E in (3.76) represents the study period number.

To estimate the unknown system parameter vector θ, I can use the LS method to minimize it in (3.77).

$$V_N(\theta) = \frac{1}{N} \sum^{i=1N} \varepsilon(t)^2 \tag{3.77}$$

VN is the LS. This results in the LS estimation as shown in (3.78).

$$\hat{\theta}_{LS}(N) = \theta + \hat{R}_{\Phi}^{-1}(N)\hat{r}_{\Phi}(N) \tag{3.78}$$

Θ is the system covariance. The square difference matrices $R(N)$ and $r(N)$ are shown in (3.79).

$$\begin{cases} R(N) = \dfrac{1}{N} \displaystyle\sum_{i=1}^{N} \phi_1 \phi_1^T \\ r(N) = \dfrac{1}{N} \displaystyle\sum_{i=1}^{N} \phi_1 y(t) \end{cases} \tag{3.79}$$

$R(N)$ and $r(N)$ are system covariance matrices. For the sake of calculation, the following four matrices are introduced in (3.80).

$$\begin{cases} Q_1 = \begin{bmatrix} I_n \\ 0 \end{bmatrix} \in R^{(n+m)*n} \\ Q_2 = \begin{bmatrix} I_n \\ 0 \end{bmatrix} \in R^{(2n+m)*n} \\ Q_1 = \begin{bmatrix} I_{n+m} \\ 0 \end{bmatrix} \in R^{(2n+m)*(n+m)} \\ Q_1 = \begin{bmatrix} 0 \\ I_{n+m} \end{bmatrix} \in R^{(2n+m)*(n+m)} \end{cases} \tag{3.80}$$

In (3.80), In and In $+_m$ are the one-bit matrices of n * n and (n + m) * (n +m).

The asymptotic analysis table of θ(N) for the LS approximation is shown in (3.81).

$$\begin{cases} \lim_{N\to\infty} \hat{\theta}_{LS}(N) = \theta + R_{\Phi}^{-1} \\ R_{\Phi} = E\left[\Phi_1\Phi_1^T\right] = \lim_{N\to\infty} \hat{R}_{\Phi}(N) \\ r_{\Phi} = E\left[\Phi_1\varepsilon(t)\right] = \begin{bmatrix} E[y_1\varepsilon(t)] \\ E[u_1\varepsilon(t)] \end{bmatrix} = \left[r_{\varepsilon}\ 0\right] = Q_1 r_{\varepsilon} \end{cases} \tag{3.81}$$

With (3.80) and (3.81), E[u1ε(t)].=.0 is get. With (3.18), two important pieces of information are get: (1) the LS estimation criterion tends to be biased due to the noise ε(t). (2) N × 1 noise matrix aberration *r* determines the noise deviation of dry disturbance. In other words, the correlation between the regression and ε is determined by the deviation, because the self-correlation of $\varepsilon(t)$ exists in the regression. It can be seen from (3.80) that even if $v(t)$ is white noise, $\varepsilon(t)$ is also self-dependent.

First, r,n regression vector η of size n × 1 as the delay of the input signal in (3.82).

$$\eta_1^T = \left[u(t-m-1)\cdots u(t-m-n)\right] \tag{3.82}$$

vector η is the delay of the input signal. At the same time, new definitions for the classification of systematic parameter vectors and numerical data regression vectors as (3.83):

$$\begin{cases} \Theta = L\theta \\ \Phi_t^T = \left[y_t^T : u_t^T - \eta_1^T H_1^T : \eta_t^T\right] \end{cases} \tag{3.83}$$

In (3.23), H_1 is shown as the $n \times m$ wt matrix; L shows the full-rank matrix of (2n + m) (n + m) showed in (3.84).

$$L = \begin{bmatrix} I_n & 0 \\ 0 & I_m \\ 0 & H_1^T \end{bmatrix} \tag{3.84}$$

According to (3.80) and (3.81), (3.85) can be get.

$$\Phi_t^T\Theta = \Phi_t^T\theta \tag{3.85}$$

That is, (3.86) can be get.

$$y(t) = \Phi_t^T\Theta + \varepsilon(t) \tag{3.86}$$

Similar to (3.26), the LS estimation criterion of the number vector θ of the system of linear regression model can be get in (3.87):

$$\begin{cases} \hat{\Theta}_{LS}(N) = \Theta + \hat{R}_{\Phi\Phi}^{-1}(N)\hat{r}_{\phi y}(N) \\ \hat{R}_{\Phi\Phi}(N) = \frac{1}{n}\sum_{i=1}^{n}\Phi_t\Phi_t^T \\ \hat{r}_{\Phi y}(N) = \frac{1}{n}\sum_{i=1}^{n}\Phi_t y(t) \end{cases} \tag{3.87}$$

Equation (3.87) is vector η as the delay of the input signal. Similar to (3.87), (3.88) can be get.

$$\lim_{N\to\infty} \Theta_{LS}(N) = \hat{\Theta} + R_{\Phi\Phi}^{-1}Q_2 r_{y\varepsilon} \tag{3.88}$$

Then introduce a matrix H of $(2 \times n + m) * n$ in (3.89).

$$H^T = \begin{bmatrix} 0^T & -H_1^T & I_n \end{bmatrix} \tag{3.89}$$

The next step of Equation (3.89) is (3.90).

$$H^T\Theta = 0 \tag{3.90}$$

Multiply (3.90) by H^T from the left and root to obtain the following (3.91).

$$H^T\left(\lim_{n\to\infty} \hat{\Theta}_{LS}(N)\right) = H^T R_{\Phi\Phi}^{-1} Q_2 r_{y\varepsilon} \tag{3.91}$$

If the inverse $H^T R_{\Phi\Phi}^{-1} Q_2$ exists, then the above expression can be transformed into the lower form.

$$r_{y\varepsilon} = \left(H^T R_{\Phi\Phi}^{-1} Q_2\right)^{-1} H^T\left(\lim_{n\to\infty} \hat{\Theta}_{LS}(N)\right) \tag{3.92}$$

In (3.92), when N tends to infinity, as shown in (3.93).

$$r_{y\varepsilon}(N) = \left(H^T R_{\Phi\Phi}^{-1}(N)\, Q_2\right)^{-1} H^T \Theta_{LS}^{\wedge}(N) \tag{3.93}$$

$r_{y\varepsilon}(N)$ and the resultant bias compensation method, the unbiased estimation of the new system parameter direction θ can be get in (3.94).

$$\hat{\Theta}_{\mathrm{BELS}}(N) = \hat{\Theta}_{LS}(N) - \hat{R}_{\Phi\Phi}^{-1}(N)\, Q_1 r_{y\varepsilon}(N) \tag{3.94}$$

The unbiased estimator of the parameter vector θ of the original system can be obtained by using the lower common formula in (3.95).

$$\hat{\Theta}_{\mathrm{BELS}}(N) = Q_3^T \hat{\Theta}_{\mathrm{BELS}}(N) \tag{3.95}$$

This section mainly introduces the recursive procedure of a deviation compensation LS method proposed above. First, it introduces the recursive formula of its off-line form, as follows:

Step 1: the minimum square criterion for calculating the parameter direction θ of the new system:

$$\hat{\Theta}_{LS}(N) = \Theta + \hat{R}^{-1}_{\Phi\Phi}(N)\hat{r}_{\Phi y}(N) \tag{3.96}$$

$\hat{R}^{-1}_{\Phi\Phi}(N)$ and $\hat{r}_{\Phi y}(N)$ in terms of the number $\{u(t), y(t), t=1, \ldots, N\}$ are computed in (3.97).

$$\begin{cases} \hat{R}_{\Phi\Phi}(N) = \dfrac{1}{n}\sum_{i=1}^{n} \Phi_t \Phi_t^T \\ \hat{r}_{\Phi y}(N) = \dfrac{1}{n}\sum_{i=1}^{n} \Phi_t y(t) \end{cases} \tag{3.97}$$

Step 2: the noise covariance aberrant $r_{y\varepsilon}$ forward estimation.
Step 3: calculation of the LS of θ row deviation compensation in (3.98).

$$\hat{\Theta}_{BELS}(N) = \hat{\Theta}_{LS}(N) - \hat{R}^{-1}_{\Phi\Phi}(N)\, Q_1 \hat{r}_{y\varepsilon}(N) \tag{3.98}$$

Step 4: the estimation and calculation of the LS method for compensating the deviation of the parameter direction θ of the original system in (3.99).

$$\hat{\Theta}_{BELS}(N) = Q_3^T \hat{\Theta}_{BELS}(N) \tag{3.99}$$

For the following trace of the parameters of the time-invariant system, the linear implementation of the lower bias compensation LS method can be used, and the linear recursive procedure of the LS method is as follows:

Step 1: assign the initial value $k = 0$; $P(0) = \alpha 2I$ (α is a large real number); $\hat{\Theta}(0) = \beta$ (β is sufficiently small in real direction)
Step 2: use the lower male group to calculate $\hat{\Theta}_{BELS}(k)$

$$\begin{cases} \hat{\Theta}_{LS}(k) = \hat{\Theta}_{LS}(k-1) + p_k \Phi_k \left(y(t) - \Phi_t^T \hat{\Theta}_{LS}(t-1)\right) \\ P_t = P_{t-1} - \frac{P_{t-1}\Phi_t\Phi_t^T P_{t-1}}{1+\Phi_t^T P_{t-1}\Phi_t} \\ \hat{r}_{y\varepsilon}(t) = \left(H^T t P_t Q_2\right)^{-1} H^T \hat{\Theta}_{LS}(t) \\ \hat{\Theta}_{LELS}(t) = \hat{\Theta}_{LS}(t) - t P_t Q_2 \hat{r}_{y\varepsilon}(t) \\ \hat{\theta}_{BELS}(t) = Q_3^T \hat{\Theta}_{LELS}(t) \end{cases} \tag{3.100}$$

Step 3: calculate θ

$$\hat{\theta}_{BELS}(t) = Q_3^T \hat{\Theta}_{LELS}(t) \tag{3.101}$$

We compute and run at $\hat{\theta}_{BELS}(t)$. If we do not, we set $k = k + 1$ to finish the calculation.

First, we prove Step 4 of the LS method of offset compensation in off-line form. We can see that $\hat{\theta}_{BELS}(N)$ is composed of the front $m + n$ elements of $\hat{\Theta}_{BELS}(N)$ theta. When both sides of (3.102) are the same, multiply HT from the left to get.

$$H^T\hat{\theta}_{BELS}(N) = H^T\hat{\Theta}_{LS}(N) - H^T\hat{R}_{\Phi\Phi}^{-1}Q_2\hat{r}_{y\varepsilon}(N) \tag{3.102}$$

Combining (3.101) and (3.102), (3.103) can get.

$$H^T\hat{\theta}_{BELS}(N) = 0 \tag{3.103}$$

For example, the vector $\hat{\theta}_{BELS}(N)$ of $(2n + m) * 1$ is replaced by three vectors of the lower in (3.104).

$$\hat{\Theta}_{BELS}(N) = \left[\hat{b}_{BELS}(N)\ \hat{b}_{BELS}^T\ \hat{\beta}_{BELS}^T(N)\right] \tag{3.104}$$

$\hat{\Theta}_{BELS}(N)$ our most n elements $\hat{\beta}_{BELS}^T(N)$, which is equal to the product of the weight moment array H^1 and $\hat{b}_{BELS}^T$ compute. $\hat{\Theta}_{BELS}(N)$ can be expressed in (3.105).

$$\hat{\Theta}_{BELS}(N) = L\hat{\theta}_{BELS}(N) \tag{3.105}$$

Equation (3.105) shows that the variables θ and Θ are always computed because of the LS method of compensation $\hat{\theta}_{BELS}(N)$ and $\hat{\Theta}_{BELS}(N)$ theta preserve a simple relation. We compute that $\hat{\Theta}_{LS}(N)$ theta is always biased, so we compute it in (3.105) without the requirement.

Second, the LS method of offset compensation has very strong phase tolerance, which is based on the following theorem.

Theorem: In the case of five false assumptions, if the parameter estimation $\hat{\theta}_{BELS}(N)$ is obtained by the LS method of bias compensation of the correction, then (3.106) can be get.

$$\lim_{N\to\infty} \hat{\Theta}_{BELS}(N) = 0 \tag{3.106}$$

It is based on the conclusion of (3.105) and from (3.106) and the following article in (3.107).

$$\begin{cases} \lim\limits_{N\to\infty} \hat{R}_{\Phi\Phi}(N) = R_{\Phi\Phi} \\ \lim\limits_{N\to\infty} \hat{r}_{\Phi y}(N) = r_{\Phi y} \end{cases} \tag{3.107}$$

From (3.106) and (3.107), (3.108) can be obtained.

$$\lim_{N\to\infty} \hat{\Theta}_{LS}(N) = \Theta + R_{\Phi\Phi}^{-1}Q_2 r_{y\varepsilon} \tag{3.108}$$

Equation (3.109) can be obtained from the root data (3.106)–(3.108).

$$\lim_{N\to\infty} \hat{r}_{LS}(N) = \left(H^T R_{\Phi\Phi}^{-1} Q_2\right)^{-1} H^T \Theta_{LS} = r_{y\varepsilon} \tag{3.109}$$

In (3.109), when N→∞, (3.106) and (3.107) can be obtained as follows in (3.110).

$$\begin{aligned} &\lim_{N\to\infty} \hat{\Theta}_{BELS}(N) = \lim_{N\to\infty} \hat{\Theta}_{LS}(N) - \lim_{N\to\infty} R_{\Phi\Phi}^{-1}(N)\, Q_2 r_{y\varepsilon}(N) \\ &= \left(\Theta + R_{\Phi\Phi}^{-1} Q_2 r_{y\varepsilon}\right) - R_{\Phi\Phi}^{-1} Q_2 r_{y\varepsilon} = \Theta \end{aligned} \tag{3.110}$$

Equation (3.110) is the recursive formula of the compensated LS method. So the compensating RLS method is unbiased and consistent.

3.6.3 Forgetting factor RLS method

Establishing a reasonable, effective battery equivalent model is of great significance, especially in the determination of related parameters in the ECM, and is a key role in the accurate estimate of SOC [135]. The lithium battery is in the process of use, and the internal complex chemical reaction occurs inside, and the reaction process is easily affected by the experimental environment. In the battery SOC estimation, a method of equipping experiment data fitting functions is generally employed to determine the value of each parameter in the battery model, but using this method will result in a large error in the estimation result.

The battery has time-varying characteristics in use, and the battery model parameters are constantly changing under different SOC, current, temperature, and aging conditions. When using the battery model for online estimation, if the model parameters can be adaptively adjusted and corrected according to the battery state, the estimation accuracy is improved. Therefore, the RLS method with the forgetting factor can be used to identify the battery model parameters. According to the LS theory, the model expression needs to be discretized first. For the second-order RC ECM, the expression of battery terminal voltage is obtained through the Laplace transformation, as shown in (3.111).

$$U_{OCV}(s) - U_t(s) = \left(R_0 + \frac{R_1}{1+\tau_1 s} + \frac{R_2}{1+\tau_2 s}\right) I(s) \tag{3.111}$$

In (3.111), the mathematical relationship can be expressed with the optimation treatment, and the transfer function is obtained accordingly, as shown in (3.112).

$$G(s) = \frac{U_{OCV}(s) - U_t(s)}{I(s)} = R_0 + \frac{R_1}{1+\tau_1 s} + \frac{R_2}{1+\tau_2 s} \tag{3.112}$$

In (3.112), R_0 is the ohmic resistance, R_1 is the polarization resistance, R_2 is the surface effect resistance, τ_1 and τ_2 are time constants. By using the bilinear transformation equation, the obtained results are shown in (3.113).

$$s = \frac{2}{T} * \frac{1 - z^{-1}}{1 + z^{-1}} \tag{3.113}$$

In (3.113), T is the sampling period. Then, (3.113) is discretized into (3.114), the purpose of this step is to simplify the parameters, and the result of this step is shown in (3.114).

$$G\left(z^{-1}\right) = \frac{\alpha_3 + \alpha_4 z^{-1} + \alpha_5 z^{-2}}{1 - \alpha_1 z^{-1} + \alpha_2 z^{-2}} \tag{3.114}$$

In (3.114), α_1–α_5 are corresponding constant coefficients, and then the discretization equation can be obtained, as shown in (3.115).

$$\begin{aligned} y(k) &= U_{OCV}(s) - U_t(s) \\ &= \alpha_1 y(k-1) + \alpha_2 y(k-2) + \alpha_3 I(k) + \\ &\alpha_4 I(k-1) + \alpha_5 I(k-2) \end{aligned} \tag{3.115}$$

In (3.115), $I(k)$ and $y(k)$ are the input and output of the system, respectively. And then, (3.116) can be expressed, it is also shown in (3.116).

$$\begin{cases} \varphi(k) = [y(k-1), y(k-2), I(k), I(k-1), I(k-2)] \\ \theta(k) = [\alpha_1, \alpha_2, \alpha_3, \alpha_4, \alpha_5] \\ y(k) = \varphi^T(k)\,\theta(k) \end{cases} \tag{3.116}$$

In the RLS identification method, $\theta(k)$ is taken as the coefficient to be identified, which is then used to deduce the parameters of the battery model accordingly. The RLS method has the attitude of infinite memory, that is, with the increase of recursion times k, old data accumulate continuously, which makes it difficult for new data to play the role of correction, which affects the effect of parameter estimation, especially in time-varying systems. Because of the imbalance between the old and new data, the newly estimated parameters cannot accurately reflect the characteristics of the current system. To avoid the above situation, a forgetting factor of λ is introduced in the RLS algorithm. The function of the forgetting factor is to give a smaller weight to the data with a longer-running time in the recognition process, while the latest observation data occupy a larger weight. The genetic factor is introduced to weaken the influence of the old data and enhance the feedback effect of the new data. The steps of the LS algorithm with the forgetting factor are described as follows.

Steps: (1) Initializing parameters can be realized, such as $\theta(k)$ and covariance matrix $P(k)$; (2) the calculation of $\theta(k+1)$ is conducted at the next time point; (3) calculating covariance matrix; (4) calculating a gain matrix; (5) calculating covariance matrix; (6) circulating Steps (2)–(5) as shown in (3.117).

$$\begin{cases} \hat{\theta}(k+1) = \hat{\theta}(k) + K(k+1)[y(k+1) - \varphi^T(k+1)\hat{\theta}(k)] \\ e(k+1) = y(k+1) - \varphi^T(k+1)\hat{\theta}(k) \\ K(k+1) = P(k+1)\varphi(k+1)[\lambda + \varphi^T(k+1)P(k)\varphi(k) + 1]^{-} \\ P(k+1) = \frac{1}{\lambda}[1 - K(k+1) * \varphi^T(k+1)P(K)] \end{cases} \tag{3.117}$$

To overcome the phenomenon of data saturation, the forgetting factor λ $(0 < \lambda < 1)$ is introduced into the traditional RLS algorithm to weaken the influence of old data and enhance the feedback effect of new data. The basic principle of the forgetting factor recursive least square (FFRLS) method is to minimize the error sum of squares. The improved objective function is shown in (3.118).

$$J = \sum_{k=n+1}^{N} \lambda^{N-k} \left[e(k)\right]^2 \tag{3.118}$$

Due to the "filter saturation" phenomenon, the value of the gain K and P will become smaller and smaller with the increase of the number of algorithm data iterations. In (3.118), the bad impact is improved. Then, the recursive equation of the FFRLS method is modified, as shown in (3.119).

$$\begin{cases} \hat{\theta}(k+1) = \hat{\theta}\left(k\right) + K(k+1)[y(k+1) - \phi^T(k+1)\hat{\theta}(k)] \\ K(k+1 = P\left(k+1\right)\phi\left(k+1\right)\left[\phi^T\left(k+1\right)P\left(k\right)\phi\left(k+1\right) + \lambda\right]^{-1} \\ P\left(k+1\right) = \lambda^{-1}\left[I - K\left(k+1\right)\phi^T\left(k+1\right)\right]P\left(k\right) \end{cases} \tag{3.119}$$

In (3.119), the closer λ is to 1, the better the simulation result. Generally, the value is greater than 0.9. The smaller the value of λ, the smaller the tracking effect of the algorithm, but it causes fluctuations in the algorithm. When $\lambda = 1$, it is the standard LS recursive algorithm. The ECM and the battery charge–discharge experiment are inseparable. Through the charge–discharge characteristic experiment and the HPPC experiment, the performance parameters of the lithium-ion battery during operation are studied. Considering the accuracy of the ECM and the difficulty of identifying and estimating parameters such as SOC, resistance, and capacitance, the second-order RC ECM is finally selected as the lithium-ion battery ECM. This model consists of a static ohmic resistance R_i and two RC circuit. The dynamic response is composed in series, as shown in Figure 3.44.

According to KVL, the KVL equation of the circuit can be obtained, as shown in (3.120). In (3.120), U_b is the OCV of the battery, R_i is the ohmic resistance, and both R_1 and R_2 represent the polarization internal resistance of the batteries.

$$U_b = U + U_i + U_1 + U_2 \tag{3.120}$$

The LS parameter identification method with the forgetting factor is applied to the identification of the equivalent model parameters of the lithium-ion battery, and the parameters in the second-order RC ECM are identified. The battery model converts into a mathematical form of LS, as shown in (3.121).

$$U_b = \left(\frac{R_1}{R_1C_1s + 1} + \frac{R_2}{R_2C_2s + 1} + R_i\right)I + U \tag{3.121}$$

Considering the time constant $\tau_1 = R_1C_1$, $\tau_2 = R_2C_2$, R_1 is the polarization resistance, R_2 is the surface effect resistance, R_i is the ohmic resistance, then (3.122) can be rewritten, as shown in (3.122).

$$
\begin{aligned}
&\tau_1\tau_2 U_b s^2 + (\tau_1 + \tau_2)\, U_b s + U_b \\
&= \tau_1\tau_2 R_i I_s^2 + \left[R_1\tau_2 + R_2\tau_1 + R_i\,(\tau_1 + \tau_2)\right] I_s \\
&+ (R_1 + R_2 + R_i)\, I + \tau_1\tau_2 U_s^2 + (\tau_1 + \tau_2)\, U_s + U
\end{aligned} \tag{3.122}
$$

Supposing $a = \tau_1\tau_2, b = \tau_1 + \tau_2, c = R_1 + R_2 + Ri, d = R_1\tau_2 + R_2\tau\tau_1 + R_i(\tau_1 + \tau_2)$ can be simplified, as shown in (3.123).

$$
\begin{aligned}
&U_b\,(k) - U\,(k) = \frac{-bT - 2a}{T^2 + bT + a}\left[U\,(k-1) - U_b\,(k-1)\right] \\
&+ \frac{a}{T^2 + bT + a}\left[U\,(k-2) - U_b\,(k-2)\right] + \\
&\frac{cT^2 + dT + aR_i}{T^2 + bT + a} I\,(k) + \frac{-dT - 2aR_i}{T^2 + bT + a} I\,(k-1) \\
&+ \frac{aR_i}{T^2 + bT + a} I\,(k-2)
\end{aligned} \tag{3.123}
$$

And then, the calculation relationship can be obtained, as shown in (3.124). In (3.124), the parameters are described as shown here

$$
\begin{cases}
U_b(k) - U(k) = k_1\left[U\,(k-1) - U_b\,(k-1)\right] \\
+k_2\left[U\,(k-2) - U_b\,(k-2)\right] \\
+k_3 I\,(k) + k_4 I\,(k-1) + k_5 I\,(k-2) \\
k_1 = \dfrac{-bT - 2a}{T^2 + bT + a},\ k_2 = \dfrac{a}{T^2 + bT + a}, \\
k_3 = \dfrac{cT^2 + dT + aR_i}{T^2 + bT + a},\ k_4 = \dfrac{-dT - 2aR_i}{T^2 + bT + a}, \\
k_5 = \dfrac{aR_i}{T^2 + bT + a}
\end{cases} \tag{3.124}
$$

Equation (3.124) is substituted into the RLS method, taking $\theta = \left[k_1\ k_2\ k_3\ k_4\ k_5\right]^T$ as the direct identification parameter, and then deriving the circuit model parameters R_i, R_1, C_1, R_2, C_2 from the identification results of these parameters, and deriving as follows. Setting $k_0 = T^2 + bT + a$, the parameter values can be obtained, as shown in (3.125).

$$
\begin{cases}
k_0 = \frac{T^2}{k_1 + k_2 + 1} \\
a = k_0 k_2 \\
b = \frac{-k_0(k_1 + 2k_2)}{T} \\
c = \frac{k_0(k_3 + k_4 + k_5)}{T^2} \\
d = \frac{-k_0(k_4 + 2k_5)}{T} \\
R_i = \frac{k_5}{k_2}
\end{cases} \tag{3.125}
$$

In (3.125), initialing $a = \tau_1\tau_2$, $b = \tau_1 + \tau_2$, and $\tau^2 - b\tau + a = 0$, the parameters values of t_1 and t_2 can be obtained, as shown in (3.126).

$$\tau_1 = \frac{b + \sqrt{b^2 - 4a}}{2}, \tau_2 = \frac{b - \sqrt{b^2 - 4a}}{2} \tag{3.126}$$

In (3.126), from $c = R_1 + R_2 + R_i$ and $d = R_1\tau_2 + R_2\tau_1 + R_i(\tau_1 + \tau_2)$, the values of other parameters can be obtained, as shown in equations from (3.127).

$$\begin{cases} R_1 = \frac{(\tau_1 c + \tau_2 R_i - d)}{(\tau_1 - \tau_2)} \\ R_2 = c - R_1 - R_i \\ C_1 = \frac{\tau_1}{R_1} \\ C_2 = \frac{\tau_2}{R_2} \end{cases} \tag{3.127}$$

From (3.127), the values of the various parameters are calculated for the ECM.

3.6.4 *Improved multi-innovation LS (MILS) method*

At present, there are four kinds of lithium battery estimation methods: direct generation, model method, data-driven method, and hybrid estimation method. The model method and hybrid estimation method establish the battery ECM according to the charge and discharge characteristics of the battery, then identify the parameters of the circuit components in the model according to the needs, and finally select the appropriate algorithm to estimate the SOC value of the battery. The accuracy of parameter discrimination will directly affect the accuracy of battery model output voltage and further affect the estimation accuracy of battery SOC.

At present, the research methods of battery model parameter discrimination mainly use off-line and online parameter discriminations. The off-line identification process of ECM parameters is relatively simple and widely used [136]. However, the actual use of the battery is greatly disturbed by the outside world, and the battery parameters will change with the increase of cycle times and the change of operating ambient temperature. The model parameters determined by off-line parameter discrimination are constants, which will remain unchanged once determined, so the accuracy of off-line parameter discrimination model will gradually decline in the process of battery use.

In recent years, the research on online parameter discrimination methods for battery models has become a hot topic. The common online parameter discrimination methods include the RLS method, maximum likelihood estimation method, Kalman filter method, particle swarm optimization algorithm, and so on. Although the RLS method can discriminate the parameters of the model, it does not consider the problem of data saturation. Although the RLS method with forgetting factor solves these problems, it does not consider the contradiction between the convergence and stability of fixed forgetting factor parameters. In recent years, multi-sensor information fusion technology is widely used in many fields, including military, global positioning system (GPS), remote sensing, robotics, image and signal processing, etc. Multi-innovation identification theory is based on the idea of expanding scalar information to vector information, extending information vector to information matrix and the multi-innovation identification method can improve the accuracy of parameter estimation.

The LS has the characteristics of easy understanding and fast convergence. It has been widely used in the field of system identification. For linear systems, the mathematical description expressions are shown in (3.128) and (3.129).

$$A\left(z^{-1}\right)Y(k)=B\left(z^{-1}\right)U(k) \tag{3.128}$$

$$\begin{cases} A(z^{-1})=1+a_1z^{-1}+a_2z^{-2}+\cdots+a_nz^{-n} \\ B(z^{-1})=1+b_1z^{-1}+b_2z^{-2}+\cdots+b_nz^{-n} \end{cases} \tag{3.129}$$

U(k) is the system input and $Y(k)$ is the system output. Then the discrete transfer function can be obtained, as shown in (3.130).

$$G\left(z^{-1}\right)=\frac{B\left(z^{-1}\right)}{A\left(z^{-1}\right)}=\frac{1+b_1z^{-1}+b_2z^{-2}+\cdots+b_nz^{-n}}{1+a_1z^{-1}+a_2z^{-2}+\cdots+a_nz^{-n}} \tag{3.130}$$

The corresponding difference equation is shown in (3.131).

$$\begin{aligned} y(k) &= -a_1y(k-1)-a_2y(k-2)-\ldots-a_ny(k-n) \\ &+ b_1u(k-1)+b_2u(k-2)+\ldots+b_nu(k-n)+e(k) \end{aligned} \tag{3.131}$$

The parameters contained are the parameters to be estimated, and $e(k)$ is the error of the vector equation. Order

$$\phi(k)=\left[-y(k-1),-y(k-2),\ldots,-y(k-n),u(k-1),u(k-2),\ldots,u(k-n)\right] \tag{3.132}$$

Thus, (3.133) can be rewritten, as shown in (3.134).

$$\theta=\left[(a_1,a_2,\ldots,a_n),(b_1,b_2,\ldots,b_n)\right] \tag{3.133}$$

$$y(k)=\Phi(k)\,\theta^T+e(k) \tag{3.134}$$

The input and output of the system are extended to N dimensions, where $k = 1, 2, N + N$, the matrix expression can be obtained as shown in (3.135).

$$Y=\phi\theta^T+e(k) \tag{3.135}$$

The calculated value of the system observation is shown in (3.136).

$$\phi=\begin{bmatrix} -y(n) & -y(n-1) & \ldots & -y(1) & u(n) & \ldots & u(1) \\ -y(n+1) & -y(n) & \ldots & -y(2) & u(n+1) & \ldots & u(2) \\ \vdots & \vdots & \vdots & \vdots & \vdots & \vdots & \vdots \\ -y(n+N) & -y(n+N-1) & \ldots & -y(N) & u(n+N) & \ldots & u(N) \end{bmatrix} \tag{3.136}$$

$$Y=\left[y(n+1)\,y(n+2)\ldots y(n+N)\right]^T \tag{3.137}$$

$$e=\left[e(n+1)\,e(n+2)\ldots e(n+N)\right]^T \tag{3.138}$$

The objective function is $J(\theta)$. The purpose of the LS method is to find θ.When there is $\theta = \theta$, $J(\theta)$ takes the minimum value. The objective function and estimated parameter values of the system are shown in (3.139).

$$J(\theta) = \sum_{i=1}^{N} (Y - \phi\theta)^2 = \sum_{i=1}^{N} e^2 (n + i) = (Y - \phi\theta)^T (Y - \phi\theta) \tag{3.139}$$

Therefore, the first derivative θ of (3.140) can be obtained, sometimes the first derivative $\theta = \theta$ is zero.

$$\frac{\partial J}{\partial \theta} = \frac{\partial}{\partial \theta} \left[(Y - \phi\theta)^T (Y - \phi\theta) \right] = 0 \tag{3.140}$$

The LS estimation value of the system can be obtained, as shown in (3.141).

$$\hat{\theta} = (\phi^T\phi)^{-1} \phi^T Y \tag{3.141}$$

In the actual simulation process, it is necessary to continuously input and output the latest experimental data to improve the accuracy of parameter estimation in the continuous iterative process until satisfactory accuracy is achieved. The specific calculation process is shown in (3.142)–(3.144).

$$\hat{\theta}(k+1) = \hat{\theta}(k) + K(k+1) \left[y(k+1) - \phi^T (k+1) \hat{\theta}(k) \right] \tag{3.142}$$

$$K(k+1) = P(k+1) \phi(k+1) \tag{3.143}$$

$$P(k+1) = P(k) - \frac{P(k+1) \phi(k+1) \phi^T (k+1)}{1 + \phi^T (k+1) P(k) \phi(k+1)} \tag{3.144}$$

where $\theta(k)$ is the estimated value of parameters at time k, $\phi^T (k+1) \theta(k)$ is the calculated value of system observation at time $k + 1$, and $y(k+1)$ is the actual observation value at time $k + 1$. In each iteration, the algorithm uses the difference between the calculated value of the systematic observation and the actual observation value, as well as the Kalman gain K to correct the final estimate. The initial value of $\theta(k)$ can be any value. $P(0) = \alpha I$ is suitable to take a larger value, and I is the unit matrix.

The unit lag operator q^{-1} is introduced and the state space model can be transferred into an *ARMA* model, the derivation formulas are shown in (3.145)–(3.147).

$$A(q^{-1}) y_i(t) = D_i(q^{-1}) \varepsilon_i(t) \tag{3.145}$$

$$A(q^{-1}) = detdet\ (I_n - q^{-1}\Phi) = 1 + a_1 q^{-1} + \ldots + a_n q^{-n} \tag{3.146}$$

$$D_i(q^{-1}) \varepsilon_i(t) = Hadj(I_n - q^{-1}\Phi) w(t-1) + det(I_n - q^{-1}\Phi) v_i(t) \tag{3.147}$$

where *adj* means the adjoint of the matrix; it defines the determinant of the matrix. $\varepsilon_i(t)$ is a zero-mean white noise with variance σ_i^2; $D_i(q^{-1})$ is a monic polynomial with order n as (3.148).

$$D_i(q^{-1}) = 1 + d_i^{(i)} q^{-1} + \ldots + d_n^{(i)} q^{-n} \tag{3.148}$$

This *ARMA* model can be written in vector form, as shown in (3.149).

$$y_i(t) = \varphi_i^T(t)\,\theta_i + \varepsilon_i(t) \tag{3.149}$$

$\theta_i \in R^n$ is the parameter vector to be identified. $\varepsilon_i(t)$ is the white noise and cannot be used in the identification process, so it has to be replaced by (3.150).

$$\hat{\varepsilon}_i(t) = y_i(t) - \Phi_i^T(t)\,\theta_i \tag{3.150}$$

By using $\{y_i(t), y_i(t-1), \ldots, y_i(0), \ldots, y_i(1-n)\}$, an MILS identification algorithm can be obtained to estimate the unknown parameter vector θ_i of the system.

$$Y_i(p,t) = \begin{bmatrix} y_i(t) \\ y_i(t-1) \\ \cdots \\ y_i(t-p+1) \end{bmatrix} \in R \tag{3.151}$$

$$\xi_i(p,t) = \begin{bmatrix} \varepsilon_i(t) \\ \varepsilon_i(t-1) \\ \cdots \\ \varepsilon_i(t-p+1) \end{bmatrix} \in R^p \tag{3.152}$$

$$\Phi_i(p,t) = [\varphi_i(t), \varphi_i(t-1), \ldots\ldots, \varphi_i(t-p+1)] \tag{3.153}$$

The matrix equation is obtained as (3.154).

$$Y_i(p,t) = \varphi_i^T(p,t)\,\theta_i + \zeta_i(p,t) \tag{3.154}$$

Then the criterion function $J(\theta_i)$ can be written as (3.155).

$$\Phi_i(\theta_i) = \sum_{i=1}^{n} [Y_i(p,t) - \Phi_i^T(p,t)\,\theta_i]^T [Y_i(p,t) - \Phi_i^T(p,t)\,\theta_i] \tag{3.155}$$

$$z_t^i = \begin{bmatrix} Y_i(p,1) \\ Y_i(p,2) \\ \cdots \\ Y_i(p,t) \end{bmatrix} \tag{3.156}$$

letting

$$H_t^i = \begin{bmatrix} \phi_i^T(p,1) \\ \phi_i^T(p,2) \\ \cdots \\ \phi_i^T(p,t) \end{bmatrix} \tag{3.157}$$

The criterion function $J(\theta_i)$ can be rewritten as (3.158).

$$J(\theta_i) = (Z_t^i - H_t^i\theta_i)^T (Z_t^i - H_t^i\theta_i) \tag{3.158}$$

Letting $\frac{\partial J(\theta_i)}{\partial \theta_i}|\theta_i = \hat{\theta}_{RMLS} = 0$, we have the recursive MILS identification algorithm as shown in (3.159)–(3.161).

$$\hat{\theta}_i(t) = \hat{\theta}_i(t-1) + P_i(t)\Phi_i(p,t)[Y_i(p,t) - \Phi_i^T(p,t)]\,\hat{\theta}_i(t-1) \tag{3.159}$$

$$S_i^{-1}(t) = S_i^{-1}(t-1) + \Phi_i(p,t)\,\Phi_i^T(p,t) \tag{3.160}$$

$$\hat{\varepsilon}_i(t) = y_i(t) - \varphi_i^T(t)\,\hat{\theta}_i(t-1) \tag{3.161}$$

where $S_i(0) = \alpha I_n, \alpha \gg 1$. $p \geq 1$ is the length of the new message, and $\hat{\theta}_i(t)$ is the estimated value of θ_i at time t.

Because of the usage of multi-innovation, the proposed method has higher accuracy than that of the common LS method. But it still has the drawback that estimation accuracy and accuracy and convergence speed of the MA process part is slow. In order to get a better estimation performance, the output data can be utilized repeatedly. By setting the estimation $\hat{\theta}_i(N)$ and $S_i(N)$ as the initial value of the second round circulation, which means $\hat{\theta}_i(0) = \hat{\theta}_i(N)$, $S_i(0) = S_i(N)$, the improved recursive MILS estimation $\hat{\theta}_i(t)$ can be calculated.

According to the different mechanisms of the battery model, it can be divided into the simple electrochemical model, an intelligent mathematical model, and an ECM. The modeling of lithium batteries can be divided into mechanism modeling, experimental modeling, and hybrid modeling. There are three kinds of electrical characteristic models: black-box model, electrochemical mechanism model, and ECM. Equivalent modeling of lithium battery is the basis of SOC estimation, and its accuracy will directly affect the accuracy of SOC estimation at present, the battery model can be divided into different types according to the different research mechanisms and objects, the most common are the electrochemical model and the ECM. The lithium battery is used more and more widely. According to the need of the battery application, the researcher has carried out comprehensive research on the battery model. Battery models can be divided into four categories: electrochemical model, thermal model, coupled model, and performance model. In the process of use, the lithium battery will show polarization and hysteresis effect, which is consistent with the characteristics of the resistance–capacitance model, so the ECM and state-space expression can be made. Using this model, the nonlinear dynamic characteristics can be vividly described, which is suitable for the simulation study. The ECM is a kind of battery performance model, which is composed of circuit elements such as a resistor, capacitor, and voltage source to simulate the dynamic characteristics of the battery. The ECM is easy to be analyzed by circuit and mathematical methods and has become the most widely used battery model, such as the Rint model, RC model, Thevenin model, and PNGV model.

The significance of the battery model lies in the establishment of clear mathematical relations and then can get important input and output. The various characteristics of the battery in work can be estimated by establishing an accurate battery model, and the feasibility and validity of the battery model can be analyzed and tested by using specific software, finally, the accuracy of the model is verified by comparing it with the experiment. The main factors considered in the modeling are system terminal voltage, battery working current, SOC, environment temperature, battery internal resistance and electromotive force, etc. The accuracy of SOC estimation mainly depends on the accuracy of the battery model. By studying the properties and parameters of circuit components in the ECM, the goal of battery management can be achieved more accurately and effectively. Lithium batteries are widely used in many fields because of their high voltage, large capacity, weak self-discharge, small mass, and small volume. The increasing application of lithium batteries in various fields and the strengthening of process complexity require more and more perfect management systems (BMS), in which the study of battery equivalent model and working characteristics is very important, so it is very important to build an accurate battery model.

Researchers have done a lot of work and obtained many effective methods and conclusions, which provide a more solid foundation for the application of lithium batteries. The dynamical system model describes the fixed parameters in the classical model as a variable that changes dynamically with the charge state and temperature to achieve a more accurate description of the lithium battery performance.

Parameter identification is the process of processing the experimental data after the experiment. There are many methods to identify the parameters of the battery model, which can be divided into two categories: one is to calculate the parameters of the model by calculating formulas, the other is to calculate the parameters of the battery model by calculating formulas, this method is suitable for a specific model; the other is to get the model parameters by data fitting. This method is suitable for most models. The first type can refer to the above formula to calculate parameters, such as the RC equivalent model. The second method, such as the PNGV equivalent model, is to perform HPPC experiments to obtain the experimental data. PPC is used to test the dynamic power characteristics of the battery under normal operation by charging and discharging pulse current. The data were obtained from the HPPC experiments of SOC with equal intervals, and then the parameters were obtained by using multiple linear regression or LS curve fitting. The regression including two or more independent variables is called multiple linear regression. The basic principle and basic calculation process of multivariate linear regression are the same as that of monadic linear regression, but because of the number of independent variables, the calculation is rather complicated.

For a specific ECM, the reliability and accuracy of the model are directly determined by the parameter identification of the model. So it is very important to accurately identify the parameters of the ECM of lithium battery. The main process of parameter identification includes the acquisition of identification data, calculation of cell ohmic resistance, and implementation of the identification method (at present, the main method is the LS algorithm). At present, the parameter identification mechanism can be divided into offline identification, online identification, and intelligent identification.

Most of the offline identification methods are based on offline data, such as linear fitting, LS, and so on. The parameters identified by the off-line identification scheme are generally fixed parameters that cannot reflect the actual operating conditions, and this kind of identification result is only suitable for the case of little change in battery operating conditions.

Online identification algorithm is the mainstream of the current battery parameter identification algorithm, which can be divided into two categories: one is the main derivative algorithm of LS, it includes RLS, deviation compensation RLS, and LS algorithm with forgetting factor.

LS are a classic data processing technique. In calculating the orbits of planets and comets, gauss estimates six parameters that describe the motions of celestial bodies, based on observations made by the telescope. GAUSS believes that the best value for an unknown parameter when extrapolating from observations would be something like this, it minimizes the sum of the square of the difference between the observed and calculated values multiplied by the value that measures their accuracy. That was the original LS idea. Since then, the LS has been used to solve many practical problems. Various LS algorithms have been developed to modify the LS for different applications.

3.7 Conclusion

The significance of battery model is to establish a clear mathematical relationship, and then we can get important input and output. Various characteristics of the battery in the work can be estimated by establishing an accurate battery model, analyzing and testing the feasibility and effectiveness of the battery model by applying specific software, and finally verifying the accuracy of the model by comparing with the experiment. Parameter identification is the process of processing the experimental data after the experiment. Most of the offline identification methods are based on offline data, such as linear fitting, LS method, and so on. Online identification algorithm is the mainstream algorithm of battery parameter identification at present, which can be divided into two categories: one is the derivative algorithm of LS method, including RLS method, deviation compensation RLS method, and LS derivative algorithm with forgetting factor. In the field of system identification, LS method is a basic estimation method. The LS method can be used in both dynamic and static systems; it can be used in both linear and nonlinear systems; it can be used for offline estimation and online estimation.

Chapter 4

Use of artificial intelligence for utility-scale battery systems

4.1 Selection criteria for choice of AI techniques

4.1.1 Common AI methods for utility-scale battery systems

Artificial intelligence is a new technological science of research and development for simulation, extension, and expanding human intelligence [137]. Its intelligent algorithm is established based on neurological calculation, fuzzy calculation, and evolutionary calculation of these three major developments relatively mature intelligent algorithm branches and has formed a new scientific algorithm through organic fusion between different algorithms, as well as is a new stage of intelligent theory and technology development. Artificial intelligence has the advantages of efficient and optimized performance, no problem special information, etc., widely used in new energy technology, engineering optimization, pattern identification, intelligent control, network intelligent automation, and other fields [138]. The artificial intelligence algorithm is commonly used in public utility battery systems including neural network (NN), deep learning (DL), support vector machine (SVM), fuzzy logic (FL), linear regression (LR), Gaussian process regression (GPR), perception machine (PM), etc. Next, the focus is on the previous six artificial intelligence algorithms that are more prominent and widely used.

(1) NN method

The basic structure of an NN through mathematics algorithm to simulate people's thinking, that is, an output model is obtained by inputting a plurality of nonlinear models and weighted interconnections between different models. NN is mainly divided into two categories: feedforward NN and feedback NN. By the feedforward NN, each neuron is divided into different groups according to the reception information, and each group can be seen as a neural layer. The neurons in each layer receive the output of the front layer neurons and output to the next layer of neurons. The information in the entire network is propagated in a direction, with no reverse information propagation, which can be represented by a directionless ring map. In the feedback NN, the neurons can not only receive signals from other neurons but they can also receive their own feedback signals. Compared to feedforward NNs, the neurons in the feedback NN have a memory function, which has different states at different times. The information propagation in the feedback NN may be a one-way

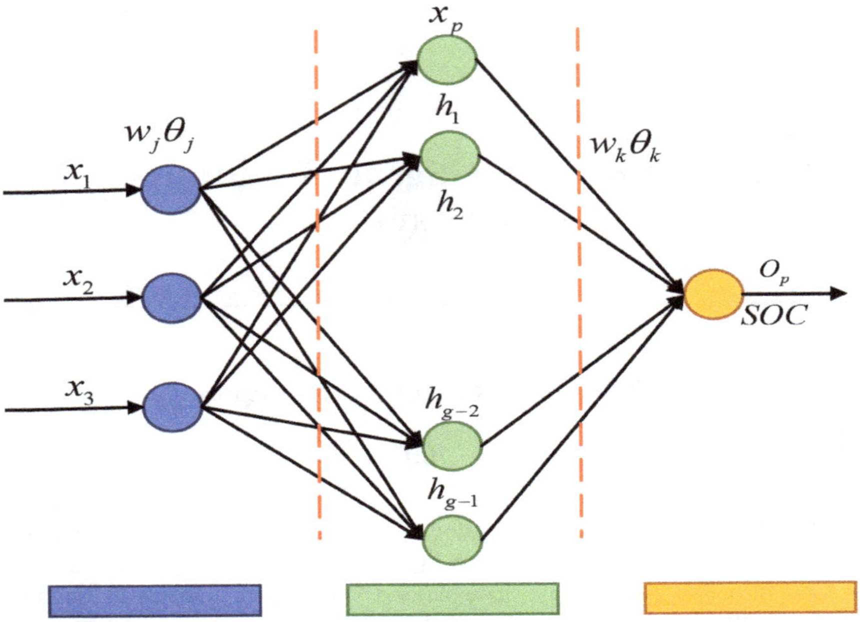

Figure 4.1 *The general architecture of the three-layer NN for state-of-charge (SOC) estimation*

or two-way propagation, so it can be represented by a channel or notes. NN consists of a three-layer formation, as shown in Figure 4.1.

In Figure 4.1, x_1, x_2, and x_3 are the three input layers, which take the vector of instantaneous current, voltage, and temperature values, respectively. h_1, h_2, h_3, and h_4 are the hidden layers. The output layer is the instantaneous SOC value. By training the NN with the input–output pairs, it can form a non-linear map that accurately models the input–output relationship without any prior knowledge of the internal structure of the battery. The relationship between the input layer and the output layer is developed using a suitable number of hidden layers, hidden neurons, and activation functions. SOC in the output layer can be expressed in (4.1):

$$SOC_i = f_i\left\{\sum_k W_{j,k}O_j + \theta_{j,k}\right\} \tag{4.1}$$

where $W_{j,k}$ and $\theta_{j,k}$ denote the weight and bias from the hidden layer to the output layer, respectively. O_j is the output of the hidden layer, and f_i represents the activation function.

According to the universal approximation theorem, a feedforward NN with a single hidden layer is capable of approximating any continuous functions. This property of NN has been exploited by researchers to solve many non-linear mapping problems in various fields that are too complex to model mathematically. Various configurations of NN are reported in the literature including backpropagation NN

(BPNN), radial basis function NN (RBFNN), recurrent NN (RNN), and wavelet NN. However, the fundamental working concept of these variations remains similar.

(2) Deep Learning

DL is to learn from the dealer's complex structure inexperienced data, that is, by constructing a computational model containing multiple processing layers or creating a plurality of levels of abstract layers to represent data. Machine learning and depth learning are different. The former is to find rules that map the input data to the target result, and the latter is a mapping that enters data to target results through a series of data transformations (layers). In DL, the transformation of each layer is implemented by a set of weights. Therefore, the means of learning are finding a set of weight values for all layers of the NN, so that the network can correctly correspond to its goals.

DL has made great strides in many software disciplines including computer vision, speech processing, natural language processing, robotics, bioinformatics, chemistry, video games, search engines, online advertising, and finance to name a few. The term deep lies in the number of computational layers in the NN. It is still an ongoing debate on the true definition of DL. However, it is a widely accepted notion that any NN with more than two hidden layers is considered a deep network. Riding on this premise, this study proceeds to categorize NNs with more than two hidden layers as DL methods.

At present, there exist several common DL network architectures. Among the notable ones are deep NN (DNN), deep convolution NN (DCNN), deep recurrent NN (DRNN), long short-term memory (LSTM), and many more. These architectures have distinct advantages and drawbacks depending on the application domain. For example, DCNN is extremely effective for image-related problems such as face recognition, traffic sign identification, and cancer cell detection from medical scans. On the other hand, DRNN and LSTM are usually used in problems related to sequential data, such as speech recognition, machine translation, or time series prediction.

To date, DL methods have been applied to battery prognostic and health management concerning SOH and remaining useful life prediction. The DNN model created by one input layer, two or more hidden layers, and an output layer is shown in Figure 4.2.

The input nodes consist of the instantaneous voltage $V(k)$ and current $I(k)$ as well as the average voltage $V_{avg}(k)$ and current $I_{avg}(k)$ over a time window. The output node is the estimated SOC at time k. L denotes the number of layers in the model. N denotes the number of neurons in a layer.

(3) Support vector machine

SVM is a two-point model; it uses nonlinear mapping to perform data mapping in some areas and applies a linear algorithm in the function space, as well it is usually used to solve the rotation classification and return task. The basic model of SVM is the maximum linear classifier defined on the feature space, and the maximum space is the most different from the PM; it also includes nuclear techniques, which make it a substantially nonlinear classifier. The learning strategy of SVM is the maximization of intervals, it can be formally viewed as a problem that solves the problem of convex quadratic programming, and it is also equivalent to the minimization

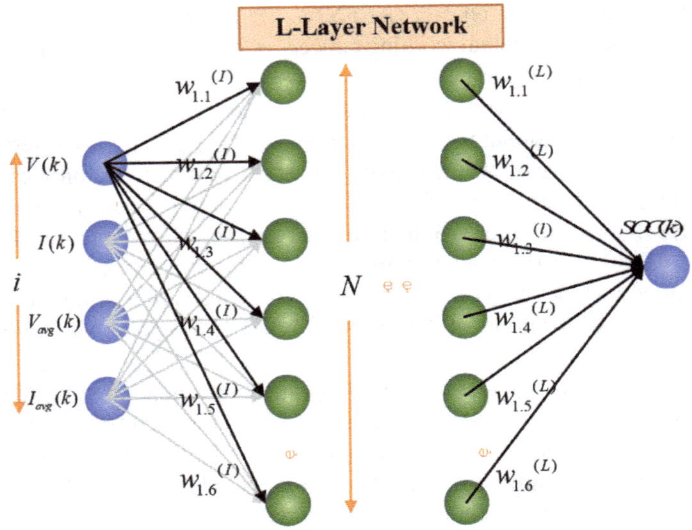

Figure 4.2 DL network architecture for SOC estimation

problem of regularized hinge loss functions. The learning algorithm for SVM is the optimization algorithm for solving the convex quadratic programming.

The basic idea of SVM is to solve the separation super plane that can correctly divide the training data set and the geometric spacing is the largest. SVM attempts to construct hyperplanes in high-dimensional space to separate data of one class from another. An optimal separation boundary is achieved when the distance from the hyperplane to the nearest data point of any class is maximized. Figure 4.3 illustrates an example of hyperplane to separate distinct classes.

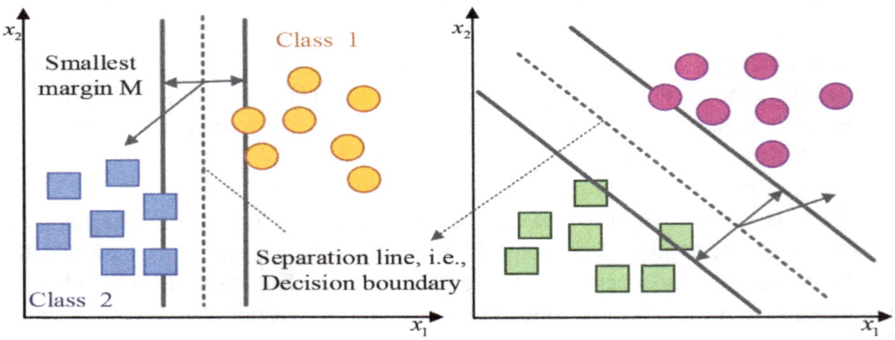

Figure 4.3 SVM separates data in the hyperplane(s) to achieve the best separation

In Figure 4.3, the right image is the result of classifying classes 1 and 2 without SVM, and the left image is the result of classifying classes 1 and 2 using SVM. M is the separate margin. SVM algorithm attempts to construct hyperplane(s) that separates one class from another such that the separate margin is the largest. The image on the right shows a large separate margin M than the image on the left.

(4) Fuzzy logic

FL is a logic that uses membership degrees replacement Boolean true value. Unlike classic binary logic, it does not use the distinct two-valued to express all propositions but uses membership degrees to express, more suitable for describing the incapacity of statements in actual life. The basic starting point of the fuzzy theory is to cancel the opposite between two values, and the membership degrees are used to represent the transition state between the two values. This not only provides the convenience of making inaccurate but effective descriptions but also provides theoretical tools to put fuzzy reasoning and fuzzy decisions that meet human thinking habits into the computer.

FL is a computing approach that offers flexibility in a statement. This approach facilitates the concept of partial truth, where the truth value may range from completely true, partially true, to completely false depending on the affiliate value it takes from 0 to 1. The greater the membership degrees, the higher the degree of authenticity; on the contrary, the smaller the membership degrees, the lower the degree of true. Instead of the conventional two-valued true or false logic, FL introduces the concept of many-valued logic. The FL interface system is structured using fuzzification, fuzzy rule base, inference engine, and defuzzification.

The focus of FL is to map the input space to the output space, and the main mechanism to do this is a series of if-then statements (called rules). All rules are parallel assessments, and the order of rules is not important. The rules are useful because they reference variables and describe the adjectives of these variables. Before building a system that explains the rules, it must define all the terms used and describe their adjectives. Figure 4.4 provides a roadmap for a fuzzy reasoning process.

In Figure 4.4, the figure on the right is the roadmap of fuzzy inference in general. After the input items are adjusted according to a series of rules set by the expert database, the output values that conform to the rules are obtained. The picture on the left is a specific example.

(5) Linear regression

One type of LR to handle is as follows: given a set of input samples and the target value corresponding to each sample, in the system, the functional relationship between the target value and the input value needs to be found, when there is one. When the new sample arrives, it can predict how the corresponding target value is. LR and linear classification (LC) are very similar, but different, the target value of LR is a continuous variable, and the target value of LC is a discrete variable. LR includes a unit LR and a plurality of LR.

The LR algorithms will be applied if the output is a continuous variable. In contrast, grading algorithms are applied when the output is broken up into sections

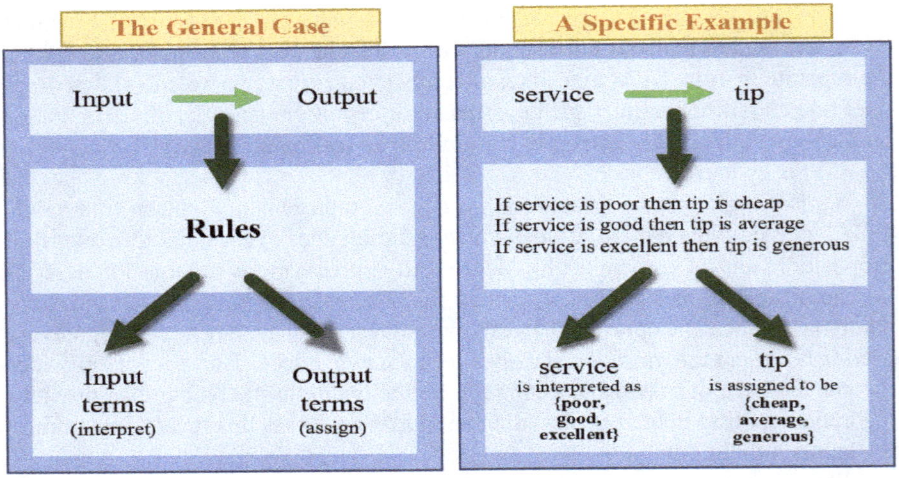

Figure 4.4 The roadmap of the fuzzy reasoning process

like pass/fail, good/average/bad, and so on. We have different regression algorithms or classifying behavior, the LR algorithm being the fundamental regression algorithm.

(6) Gaussian process regression

GPR is usually used for the regression problem of low-dimensional and small samples, but there are also extension algorithms suitable for large samples and high-dimensional situations. In the regression prediction, we usually predict just a single point value; the GPR can be understood as probability prediction, which can also predict the upper and lower limits while predicting accuracy point, adding more reference values to predict. GPR model assumptions include the two parts of the returning residual and the Gaussian process prior art, which is solved by Bayesian inference. GPR is a probability model with general use and analytical. Based on the convenience of the Gaussian process and its nuclear functions, GPR is widely used in the field of time series analysis, image processing, and automatic control.

The difference in the sample between the Gaussian process and the general machine learning is that there is a correlation between the features in the Gaussian process, which is reflected by the covariance matrix. For example, in some time series models, the time series of each variable output will reflect a correlation before and after time, such as a smooth transition, and the output of those models is ideal for the simulation of the Gaussian process. The covariance between individual dimensions can be simulated through the kernel, and the common kernel includes RBF kernel, periodic kernel, linear kernel, RBF kernel, and more.

GPR is a non-parametric Bayesian regression method that generates waves in machine learning. GPR has various advantages, such as that it works well on small data sets and can provide predictive uncertainty measurements.

4.1.2 AI technology evaluation index

The development of artificial intelligence has injected new impetus into social progress. Machine learning technology represented by deep convolutional NNs has triggered a new round of technological revolution and has had a profound impact in many fields such as transportation, medical care, finance, manufacturing, and security. However, there are certain hidden dangers in the development of artificial intelligence technology [139]. In recent years, relevant studies at home and abroad have shown that intelligent algorithms based on DNNs are extremely susceptible to noise interference and cause unpredictable errors and may even be misled and cause serious problems. Therefore, how to evaluate the quality of machine learning models and ensuring the security of artificial intelligence technology has become an urgent problem to be solved.

T/CESA 1036–2019 "Quality Elements and Test Methods of Information Technology Artificial Intelligence Machine Learning Models and Systems" was led by Beihang University and jointly with China Electronics Standardization Institute and other units [140]. This standard specifies the quality elements of machine learning models and systems, provides a quality test index system and corresponding test methods for machine learning models and systems, and applies to the design, development, and quality assessment of machine learning models and systems.

This standard proposes a quality index system for machine learning models, specifies the quality elements of standard datasets for machine learning model quality, provides general quality elements of models and the quality elements involved as software and hardware system modules, and provides machine learning model-based quality elements. Evaluation methods are used for evaluating quality elements. The evaluation index mainly includes three elements: the evaluation index system, standard-quality elements, and standard measurement methods.

(1) Evaluation index system

The evaluation index system refers to the functional characteristics of machine learning models and systems, including the understanding of the life cycle of machine learning models and systems, as well as common evaluation indicators of existing machine learning models and systems, and has field versatility [141]. The current machine learning model has the characteristics of strong data dependence. The model generally obtains knowledge from known data and acts on unknown data to make predictions [142]. The quality of the data restricts the quality of the machine learning model to a certain extent. Therefore, the quality elements of machine learning models and systems proposed by this standard are based on the metric requirements for standard datasets. Based on unified data standards, because of the general quality elements of machine learning models and the quality requirements involved in software and hardware systems, combined with the security challenges of artificial intelligence security, it further divides four main index characteristics such as functionality, reliability, efficiency, and maintainability. The schematic diagram of the whole quality factor index system is shown in Figure 4.5.

In Figure 4.5, using a standard dataset as a basic guide is the basis for model quality assessment. It includes four major indicators: Feature, Reliability, Efficiency, and

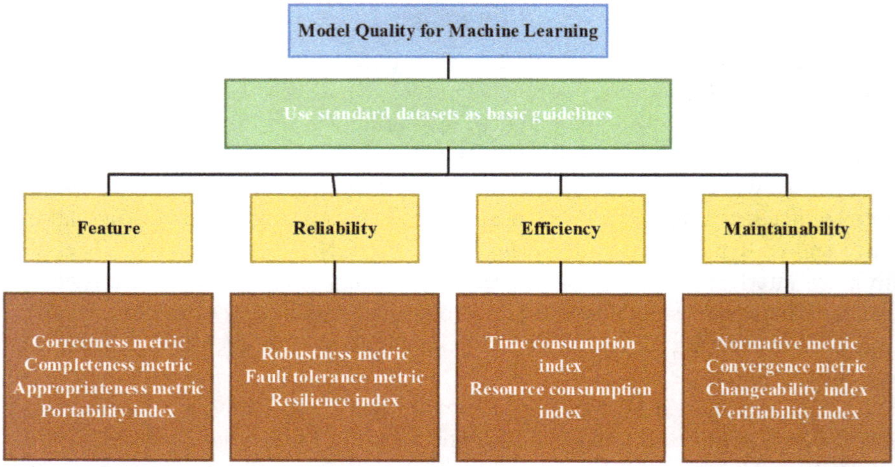

Figure 4.5 Quality indicator system for machine learning models

Maintainability. Where, Feature includes four evaluation indicators: Correctness metric, Completeness metric, Appropriateness metric, and Portability index. Reliability includes three evaluation indicators: Robustness metric, Fault tolerance metric, and Resilience index. Efficiency includes two evaluation indicators: Time consuming and Resource consumption index. Maintainability includes four evaluation indicators: Normative, Convergence metric, Changeability index, and Verifiability index. The above evaluation indexes included in the evaluation index system are important criteria for evaluating the performance characteristics of a model.

(2) Standard quality elements

Based on the feature, machine learning models learn from known data and act on unknown data to make predictions. The standard first considers the dataset quality elements and, on this basis, considers the functionality, reliability, efficiency, and maintainability quality elements.

1) Standard dataset quality elements

Generally, the life cycle of a machine learning model consists of three stages, namely, the model training stage, the model validation stage, and the model testing stage. Machine learning models at different stages have different requirements for datasets.

2) Functional quality elements

Since the application of quality standards is to evaluate the quality of machine learning models, indicators should have the ability to measure the basic functions of machine learning models. Specifically, it includes the correctness metric element, the completeness metric element, the appropriateness metric element, and the transferability index of the model. Among them, the correctness is expressed as the correctness of the machine learning model's operation process and results for the specified tasks and user goals, completeness indicates the degree of coverage of the machine learning model for specified tasks and user goals, appropriateness is used to define and evaluate the reasonable degree of the machine learning model chooses different components to fulfill the requirements, and transferability is used to define and evaluate the transferability of machine learning models.

3) Reliability quality elements

The purpose of quality standards is to help improve the reliability and security of machine learning models, so metrics can measure how reliable a machine learning model is. It is mainly divided into three sub-features: robustness, fault tolerance, and easy recovery. Among them, robustness is used to define and evaluate the ability of machine learning models to avoid failures caused by hazards such as anomalies and extreme conditions, fault tolerance is used to define and evaluate machine learning models to maintain user-expected performance levels in the event of failures, and recoverability is used to define and evaluate the ability of a machine learning model to regain its specified function and restore the affected data within a certain period when it fails.

4) Efficiency quality factors

The efficiency of the machine learning model reflects the quality of the machine learning model. It is used to define and evaluate the ability of the machine learning model to complete the work relative to the resources used. It can be further divided into time consumption requirements and resource consumption requirements.

5) Maintainability quality factor

Machine learning models have the possibility of long-term use, so quality standards measure the maintainability of machine learning models, and are used to evaluate the ease of maintenance of machine learning models and their systems. Sex and other four sub-characteristics are used to evaluate the ease of maintenance of machine learning models and their systems. It is mainly divided into four sub-features: normative, convergent, changeable, and verifiable. Among them, normative is used to define and evaluate whether the training, operation, and maintenance of machine learning models and systems meet the normative standards of the model, convergence is used to define and evaluate whether the training process of the machine learning model can quickly converge to achieve the expected performance, changeability is used to define and evaluate the degree of difficulty for maintainers or users to modify and validate a machine learning model system, and verifiability is used to define and evaluate whether the computational process and computational results of machine learning models and systems are easy to understand and verify.

(3) Standard measurement method

The specific test method of the quality standard is the operation specification for evaluating the quality of machine learning models and systems. It is derived from the quality factor index system of the quality standard and is expressed as a series of specific indicators, measurement elements, and methods, which are used for quantitative or qualitative evaluate the quality of specific machine learning models and systems. Test methods and quality elements are the two major elements of measuring quality standards.

The measurement method of the quality standard is based on the machine learning model and the quality elements of the system to expand the specific indicators, which are divided into five aspects. Focusing on the quality requirements of standard datasets, three methods of dividing characteristics, such as "Hold-out," "Cross Validation," and "Bootstrapping," are proposed, and basic characteristic indicators and measurement elements such as data volume and data distribution; focusing on functional quality elements, put forward indicators and measurement elements such as accuracy rate, requirement coverage ratio, parameter setting, and migration performance change; focusing on reliability quality elements, indicators and methods such as resistance to confrontation, hazard detection, hardware failure interference, average time required to repair software systems and data, and average restart time are proposed; focusing on the factors of efficiency and quality, indicators such as calculation amount and storage occupancy of model training are proposed; focusing on maintainable quality elements, methods such as formal verification methods, correlation analysis, and causal analysis are proposed.

Next, this book mainly introduces some important indicators related to the aspects mentioned above.

1) Accuracy

The proportion of all predicted correct samples (including positive and negative) in all samples.

2) Precision

Also known as precision, the proportion of correctly predicted positive samples to all predicted positive samples. It is different from accuracy.

3) Mean absolute error (MAE)

MAE is the average of the absolute errors, which can better reflect the actual situation of the predicted value error. MSE and MAE are suitable for when the error is relatively obvious, and the large error also has a relatively high weight. MAE is a linear indicator, and all individual differences are equally weighted on the average, so it can highlight outliers more than MSE. Let $x\left(i\right)$ be the real value and $x'\left(i\right)$ be the predicted value, then the calculation formula of MAE is shown in (4.2):

$$MAE = \frac{1}{n}\sum_{i=1}^{n}|x(i) - x'(i)| \tag{4.2}$$

4) Root-mean-square error (RMSE)

The arithmetic square root of mean squared error, RMSE, refers to the sample standard deviation of the difference between the actual value and the predicted value, mainly used when the error is not obvious. Let $x\left(i\right)$ be the real value and $x'\left(i\right)$ be the predicted value, then the calculation formula of RMAE is shown in (4.3):

$$RMSE = \sqrt{\frac{1}{n}\sum_{i=1}^{n}|x(i) - x'(i)|} \tag{4.3}$$

5) Standard deviation

The standard deviation (SD) is the arithmetic square root of the variance, which reflects the degree of dispersion of a data set. Two sets of data with the same mean may not necessarily have the same standard deviation. SD includes population standard deviation, sample standard deviation, and standard error, as shown in (4.4)–(4.6), respectively:

$$\sigma = \sqrt{\frac{1}{n}\sum_{i=1}^{n}(x_i - \mu)^2} \tag{4.4}$$

$$S = \sqrt{\frac{1}{n-1}\sum_{i=1}^{n}(x_i - \bar{x})^2} \tag{4.5}$$

$$\sigma_n = \frac{\sigma}{\sqrt{n}} \tag{4.6}$$

where x_i is the predicted value, μ is the mathematical expectation, and $\bar{x}$ is the mean of the data.

6) R-squared

R-squared uses the least-squares method for parameter estimation. R-squared is the ratio of the regression sum of squares to the total deviation sum of squares, indicating the proportion of the total deviation sum of squares that can be explained by the regression sum of squares, the larger the ratio, the more accurate the model, and the more significant the regression effect. R-squared is between 0 and 1. The closer it is to 1, the better the regression fitting effect. It is generally believed that the model with more than 0.8 has higher goodness of fit.

7) Recall

Also known as recall, the proportion of correctly predicted positive samples to those marked as positive.

8) True positive rate (TPR)

The proportion of actual positive instances in the positive class predicted by the classifier to all positive instances.

9) False positive rate (FPR)

The proportion of actual negative instances in the positive class predicted by the classifier for all negative instances.

10) Significance detection and evaluation of F1 index, $F\beta$

It takes into account both the precision and recall of the classification model. The F1 score can be regarded as a weighted average of model precision and recall, with a maximum value of 1 and a minimum value of 0.

11) Significance detection and evaluation of PR curves

The PR curve is the precision–recall curve. The closer the PR curve is to the upper right corner, the better the model, that is, the larger the area of the PR curve integral, the better the model.

12) ROC curve

It is the TRP and FPR curves. In contrast to the PR curve, the closer the ROC curve is to the upper left corner, the better the performance of the classifier.

4.2 Monitoring of utility-scale batteries development with AI

4.2.1 Development status

The double pressure of the energy crisis and environmental protection has greatly promoted the rapid development of new energy vehicles mainly based on electric vehicles. Oil, gas, and coal are expected to last only about 40, 67, and 190 years, and the automotive industry currently consumes almost a third of all oil sold. In order to strengthen the support for "energy conservation and environmental protection," the development of energy conservation and the new energy vehicle industry represented by electric vehicles has been strongly supported by all countries. Establish a sound electric vehicle power system technology system and industrial chain technology system, realize the industrialization of all kinds of electric vehicles, and promote the strategic emerging industry of new energy vehicles into the rapid growth period. Compared with traditional cars, electric vehicles not only have the advantages of being clean, energy-saving, pollution-free, and low noise but also have the disadvantages of short-range and high cost [143]. The core of the development of electric vehicles is to solve the problem of high cost and short range, which is mainly limited by the performance and cost of the power battery for electric vehicles.

In other words, the technical bottleneck of power batteries and its management system directly restricts the large-scale promotion and application of electric vehicles. In view of this problem, Germany, the United States, Japan, South Korea, and other countries have launched special studies on key technologies of power batteries in order to promote the commercial development of electric vehicles. A power battery provides the power source for an automobile. According to different use modes, it can be divided into a transportation power supply, electric energy storage power supply, mobile communication power supply, and military aerospace power supply, as shown in Figure 4.6.

The electrode material can be divided into lead-acid battery, nickel-cadmium battery, nickel-hydrogen battery, and lithium-ion battery. Lithium-ion battery has gradually become an important development direction of new energy vehicle power source because of its high energy density, long cycle life, wide temperature adaptability, small self-discharge rate, green, and other characteristics [144]. Due to the different battery materials, lithium-ion battery can be subdivided into lithium-ion manganese acid battery, lithium-ion phosphate battery, ternary system battery, and lithium-ion titanate battery.

As the main power source of electric vehicles, power batteries need to be grouped in series and parallel in actual use to meet the requirements of voltage, capacity, power, and other aspects. However, with the increase of the number of batteries, the degree of battery inconsistency is aggravated, and different degrees of aging occur over time, leading to the decline of the performance of the entire power battery system, increasing the possibility of system failure, resulting in high operation and maintenance costs. In order to ensure the safety and reliability of the power battery system, optimize the energy management of electric vehicles, reduce the use cost of power battery, and extend the effective life of the battery system, it is

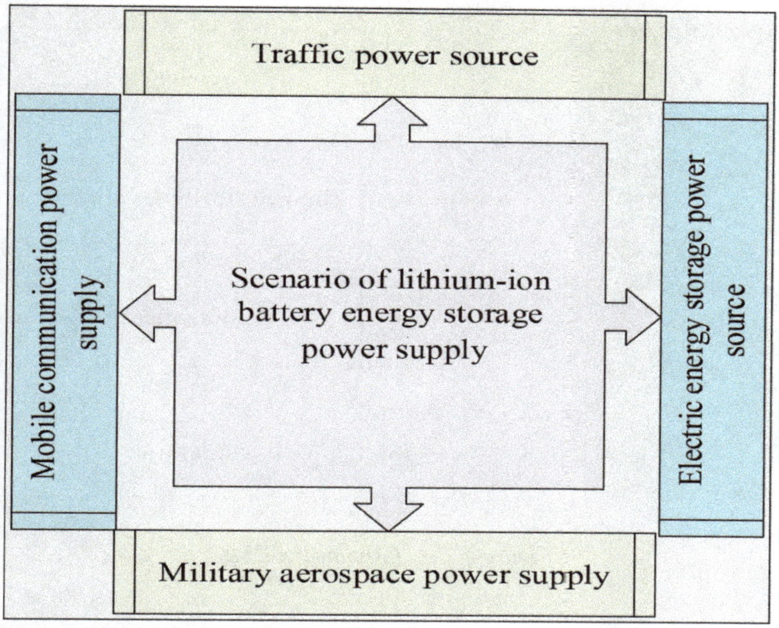

Figure 4.6 *Lithium-ion battery power suppty application*

necessary to adopt the battery management system (BMS) provides comprehensive, systematic, efficient, and detailed management of battery cells and battery strings. BMS consists of sensors, actuators, controllers with embedded algorithms, and signal cables to manage battery strings. The BMS is shown in Figure 4.7.

At present, artificial intelligence and other algorithms are widely used in battery state estimation, including charge state, energy state, power state, health state, and battery charge and discharge control and balance control. In the field of charge state estimation of power lithium-ion battery, the application of artificial intelligence algorithm is relatively comprehensive. Compared with the method of battery model and filter, the algorithm of data-driven SOC prediction can accurately predict battery

SOC only by knowing battery voltage, temperature, current, and other parameters, among which the parameters of the data-driven model need to be adjusted by themselves. Some scholars can accurately estimate the state of charge by optimizing the existing reverse NN and RBFNN. In recent years, some scholars have used a new method of DNN for battery SOC estimation [145]. By applying the driving cycle load of various ambient temperatures to lithium-ion batteries, DNN can adapt to various ambient temperature conditions.

BP network has been widely used in various fields since 1986, and it has also been applied in the prediction of battery SOC. In BP network, as the number of input and output neurons is determined, the number of hidden layer neurons should be determined according to the following empirical formula as shown in (4.7):

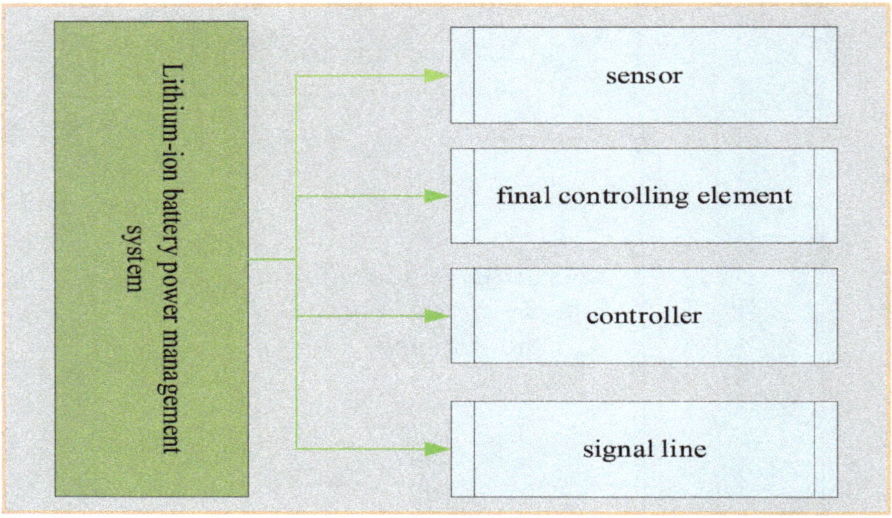

Figure 4.7 Lithium-ion BMS

$$\begin{cases} m = \sqrt{n+1} + \alpha \\ m = \log_2 n \\ m = \sqrt{nl} \end{cases} \tag{4.7}$$

where m represents the number of nodes in the hidden layer, N represents the number of nodes in the input layer, and L represents the number of nodes in the output layer and represents the number between 1 and 10. Finally, $m = 11$ is obtained. The BPNN model used to estimate the charged state of lithium-ion battery is shown in Figure 4.8.

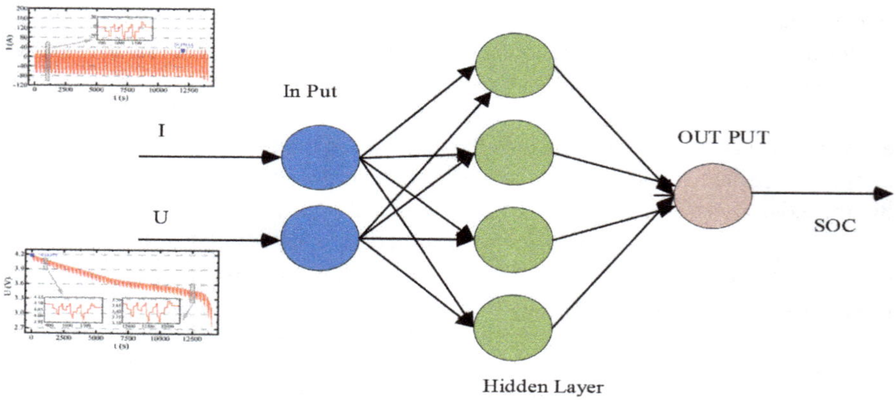

Figure 4.8 The BPNN model

Table 4.1 *The main parameters in a BP network*

Parameter name	Description	Value
Net.tarinparam.goal	Target minimum error	0.001
Net.tarinparam.epochs	Number of training	20,000
Net.tarinparam.lr	Learning rate	0.01

Based on RandomForest (RF) SOC prediction, RF is a decision tree algorithm using random sampling, which belongs to bagging integration in a broad sense. Its

After determining the basic structure of BP network, it is necessary to determine the training function of the model. Commonly used in MATLAB® are TraingD-gradient descent method, which has a slow convergence speed, TraingDA-adaptive learning rate gradient descent method, and TraingDM adaptive learning algorithm with momentum term, which has a fast convergence speed. Table 4.1 shows the main parameters in BP network training.

biggest feature is to build different trees through random sampling. Then for the classification problem, the majority vote of multiple trees is carried out, and the final result is the one with the most occurrence times. For the regression problem, the mean value is taken for the final prediction result of each tree, and the prediction of battery SOC belongs to the regression problem. The main advantages of RF are as follows:

1. Random samples in multiple trees can carry out a high degree of parallel calculation, with high accuracy and efficiency.
2. In RF algorithm, multiple trees are computed in parallel, resulting in small variance and strong generalization ability.
3. Insensitive to missing data sets, able to deal with missing values, mainly including violent filling (mode filling) and similarity matrix filling.
4. It can better deal with unbalanced data and still has high accuracy in the case of unbalanced data.
5. Simple implementation, high code integration.

RandomForest has a well-integrated Sklearn machine learning package in Python, which was used to train battery data sets with Random Forestsor from the SkLearn Ensemble integration module. However, as the algorithm has been proved to overfit in noisy data sets, the specific parameter selection results are obtained through several experiments, which ensures the accuracy of the final result prediction to a certain extent.

XGBoost, as a boosting tree, has the advantages of parallel operation, high accuracy, and easy implementation. Therefore, XGBoost has become one of the most commonly used decision tree models. Compared with gradient lifting decision tree (GBDT), the GBDT only performs first-order expansion in loss function. Due to the advantages of XGBoost second-order Taylor expansion, it is closer to the

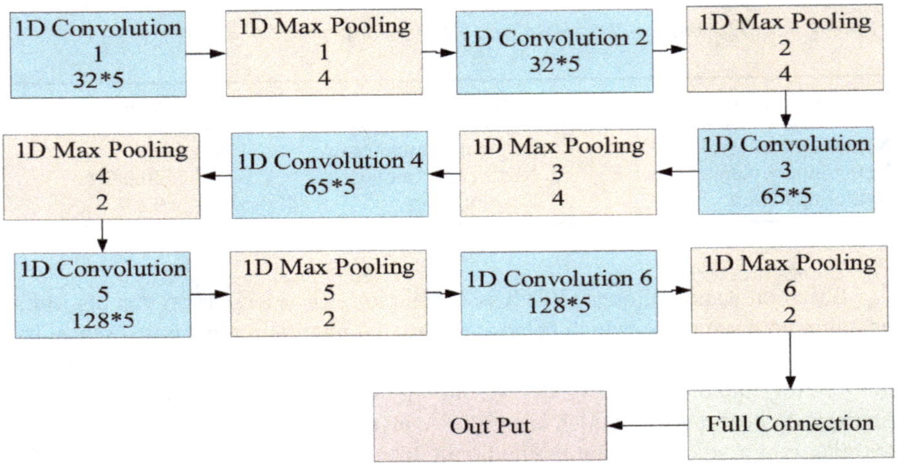

Figure 4.9 The structure of 1D-CNN

actual loss error. At the same time, XGBoost adopts column sampling in the objective function and introduces regular terms in the objective function to ensure that the model is not easy to overfit. The parameters in XGBoost mainly include general parameters, Booster parameters, and learning target parameters.

Lithium-ion battery state prediction method is based on the CNN-LSTM model. The nonlinear relationship between SOC and discharge data of lithium battery was obtained by using CNN and LSTM NN and the long-term dependence of SOC sequences.

CNN is a DNN with convolution structure, which can extract higher-level and more abstract data features from original data and is widely used in image classification, face recognition, and other fields. Among them, one-dimensional CNN is mainly applied to time series data. Figure 4.9 shows the structure of one-dimensional CNN.

RNN takes the output of the previous moment as the input of the current moment and maintains a hidden state containing historical features of data in the hidden layer, so it has a great advantage in time series prediction. Figure 4.10 shows the network structure of the RNN model.

Although RNN adopts a very deep feedforward network structure, theory and experience show that the prediction effect of RNN is very poor when the sequence is long. LSTM NN, as a variant of RNN, solves the problems of gradient

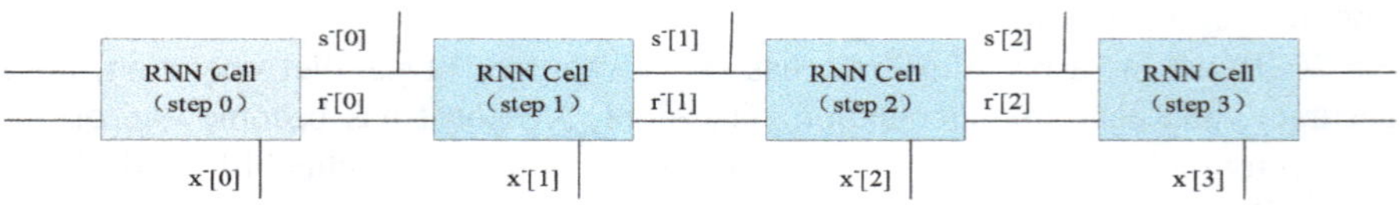

Figure 4.10 The structure of RNN unit

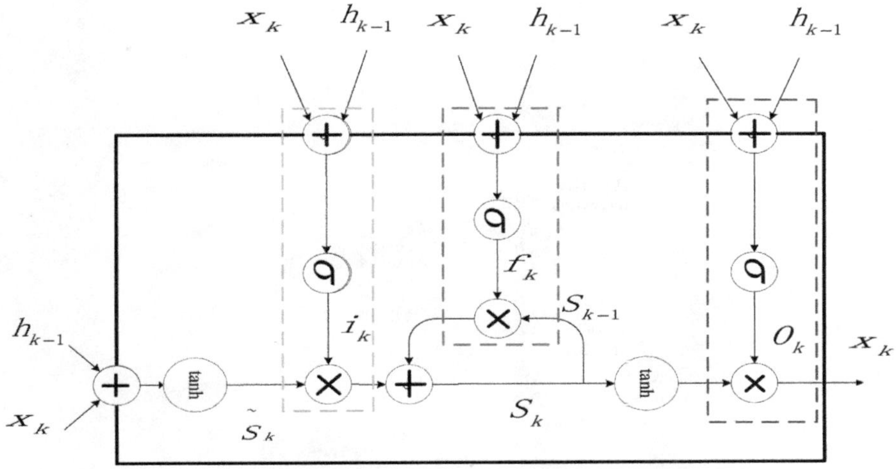

Figure 4.11 The structure of LSTM unit

disappearance and gradient explosion that are easy to occur in the training process of RNN. Figure 4.11 shows the internal structure of the LSTM unit. By introducing the design of gate structure, LSTM NN has the ability to recall past information, which is suitable for solving problems with long time series, such as battery SOC estimation.

These artificial intelligence algorithms cannot only be applied to the charge state, energy state, power state, health state, life prediction, charge and discharge control, and equalization control of lithium-ion battery but also be applied to the parameter identification of equivalent model of lithium-ion battery.

4.2.2 Development prospect

The artificial intelligence method takes data as the starting point and studies online and offline data by using GUI objects, using statistical theory and data mining, pattern recognition technology as theoretical guidance and application of NN, logistic regression, genetic algorithms, Bayesian methods, SVM, and other machine learning methods to analyze data contains information, summary data to study the deep relationship, Finally, it realizes various expected functions such as forecast, evaluation, scheduling, monitoring, diagnosis, decision-making, and optimization based on data. The schematic diagram of artificial intelligence is shown in Figure 4.12.

With the development of mobile Internet, cloud computing, big data, the Internet of Things, and social technology, everything can be digitized. The world is gradually entering the stage of a data society, and the core significance of data is increasingly prominent. The generation of massive data makes people gradually realize that there is a lot of effective information in data. With the help of artificial intelligence data analysis technology, data resources will be utilized to the maximum to help people make inferences and decisions in various fields of work and life. The ultra-high

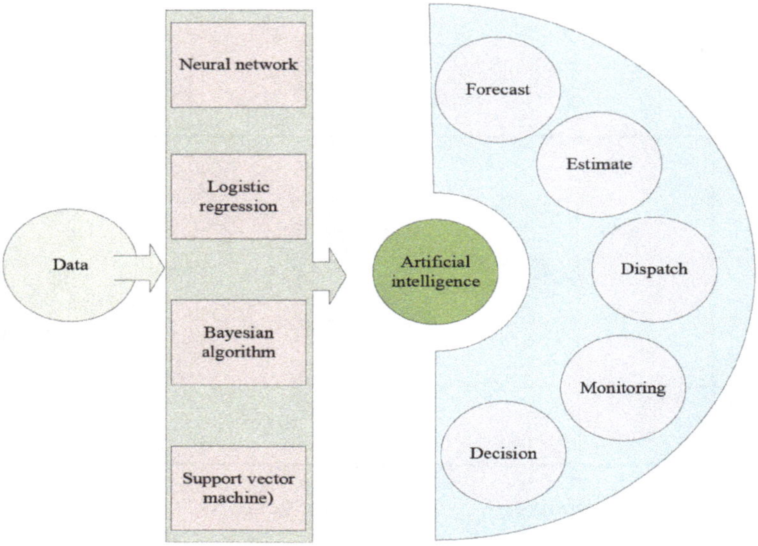

Figure 4.12 Diagram of artificial intelligence

performance of DL makes it the most popular research direction nowadays, and data are the cornerstone of DL. It can even be said that in the era of DL, having data take over the high ground of artificial intelligence. The artificial intelligence method has been applied and developed in many fields and is the hotspot of many research directions including computer science.

Traditional component or system modeling usually starts from mechanism research and builds a corresponding physical model, circuit model, chemical model, etc., based on the theoretical analysis of the internal characteristics of the studied object. This kind of mechanism model can essentially reflect the natural properties of the object being studied, but it is often difficult to apply or popularize in engineering practice. On the one hand, the mechanism model requires a deep understanding of the research object, which requires a large amount of preliminary research investment. The existing research results cannot meet the requirements of establishing a perfect mechanism model. On the other hand, the complexity of the machine model is high, the efficiency of operation is low, and the model parameters may be difficult to obtain, because in practice, the lithium battery system is nonlinear and extremely complex. The schematic diagram of the lithium-ion battery model is shown in Figure 4.13.

The artificial intelligence data-driven method starts from the data itself, studies and analyzes the historical data of research objects with the help of data-driven theory and technology, converts the original data into relevant information and behavior model, realizes the modeling of components, systems, or processes, and is used in state estimation and prediction. Data-driven method's advantage is that its internal mechanism does not require the depth of understanding of the object of study but,

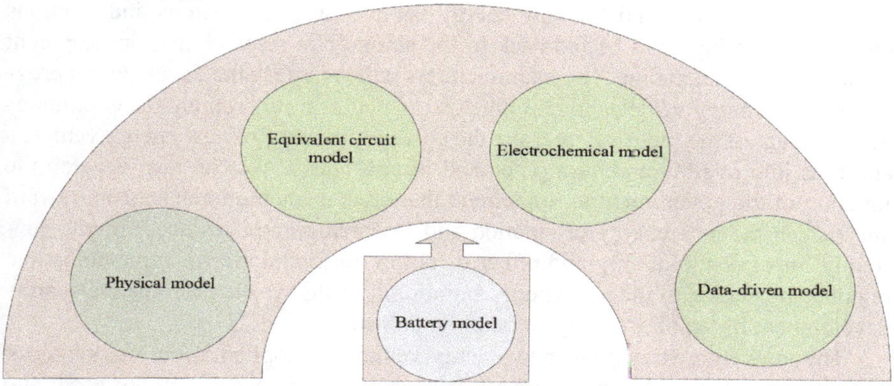

Figure 4.13 Lithium-ion battery model

under the guidance of the theory, such as statistics, data mining using NNs, SVMs, and other machine learning algorithms, through the study of data and information mining, summarizes the deep relationship between data variables, driver input, and output variables of the implicit relationship between the model established. Data-driven research can provide details that the prediction may ignore, assumptions on which the model depends, and factors known or unknown from various aspects. Therefore, the artificial intelligence data-driven method is to make full use of the existing data information, to achieve the maximum application of data resources.

With the rapid development of the mobile Internet, the amount of data shows exponential growth. However, due to the limited computing capacity of artificial intelligence technology, large-scale parallel computing and processing cannot be completed. However, the outstanding computing and processing capacity of big data platforms can solve this problem and provide basic support for artificial intelligence.

The combined application of big data and artificial intelligence technology has become a new wave of development. It has the advantages of intelligence, convenience and efficiency, and flexible and diverse technology types. It has been applied in various fields of life, achieved good results, and promoted the innovation and reform of various industries.

At present, the development of the new energy vehicle industry is jointly dominated by national policies and the industrial market. In this context, to develop the new energy automobile industry, on the one hand, it is necessary to fully implement the market-oriented model; on the other hand, it is necessary to strengthen the government function to regulate the market. We can use large data, artificial intelligence judgment and decision, machine learning, NN, and other technologies to create a human–computer interaction integration platform associated with the government, the public, new energy vehicle research and development manufacturers, and after-sales maintenance service departments, and develop a platform with intelligent interaction and real-time feedback functions.

The public's demand for new energy vehicles and suggestions and opinions on its shortcomings will be fed back to the automobile research and development manufacturers in real time. The manufacturers will announce the development progress of new energy vehicles in real time on the platform, to promote the continuous development of the industry, improve the cost performance of new energy vehicles, and thus improve the purchasing demand of consumers. We can also develop the function of the government to implement the effect evaluation mechanism, reward and punish the new energy production and research enterprises accordingly, constantly improve the quality and effect of policy formulation and implementation, regulate the market of the new energy automobile industry, and promote the healthy development of the new energy automobile industry.

The marketing strategy of new energy vehicles is applied in the marketing of new energy vehicles, big data and artificial intelligence technology are used, and the interaction between automobile manufacturers and users is formed. The system can be developed to judge the fit degree of marketing promotion objects and target consumer groups through multiple dimensions, accurately grasp the market situation changes, extract and summarize data information, and optimize marketing programs. On the other hand, we can also make use of the learning characteristics of artificial NNs to study and simulate the evaluation thought of management experts to carry out marketing forecasting.

New energy automotive technology research and development of application: on the research and development of new energy automotive technology, using the technology of data to data mining and knowledge discovery, and integrated intelligent expert system in artificial intelligence, developing integrated technology platforms as a new energy technology research and development with a carrier, improve the detection, the ability to solve the problem, optimization of continuous innovation, and promote the innovation in technology research and development; in the new energy vehicle design, the use of artificial intelligence machine vision and recognition, expert system, automatic programming, and other methods to assist the overall design of the car; in key parts technology, lithium battery technology, fuel cell technology, charging station design, and other aspects, the combination of big data artificial intelligence expert system, robotics, automatic planning, the use of artificial intelligence algorithm for

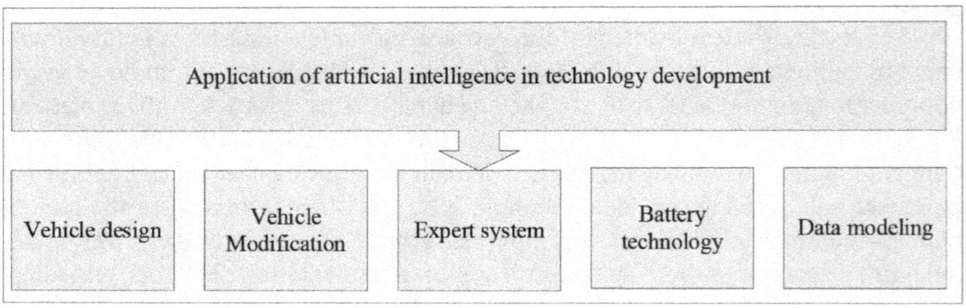

Figure 4.14 Application of artificial intelligence in new energy

combinatorial optimization, data modeling, technology research and development, problem-solving, as shown in Figure 4.14.

The application of new energy vehicle manufacturing in the manufacturing of new energy vehicles and the use of artificial intelligence technology can largely improve product quality and reduce production costs. For example, machine vision and other cognitive technologies are used to automatically program high-performance sensors, with sophisticated hardware design, to form a new industrial robot, and to design automatic machine processes. In the production and operation of the new energy automobile industry, it can also replace technicians to flexibly handle different tasks in a high-risk and uncontrolled environment. In terms of product inspection, the machine vision technology of artificial intelligence is used to carry out intelligent inspection, measurement, and scanning of products, and the result data are sent back to the airborne computer control console to evaluate the status of equipment automatically, to continuously improve product quality and benefit return.

The maintenance service during the use of new energy vehicles is also a part that cannot be ignored. Artificial intelligence mainly applies the prediction technology of artificial intelligence in the maintenance area, self-healing regulation, and intelligent damage identification for fault diagnosis, maintenance, and automobile equipment life prediction. The main applications are as follows: The intelligent diagnosis system is developed by using artificial intelligence technology, which can automatically analyze and judge the faults of the system, locate the cause of the faults accurately and quickly, and provide solutions. Artificial intelligence in vision and pattern recognition is used to produce a new type of robot and in the implementation of special complex environment repair tasks. The artificial intelligent control system is developed by using the expert control system, which is connected to the Internet network to realize intelligent remote monitoring and timely obtain the condition of equipment for maintenance.

The application of big data and artificial intelligence technology in different aspects of the new energy vehicle industry and the selection of different technical methods according to the actual situation will greatly improve efficiency, quality, and results, but it is inevitable to consider the shortcomings of the current big data and artificial intelligence technology. The problems of artificial intelligence application in new energy are shown in Figure 4.15.

Second, broadband network speed problem: the adoption of artificial intelligence technology needs to process massive data, artificial intelligence DL method

First, potential safety problems: when artificial intelligence technology is used to assist car driving and automatic driving, the control of intelligent devices needs to be further considered. There are potential problems of personal safety, regulations, supervision, and other aspects that need to be stipulated and implemented. In addition, due to resource sharing, the application of big data platforms, and the use of big data analysis technology, there are data security issues of data privacy leakage, and preventive measures need to be taken.

also needs to have a powerful computing platform to do support, and these need to develop high broadband network speed and the high processing capacity of the computing platform to support.

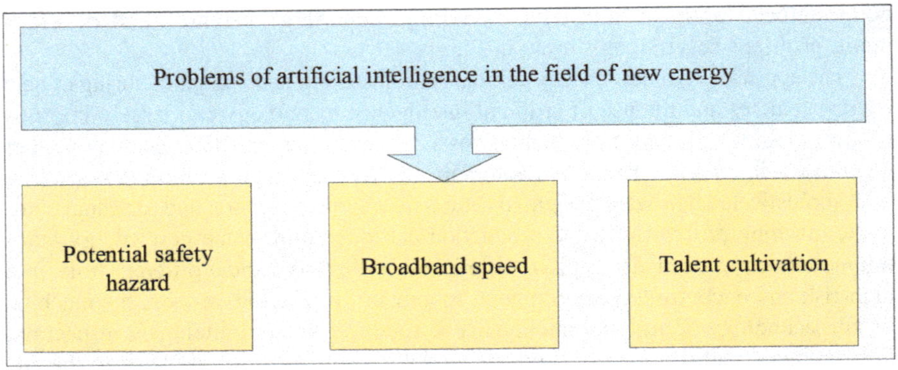

Figure 4.15 Problems existing in the application of artificial intelligence

Third, talent training: at present, the big data artificial intelligence technology is developing rapidly, but there is still a shortage of talents proficient in this field. The society and the government must promote the training of relevant talents to promote the development and maturity of technology.

Believe that with the constant progress of the society and the rapid development of science and technology, big data and artificial intelligence technology will become increasingly extensive application in the field of new energy vehicles, more fit, prompting mature industry rapid development; the new energy vehicles gradually replace traditional automobile become mainstream, to achieve the purpose of environmental protection, energy saving, and emission reduction to the sustainable development of human society.

4.3 Basic parameters in AI-based status monitoring

4.3.1 Voltage for input and correction

In the lithium-ion battery balance management unit, the accuracy of lithium-ion battery voltage sampling affects the decision-making and judgment of the control module [146]. Therefore, it is very important to design a fast and accurate battery voltage sampling circuit for the lithium-ion battery balance management unit, and the accuracy directly affects the control module's judgment on balance. The balance management unit takes the LF198 chip as the core to make the battery voltage acquisition circuit. Through the measurement results, the source of the detection error is analyzed, and the error compensation scheme is designed, which meets the precision requirement of single-battery voltage detection. The balance management unit takes the LF198 chip as the core to make the battery voltage acquisition circuit. Through the measurement results, the source of the detection error is analyzed, and the error compensation scheme is designed, which meets the precision requirement of single-battery voltage detection.

In the lithium-ion battery pack, because lithium-ion batteries are connected in series, the negative poles of the batteries are not in common, which also causes some difficulties in collecting the voltage of every single battery [147]. In the past research and design, there are many different methods for voltage detection of lithium-ion batteries, such as common-mode battery cell voltage detection method, direct voltage detection method, optocoupler isolation detection method, and integrated chip detection method. In the balanced topology of a bi-directional DC-AC converter, the planar transformer with the turn ratio of 1 is adopted, and the voltage of series lithium-ion batteries is grounded on the secondary side by the transformer, thus simplifying the detection of single-battery voltage of lithium-ion battery pack [147]. In the balanced topology of a bi-directional DC-AC converter, the planar transformer with the turn ratio of 1 is adopted, and the voltage of series lithium batteries is grounded on the secondary side by the transformer, thus simplifying the detection of single battery voltage of lithium-ion battery pack.

In this voltage unit design, based on LF198 sample-and-hold chip, a voltage detection circuit is designed. Figure 4.15 shows the detection circuit corresponding to a single battery. It is different from the previous voltage detection methods of the lithium-ion battery pack [148]. The detection circuit realizes the independent detection of lithium-ion battery voltage through the transformer, and the detection circuits are independent of each other and do not affect each other, thus ensuring the accuracy and reliability of the detection. The input of the sample-and-hold is connected to the secondary side of the planar transformer. After the battery voltage is sampled, it is sent to the MUX through a low-pass RC filter circuit, then enters the 12-bit AD circuit for digital processing, and then is transmitted to the control module and the host computer through the 1,553B data bus.

As can be seen from Figure 4.16, the structure of the battery voltage sampling circuit is relatively simple, and the main component is the external holding capacitor C_1. The selection of the holding capacitor has an important influence on the battery

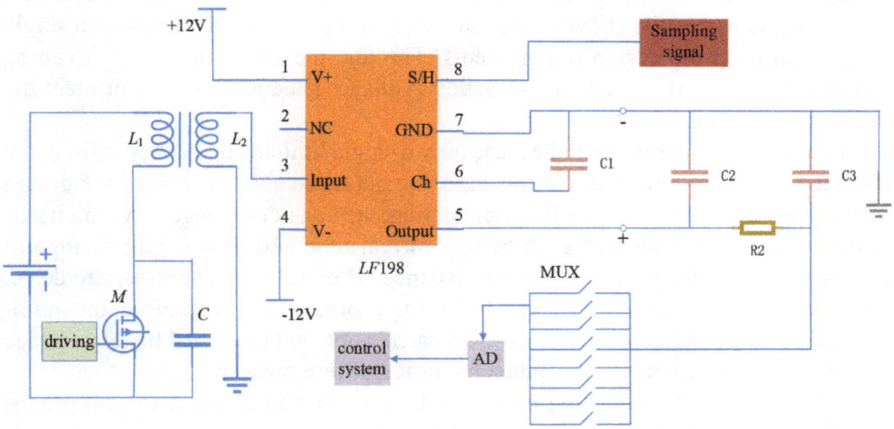

Figure 4.16 Second-order Thevenin model

detection accuracy. The value of C_1 is not fixed, and the size of the holding capacitor used will be different under different sampling frequencies. In the actual battery voltage detection process, the value of C_1 can be determined by experimental methods. In the process of voltage detection, after many tests, when C_1C_1 is 300 pF, the accuracy of voltage detection is the highest, so in this voltage detection circuit, the holding capacitor used is 300 pF.

The sampling signal also has an important influence on the accuracy of the sample-and-hold chip. In this design, due to the bi-directional DC–AC converter equalization topology, the voltage on the secondary side of the transformer is the battery voltage during the on-state of the switch, but the AC voltage will appear on the transformer during the off-state of the switch [149]. Therefore, the sampling signal should be designed to detect the battery voltage during the on-state of the switch, and the sampling signal should be synchronized with the equalization signal. When the equalization frequency changes, the sampling signal can also change accordingly. Thereby the sampling signal MON CLK is also generated by the phase-locked loop. As for the design of the sampling signal, it is required that not only the rising edge and falling edge of the sampling signal must be located in the conduction interval of the switch tube but also the high-level interval of the sampling signal should be located in the conduction interval of the switch tube. If the MON CLK period is longer than the BAL CLK period, during the sampling period, due to the tracking function of the sample-and-hold, AC signals will enter the sample-and-hold, but the actual measurement shows that the accuracy of battery voltage detection is low in this case.

First, measure the detection results of a single battery without accessing the Share bus and determine whether the designed voltage detection circuit can meet the design requirements. To ensure the accuracy of the test, an Agilent 34,401A digital multimeter is used for the test, and the test results are shown in Table 4.2. It can be found from Table 4.2 that the detection error of the detection circuit can be controlled within 5 m V within the measured battery voltage range of 3.0–5.1 V. In the actual experiment, the voltage value of the lithium-ion battery used is standard 4 V, which is within the above detection range, so the voltage detection of a single battery can meet the design requirements. Through the above detection, it can be concluded that the designed LF198 battery voltage detection circuit can meet the measurement requirements.

In the above measurement, because only a single lithium-ion battery is used, the balanced work is not realized. In this case, the detection accuracy meets the design requirements. However, in practical application, the balance management unit needs to collect the information of each battery in real time and monitor the change of battery voltage in the process of battery balance. Therefore, it is necessary to detect the battery voltage during the battery balancing process [149]. Therefore, the multichannel battery balancing experiment was carried out, and the actual battery voltage and voltage detected during the balancing process were measured.

First, two battery balancing experiments were carried out, and several groups of different initial voltage values were set. In many experiments, relevant data were recorded. Assuming that the actual battery voltage of the *n*th battery is u_n, and the

Table 4.2 Test results of cell voltage

Battery voltage (V)	Sampling output (V)	Error (mV)
3.019	3.023	4
3.298	3.302	4
3.547	3.550	3
3.846	3.849	3
4.025	4.027	2
4.301	4.302	1
4.494	4.494	0
4.833	4.831	2
5.031	5.028	3

sampled value of the voltage of the *n*th battery is u_n, the detection results in the process of equalization are shown in Table 4.3.

As shown in Table 4.3, after starting the equalization work, when there is energy flow in the Share bus, the detection error becomes unstable. At the same time, it can be found that the greater the difference between the actual battery voltages, the greater the error. The voltage sampling value of the discharged battery is lower than the actual battery voltage, and the voltage sampling value of the rechargeable battery is higher than the actual battery voltage.

4.3.2 Capacity for internal state parameters

The total amount of charge released by the battery under certain discharge conditions is called the battery capacity, expressed by the symbol C, usually in units of Ah or mAh. Battery capacity is one of the most important performance indicators of battery performance, which is related to battery life. Typically, lithium-ion batteries are discharged in the laboratory using a constant current or constant voltage method, i.e., a constant current and constant voltage charge are given to a battery and then discharged at a constant current; with the highest battery voltage put to the lower protection voltage of 3.0 V, the capacity released is the truest capacity of the battery [150]. The storage capacity of a lithium-ion battery can be calculated by measuring the energy released by the battery throughout the discharge process or the product of the discharge current and time.

Table 4.3 Two-way balanced voltage detection results

u_1 (V)	u_1 (V)	Error (mV)	u_2 (V)	u_2 (V)	Error (mV)
3.844	3.822	–22	3.696	3.714	18
3.850	3.832	–18	3.741	3.755	14
3.853	3.834	–19	3.738	3.753	15
3.863	3.853	–10	3.793	3.799	6
3.873	3.864	–9	3.810	3.804	6

Battery capacity is divided into theoretical capacity, rated capacity, actual capacity, and remaining capacity according to different conditions.

(1) Theoretical capacity

The theoretical capacity is the amount of power that can be supplied assuming that all the active materials on the electrodes are involved in the electrochemical reaction of the battery [150]. The theoretical capacity can be accurately calculated from the amount of electrode active material in the battery reaction equation and the electrochemical equivalent of the active material calculated according to Faraday's law. The calculation processing equation is shown in (4.8).

$$Q = \frac{zmF}{M} \tag{4.8}$$

Faraday law states that when current passes through the electrolyte solution, the amount of chemically reacted substance on the electrode is proportional to the amount of electricity passed. In (4.8), Q is the amount of electricity passed in the electrode reaction (Ah); z is the electronic measurement coefficient in the electrode reaction equation; m is the mass of the reactive material (g); M is the molar mass of the active substance (g/mol); and F is the Faraday constant of about 965,000 C/mol or 26.8 Ah/mol. It can also be understood that the active material with mass m can release electricity Q after a complete reaction. The electric quantity Q is the theoretical capacity (C_0) of the electrode active material, so it can also be written in the form of (4.9):

$$C_0 = 26.8z\frac{m}{M} = \frac{1}{K}m \tag{4.9}$$

where K is the electrochemical equivalent of the active material and $K = M/26.8z$ (g/Ah) is the mass of active material required to obtain 1 Ah of electricity. Equation (4.9) is the calculation equation for the theoretical capacity of the electrode active material.

(2) The rated capacity

The rated capacity (C_g) of the battery is the capacity of the battery under certain discharge conditions (temperature, discharge rate, termination voltage, etc.), by the standards set by the state or relevant departments, and the minimum capacity of the discharge to the termination voltage is the capacity of the battery as specified in the design and production.

(3) The actual capacity

The actual capacity (C) refers to the total amount of electricity released by the battery under actual application working conditions. It is equal to the integral of the discharge current and the discharge time, which is in the lower right corner of the capacity C and is expressed by Arabic numerals, such as $C_5 = 50$ Ah, which means that the discharge rate is completed in 5 hours. And the obtained capacity is 50 Ah. The actual capacity calculation method is described as follows.

The calculation process is shown in (4.10) when discharging with a constant current:

$$C = It \tag{4.10}$$

The calculation process is shown in (4.11) when discharging with variable current:

$$C = \int_0^T I(t)\, dt \tag{4.11}$$

where I is the discharge current as a function of the discharge time t. T is the total discharge time. The battery's actual capacity depends on the amount of active material in the battery and the utilization rate of the active material. The more active material there is, the higher the utilization rate of the active material, the greater the capacity of the battery, and vice versa. Due to the existence of internal resistance and other various reasons, the active material cannot be fully utilized, that is, the utilization rate of the active material is always less than 1. Therefore, the actual capacity and the rated capacity of the chemical power supply are always lower than the theoretical capacity. The actual capacity of the battery is closely related to the discharge current. During high-current discharge, the polarization of the electrode and internal resistance increases, but the discharge voltage drops quickly, with the battery energy efficiency, so the actual discharge capacity is always lower. Correspondingly, the discharge voltage drops slowly under low current discharge conditions, and the battery's actual capacity is often higher than the rated capacity.

(4) The remaining capacity

The remaining capacity refers to the remaining usable capacity of the battery after being discharged at a certain discharge rate. The estimation and calculation of the remaining capacity are affected by factors such as the battery's discharge rate and discharge time, the degree of battery aging, and the application environment. Therefore, there are certain difficulties in the accurate estimation.

Many factors affect the capacity of a battery; the following are common.

(1) Discharge multipliers

With the increase of the discharge rate, the lithium-ion battery will continue to show the discharge capacity decay phenomenon; the reason is due to the serious polarization, the discharge voltage is reduced to the discharge cut-off voltage in advance, that is, the discharge time is shortened, the discharge is not sufficient, and the negative electrode $Li+$ is not completely de-embedded. When the discharge rate of the battery is between 1.5 and 3.0, the discharge capacity starts to show signs of a rebound in different degrees. As the reaction continues, the temperature of the battery itself rises significantly with the increase of the discharge multiplier, which strengthens the thermal movement of $Li+$ and accelerates the diffusion rate, resulting in faster $Li+$ disembedding and a rebound of the discharge capacity. It follows that the combined effect of the large discharge multiplier and the temperature rise of the cell itself leads to the non-monotonic phenomenon of the battery.

(2) Temperature

Temperature mainly affects the activity of the internal cell substance and the performance of the electrolyte. Both too high and too low temperatures have a large impact on the capacity of the battery. Lithium-ion battery activity decreases significantly at low temperatures, the ability to embed and release lithium decreases, the internal resistance of the battery, the polarization voltage increases, the available capacity decreases, the battery discharge capacity decreases, the discharge plateau is low, the battery is more likely to reach the discharge cut-off voltage, the available capacity of the battery decreases, and the energy utilization efficiency of the battery decreases.

Similarly, when lithium-ion batteries are exposed to elevated temperatures, the lithium ions between the positive and negative electrodes become active, resulting in a reduction in the internal resistance of the battery and a longer stabilization time for the internal resistance, increasing the amount of electron narrowing in the external circuit and more efficient use of capacity. However, when the battery is exposed to high temperatures for a long time, the stability of the cathode lattice structure deteriorates, the safety of the battery is reduced, and the life of the battery is significantly shortened.

(3) Termination voltage

When the battery is discharged to a certain voltage value, the voltage drops sharply and the energy gained is very small, which can cause considerable damage to the battery if discharged deeply for a long time. Therefore, the discharge must be terminated at a certain voltage value, and the cut-off voltage is called the discharge termination voltage. The cut-off voltage is called the end-of-discharge voltage. Setting the end-of-discharge voltage is of great significance for extending the life of the battery.

(4) The geometry of the electrode plate

When the amount of active substance is certain, the geometric area of the electrode plate in direct contact with the electrolyte increases, and the battery capacity increases, so the geometric size of the electrode plate has a significant impact on the battery capacity.

First, when the amount of active substance is determined, the capacity of the battery decreases with the increase of the thickness of the electrode plate, the thicker the plate, the smaller the contact area between sulfuric acid and the active substance, the lower the utilization rate of the active substance, and the smaller the capacity of the battery.

Second, there is a large difference in the utilization rate of active substance in the battery between the upper and lower parts of the plate. Experiments confirmed that the current density of the upper part of the plate is about 2–2.5 times higher than the current density of the lower part at the beginning of the discharge, and this difference gradually decreases as the discharge time goes on, but the current density of the upper part is higher than that of the lower part.

Finally, the larger the geometric area of the plate, the higher the utilization rate of the active substance and the higher the capacity of the battery. In the case of the same battery shell and the same quality of active substance, the use of thin plates increases the number of plates, which increases the effective reaction area of the plates, thus increasing the utilization rate of the active substances and increasing the capacity of the battery.

4.3.3 Internal resistance for state parameters

Inside a lithium-ion battery, lithium-ions move from one electrode to another through the electrolyte, and the factors that hinder the movement of ions in the process constitute the internal resistance [151]. Due to the battery's internal resistance, the terminal voltage of the battery is lower than the electromotive force and open-circuit voltage during discharge, and the terminal voltage of the battery is higher than the electromotive force and open-circuit voltage during charging. The essence of electric current is the directional movement of electric charges, and the motion of the electrons is opposed by the material itself and the magnetic field. Internal resistance is one of the important parameters of lithium-ion batteries that are not always static. The magnitude of the internal resistance directly affects the battery operating voltage, operating current, and battery capacity. In the process of battery charge–discharge, changes in temperature, charge–discharge rate, charge–discharge time, electrolyte concentration, and active material quality cause changes in the batteries.

The internal resistance of the battery includes the ohmic resistance R_0 and the polarization of internal resistance R_p. The ohmic resistance comes from the resistance of the electrolyte, diaphragm, and the contact resistance of various parts. During the internal oxidation–reduction reaction of the battery, an electric field is generated [152]. Under the action of the electric field, the dielectric generates polarization charges due to the polarization effect. The polarization resistance is the resistance of the polarization charge to the current. During the charge–discharge process of lithium-ion batteries, the internal resistance generated by the polarization reaction is mainly electrochemical and concentration polarization. The nature of the active material, the structure of the electrode, the manufacturing process of the battery, and the operating conditions of the battery lead to the difference in the polarization internal resistance. The main influencing factor of the polarization internal resistance is the operating condition of the batteries.

The battery is polarized during the charging and discharging process, and the polarization of lithium-ion batteries can usually be divided into three categories: ohmic polarization, electrochemical polarization, and differential polarization. Each of the several types of polarization has a different response rate. There are many factors affecting the degree of polarization, but in general, the higher the charge and discharge current density, the greater the polarization.

(1) Ohmic polarization

The polarization caused by the ohmic internal resistance of a lithium-ion battery is called ohmic polarization, and ohmic polarization occurs instantaneously. The ohmic internal resistance (R) of a battery consists of the electrode material, the electrolyte, the diaphragm resistance, and the contact resistance of the parts.

(2) Electrochemical polarization

Electrochemical polarization is generally considered to be on the microsecond scale and is caused by the polarization of the electrochemical reaction on the positive and negative electrodes at a rate less than the speed of electron movement.

(3) Differential concentration polarization

Concentration polarization is generally considered to be secular and is caused by the polarization of the lithium ions involved in the reaction diffusing in the solid phase at a rate less than the electrochemical reaction rate.

Ohmic resistance is divided into three main components, one is ionic impedance, the second is electronic impedance, and the third is contact impedance. The smaller the internal resistance of the lithium battery, the more specific measures need to be taken to reduce ohmic resistance for these three components.

(1) Ionic impedance

The ionic impedance of lithium batteries is the resistance to the transfer of lithium ions within the battery. In lithium batteries, the rate of lithium-ion migration plays an equally important role as the rate of electron conduction. Ion impedance is mainly influenced by the positive and negative electrode materials, the diaphragm, and the electrolyte [153]. To reduce the ionic impedance, first, we need to ensure that the anode and cathode materials and the electrolyte have good wettability, and second, the electrolyte should have the right concentration, viscosity, and conductivity, to ensure that the electrolyte fully wets the diaphragm, and at the same time, there should not be a residual amount of electrolyte left in it, which reduces the efficiency of the electrolyte.

(2) Electronic impedance

Electronic impedance is influenced by many factors and can be improved by materials and processes, such as positive and negative electrode plates, diaphragms, and collector substrates.

(3) Contact impedance

The contact resistance is formed between the contact of the copper and aluminum foil with the living substance and requires a focus on the bonding of the positive and negative pastes.

The total internal resistance R of the battery is equal to the ohmic resistance R_0 plus the polarization internal resistance R_p. The calculation equation is shown in (4.12):

$$R = R_o + Rp \tag{4.12}$$

In (4.12), different types of lithium-ion batteries show different internal resistance characteristics [154]. Even the same type of lithium-ion battery, due to the difference in internal material composition, operating environment, and aging degree, also shows different internal resistances. The internal resistance is an important indicator to measure the battery performance. Therefore, the research on the change characteristics of the internal resistance of the battery is of great significance for equivalent modeling and SOC estimation. During the battery cycle discharge experiment, the partial voltage change curve is shown in Figure 4.17.

In Figure 4.17, when the battery is connected to or disconnected from the external circuit, the terminal voltage of the battery drops or rises rapidly for dU_R and dU_{R1} in the figure. It is due to the ohmic resistance of the lithium-ion battery itself, causing the voltage to change drastically at the beginning or end of discharge, which is called the internal resistance effect. In the first stage, the terminal voltage drops rapidly and then slowly decreases, or when the discharge ends, the voltage slowly rises (dU_1 and dU_1' in the figure), which is caused by the polarization effect of the lithium-ion batteries. When the battery is fully allowed to rest, the internal balance of the battery is reached, and the polarization effect disappears. The polarization effect produces polarized internal resistance, internal structure, and working conditions

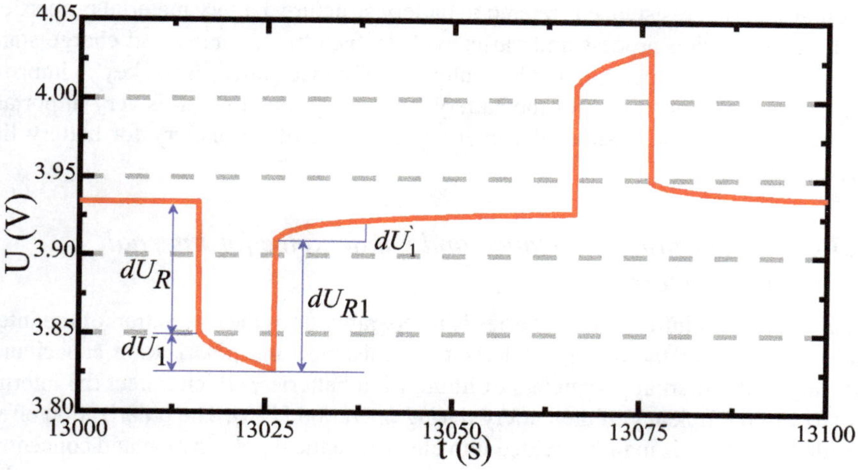

Figure 4.17 *The voltage changes curve of the cyclic discharge-shelving experiment*

that affect the polarization internal resistance of the batteries. However, the discharge current and the operating temperature have a particularly obvious effect on the polarization internal resistance of the batteries. It is due to its effect on the movement of lithium ions inside the battery during operation.

Internal resistance is most sensitive to temperature, the lower the temperature, the slower the ion transport within the battery, and the greater the internal resistance of the battery. Battery impedance can be divided into bulk phase, SEI membrane, and charge transfer impedance. Internal resistance, bulk phase impedance, and SEI membrane impedance are mainly influenced by the electrolyte ionic conductivity, and the trend of change at low temperature is the same as the trend of change in electrolyte conductivity. Compared to the increase in bulk phase impedance and SEI membrane impedance at low temperatures, the charge response impedance increases more significantly with decreasing temperature, and below –20°C, the charge response impedance accounts for almost 100% of the total internal resistance of the battery.

When the battery is in different SOC, its internal resistance size is different, especially the DC internal resistance directly affects the power performance of the battery, which in turn reflects the battery performance in the actual state: the DC internal resistance of the Lithium-ion battery increases with the increase of the battery discharge depth DOD, the internal resistance size is unchanged in the discharge interval of 10–80%, and generally the internal resistance increases significantly in the deeper discharge depth.

Internal resistance is an important parameter for measuring lithium-ion power performance and assessing battery life. The higher the internal resistance, the poorer the multiplier performance of the battery and the faster it increases in storage and cycling. Internal resistance is related to battery structure, battery material properties, and manufacturing process and varies with ambient temperature and charge state. Therefore, the development of a low internal resistance battery is the key to improving the power performance of the battery, and at the same time, it is very important to grasp the variation law of the internal resistance of the battery for battery life prediction.

4.3.4 Polarization resistance and capacitance for internal parameters

The research on lithium-ion batteries is inseparable from the estimation of the internal parameters of the battery. Polarization resistance and polarization capacitance are important internal parameters of lithium-ion batteries, which reflect the internal polarization efficiency of the battery during operation [155]. The polarization effect inside the battery is mainly divided into electrochemical polarization and concentration polarization.

During the electrolysis process, the concentration of an ion near the electrode changes due to the electrode reaction, and the speed of ion diffusion in the bulk solution cannot keep up with this change, resulting in a concentration gradient between the concentration of the solution near the electrode and the bulk solution. The resulting change in electrode potential is called concentration polarization.

The electrode reaction is always carried out in several steps. If the reaction rate of one step is slow, higher activation energy is required. To make the electrode reaction proceed smoothly, the additional voltage applied is called electrochemical overpotential, which is also known as activation overpotential [156]. This polarization phenomenon is called electrochemical polarization.

In recent years, with the continuous development of electric vehicles, the requirements for the high-rate charge–discharge performance of power batteries are getting higher and higher. In practical applications, due to the internal polarization, the higher the discharge rate of the power battery, the lower the electric voltage, and the lower the energy that can be released. According to the formation mechanism of polarization, polarization can be decomposed into electrochemical activation polarization, concentration polarization, and ohmic polarization. Therefore, how to accurately and quantitatively analyze various types of polarization inside the battery has received extensive attention from electrode materials and battery optimization designers.

Different types of polarization can be measured by experimental methods, such as electrode concentration polarization that can be measured by galvanostatic intermittent titration-voltage relaxation technique, static method, and so on, but the measurement process generally requires tedious experiments. With the development of computer technology and the maturity of the multi-physics theory inside the battery during the discharge process, the rapid development of battery numerical simulation can effectively solve this problem. Some researchers established a microstructure model of $LiCoO_2$ electrode to investigate the effect of microstructure on electrode polarization during charging and discharging, indicating that concentration polarization mainly depends on the spatial position, twist rate, and porosity; activation polarization is affected by porosity, state of charge, and spatial position which have a greater influence [157]. Besides, the difference of activation overpotential at different spatial positions at the end of discharge increases, and the difference at the end of charge decreases.

Although the microstructure model can better study the distribution characteristics of the internal polarization of the battery in the spatial position, the calculation efficiency of the model is low, and the performance requirements of the computer are high, which is not conducive to its popularization and application. Some researchers discussed $LiMnO_2$ based on a quasi-two-dimensional electrochemical model. Some researchers discussed the influence of electrode thickness of graphite battery and particle size of active material on concentration polarization. And some researchers investigated the effects of discharge rate, electrolyte diffusion coefficient, and electrode thickness on concentration polarization in all-solid-state lithium-ion batteries.

In addition to characterizing the polarization effect of the battery by electrochemical means, constructing an equivalent circuit model is also an important method to characterize the internal polarization effect of the battery. In the equivalent circuit model, the Thevenin model and the Partnership for New Generation of Vehicles (PNGV) model both use a resistance–capacitance (*RC*) loop to characterize the polarization effect of the battery, the Thevenin model is shown in Figure 4.18.

In Figure 4.18, U_{OC} is the open-circuit voltage, R_0 is the ohmic internal resistance, R_P is the polarization resistance, and C_P is the polarization capacitance. The parallel circuit of R_P and C_P describes the polarization process. U_L is the closed-circuit voltage after the battery is connected to the external circuit. This model takes into account the polarization process of the battery, and the first-order model constructed is relatively simple to calculate. In the process of obtaining the state-space equation, first, according to the basic operating characteristics of the capacitor components, the relationship between the current flowing through the battery polarization capacitor C_P and its two closed-circuit voltages is shown in (4.13):

$$I_{Cp}(t) = C_P \frac{dU_{Cp}}{dt} \tag{4.13}$$

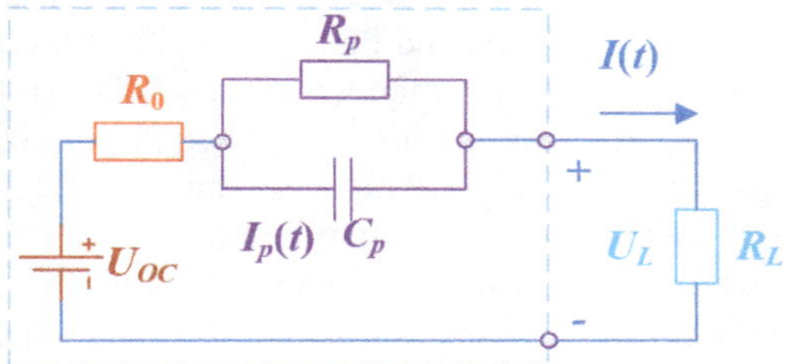

Figure 4.18 Thevenin equivalent model structure diagram

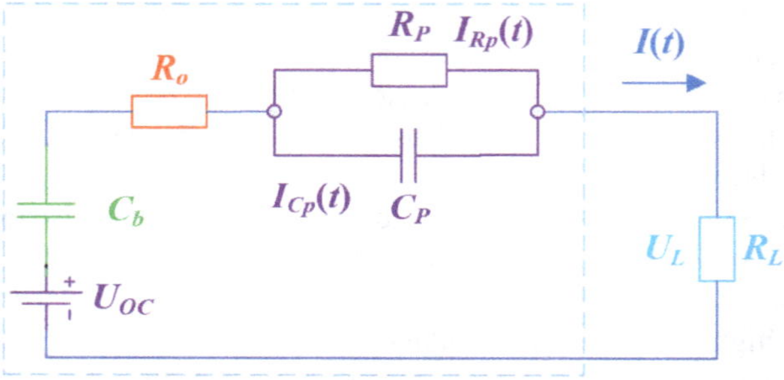

Figure 4.19 Lithium-ion battery PNGV model

Based on the analysis of the equivalent circuit composition, through the application of the knowledge of circuit science, according to Kirchhoff voltage laws, the voltage relationship in the equivalent circuit of Figure 4.19 can be expressed in (4.14):

$$U_L = U_{OC} - R_0 I_{Cp}(t) - R_0 \frac{U_{Cp}(t)}{R_P} - U_{Cp}(t) \tag{4.14}$$

The parameter identification of this model combined with the hybrid pulse power characterization (HPPC) test is the standard battery model in the PNGV Test Manual. Increase the capacitance C_b to describe the open-circuit voltage change; the model structure is shown in Figure 4.19.

In Figure 4.19, U_{OC} is the ideal voltage source, R_0 is the ohmic internal resistance, R_P is the polarization resistance, and C_P is the polarization capacitance. The parallel circuit of R_P and C_P is used to reflect the production of the battery polarization process, thereby eliminating the influence of the difference in the production process and the working environment. There are differences between cells such as voltage, capacity, and temperature, which will become serious as the battery ages. This model does not consider the self-discharge effect of the battery and the influence of temperature on the battery. By analyzing the illustrated equivalent circuit model, the equivalent circuit model can be obtained as shown in (4.15):

$$U_L = U_{OC} - C_b \left(\int i(t)dt - R_0 I_L - R_p I_p \right. \tag{4.15}$$

To better characterize different polarization effects, subsequent researchers made improvements based on the Thevenin and PNGV model and proposed an improved dual-polarization (DP) model. The DP model uses two *RC* loops to characterize the electrochemical polarization effect and the concentration polarization effect of the lithium-ion battery, respectively. The DP model is shown in Figure 4.20.

In the research of the polarization resistance and capacity, the internal parameters estimation of the lithium-ion batteries by some tests are used by many researchers. To realize the polarization resistance and capacity estimation of the batteries, the HPPC test and pulse discharging test are often used as the internal parameter identification method. The commonly used HPPC experiment is shown in Figure 4.21.

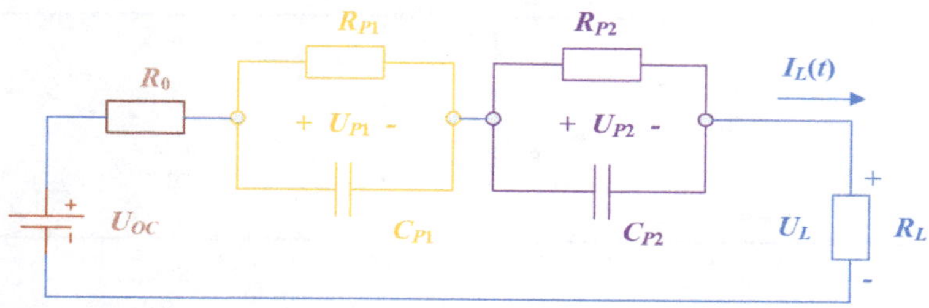

Figure 4.20 Lithium-ion battery DP model

The voltage drop and rise are caused by the discharging and charging current. The current corresponding to U_1–U_4 in the figure is the current from t_1 to t_1. In the changing stage of the current, the battery is discharged at a current of I_m for 10 seconds, and the corresponding voltage changes are U_1–U_4. When the battery is discharged, the battery voltage instantly drops from U_1 to U_2, and after the battery discharge ends, the voltage instantly rises from U_3 to U_4. The appearance of these two stages is mainly due to the presence of ohmic internal resistance in the battery.

In lithium-ion battery PNGV model in Figure 4.19, the U_2–U_3 phases are mainly caused by the polarization reaction of the lithium-ion batteries. Therefore, according to this voltage and the corresponding current change, the polarization resistance and polarization capacitance inside the battery can be studied. Taking the Thevenin model as an example, the zero-state response expression is shown in (4.16):

$$U_L = U_{OC} - U_0 - U_p \tag{4.16}$$

wherein U_{OC} is the open-circuit voltage of the lithium-ion batteries. U_0 is the voltage that is used to represent the Ohmic effect. U_P represents the polarization effect. According to the zero-state response, the polarization effect of the Thevenin model can be shown as (4.17):

$$U_p = IR_p(1 - e^{-\frac{t}{\tau}}) \tag{4.17}$$

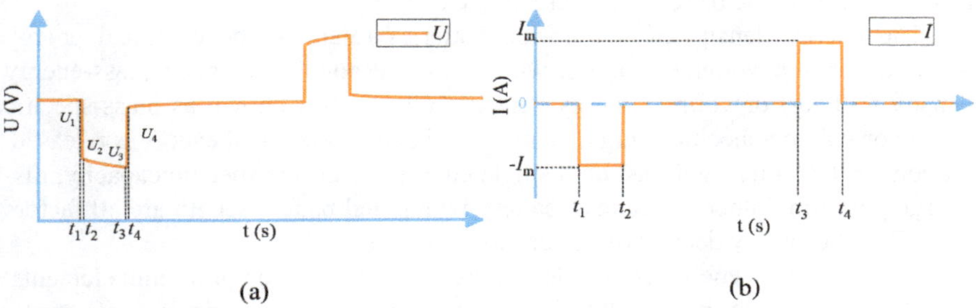

Figure 4.21 The voltage and current of the HPPC test: (a) HPPC test voltage and (b) HPPC test current

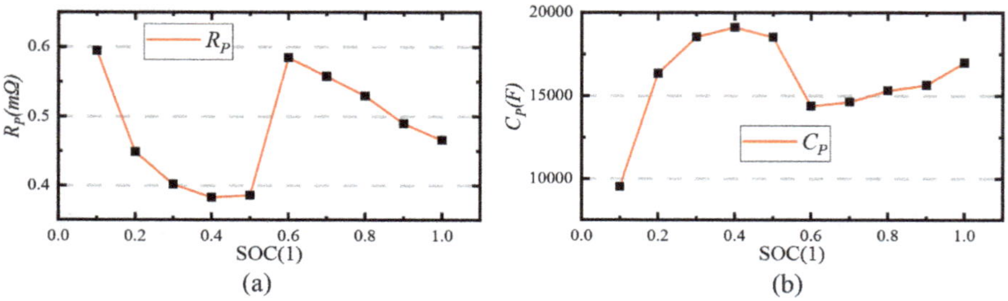

Figure 4.22 Polarization resistance and capacity identification results: (a) polarized resistance R_P *and (b) polarization capacitance* C_P

wherein τ is the time constant of the *RC* loop in the model, t is the sampling time, and I is the load current. According to the voltage curve U_2–U_3 and the corresponding current, the parameters in (4.17) can be obtained by curve fitting to realize the identification of model parameters. The fitting curve is obtained by the least square polynomial fitting method. If the identified parameters are obtained and drawn in the coordinate system, SOC is used as the independent variable, and the parameter is used as the dependent variable, the scatter points of the relationship between R_P, C_P, and SOC are obtained. The polarization resistance and polarization capacity are shown in Figure 4.22.

Figure 4.22a and b shows the changing R_P and C_P with SOC. It can be observed from the figure that the internal polarization parameters fluctuate within a certain range with SOC changes.

4.3.5 Energy density for correction

Lithium-ion batteries are widely used in production and life due to their advantages of high specific energy, long life, high voltage, and low self-discharge rate, especially in the field of vehicle engineering, with lithium-ion power battery packs as the main energy storage system. New energy vehicles are gradually replacing traditional fuel-powered vehicles [158]. For lithium-ion batteries, their energy density is an important parameter to evaluate the performance of a battery, and energy density can be used to reflect the battery's power storage capacity.

The energy density of lithium-ion batteries refers to the electrical energy released per unit volume or mass, which is correspondingly called mass-energy density (Wh/kg) or volume energy density (Wh/L), also known as mass-specific energy or volume-specific energy. Generally, the more electrical energy is released, the longer the battery will last on a single charge. Electrode specific capacity, discharge platform, battery thickness, battery width, and battery length are all factors that affect the energy density of lithium-ion batteries.

To improve the energy density is to increase the proportion of lithium elements, and at the same time, let as many lithium ions run out from the positive electrode, move to the negative electrode, and then return to the positive electrode from the original number of the negative electrode, which is shown in Figure 4.23.

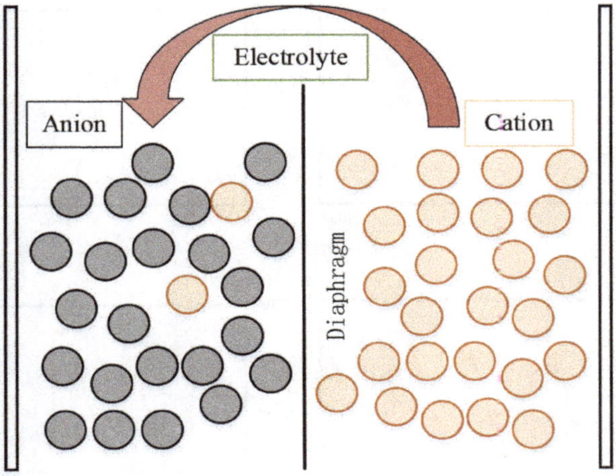

Figure 4.23 *Lithium-ion battery working theory*

At present, the methods to improve the energy density of lithium-ion batteries mainly include increasing the proportion of positive active materials, increasing the proportion of negative active materials, increasing the proportion of negative active materials, increasing the proportion of negative active materials, and decreasing battery volume.

The data show that battery energy density is the product of battery capacity and discharge platform, battery thickness, battery width, and battery length. In other words, electrode-specific capacity, discharge platform, battery thickness, battery width, and battery length are all factors that affect the energy density of lithium batteries. The calculation equation of the energy density is shown in (4.18):

$$E = C \times U \times H/L/W \tag{4.18}$$

wherein E represents the energy density, U is the voltage of lithium-ion batteries, and H, L, and W are the battery thickness, battery width, and battery length, respectively. Under the condition that other conditions remain unchanged, as long as one of the battery capacity, discharge platform, battery thickness, battery width, and battery length changes, the energy density of lithium batteries will also change [159]. After the lithium-ion battery is produced, the volume, length, width, and height of the battery do not change. Therefore, the main factor affecting the energy density of a lithium-ion battery during use is the change in battery capacity. In the study of energy density correction of lithium-ion batteries, the study of battery capacity plays a crucial role.

Battery capacity refers to the amount of charge that the battery can store and is one of the important performance indicators to measure battery performance. The battery capacity test methods mainly include constant current discharge test, constant resistance discharge test, constant voltage discharge test, continuous discharge test, and intermittent discharge test methods. The capacity of the battery can

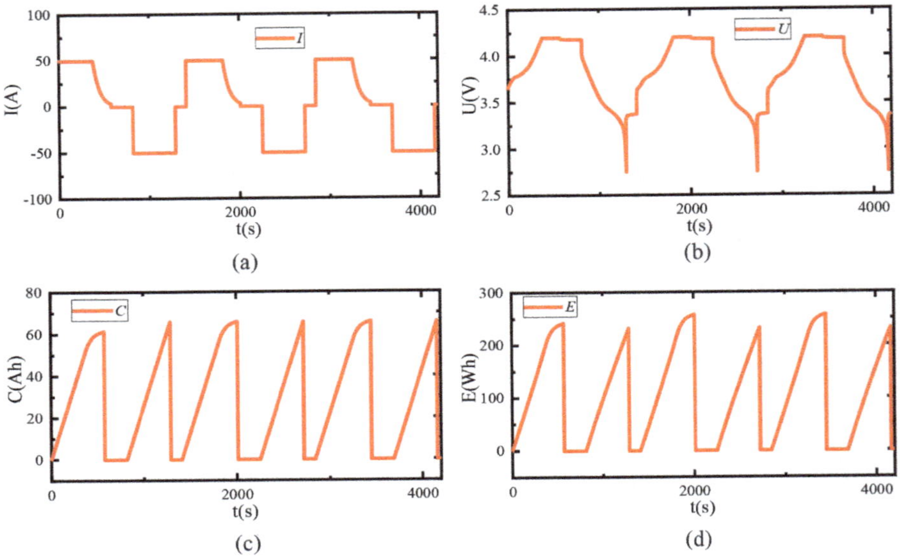

Figure 4.24 Lithium-ion battery capacity test: (a) current, (b) voltage, (c) capacity, and (d) energy

be calculated according to the discharge time and the current. The purpose of the capacity test experiment is to test to obtain the maximum usable capacity of the lithium-ion battery in its current state. At present, lithium-ion battery capacity test experiments are mainly based on constant current discharge tests. First, the lithium-ion battery is charged with a constant current and constant voltage until the battery reaches a fully charged state, and then the battery is discharged with a constant current until the voltage of the battery drops to the cut-off voltage, so that the discharge capacity of the battery can be measured. The discharge capacity test experiment of the ternary lithium-ion battery is shown in Figure 4.24.

According to the capacity test of the lithium-ion battery, the capacity and the energy of the battery can be obtained under certain conditions. However, the capacity of the lithium-ion battery varies with the ambient temperature, aging, and discharge current of the battery. Therefore, it is necessary to consider these factors that affect the capacity of lithium-ion batteries for the energy density correction research of lithium-ion batteries.

Temperature is one of the most important parameters affecting battery performance. The low temperature will reduce the ion diffusion rate and chemical reaction rate inside the battery, which will lead to an increase in the internal resistance of the battery and a decrease in the discharge capacity and the output voltage. Low temperature will also lead to lithium precipitation in the negative electrode, which seriously affects the life and safety of lithium batteries. With the increase in temperature, although the ion diffusion rate and electrochemical reaction rate in the battery are accelerated to improve the performance of the battery, it will also lead to the acceleration of the side reaction rate and accelerate the aging of the battery. More seriously, when the battery temperature exceeds a certain threshold, a series of chain reactions such as the decomposition of the SEI film, the melting of the diaphragm, and the reaction between the electrode and the electrolyte will occur successively, resulting in a large amount of heat, thermal runaway, thermal diffusion, and even explosion, pressure, and other disasters. Therefore, at the battery management level, the thermal management system that controls the battery temperature is an important guarantee to avoid battery charge and discharge performance degradation, prolong battery life (relieve the reduction of vehicle maximum cruising range and available power after battery aging), and improve battery thermal safety. So, the research on energy density correction for the battery should consider the effect of temperature.

4.4 Conclusion

In the battery system, the application of artificial intelligence will have a profound impact on it. With the use of lithium-ion batteries, more and more data are accumulated, so the use of artificial intelligence methods can achieve accurate battery parameter prediction for battery systems. This chapter mainly introduces artificial intelligence that is an application in utility-scale battery systems. First, this chapter analyzes the application of commonly used artificial intelligence methods in battery systems from the selection criteria for selecting artificial intelligence technologies, which include the common artificial intelligence methods for utility-scale battery systems and the artificial intelligence technology evaluation index. Besides, this chapter describes the monitoring of utility-scale battery development with artificial intelligence, which includes the development status and prospects. Last, the basic parameters in AI-based status monitoring are analyzed in this chapter, such as voltage for input and correction, capacity for internal state parameters, internal resistance for internal state parameters, polarization resistance and capacitance internal state parameters, and energy density for correction. For the application of artificial intelligence methods in lithium-ion battery systems, the first part is the selection criteria for artificial intelligence technology. This part analyzes the application of some existing artificial intelligence methods in lithium-ion battery systems, including utility-scale applications. Common artificial intelligence methods are explained, such as some machine-learning artificial intelligence methods, data-driven methods, etc., through the introduction of these methods, the application prospects of artificial intelligence methods in lithium-ion battery systems, and the resulting benefit.

Chapter 5
AI-based state-of-charge estimation

5.1 Overview of SOC estimation

5.1.1 Definition of SOC

The state of charge (SOC) of a battery is usually defined as the Q_n ratio of its current capacity $Q(t)$ to its nominal capacity, which also indicates the maximum amount of charge that can be stored in the battery. The main task of the battery management system (BMS) is to monitor the working condition of the battery, estimate the SOC of the battery and the health condition of the battery, and complete the functions of heat management, charge and discharge control, CAN communication, balance detection, and fault diagnosis, and LCD. There are a variety of SOC definitions [160]. At present, the internationally uniform definition is from the point of capacity, that is, charged state SOC is the residual capacity of lithium iron phosphate battery, which is equal to the ratio of the residual capacity to the rated capacity. The formula is as shown in (5.1):

$$\mathrm{SOC}\,(t) = \mathrm{Q}\,(t)\,/\mathrm{Qn} \tag{5.1}$$

The SOC of the lithium battery is simply how much electricity is left in the battery; SOC is the most important parameter in BMS, and because everything else is SOC-based, its accuracy and robustness are extremely important [161]. Accurate SOC estimation can reflect important information, such as battery performance, battery remaining life, etc., which ultimately leads to effective management and utilization of battery power and energy. In addition, SOC estimation can be used to adjust for reduced battery life, explosion or fire, accelerated aging, and permanent damage to the cell structure due to over-discharge or overcharge of the battery. Therefore, accurate SOC indication is important for user convenience and to ensure battery efficiency, safety, and life. An accurate SOC estimation is our basic consideration for eliminating thermal failure and regulating cell equalization.

Without an accurate SOC, no amount of protection can keep BMS from working properly, as lithium-ion batteries are often in a protected state and cannot extend their life. The estimation accuracy of SOC is also very important. The higher the accuracy, the higher the range for the lithium-ion battery of the same capacity. Therefore, high-precision SOC estimation can effectively reduce the cost of a lithium batteries. Most SOC estimation techniques require very precise measurements of the battery's chemical composition (type of electrolyte), its operating conditions, and cell variables (voltage, current, etc.) and are therefore only suitable for laboratory rather than real-world applications [162]. In addition, a given SOC estimation method is more suitable or applicable to a particular application than other methods. Thus, the SOC of a battery is not a state that can be measured directly but rather estimated from available parameters such as voltage, current, or surface temperature.

The SOC of a lithium-ion battery is mainly related to the battery voltage, current, and internal state parameters. As the most advanced and flexible battery technology, lithium-ion battery has demonstrated its advantages over other standard types of batteries (lead-acid, cadmium nickel, etc.), such as high capacity, high energy density, low self-discharge rate, long life, more charge and discharge cycles, and lower operation and maintenance requirements [163]. As an important threshold of charging and discharging, SOC can regulate and protect the lithium battery pack.

To give full play to the power performance of the battery system, improve its safety in use, prevent the battery from overcharging and over-discharging, prolong the service life of the battery, optimize driving, and improve the service performance of the electric vehicle (EV), the SOC of the battery is estimated accurately by the BMS system. SOC is an important parameter to describe the charge and discharge capacity of lithium iron phosphate batteries.

By definition, the SOC value is directly determined by two parameters, the maximum available capacity and the current capacity of the lithium battery. The voltage at the battery end and the loop current are the most important parameters to be used in all kinds of algorithms However, over-discharge and overcharge can lead to permanent damage to the lithium-ion battery cell, which can catch fire or even explode. Therefore, accurate estimation of battery SOC can prevent the battery from being frequently charged and discharged, thus saving the service life of the battery. As an important parameter to describe the state of the battery, it is a difficult problem and hotspot to estimate the state of the battery accurately. Therefore, if the SOC can be accurately estimated, it will play an important role in the research and development of batteries. There are many more accurate methods for estimating SOC, and the following are some of the main ones.

1. Discharge test method

The power battery is discharged to the cut-off voltage of the battery at a constant discharge rate (usually 0.3 or 1 C), and the constant current (0.3 or 1 C) is multiplied by the time of discharge to obtain the output of the battery, that is, the SOC value of the battery. This method is often used as a reference standard for battery capacity testing and is suitable for various types of power batteries. However, the test conditions of this method are strict, the current is constant, and the measurement is accurate, so it can only be limited to the data measured in the laboratory environment. In practical application, the current is not only constant, and the current is often forced to interrupt, but also needs to consume too much time, there is no stable and reliable ideal environment in the laboratory, so this method is not suitable for practical application.

2. The open-circuit voltage method (OCV)

The OCV method is based on the relationship between the OCV and the concentration of lithium ion in the battery. In the actual operation, it is necessary to discharge the battery at a fixed discharge rate (usually 1 C) after the battery is charged, until the battery cut-off voltage when the discharge stopped, according to the discharge process obtained OCV and SOC relationship curve. When the battery is in the actual working state, we can find the current SOC of the battery by looking up the OCV–SOC relationship table. Although this method is effective for all kinds of batteries, it also has its defects, before measuring OCV, the target battery must be left standing for more than 1 hour, so that the electrolyte inside the battery is evenly distributed to obtain a stable terminal voltage; however, the actual SOC may be quite different, and the measurement results cannot be completely accurate by using this method for a long time. Therefore, the OCV method, like the discharge test method, is not suitable for dry-running battery SOC estimation.

3. The current integral method

This method, also known as the ampere-hour measurement method, is one of the most widely used SOC estimation methods in the BMS, the SOC of the battery is estimated by the accumulated charge or discharge, and the SOC is compensated according to the discharge rate and the temperature of the battery; then the SOC of the battery after t time is shown in (5.2):

$$\mathrm{SOC}\,(t) = \mathrm{SOC}\,(0) + \frac{\int_0^t I_L\,(t)\,dt}{C_n} \tag{5.2}$$

Compared with other SOC estimation methods, the current integral method is relatively simple and reliable and can estimate the SOC value of the battery dynamically, so it is widely used. However, this method has two limitations: First, the current integral method needs to obtain the initial SOC value of the battery in advance, and the current flowing into or out of the battery must be accurately collected so that the estimation error is as small as possible. This method only takes the external characteristics of the battery as the basis of SOC estimation; to some extent, it ignores the influence of the self-discharge rate and aging degree.

4. Artificial neural network (ANN) method

The neural network method is a new algorithm to simulate the human brain and its neurons to deal with the nonlinear system, the SOC value in operation can be obtained by pre-extracting a large number of input and output samples from the target battery and putting them into the system built by this method. When an ANN is used to calculate the SOC of the battery, some parameters such as open circuit, charge/discharge ratio, ambient temperature, and surface temperature of battery are usually used as input variables. Some researchers take the attenuation of battery capacity as the input variable and apply it to different size batteries and different capacity attenuation AV batteries; the results show a certain degree of accuracy. This method has many advantages in battery SOC estimation, but it also has some problems such as large computation, time-consuming, large storage space, and training samples.

5. Fuzzy logic method

This is based on the theory of fuzzy logic reasoning, to imitate the way of thinking system, to establish an accurate mathematical model of the object is difficult to implement a control strategy given a description control. Unlike classical logic, which requires an accurate mathematical model and precise numerical values, fuzzy logic allows for modeling using abstract concepts derived from our knowledge and life experience, using subjective concepts such as big, small, very hot and bright red, long time, fast, or slow. Compared with the numerical system, this algorithm is relatively simple to develop, and then the system output is mapped to the precise numerical range to characterize the system characteristics.

Fuzzy control is widely used in automatic control systems. Using this algorithm, we can get a lot of information that can characterize battery performance, estimate battery SOC or Soh more accurately, and can carry on a more comprehensive analysis monitoring the battery condition.

6. System filtering method

Compared with the methods mentioned above, the system filter algorithm has the advantages of closed-loop control and real-time performance, so it is widely used in battery SOC estimation, the most commonly used system filtering algorithm is the Kalman filter algorithm. The Kalman filter was first proposed in the 1960s by the Hungarian-American mathematician Emmerich Kalman, who incorporated state-space analysis into filtering theory; as an optimal state estimation method, the Kalman filter can be applied to the dynamical system affected by random disturbances. To be precise, the Kalman filter is essentially a recursive algorithm that provides the best linear, unbiased, and minimum error variance estimation of the system state from noisy discrete observations in real-time.

Therefore, the Kalman filter algorithm is not only suitable for dynamic stochastic processes but also can be used to achieve recursion by predicting new states and their uncertainties and then calibrating the predicted values with new measurements, very suitable for multi-input systems. Therefore, for computer operations, the amount of computation and storage of Kalman filter is much less than that of traditional methods and can meet the requirements of real-time performance; at present, the SOC of lithium-ion power battery has been widely used in engineering practice. It is a non-direct measurement variable, which cannot be obtained directly by sensor devices, and can only be estimated by the combination of measurable variables and the control algorithm; in addition, the working process of a lithium-ion power battery is a complex electrochemical reaction process, the inner state is difficult to predict, and it is also affected by the external environment. The comparison of estimation methods is shown in Table 5.1.

The methods used to estimate SOC in EVs are based on the traditional method, that is, adding some influence factors to the time integral, and the disadvantage is that the estimation result of SOC has a great error, at present, the SOC estimation technology applied in the BMS is not very mature.

Table 5.1 Comparison of estimation methods

Estimation method	Merits	Faults
Discharge test method	Accurate and reliable	Need interruption
OCV method	Numerical approximation of electromotive force	Take a long time to set
Current integral method	High accuracy	Mass computation
ANN method	Model is simple	Inaccurate
Fuzzy logic method	Apposite theory	Measurement difficulty
System filtering method	Suitable for nonlinear models	Algorithmic complexity

In various power system applications, such as battery energy storage system (BESS) and EV, estimating and controlling SOC is very important. BESS is also a potential solution to the intermittency problem of renewable energy when it comes to the high penetration of renewable energy in the distribution network. However, the operation of the BESS system is easily affected by dynamic disturbances such as unbalanced load dynamics, for charging applications and single-phase distributed generation. Therefore, to develop a BESS that includes these dynamic aspects and involves aspects such as modeling, simulation, control scheme development, and standard test procedures, an adequate battery model is required to simulate real battery characteristics. With the development of the EVs, the BMS has been widely used.

The most commonly used battery models are equivalent circuit battery models (ECMs) and electrochemical battery models. ECMs are more flexible and convenient to calculate than complex electrochemical models that take into account the chemical processes and degradation of cells, which is supported by time-variant spatial partial differential equations (PDEs). One of the most commonly used electrochemical models is single particle model (SPM), which designs a state estimator for PDEs in a backward way. However, since the SPM model captures fewer cell dynamics and this method requires high computational cost and time, we only consider the ECM model here.

In general, ECM models include the Rint model, the Thevenin model, the RC model, Partnership for a New Generation of PNGV Vehicles, and the improved Thevenin model. In these ECMs, the second-order RC models are composed of polarized capacitors to represent the response of transient behavior, as shown in Figure 5.1.

The methods widely used are the discharge test method, ampere-hour method, OCV method, internal resistance method, Kalman filter method, linear model method, and neural network method. The wide voltage platform and severe polarization of lithium batteries are not conducive to SOC estimation, but the SOC of lithium batteries has an important effect on the battery pack inconsistency and battery life. The cycle life of the lithium-ion battery is consistent with SOC, and the cycle life of a lithium-ion battery increases obviously with the decrease of SOC.

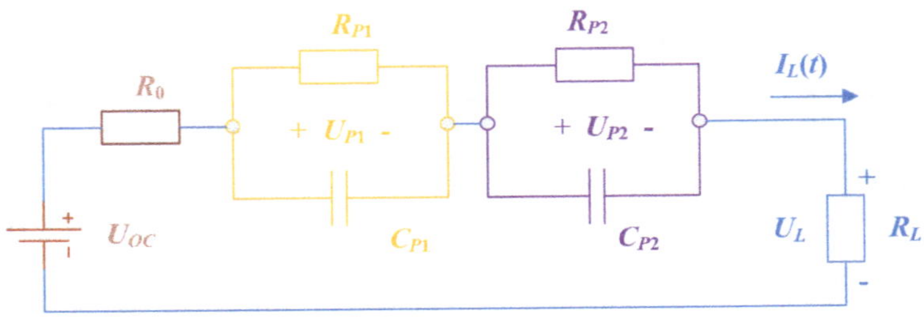

Figure 5.1 Second-order Thevenin model

To accurately estimate SOC, cell models must be able to accurately represent both static and dynamic reactions. As the accuracy of the battery model increases, so does the cost and time of calculation. Therefore, the advantages of the second-order RC model appear in the compromise choice between model accuracy and computational efficiency. The other models mentioned in this chapter have high computational efficiency but much lower accuracy. In the second-order RC model, a voltage source represents the OCV of the cell. Therefore, to obtain an accurate battery model, it is necessary to adopt an appropriate SOC estimation method. We should consider the precision and complexity of SOC estimation methods.

The SOC estimation algorithm usually runs in the BMS and adjusts the energy in the battery pack according to the voltage, temperature, SOC, and SOH status of individual cells. The main function of BMS is to maintain a safe operating environment for the battery system and protect the battery system from damage. Although battery SOC estimation is a key function of BMS, its accuracy and online estimation are still a challenge due to the nonlinear and complex electrochemical processes in the battery.

Although there are many kinds of estimation methods for battery SOC, there are some defects in each method; it is difficult to meet the requirement of SOC real-time online and high-precision estimation. The future research of SOC estimation methods, electric experts believe will be improved from the following four aspects, first, through a large number of experiments, the establishment of a rich database, so that SOC estimation has evidence, evidence-based; second, rely on hardware technology, to improve the measurement accuracy of current and voltage, and to ensure the accuracy of the basic data used to estimate SOC, third, to introduce accurate battery models to characterize more truly the dynamic and dynamic characteristics of the battery in use; and finally, to synthesize various algorithms, different correction methods are introduced for different stages of SOC to reduce the error in different states and improve the estimation accuracy.

5.1.2 Main affecting factor analysis

In EVs and energy stations, the battery pack exhibits a high degree of nonlinearity, and the SOC is affected by many factors such as discharge rate, battery temperature, self-discharge rate, and cycle times, which brings great difficulties to the accurate SOC estimation.

1. Charge–discharge rate

The current of charge and discharge is the rate of charge and discharge. During the same period, the amount of charge and discharge of lithium batteries will change with different charge rates [164]. Generally, the amount of power that the battery can discharge is inversely proportional to the rate of charge and discharge. Under the condition that other influencing factors remain unchanged, the discharge capacity of the battery decreases as the discharge rate increases. This is because the active material inside the battery has a limited depth of action with the thickness of the electrode. When a large current is discharged, the greater the discharge rate, the shallower the depth of action, and the lower the utilization rate, so the battery capacity is smaller, and vice versa, the greater the capacity.

The charge–discharge current can be described by the charge rate and discharge rate of lithium-ion batteries, which is represented by C. The charge–discharge rate refers to the ratio of the working current to the rated capacity of the battery during the charging and discharging process or working process [165]. Within the same duration, the charge–discharge amount of the battery will vary with different rates of charge. The battery is charging, discharge mode, the number of cycles, and depth of discharge (DOD) will affect the aging process of the battery.

The C-rate is to regulate the charge–discharge rates of the battery. The rated charge–discharge rate of the battery is usually 1 C, which means that the fully charged battery rated at 1 Ah can supply 1 A for 1 hour The same battery is discharged at 0.5 C and should deliver 500 mA for 2 hours, and 2 C should deliver 2 A for 30 minutes. Losses at accelerated discharge decrease the time of discharge, and these losses also impact the time of loading. The C-rate of 1 C is often referred to as 1-hour release such as the 2-hour release of 0.5 C or C/2 and the 5-hour release of 0.2 C or C/5. Any high-performance low-stress batteries can be charged and discharged above 1C.

2. Battery temperature

The SOC estimation accuracy is always affected by the external temperature because the external temperature will affect the electrochemical reaction inside the lithium battery. The higher the temperature, the higher the degree of the electrochemical reaction, and the energy released by the battery will also increase. A large amount of high-temperature gas will damage the battery, aggravate the aging speed of the battery, and affect the life of the battery. Therefore, in actual work, only by finding a suitable working temperature point can the performance of the battery be effectively exerted and the charge and discharge can be effectively performed.

The battery power and active material utilization rate increase with the rise of battery temperature, which is mainly caused by changes in electrolyte temperature performance.

When the battery temperature rises, the viscosity of the electrolyte decreases, and the activity increases, which increases the ion diffusion movement ability, and finally an increase in the utilization rate of the active material, and the actual available power of the battery increases. Conversely, when the battery temperature drops, the active material utilization rate decreases and the actual available power decreases. Therefore, the actual available power of the battery is directly proportional to the battery temperature. In practical application, the charge–discharge working temperature range of lithium iron phosphate polymer battery is 0–45°C.

The battery capacity indicates how much energy it can hold, which decreases with the decrease of temperature and increases with the increase of temperature. Therefore, the battery fails on the cold winter morning, even though it worked the previous day perfectly. If the batteries work in a cold environment most of the year. In that case, the decreased power of the lithium-ion battery must be taken into consideration. The typical operating temperature of a battery is 25°C (77°F). At approximately –30°C (–22°F), the battery power will be reduced to 50%. At freezing, the battery power is decreased by 20%. Capacity is improved at higher temperatures at 122°F. The capacity of the battery will be around 12% higher.

The SOC estimation accuracy is always affected by the external temperature because it affects the electrochemical reaction inside the battery. The degree of the electrochemical reaction is high when the temperature is high. The energy released by the battery also increases, which will produce a large amount of high-temperature gas, damage the battery, accelerate the aging speed of the battery, and affect the life of the battery.

3. Number of cycles

The aging of the battery is due to changes in the internal structure of the battery. The heating and cooling of the battery and the continuous charging and discharging will lead to changes in the internal structure of the battery, which in turn affects the degree of battery aging. During the aging process of the battery, the SOC value will also decrease, which will eventually lead to the battery being scrapped and unusable. Generally, the charging and discharging method of the battery, the number of cycles of use, and the DOD will affect the aging degree of the battery. The aging of the battery is due to the change in its internal structure, the rise in temperature, and the batteries dropping. The constant charging and discharge will lead to a change in the internal structure, influencing the aging degree. The SOC value will also decrease during the battery aging process, which will lead to the failure of batteries to operate. Generally, the battery charging–discharge mode, the number of cycles, and DOD can affect the aging of batteries.

The everyday use of batteries in HEVs implies a paradigm of accurate battery aging and battery life. The aging of batteries can be divided into two parts, including the aging of calendars and cycle one. Calendar aging refers to the endless amount of lost storage ability. In other words, battery capacity is responsible for the deterioration. Cycle aging is related to the influence of cycles (charge or discharge) for battery use times. It occurs when the battery is either loaded or unloaded. This is directly due to the level, the temperature of the usage pattern, and current battery specifications. As a result, its performance steadily deteriorates during battery life

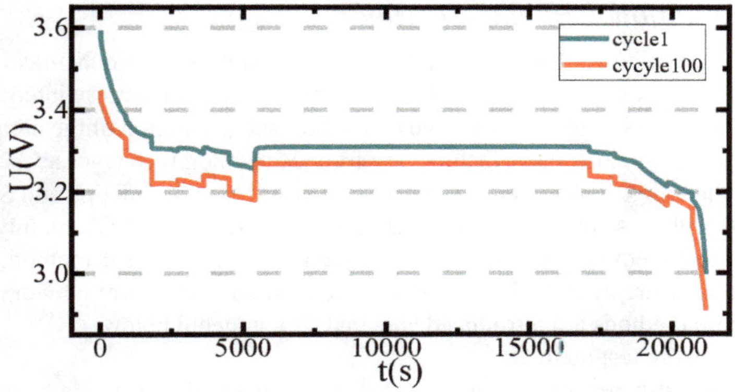

Figure 5.2 *The schematic diagram of terminal voltage under different numbers of cycles*

because its electrochemical components are deteriorating and the performance of EVs and fuel efficiency deteriorates.

After a period of use, the standard total capacity of lithium-ion batteries changes to a certain extent. At first, the power increases, and it remains the same for a period, and then the power gradually decreases. For lithium iron phosphate batteries, the cycle life of the battery is generally expressed by the number of charge–discharge times when the total discharge capacity drops to 80% of the nominal capacity. The schematic diagram of terminal voltage under different numbers of cycles is shown in Figure 5.2.

4. Self-discharge rate

When the battery is not in use, some slow chemical reactions are also taking place inside it. As the essence of battery charge–discharge is an electrochemical reaction and the electrochemical reaction inside the battery increases with the rise of temperature, when the temperature is high, the intensity of self-discharge inside the battery is large, which will lead to the automatic drop of SOC.

Self-discharge, also known as charge retention capacity, refers to the storage capacity retention capacity of the battery in an open-circuit state under certain environmental conditions. Under the self-discharge of the battery, the SOC value decreases as the storage time increases. In general, the self-discharge rate can be expressed as the percentage of capacity reduction per unit of time, and the expression is shown in (5.3)

$$I_{sdr} = \left(C_a - C_b\right) / \left(C_a * T\right) * 100\% \tag{5.3}$$

In (5.3), I_{sdr} is the self-discharge rate, C_a is the battery power before storage, C_b is the battery power after storage, and T is the battery storage time. The self-discharge rate of lithium-ion batteries is related to many factors such as the number of cycles, ambient temperature, and storage time. It is generally calculated and derived by experimental methods.

5.1.3 Traditional estimation method

The discharge of power batteries in EVs is a very complex electrochemical reaction process. Factors such as discharge rate, ambient temperature, battery internal resistance, and self-discharge rate affect the estimation and prediction of the battery SOC value. At the same time, these factors vary from each other. It changes as the number of cycles increases, which increases the difficulty of battery modeling and SOC estimation algorithm. At present, the estimation methods of battery SOC mainly include discharge test method, OCV method, ampere-hour measurement method, internal resistance measurement method, load voltage method, and neural network method [166]. These methods are introduced and analyzed in detail below.

1. Discharge test method

The discharge test method is an experimental method that uses a constant current to continuously discharge the battery until the battery terminal voltage reaches the discharge cut-off voltage [167]. It is the most reliable SOC estimation method and applies to all types of batteries. However, it also has several defects, which greatly limit its application. First, it takes a lot of measurement time. Only after the entire discharge test is over, the SOC value can be calculated for each time point, and the real-time SOC estimation cannot be achieved. Second, the previous work of the battery must be forced to stop and switch to a constant current discharge state. Therefore, the discharge test method is not suitable for estimating the SOC estimation of power batteries with real-time estimation requirements, but it can be used for the maintenance of power batteries and the determination of battery model parameters.

This method is generally used as a calibration method for battery SOC estimation or used in the later maintenance work of the battery. In the case of not knowing the SOC value of the battery, this method is relatively simple and reliable, the result is relatively accurate, and it is also effective for different types of batteries. Due to the reliability of the performance of the discharge method, most battery manufacturers and research institutions now choose this method as the basic method for SOC determination. The inevitable disadvantage of the discharge method is that the measurement takes a lot of time, and online measurement cannot be performed. The discharge method must be cut off after the working circuit where the battery is located. Therefore, in most cases, the discharge method can only be used for factory testing and pre-application testing. The schematic diagram of voltage curve during constant current discharge is shown in Figure 5.3.

2. OCV method

The OCV method estimates SOC according to the approximately linear relationship between SOC and OCV. The actual operation process is described as follows. First, the battery needs to rest for a long enough time to ensure the stability of its OCV and other parameters. Then, the OCV of the battery is measured to obtain the SOC value according to the known approximate linear function relationship. The advantage of the OCV method is its simplicity and strong operability. However, the OCV can only be obtained after the battery is fully rested until the electrochemical reaction of the battery is stable, but it is impossible to track the SOC value in real

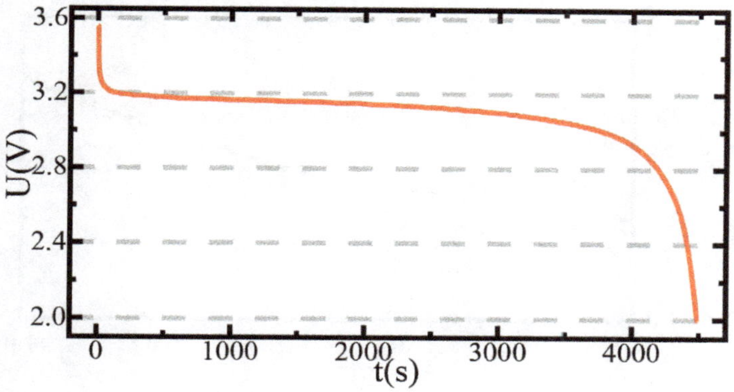

Figure 5.3 *The schematic diagram of voltage curve during constant current discharge*

time. In addition, although the functional relationship between SOC and OCV is relatively fixed in a short time, the relationship between SOC and OCV changes with the aging of the battery, which also affects the accuracy of SOC estimation. Therefore, the applicability of this method is poor.

The OCV refers to the battery's positive and negative potential difference when the external circuit current is zero and the battery reaches equilibrium after a long time of standing. The OCV of the battery has a relatively fixed mapping relationship with SOC, as shown in (5.4):

$$U_{OC} = F(SOC) \tag{5.4}$$

In (5.4), the mapping relationship can be expressed in different forms, mainly divided into discrete and continuous forms. Discrete forms, such as the look-up table method, use a series of discrete isolated points to express the one-to-one correspondence between the OCV and SOC that uses a simple piecewise linear method to express the parts between points, which also exposes the coarseness of this method. The continuous form, such as the function method, is to get the OCV–SOC curve through multiple measurements. The continuous relationship expression between discrete points is obtained by function solution or curve fitting methods.

A relatively fixed-function relationship can be used to estimate the SOC, and its obvious disadvantage is that the battery needs to be rested for a long time to measure, which will take a long time to measure. An improved method is to simulate the polarization effect of the battery by establishing a battery model and conducting multiple experimental tests on the battery's static curve after charging and discharging. According to the obtained regular formula, the battery voltage after stopping charging and discharging for a certain period is used as the OCV to estimate the SOC, to solve the defect of directly measuring the OCV. The OCV is also affected by temperature. When the SOC is low, the OCV begins to bifurcate with the temperature change. The lower the temperature, the lower the OCV. At present, in most cases, the OCV–SOC curve at room temperature is studied, which will cause large

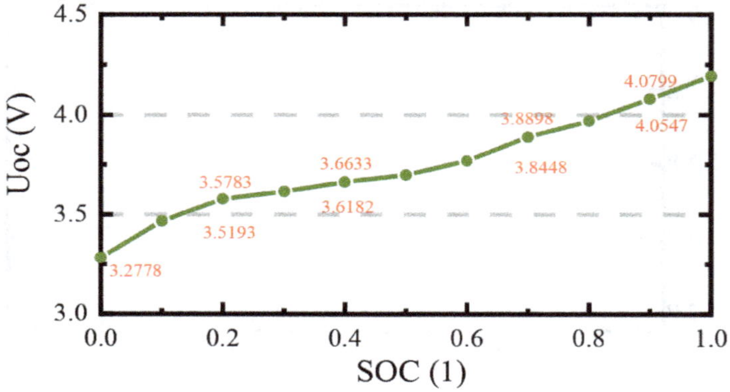

Figure 5.4 The schematic diagram of OCV–SOC curve

errors in SOC estimation at other ambient temperatures. The limitation of the OCV method is that the battery always needs to be left to stand, and the SOC cannot be estimated in real-time during the charging and discharging process.

The OCV–SOC curve of lithium batteries is relatively flat, which means that a little difference will cause a large error in SOC estimation. This method overcomes the dependence of the ampere-hour integration method on the initial value of SOC. However, due to the plateau effect of the battery voltage, a small change in voltage error outside the plateau period will cause drastic changes in SOC, and the battery needs to be left standing for a long time to obtain the OCV of the battery. These disadvantages make it less suitable for engineering applications. The schematic diagram of OCV–SOC curve is shown in Figure 5.4.

3. Ampere-hour measurement

The ampere-hour integral method, also known as the Coulomb counting method, is to calculate the total change of a certain charge–discharge period when the SOC basic value of a certain time is known in advance to obtain the SOC value of the battery at the special time point. Comparing the estimated value with the actual value, the error is obtained. Along with the time extension, the error increases obviously and cannot take the correct response strategy to the error accumulation problem, which cannot guarantee the stability of the test results. This approach has a certain dependence on the initial value. The Ah integral method is widely used in SOC estimation. Its goal is to focus on the external characteristics of the battery system, without considering the complex relationship between the electrochemical reaction and parameters in the battery.

The ampere-hour measurement method estimates the SOC of the battery by calculating the accumulated power of the battery during charging or discharging and compensates the estimated SOC value according to the temperature and the charge–discharge rate. It is currently the most commonly used and simplest SOC estimation method, and it has been successfully applied to the power estimation of electronic consumer products. If the initial state of charging and discharging is specified as SOC_0, then the SOC value of the current state can be calculated by (5.5):

$$SOC = SOC_0 - \frac{1}{Q_n}\int_0^t \eta i dt \tag{5.5}$$

In the formula, Q_n is the nominal capacity; i represents the battery current, which is positive when discharging and negative when charging; η is the battery efficiency coefficient, including the temperature influence coefficient η_r and the charge–discharge rate coefficient η_i, where η_i can be obtained by the Peukert equation. Three problems should be paid attention to when using the ampere-hour measurement method: the method itself cannot provide the initial battery value SOC_0; inaccurate current measurement will increase the SOC estimation error, which will become larger and larger after a long-time accumulation. The battery efficiency coefficient η must be considered when estimating SOC. Although the accuracy of current measurement can be solved by using high-performance current sensors, this will greatly increase the cost of the system. At the same time, to solve the problem of battery efficiency coefficient η, it is necessary to establish empirical formulas for temperature influence coefficient η_r and charge–discharge rate coefficient η_i through a large amount of experimental data. If the estimation accuracy requirements are reduced, the ampere-hour measurement method can be applied to the calculation of the remaining battery power of all EVs.

4. Internal resistance measurement method

For the internal resistance method, there is a certain functional relationship between SOC and internal resistance. The resistance method can also be called the conductivity method. It can analyze the relationship between the conductivity or internal resistance of the battery and SOC from numerous experimental data through a long-term tracking test of the conductivity or internal resistance, to predict the SOC of the battery. Since the conductance method relies only on the internal resistance of the battery to estimate SOC, the measurement accuracy of the internal resistance of the battery directly affects the accuracy of the SOC estimation. Therefore, it is necessary to ensure good contact during the measurement and try to make the contact resistance zero.

In a certain period, the functional relationship between SOC and internal resistance is relatively stable, so the accuracy of the internal resistance method depends entirely on the accuracy of internal resistance measurement. However, the internal resistance of the battery will gradually increase along with the aging process of the battery, and the fixed function relationship between SOC and internal resistance is no longer applicable, which will affect the accurate estimation of SOC.

The internal resistance measurement method refers to the method of measuring the internal AC resistance of the battery by exciting the battery with different frequencies of alternating current and calculating the SOC value using the battery static model of the remaining power and the AC internal resistance. In the later stage of battery discharge, the internal resistance measurement method has high estimation accuracy and battery adaptability and is generally used in combination with the ampere-hour measurement method. However, this method only considers two aspects of discharge current and internal resistance and does not involve factors such as ambient temperature, battery cycle times, and battery pack imbalance. In addition, the cause of battery internal resistance is very complicated, so the internal resistance measurement method. The final SOC accuracy is not high. In addition, there are great differences in AC impedance at different current frequencies, and EV power batteries cannot work at a fixed AC frequency, so the internal resistance measurement method cannot meet the requirements of EV SOC estimation.

5. Neural network method

The neural network is an artificial intelligence technology that has been developed in recent years and uses simulating brain signal processing. A neural network is a complex nonlinear system formed by a large number of simple neurons connected extensively. It can automatically summarize the collected data and obtain the internal laws of the data from it. It also has a nonlinear mapping capability and is suitable for non-causal relationships with more complex nonlinearities: accuracy identification, judgment, reasoning, and classification issues. The battery is a highly nonlinear system, and it is usually difficult to establish a reasonable and accurate mathematical model for its charging and discharging process. Therefore, given external incentives, the neural network method can use the learning ability of the neural network and the parallel structure to simulate the nonlinear characteristics of the battery to estimate the SOC value.

SOC estimation often uses a typical three-layer neural network, in which the number of neurons in the input and output layers is determined by the actual system needs, and the number of neurons in the middle layer depends on the system complexity and analysis accuracy requirements. In the neural network method, the system input includes battery voltage, ambient temperature, charging and discharging current, battery internal resistance, accumulated discharge power, and so on. Whether the input type and quantity are appropriate will directly affect the calculation and accuracy of the method model. The disadvantage of the neural network method is that a large amount of experimental data is needed to train the parameters, and the SOC estimation error is greatly affected by the training data and the training method. The schematic diagram of the three-layer model is shown in Figure 5.5.

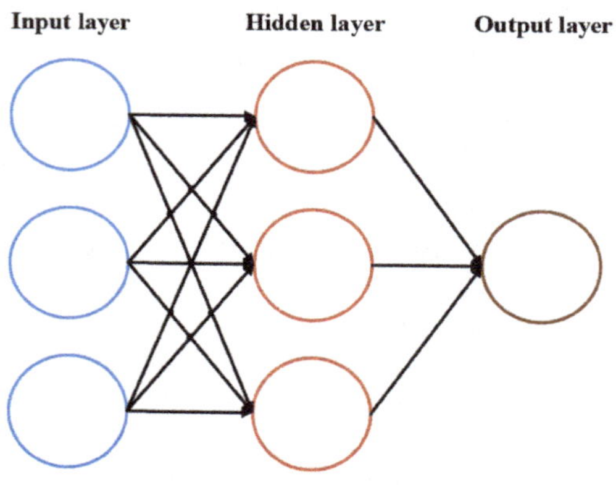

Figure 5.5 The schematic diagram of the three-layer model

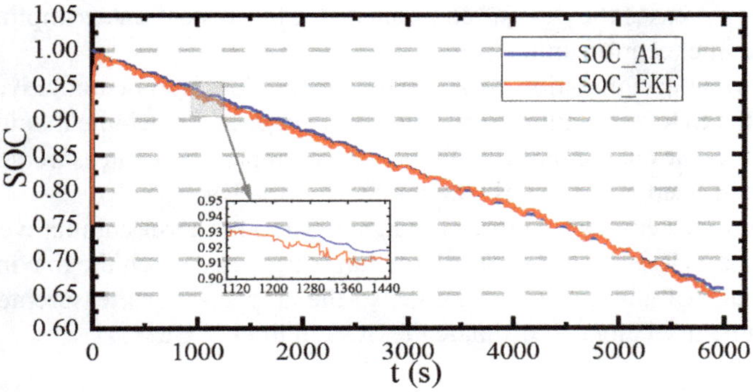

Figure 5.6 The SOC estimation results based on Ah measurement and EKF

6. Kalman filtering method

Lithium batteries are a typical nonlinear system. With the development of Kalman filter technology, the improved Kalman filter can already be used in complex nonlinear systems. For nonlinear Kalman filtering, there are mainly extended Kalman filtering (EKF) and unscented Kalman filtering (UKF). Among them, the EKF uses the partial differentiation method for linearization. It is necessary to use the partial differentiation method to generate the Jacobian matrix in each cycle of the algorithm, and the calculation is more complicated. On the other hand, when using the partial differential method in the EKF to convert the expected signal and covariance, if the system is nonlinear, the converted result may be biased, which will lead to errors and no trace.

Compared with the EKF, the transformation has some advantages. The data transformation is not only unbiased but also simple to calculate and easy to implement, so it is more suitable for the estimation of the state of the nonlinear system. Under the BBDST condition, the SOC estimation results based on ampere-hour measurement and EKF are shown in Figure 5.6.

Compared with commonly used electronic consumer products, EVs have higher requirements for power battery performance parameters. Battery packs are often under discharge conditions with large currents and severe current fluctuations. At the same time, the components of battery capacity affected by temperature are relatively increased. This puts more stringent requirements on the SOC estimation algorithm:

1. The SOC estimation result should have high accuracy. To improve the energy utilization rate of power batteries and provide accurate battery life information, the maximum estimation error of SOC should generally be controlled within 10%, and in the later period of charge and discharge, the estimated value of SOC can converge to near the true value to ensure that the battery can be completely charged and discharged without overcharging or over-discharging.
2. SOC must be estimated online in real time. The vehicle control strategy of EVs is adjusted at any time according to the current SOC value of the power battery pack, so the SOC estimation must have online real-time performance, which

greatly increases the design requirements of the system data sampling unit and algorithm execution unit.

3. The current accumulation error is reduced. The long-term cumulative error will make the SOC estimation error larger and larger and reach the maximum in the later stage of charge and discharge, so the ampere-hour measurement method cannot be used alone to estimate the SOC in the EV BMS.
4. The current cumulative error is reduced. The long-term cumulative error will make the SOC estimation error larger and larger and reach the maximum in the later stage of charge and discharge, so the ampere-hour measurement method cannot be used alone to estimate the SOC in the EV BMS.

Considering comprehensively, the Kalman filter method has a relatively large advantage in estimating the battery SOC. Compared with the neural network, the amount of calculation is smaller and the amount of preliminary data required is smaller. Compared with other algorithms, it has the ability of adaptive correction. In the actual BMS, the characteristic comparison table of the SOC estimation method is shown in Table 5.2.

In recent years, researchers have proposed some new estimation methods, such as the least squares support vector machine (SVM) regression algorithm, the adaptive theory method, the improved neural network method, and the UKF method. These new methods are to correct the errors that occur in the process of estimating SOC by the traditional estimation method and the new estimation method, so as to reduce the estimation error as much as possible, thereby improving the estimation accuracy. In practical BMSs, the methods for estimating battery SOC are traditional methods, and most of the new methods are only in the theoretical research or simulation stage.

5.2 Backpropagation-based SOC estimation method

5.2.1 Backpropagation network model structure

Lithium-ion batteries are widely used in EVs. In daily use, with the increase in the number of charge and discharge cycles of lithium-ion battery, its capacity attenuation also occurs. When the lithium-ion battery is aged to a certain extent, it not only has a significant decline in durability but also has potential safety hazards [168]. Therefore, it provides a key reference basis for the efficient utilization of lithium-ion batteries and the timely replacement of the aged lithium-ion battery. Accurately and effectively estimating the SOC of power lithium-ion batteries for EVs is of great significance for the estimation of EV mileage and the safety protection of power batteries.

The estimation of battery residual power has always been the focus and difficulty of the BMS [169]. To prolong the service life of lithium-ion battery and give full play to the discharge efficiency, the premise is the need for a high-precision estimation of lithium-ion battery residual power. China's 13th 5-year plan also takes lithium-ion battery as an important development object in the field of new energy.

Table 5.2 the characteristic comparison table of the SOC estimation method

Estimation method	Fundamental	Advantages	Disadvantages
Discharge test method	Constant current discharge to cut-off voltage	Easy to measure	Requires offline measurement and takes a long time
OCV method	Sufficient rest and use of OCV–SOC curve	Simple calculation and good accuracy	Requires prolonged storage and is affected by temperature and noise
Ampere-hour measurement	Integral of current over time	Simple and easy to calculate and widely used	Depends on initial SOC and accumulated error cannot be eliminated
Internal resistance measurement method	Calculate SOC value using static model and AC internal resistance	High estimation accuracy and battery adaptability	Considers only discharge current, internal resistance, and fixed frequency
Neural network method	Simulate the working mechanism of the human brain	Strong nonlinearity, self-learning ability, and high precision	Rely on a large amount of data and the impact of data errors is large
Kalman filtering method	Combines the battery equivalent model and uses the two links of prediction and correction	High estimation accuracy and real-time prediction correction function	Large amount of calculation and low precision

Its application scope has penetrated all walks of life, such as new energy vehicles, aviation, and household appliances. Lithium-ion battery is more and more widely used in the field of new energy, and its state detection is paid more and more attention. Whether the SOC of the lithium-ion battery can be accurately estimated is related to whether the performance of the lithium-ion battery can be brought into full play. It is of great significance to realize real-time state detection and safety control of lithium-ion battery. The complex internal structure of lithium-ion battery makes it show strong nonlinear characteristics in the process of use, which puts forward new requirements for the traditional SOC estimation method of lithium-ion battery.

The traditional SOC prediction methods of lithium-ion battery include discharge method, conductivity method, OCV method, and ampere-hour integration method. Their basis is relatively simple, but the implementation process is relatively simple, and their accuracy is often not high or adaptability is not strong. New lithium-ion battery SOC prediction methods include EKF, particle filter (PF), fuzzy logic method, and neural network algorithm (backpropagation, BP). These methods combine mathematics and computer theory, and the implementation process is complex, but they can achieve good estimation effect and strong adaptability.

The neural network method and battery neural network model are the same concepts, which are put forward with the help of the uniqueness of the neural network. The neural network method is relatively new. According to the parameters measured in the process of battery charging, it is modeled by the highly nonlinear and autonomous learning mechanism of the neural network to obtain the estimated value of SOC. The richer the types of input parameters and the larger the amount of data, the more accurate the SOC estimation. However, a neural network needs a lot of mathematical operations, which requires high speed and the ability of the CPU, and the cost is relatively expensive. The neural network method is generally complex, and the estimation error is highly dependent on training data and training methods.

The principle of using a neural network to estimate the SOC of the lithium-ion battery is as follows: the neural network is used to model the lithium-ion battery, take the external characteristic parameters of the battery such as voltage and current as of the input, and train the system through a large number of sample data. When the SOC reaches the required error range, the system is used to estimate the SOC of new input. The advantage of using a neural network to estimate battery SOC is that there is no need to establish a certain mathematical model. In addition, from many kinds of literature, it can be seen that there is a nonlinear relationship between battery SOC and parameters such as current through the battery and terminal voltage of the battery, while neural networks can well determine this nonlinear relationship through sample learning. The more sample learning data, the higher the estimation accuracy. Of course, the error of sample data will also have an impact.

At present, BP neural network algorithm has high nonlinearity and strong self-learning ability. It is widely used in SOC estimation of lithium-ion battery states.

1. The development background of BP network

ANN technology has been widely used to deal with nonlinear complex problems in various fields of social and economic development. In the development history of the ANN, multilayered perceptron (MLP) network has played a great role in its development of the ANN. It is also considered to be a real ANN model. Its emergence has set off an upsurge in people's research on ANNs. As the initial neural network, a single-layer perceptual network has the advantages of a clear model, simple structure, and a small amount of calculation.

However, with the deepening of research work, it is found that it still has some shortcomings. For example, it cannot deal with nonlinear problems. Even if the action function of the calculation unit does not use the valve function and uses other more complex nonlinear functions, it can only solve the linear separable problem. It cannot realize some basic functions, which limits its application. The only way to enhance the classification and identification ability of the network and solve the nonlinear problem is to use a multi-layer feedforward network, that is, add a hidden layer between the input layer and the output layer to form a multi-layer feedforward perceptron network.

In the mid-1980s, David Runelhart, Geoffrey Hinton, Ronald Willians, David Parker, and others independently discovered the Error BP training algorithm, which

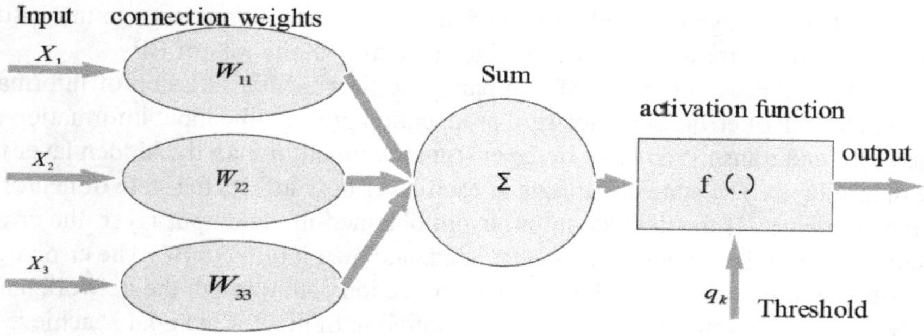

Figure 5.7 Artificial neuron model

systematically solved the learning problem of hidden layer connection weight of multilayer neural network, and gave a complete derivation mathematically. The multilayer feedforward network using this algorithm for error correction is called the BP network.

BP neural network has the ability of arbitrarily complex pattern classification and excellent multi-dimensional function mapping. It solves the exclusive OR (XOR) and some other problems that cannot be solved by a simple perceptron. In terms of structure, the BP network has an input layer, a hidden layer, and an output layer. In essence, the BP algorithm takes the square of network error as the objective function and uses the gradient descent method to calculate the minimum value of the objective function.

With the further development of the BP algorithm, it will face more challenges in the field of local convergence and optimization.

2. The principle of BP network

The neural network is a nonlinear system composed of a large number of simple units. It has the ability of nonlinear mapping, does not need an accurate mathematical model, and is good at learning useful knowledge from input and output data. The neuron is the basic unit of the neural network. The neuron model is shown in Figure 5.7. It consists of three elements: connection weights, sum, and nonlinear activation function.

Since the first neural network model was established in the early 1940s, artificial intelligence information processing technology that imitates the structure and function of human brain neurons, namely, ANN technology, has attracted the favor and attention of many scholars, including Hopfield neural network, adaptive resonance technology, and BP neural network. Among many neural network models, the BP neural network is the most widely used in various fields.

BP algorithm is a supervised learning algorithm. The purpose of its learning is to modify its weight with the error between the actual output of the network and the target vector, so that the output is as close as possible to the expected, even if the sum of squares of errors in the output layer of the network reaches the minimum. It gradually approaches the target by continuously calculating the changes in network weight and deviation in the direction of decline relative to the slope of the error function. Each change of weight and deviation is directly proportional to the influence of network error and is transmitted to each layer in the way of BP.

BP algorithm is composed of two parts, the forward transmission of information and the BP of error. In the forward propagation process, the input information is calculated and transmitted layer by layer from the input through the hidden layer to the output layer. The state of neurons in each layer only affects the state of neurons in the next layer. If the desired output is not obtained in the output layer, the error change value of the output layer is calculated, and then it turns to BP. The error signal is transmitted back along the original connection path through the network, and the weights of neurons in each layer are modified until the desired goal is achieved. Its structure is generally a classic three-layer model, including an input layer, a hidden layer, and an output layer.

The most special is the existence of its hidden layer, which can deal with nonlinear problems well. In the process of forwarding transmission, the input signal is processed layer by layer from the input layer through the hidden layer to the output layer. The neuron states between each layer affect each other. If the expected output is not obtained in the output layer, it will propagate back and adjust the weight and threshold of the network according to the error, to make the whole network close to the expected output value.

3. The model structure of BP Network

BP neural network structure is generally a typical three-layer model, which is composed of the input layer, output layer, and hidden layer. Its three-layer structure model is shown in Figure 5.2. $X_1, X_2, ..., X_m$ are input variables in the model input layer; $y_1, y_2, ..., y_n$ are the output variables of the model output layer, which is shown in Figure 5.8.

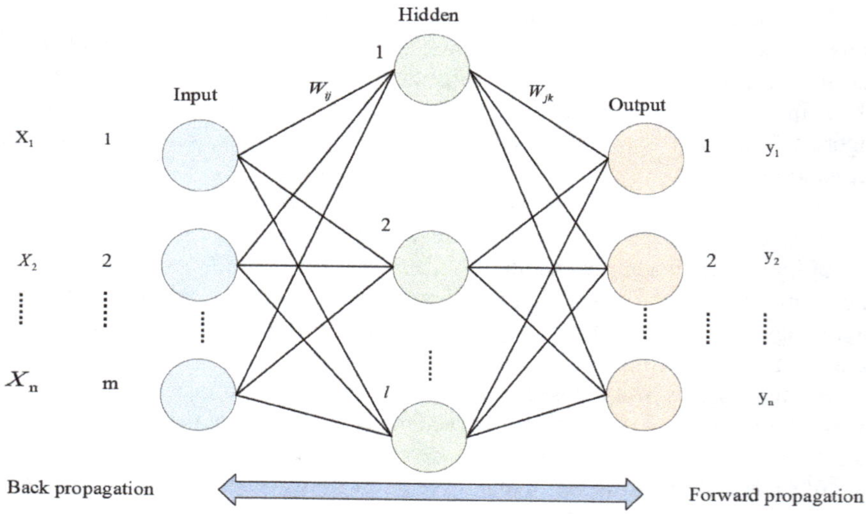

Figure 5.8 Three-layer model of neural network

According to the definition, they have the following corresponding relationship, as shown in (5.6). According to the learning and training process of the gradient-descent method, the error and correction weights are calculated. E is the error and t_k is the expected output:

$$\begin{cases} P_i = p(\sum_{i=1}^{m} \omega_{ij}X + \theta_i) \quad j = 1,2,\ldots\ldots,l \\ y_k = q(\sum_{j=1}^{l} \omega_{jk}P_i + \theta_k) \quad k = 1,2,\ldots\ldots,n \\ E = \frac{1}{2}\sum_{k=1}^{n}(t_k - y_k)^2 \quad k = 1,2,\ldots\ldots,n \\ \omega_{\alpha+1} = \omega_\alpha - \eta\frac{\partial E}{\partial \omega_\alpha} \end{cases} \tag{5.6}$$

The neural network is a complex network system formed by a large number of simple neurons connected. First, the structural level of the network (the selection of the number of neuron nodes) should be determined. According to the target, the number of neurons (input layer, output layer, and hidden layer) in the BP neural network model can be selected randomly.

Input layer modeling: the off-voltage of the battery, the current of charge and discharge, the working temperature of the battery, the amount of electricity discharged by the battery, and other factors that affect the battery charge. Therefore, this work determines that the number of neurons in the input layer is 4, that is $M = 4$.

Hidden layer model: the number of neurons in the hidden layer can be determined according to any one of the empirical of the BP neural network. The three common empirical formulas are shown in (5.7):

$$\begin{cases} l = \sqrt{m+n} + a \\ l = \log_2 m \\ l = \sqrt{mn} \end{cases} \tag{5.7}$$

is a constant between 1 and 10.

Output layer model: since the target value is the SOC value, it is determined that there is only one neuron node in the output layer. Therefore, the neural network model is established, as shown in Figure 5.9.

4. The scope of application of BP Network

Since the ANN is a nonlinear simulation technology among them, the BP neural network has the ability of highly autonomous organization, autonomous learning, autonomous adaptation, and autonomous reasoning. It can solve many nonlinear and complex practical problems. It can be applied to many fields of social economy and urban development to solve the difficulties of nonlinear influence.

BP neural network is widely used in various fields and has good mapping ability for nonlinear problems in various fields. The neural network algorithm itself overcomes the limitations and deviations caused by the excessive dependence of the traditional algorithm on the battery model, provides a new idea and method for the

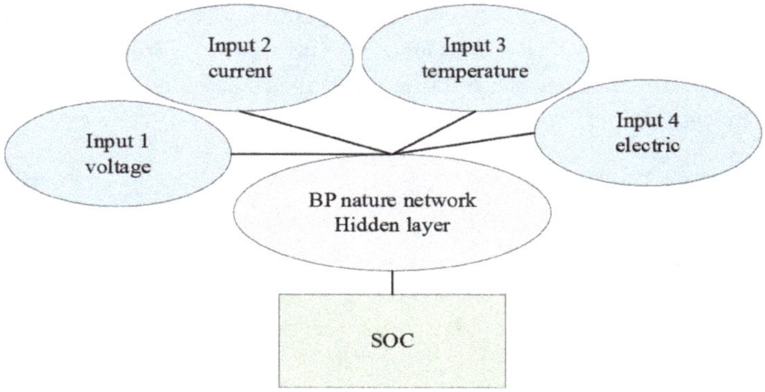

Figure 5.9 The neural network structure diagram of lithium-ion battery

SOC estimation of power battery, and provides a scientific prediction and estimation algorithm basis for the charge estimation of lithium-ion power battery of the EV. Especially in intelligent transportation, intelligent medical treatment, intelligent education, intelligent economy, and other aspects, it has very broad development prospects, provides more efficient services for all walks of life, and can provide the objective basis and effective early warning signals for decision-makers, which has important theoretical significance and practical value.

In the practical application of ANNs, most neural network models adopt the BP network and its variation form. It is also the core part of the forward network, which embodies the essence of an ANN.

BP network is mainly used in the following four aspects:

1. Function approximation: train a network to approximate a function with input vector and corresponding output vector.
2. Pattern recognition: associate it with the input vector with an undetermined output vector.
3. Classification: classify the appropriate way defined by the input vector.
4. Data compression: reduce the dimension of the output vector to facilitate transmission or storage.

5.2.2 Network information capacity optimization

As the scale of wireless networks continues to expand, the requirements of users for communication services also continue to increase. The correct calculation of network capacity is extremely important for guiding network information transmission. Due to the special characteristics of the information transmission process in wireless networks (multi-hop transmission, multi-channel fading, etc.), it is impossible to directly apply the classical Shannon formula to calculate network capacity [170]. Therefore, how to calculate the network capacity has attracted extensive attention from experts and scholars at home and abroad and has become a hot area of wireless network research.

In 2000, Gupta and Kumar proposed to define the network capacity as the product of the maximum node density allowed in the network and the bit rate at which information is successfully transmitted, as shown in (5.8):

$$C(\varepsilon) = R\lambda\varepsilon(1-\varepsilon) \tag{5.8}$$

where R denotes the information transmission rate, ε denotes the channel outage probability, $\lambda\varepsilon$ denotes the maximum node density of the network when the channel outage probability is ϵ, and $C(\varepsilon)$ denotes the network capacity.

The existing literature on network capacity calculation sets the information transmission rate as a constant [171]. The network transmission rate is an important component of network capacity and increasing the network transmission rate is an important means to increase the network capacity. Network capacity refers to the maximum number of packets that can be delivered from Alice to Bob in an average time, denoted by C. If no interference from the adversary is assumed, C is equal to the maximum flow of the network, and in addition, it is the minimum cut from the source node to the destination node (for the multicast case, C is defined as the minimum of the minimum cut from the source node to all destination nodes); the error probability refers to the probability that Bob will not be able to recover Alice's information accurately. The rate R is the average number of message symbols that can be delivered from Alice to Bob per unit time step. A rate R is said to be achievable if for any $\in 1 > 0$ and $\in 2 > 0$, there exists a coding scheme that can encode packets of packet size n satisfying a rate greater than or equal to $R - \in 2$ and an error probability less than or equal to $\in 1$. The capacity in a given attack scenario is the maximum achievable rate in that scenario.

For systems, capacity, quality, and network performance are always relative and contradictory. Capacity refers to the maximum number of users accessing the network at any given time without call failure, while quality is determined by the false frame rate FER and EC/IO, which are usually fixed by the manufacturer or operator. An important goal of network optimization is to maximize the system capacity of the network while achieving network quality and user QoS requirements.

In practice, there are many methods of information capacity optimization, and all kinds of methods can be used in a mixture. If the effect of the above methods is not obvious, consider upgrading the network capacity using carrier expansion and heterogeneous networks, etc., that require new equipment and investment; the introduction of new technologies should be demonstrated through the form of the pilot and promoted one after another, paying attention to the application scenarios and limitations of new technologies.

Network capacity optimization is one of the most important means to improve user service perception and operator revenue, and the level of network capacity optimization also plays a very important role in creating the operator's brand image. With the development of network technology and the market, the demand for network capacity optimization will become more and more urgent, and the research and application of network capacity optimization methods are of far-reaching significance. Network optimization is to detect, analyze, and adjust the system dynamically according to the actual operation status and operation quality of the system, to achieve the goal of providing the best quality of service and the best coverage under the existing system configuration, and to realize the highest network cost performance.

Capacity optimization refers to the optimization of frequency. In the current practice environment, because multiple operators are in one area, which leads to the signal interference of different operators, and the signal loss will occur in the case of the interval, so to deal with the interference problem, the following points can be used as reference:

1. Control and circumvent the frequency between devices and choose the appropriate frequency point.
2. Use smart antennas.
3. Increase the capacity of AC users.
4. Balance the load.

In wireless networks, interference problems are very common, specifically including hardware failure, external interference, and internal interference, of which co-channel interference is an important form of internal interference, closely related to coverage, frame out-of-step interference, and mode three interference; while RRU failure, antenna feeder arrester, antenna feeder interference, and other aspects of interference problems belong to the system hardware failure interference; illegal use of the wireless network frequency band, clutter, different. The problem of system blockage belongs to the system external interference. For the above problems, the following strategies should be adopted in the optimization process: find the fault point to determine the specific cause of the fault, usually RRU fault, antenna feeder fault, GPS fault, etc.; apply spectrum meter equipment to determine the scanning range of the frequency band area to accurately locate the source of interference; use the detection data and the surrounding environmental factors as the basis to determine the location of the source of interference, and then use the two-way detection technology to use; the line network is used outside the interference source to gradually narrow down the ranking area and finally determine the location of the interference source and optimize it.

Optimization methods

1. Network structure optimization

Due to the use of co-channel networking, the network has very high requirements for network structure, and unreasonable arrangements have a large impact on the network structure, and such problems affect the capacity of the system by affecting the coverage and interference level of the nearby network structure. Therefore, the optimization and adjustment of the problematic points of the network structure can help to improve the network capacity and can be considered to optimize the network structure to balance the service volume with each other and improve its capacity by adjusting the system parameters and power. The optimization process is shown in Figure 5.10.

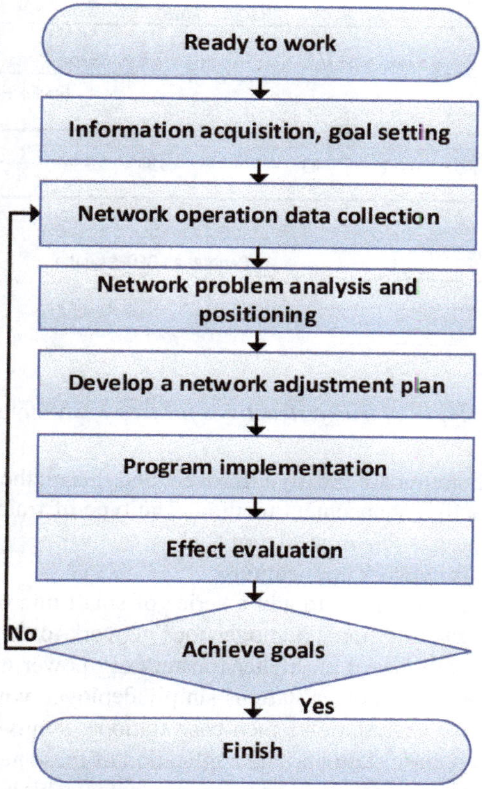

Figure 5.10 Optimize the process

2. Adjust the frequency use strategy

To improve the spectrum utilization, the network is mainly based on the same frequency grouping, but for the problem of capacity degradation mainly caused by interference, in addition to the structural adjustment, frequency adjustment can also be considered, and the heterodyne network can be used locally to avoid the problem of co-channel interference and improve its capacity.

In the process of network capacity optimization, the fractional frequency multiplexing structure is mainly considered, which can improve the utilization of spectrum resources while suppressing the interference between neighboring devices.

3. Carrier frequency expansion.

Carrier frequency expansion is to enhance the ability to absorb business by adding carrier frequency on the original basis. After expansion, the terminal selects the carrier frequency for residing and initiating business according to the field strength, and at the same time, the load balancing function is opened between the original carrier frequency and the expanded carrier frequency, so that it can be evenly carried on multiple carrier frequencies, and the user experience and the business bearing

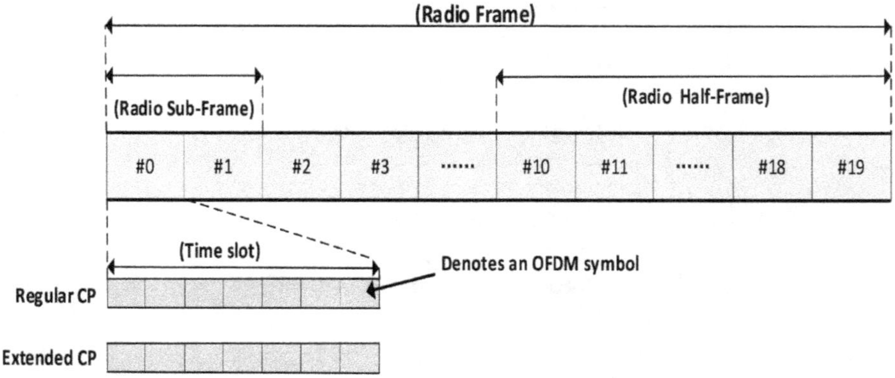

Figure 5.11 Types of frame structures for frequency-division duplexing

capacity of the base station are greatly improved, and this method is the most effective and direct capacity enhancement method. The type of frame structure for frequency division duplex is shown in Figure 5.11.

4. Heterogeneous network applications

A heterogeneous network is to add a series of small micro base stations with various functions to the traditional homogeneous network to enhance the frequency usage efficiency, usually based on higher transmitting power macro base stations, supplemented by some low-power stations simply deployed within the macro base stations, such as micro base stations, pico base stations, femto base stations, and a small number of relay base stations. The application of these new base stations can make the base stations closer to users and services, achieve high-quality coverage of network blind spots and hotspots, absorb and divert services, and increase network capacity.

Small base stations are heat-absorbing layers, heterodyne networks, and the signal quality of small base stations is superior to that of a large network which is the basic coverage layer, so it is suggested that UEs are given priority to reside on small base stations if the minimum access threshold of small base stations is met. It is suggested that the priority of small BTS is set higher than that of a large network.

5. Group scheduling optimization

Packet scheduling is the most important part of the network when it comes to congestion control. Packet scheduling handles non-real-time (NRT) packet service data and determines the initial packet time and the rate that should be used for packet transmission, which has different applications for different types of services. NRT packet data is essentially a burst service, containing one or more data calls. Packets can be scheduled by using time-division techniques, code-division techniques, or both together. Packet scheduling and load control (which includes admission control, AC) work one after the other. A higher load will result in more interference, which means that the network can admit fewer users. This affects the bit rate at which NRT packet data are distributed. Therefore, for packet scheduling, load control is an important parameter to be analyzed and optimized. In network optimization, the transmit power of the downlink and the interference power of the uplink are more important parameters. When the thresholds of both parameters are exceeded, a preventive measurement of the control load is initiated, which allows for improving capacity optimization.

6. Interference optimization

The main sources of interference in the network are UP interference, frequency pollution, interference caused by unreasonable scrambling code planning, interference from outside the system, etc. These interferences, when mixed, often cause KPI deterioration, data service rate degradation, poor user perception, and eventually dropped calls, so it is one of the important problems to be solved for network optimization. The root cause of intra-network interference is the short network channel code and scrambling code, insufficient frequency resources, coupled with the network site layout and base station coverage is not reasonable, the main aspects include distant base station downlink guide frequency (DwPTS) to the near base station uplink guide frequency (UpPTS) interference; inter-cell co-channel interference, including public channel interference and service channel interference. The main guide frequency pollution is the guide frequency strength pollution, that is, when the MS receives more than three Ec/Io guide frequencies with strength greater than T_ ADD, and because the MS's RAKE receiver can demodulate up to three path signals, so the excess strong guide frequencies form a thousand disturbances to the MS's signal demodulation, which will seriously affect the MS's demodulation of downlink signals. The frame structure type of the time division duplex is shown in Figure 5.12.

7. New technology application

As the network develops and evolves, new technologies should be attracted to the network one after another, and the throughput enhanced by way of software and hardware upgrades, such as interference coordination and Massive MIMO, and network optimization methods and concepts should be.

8. AC optimization

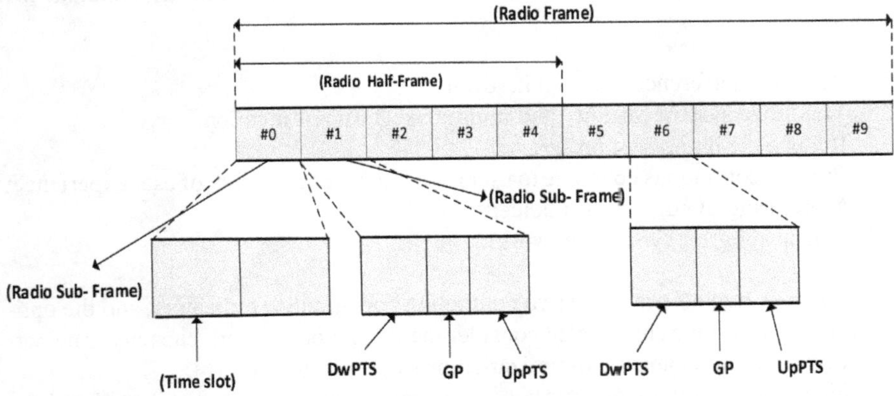

Figure 5.12 Time-division duplexing frame structure type

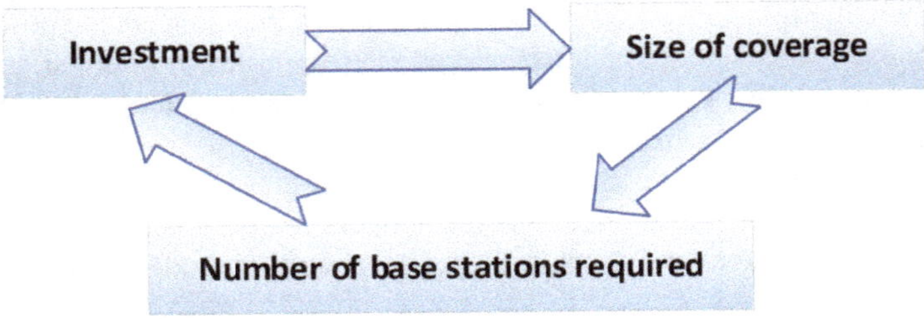

Figure 5.13 System capacity, coverage, and quality relationship

The AC function is related to the load control process. The admission process is important for both real-time and NRT services. Since the AC function is based on power and Joe spit, these parameters are important for capacity and coverage optimization. Two of the most important parameters are the transmit and receive power in the uplink and downlink directions and the orthogonality and Joe spit in both directions.

Analysis of common problems in the network optimization process

1. The relationship between coverage and capacity

In network operation, network optimization based on network coverage and its capacity is an important task for the proper operation of the network. Network coverage and capacity optimization are mainly calculated and planned by using some mathematical models. Considering the actual network operation, it is also necessary to measure the network performance indicators, including drop rate and traffic count. The drop rate can be used to indicate whether there is sufficient coverage in a certain area of the network; the traffic count can be used to detect whether the network has capacity problems. For network coverage and capacity issues, the traditional approach is to collect data through road tests and then analyze it in planning tools to find solutions. The system capacity, coverage, and quality relationship diagram are shown in Figure 5.13.

The main objectives of network coverage and capacity optimization include the following:

1. Reduce interference between hardware;
2. Maximize system capacity and improve spectrum utilization;
3. Reduce the number of probes;
4. Provide continuous coverage to users and improve the quality of user experience;
5. Monitoring of edge performance;
6. Self-healing for synergistic work.

Capacity and coverage are two conflicting performance indicators, and the optimization of coverage should also consider the impact on network capacity. The network coverage and capacity optimization process is as follows.

First, the measurement reports of base stations and users are collected and the data are analyzed to find out the coverage and capacity-related data; second, the control parameters adjusted in the network planning tool are adjusted and the coverage and capacity of the system are optimized and the parameters to be adjusted are passed to the coverage and capacity optimization module; and finally, the optimization module updates the related control parameters and continues to collect the measurement data.

The "breathing effect" is the relationship between capacity and coverage: when the load increases, the capacity increases, the required interference margin increases, and the coverage becomes smaller; conversely, when the load decreases, the capacity decreases, the required interference margin becomes smaller, and the coverage becomes larger. From the respiration effect, we can know that the optimization of capacity and coverage should be considered jointly in the network optimization, unlike the optimization of capacity and coverage in the GSM system, which is considered separately.

Coverage is directly related to performance, and an increase in coverage requires an increase in the average transmit power of the base station. If the system capacity is downlink constrained, then an increase in coverage will result in a decrease in capacity. If the system is uplink constrained, then the capacity is not affected. There are many ways to increase the coverage; the main parameters that affect the link performance (e.g., false block rate, E/N, power control reserve) directly affect the power budget and therefore the coverage. The uplink coverage can be improved by reducing the interference margin, lowering the BTS noise factor, or even increasing the antenna gain. However, processing gain and E/N are the two main factors affecting coverage. Another way to improve the E/N value is to increase multipath diversity. Two signals arriving at two antennas can be coherently combined, while receiver noise can be non-coherently combined. This technique not only provides higher gain but also gives protection against fast fading. Concepts such as antenna down-tilt can also be used for wireless network coverage enhancement.

For capacity and coverage optimization, it is very important to optimize the switching control function components in the network. One of the important parameters involved is the CPIHP transmit power, which affects coverage and should be set as low as possible, the optimal value of this parameter will determine coverage and capacity, and this parameter also affects packet scheduling. If the CPIHP value is not optimal, then either the network is not fully utilized or the number of users is higher than the planned number of users, resulting in significant interference in the system and thus degrading the network quality. These parameters also affect the call success rate and the dropped call rate. As with GSM networks, these two factors directly determine the quality of the network.

2. Network scalability

When optimizing a network, network scalability must first be considered. The network optimization engineer cannot simply assign additional frequencies to the relevant hardware, as is the case for optimizing GSM networks. A defined signal margin must be considered at the beginning of the network optimization and used in the calculations as compensation for the interference due to the increased service volume. This suggests that, from the beginning, the network needs to be built with smaller hardware or more base stations, which would increase the investment cost.

If the service volume signal margin is set too small, then more base stations must be built once the service volume increases significantly in the future.

Before optimizing the network capacity of the base station, it is necessary to carry out detailed data collection and analysis of the business of the base station and clarify the root cause of the capacity problem, in addition, to paying attention to the lack of coverage of the base station signal and coverage of blind areas, and to solve the problem by adjusting the engineering parameters or base station antenna adjustment. When a local BTS is overloaded, it is necessary to further check the soft switching ratio of the BTS, if its soft-switching ratio is very high, the parameter ratio will be switched by soft switching, if the problem cannot be solved, the downward angle, azimuth angle, and height of the BTS antenna can be adjusted. Finally, through the above measures, the local base station with excessive service load shares its service with the surrounding base stations to reduce the load on that base station.

In the network, soft-switching optimization is an important part of improving system capacity. The coverage optimization and power control described earlier not only reduce the inter-cell interference but also reduce the overlapping area of cells and, accordingly, the soft-switching ratio. The access control strategy can also improve the network performance; it can not only ensure the service quality but also maximize the system to provide high capacity so that the service distribution of the system is more rationalized.

3. Integrated multimedia services

The network must meet the needs of different services at the same time. Therefore, network optimization engineers have to consider all kinds of services. Services that do not require high communication quality; on the contrary, some services that require high communication quality have a small coverage area. In this way, the network optimization engineer cannot consider only a single radius in practice, because different services correspond to different radii. If the minimum radius, that is to say, the communication, the major network optimization software companies have developed and developed effective algorithms for this new network-integrated multimedia services. In fact, due to the characteristics of the network itself, such as fast power control, soft switching, and downlink orthogonal codes are not completely "orthogonal," network optimization is much more complex, and the number of parameters to be optimized is also very large, which is often achieved with the help of emulators network optimization.

5.2.3 Training sample collection design

The BP neural network method and the battery neural network model are the same ideas, which are based on the unique characteristics of the neural network. According to the parameters measured in the process of battery charging, the neural network is modeled by the highly nonlinear and autonomous learning mechanism, and the SOC is estimated [172]. The more kinds of input parameters, the more data, and the more accurate the SOC estimation. However, the neural network requires a lot of mathematical operations, which requires high speed and capability of the central processing unit (CPU), and the cost is high. The neural network method is generally complex, and the estimation error is highly dependent on the training data and training methods.

The principle of using a neural network to estimate the SOC of the battery is to take external characteristic parameters of the battery such as voltage and current as input, the system is trained through a large number of sample data, and then the system is taken advantage of to estimate SOC of new input when it reaches the required error range [173]. The advantage of using a neural network to estimate the SOC of the battery is that a certain mathematical model is not required.

The principal aim of the battery state estimation is to use a BP neural network for describing its characteristics. The external battery characteristic parameters are taken as input such as voltage and current. The system is trained for numerous sample data which is then utilized to estimate the state of new input when the state reaches the required error range [173]. The advantage of using a network to estimate the battery state is that there is no need to establish a certain mathematical model. Besides, it can be seen from many pieces of literature that there is a nonlinear relationship between the battery state parameters such as the current and terminal voltage. It can determine this relationship well through sample learning, in which the estimation accuracy will be higher with more sample learning data. It can achieve the best effect on dynamic parameter identification of time-varying systems.

The BP neural network is a multilayer feedforward system. The main features are as follows. The signal propagates forward, and the error propagates backward. The computing network consists of an input layer, a hidden layer, and an output layer. For a given training set, the input consists of attributes and the output is a one-dimensional real-valued variable. The calculation algorithm generally uses a three-layer BP network structure, as shown in Figure 5.14.

The algorithm has been applied to many fields. The PF algorithm is to extract some discrete random particles that are then used to implement the probability density function, instead of the mean of the extracted samples, so that no integral operation is needed. This way, it is possible to estimate the dynamic parameters of the

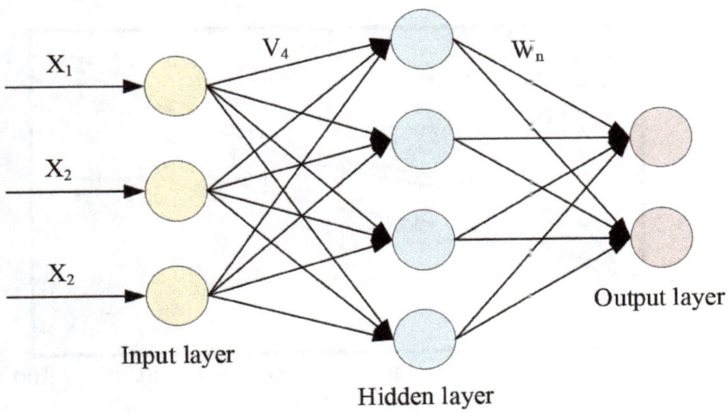

Figure 5.14 BP network structure diagram

system and get a better estimation result. The PF can be tracked down for many complex situations with high precision that achieves efficient estimation. Neural networks are used in almost all fields of classification and regression, and it contributes more to artificial intelligence. These neurons are responsible for building up a network grouping of similar datasets or classes. Hence, it is applied to both supervised and unsupervised learning, which initially shows good results. As the number of nodes is increased, the complexity goes on increasing with a bit difficult to be realized. A small number of neural networks are suited. The SVM recovers these disadvantages and can be applied to huge datasets as well. It is simple to be realized that also uses MLP properties, in which the MLP uses recurrent and feedforward networks. Properties of MLP include the approximation of continuous nonlinear functions that may not provide accurate results.

The experimental sample quality determines the final prediction effect after network training, through the above-built experimental platform for battery discharge data acquisition under different conditions. First, the battery is put at the constant temperature box (constant temperature box temperature adjustable), the wires are used to the battery and battery charging and discharging device to connect, real-time data are acquainted in the process of battery charging and discharging current and voltage, and at the same time, the signal lines are taken advantage of connecting the battery and the external temperature monitoring equipment, real-time acquisition of battery in the process of charging and discharging temperature data. In order to ensure the collection of data can have a stronger representation and can better cover the entire scope of work of the battery, this chapter collects the discharge data of the battery in different temperatures and different discharge rates, respectively. The data figure of the SOC-OCV curve of different discharge rates is shown in Figure 5.15.

The SOC of the battery OCV curve shows that in the middle of the SOC area, the change rate of OCV is small; when the battery SOC < 15%, the rate of OCV increases sharply. The discharge ratio is different, and the discharge current of the

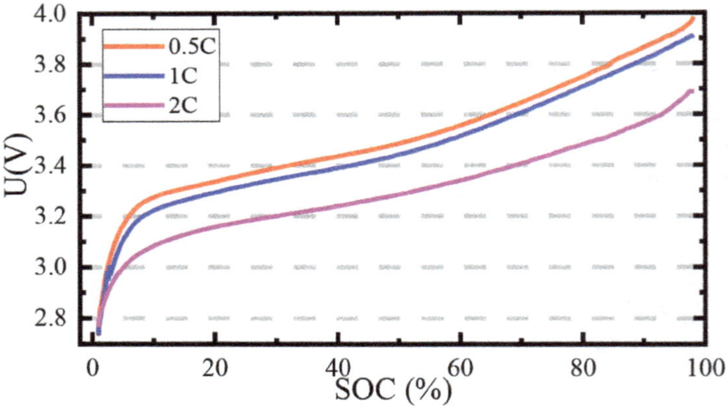

Figure 5.15 Data curve of SOC–OCV under 20°C

battery size is also different; therefore, in the process of discharging the battery, temperature rise is also different.

Deep neural network is a model type as an MLP that is trained by algorithms to learn representations of data sets without any manual design of feature extractors. It consists of a higher or deeper number of processing layers, which contrasts with a shallow learning model with fewer layers of units. The shift has allowed for more complex and nonlinear functions to be mapped, as they cannot be efficiently mapped with shallow architectures. This improvement has been complemented by the proliferation of cheaper processing units such as the general-purpose graphic processing unit and the large volume of data set to train in big data. The general-purpose graphic processing unit is less powerful than the CPU, but the number of parallel processing cores in them is more than CPU cores by orders of magnitude.

This makes general-purpose graphic processing units better for implementing deep neural networks. In addition to the BP algorithm and graphics processing unit, the adoption and advancement of machine learning (ML) and particularly deep learning can be attributed to the explosion of data or big data in the last 10 years. It will continue to impact and disrupt all areas of our lives from education, finance, governance, healthcare, manufacturing, marketing, and others. The neural network is a multilayer feedforward system, the main features of which are described as follows. The signal is forward propagating, and the error is back-propagating. The calculation network consists of an input layer, a hidden layer, and an output layer. For a training set, the input consists of attributes and outputs a one-dimensional real-valued variable.

A neural network is a series of algorithms that endeavor to recognize underlying relationships in a set of data through a process that mimics the way the human brain operates. In this sense, neural networks refer to systems of neurons, either organic or artificial. Neural networks can adapt to changing input. Consequently, the network generates the best possible result without needing to redesign the output criteria. The concept of neural networks, which has its roots in artificial intelligence, is swiftly gaining popularity in the development of trading systems.

In the world of finance, neural networks assist in the development of such processes as time-series forecasting, algorithmic trading, securities classification, credit risk modeling, and constructing proprietary indicators and price derivatives. A neural network works similarly to the human brain neural network. A "neuron" in a neural network is a mathematical function that collects and classifies information according to a specific architecture. The network is analogous to statistical approaches like curve fitting and regression analysis.

A neural network contains layers of interconnected nodes. Each node is a perceptron and is similar to multiple linear regressions. The perceptron feeds the signal produced by a multiple linear regression into an activation function that may be nonlinear. In an MLP, perceptrons are arranged in interconnected layers. The input layer collects input patterns. The output layer has classifications or output signals to which input patterns may map. For instance, the patterns may comprise a list of quantities for technical indicators about security. Hidden layers fine-tune the input weightings until the neural network margin of error is minimal. It is hypothesized

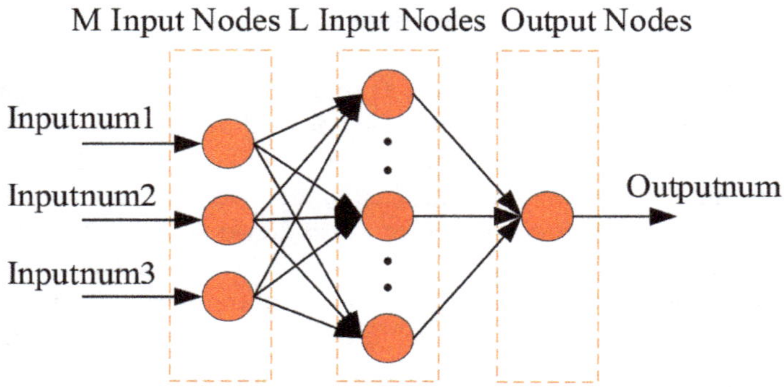

Figure 5.16 Schematic diagrams of the NN training process

that hidden layers extrapolate salient features in the input data that have predictive power regarding the outputs. This describes feature extraction, which accomplishes a utility similar to statistical techniques such as principal component analysis (PCA), as shown in Figure 5.16.

Neural networks are broadly used, with applications for financial operations, enterprise planning, trading, business analytics, and product maintenance. Neural networks have also gained widespread adoption in business applications such as forecasting and marketing research solutions, fraud detection, and risk assessment. A neural network evaluates price data and unearths opportunities for making trade decisions based on the data analysis. The networks can distinguish subtle nonlinear interdependencies and patterns other methods of technical analysis cannot. According to the research, the accuracy of neural networks in making price predictions for stocks differs. Some models predict the correct stock prices 50–60% of the time, while others are accurate in 70% of all instances. Some have posited that a 10% improvement in efficiency is all an investor can ask for from a neural network.

There will always be data sets and task classes that a better analyzed by using previously developed algorithms. It is not so much the algorithm that matters, in which it is the well-prepared input data on the targeted indicator that ultimately determines the level of success of a neural network. Wu et al. applied the bat PF (Bat-PF) to recursively update the parameters of the NN degradation model. The RUL prediction error of the proposed NN_Bat-PF method is 2 cycles in 500 prediction cycles, and the probability density function (PDF) width is 35 cycles. She *et al.* proposed a battery aging assessment method based on incremental capacity analysis and radial basis function neural network (RBFNN) models. The average prediction error of this method reaches 4.00%, the confidential interval of the derived model is 92%, and the prediction accuracy is 90%. Wu *et al.* proposed an online method to estimate the RUL of lithium-ion batteries, using importance sampling (IS) for the feedforward neural network (FFNN) input selection and setting the number of hidden neurons to 40.

The error of this FFNN + IS method in actual operation is less than 5%. Li *et al.* established an indirect prediction method based on Ellman neural network and estimated the RUL of lithium-ion batteries under vibration stress. The mean absolute error (MAE) is 0.0243, and the mean square error (MSE) is 0.0278. Tang *et al.* built a model through the input–output slope and bias correction (SBC) method and further integrated the SBC model into a four-layer feedforward migration NN and trained it via the gradient correlation algorithm. When only the first 30% of the aging trajectory is used for NN training, the RMSE of the prediction is less than 2.5%.

A hybrid neural network is proposed by combining convolutional neural networks and long-short-term memory with the false nearest neighbor method, and the accuracy of the proposed method can reach 98.21%. A method proposed to predict RUL based on EMD, deep neural network (DNN), and long-short-term memory model. Compared with the hybrid model of EMD and autoregressive integrated moving average model, the average value of RMSE is reduced effectively.

A model framework is proposed for battery SOH monitoring and RUL prediction based on temporal convolutional networks (TCNs). For RUL prediction tasks, in the case of offline data, the offline data can be de-noised through empirical mode decomposition (EMD) and then combined with some online data to fine-tune the model to achieve highly accurate RUL prediction. The average RMSE accuracy of the TCN model is 5% higher than that of traditional networks with different starting points. Compared with other models, the average RUL error is nearly eight cycles higher. An online synthesis method is proposed by combining part of the incremental capacity and ANN. The ANN model has only two inputs and has a simple structure. The model estimated MAE and RMSE for RUL are less than 4 and 6 cycles, respectively. A simple model is first created from the current battery aging data set that explains the power decay over time. This base model is then transformed by an input–output SBC method structure to capture the degradation of the target cell. To enhance its nonlinear transfer capability, the SBC model is further integrated into a four-layer NN and easily trained via the gradient correlation algorithm.

NN has many advantages, but it requires very large number of parameters to be estimated. When there is a hidden layer or multiple hidden layers, and multiple input variables, the number of parameters that need to be accurately estimated explodes. However, each parameter should be estimated consuming some of the informational budgets. It is essentially guaranteed to overfit this model. This has nothing to do with the computational feasibility of the algorithm. If a predictive model with continuous variables is built, the logistic regression model is better than NN. It will use fewer parameters. Herein, one or two variables are just used to fit the model, and then the test set is used to see if other variables-only outside the intercept-cause instability and reduce sample accuracy. Regarding the x variable, the PCA can be used to improve its estimation effect, in which only the previous one or two PCs need to be extracted.

5.2.4 Initial parameter weight design

To reduce the estimation error of the network, it is necessary to adjust its weight coefficient. The function always changes the fastest along the direction of the gradient. Then, the partial derivative is obtained for each parameter that needs to be adjusted [174]. Then, the –1 * partial derivative is used to get the value that the parameter needs to change. The output and implicit layers expression is shown in (5.9):

$$\begin{cases} o_k = f(net_k), net_k = \sum_{j=0}^{m} w_{jk}y_i, (k = 1, 2, 3, \ldots, l) \\ y_k = f(net_k), net_k = (k = 1, 2, 3, \ldots, l) \end{cases} \tag{5.9}$$

After identifying the output and input layers, when the output of the network is different from the expected output, the error E can be generated which is defined by (5.10):

$$E = \frac{1}{2}(d - o)^2 = \frac{1}{2}\sum_{k=1}^{l}(d_k - o_k)^2 \tag{5.10}$$

The training process of the calculation algorithm is to continuously modify the weight reduction error of the network. The gradient-descent method is used to update the parameters. The relationship between the hidden layer and the input error is shown in (5.11):

$$E = \frac{1}{2}(d_k - f(net_k))^2 = \frac{1}{2}\sum_{k=1}^{l}\left[d_k - f\left(\sum_{j=0}^{m} w_{jk}y_i\right)\right]^2 \tag{5.11}$$

If the partial derivative is greater than 0, it should change in the opposite direction of the partial derivative. If the partial derivative is less than 0, the change in this direction can be calculated. The variables are brought into the above equations, and they are simplified to get the function relationship, as shown in (5.12):

$$E = \frac{1}{2}\sum_{k=1}^{l}\left(d_k - f\left(\sum_{j=0}^{m} w_{jk}f(net_j)\right)\right)^2 = \frac{1}{2}\sum_{k=1}^{l}\left\{d_k - f\left(\sum_{j=0}^{m} w_{jk}f\left(\sum_{j=0}^{m} w_{jk}x_i\right)\right)\right\}^2 \tag{5.12}$$

As can be known from (5.12), the error of the network is closely related to the weight of each layer. At the same time, the learning rate η is set, the learning rate cannot be too fast or too slow. Too fast may cause the optimal solution to be exceeded. If the learning rate is too slow, it may reduce the efficiency of the algorithm. The parameter expression is shown in (5.13):

$$\begin{cases} \Delta w_{jk} = -\eta\frac{\partial E}{\partial w_{jk}} \; (j = 1, 2, 3, \ldots, l) \\ \Delta v_{ij} = -\eta\frac{\partial E}{\partial w_{ij}} \; (j = 1, 2, 3, \ldots, n; j = 1, 2, 3, \ldots, m) \end{cases} \tag{5.13}$$

wherein the negative sign indicates a gradient decrease, in which η is a constant. It reflects the learning rate varying from 0 to 1. The higher the learning rate, the higher the accuracy, but with the rapid increase of computational complexity.

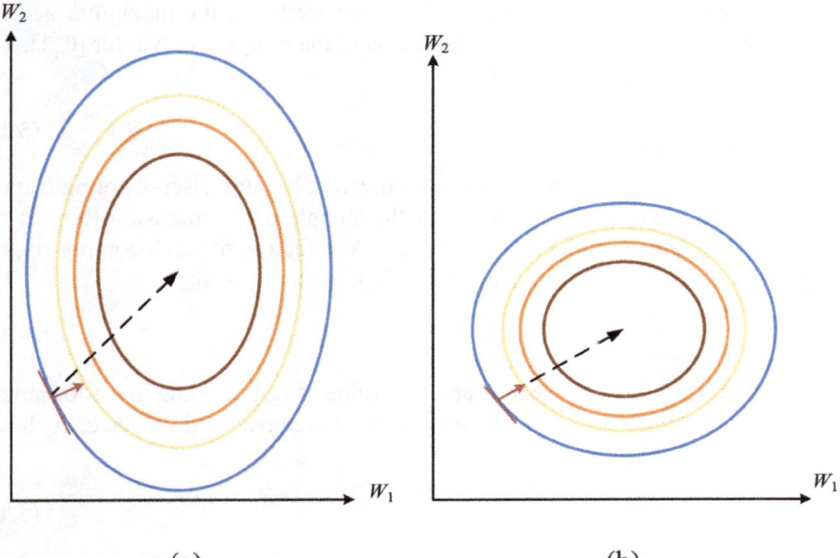

Figure 5.17 The influence of data normalization on algorithm (a) Gradient of unnormalized data (b) Gradient of normalized data

Last through constructing the experimental platform under different working condition of discharge the battery data acquisition, but in the network input, the input variables, the characterization of the physical meaning is different, and the corresponding dimension is also different, such as the change of the voltage range of 2.7–4.2 V, current changes in the range of 0–150, the change in the temperature range for −20°C to 55°C, and lose A vector of SOC range is 0–100%; the source of the characteristics of each dimension and the unit of measure are different; it can cause the eigenvalue distribution difference over big, if not normalized processing; it is difficult to compare and also has no unified evaluation standard; samples of each component in the network training to the training of the network each have the same effect as a result.

Therefore, the data in line with normalized different dimensions can be not only the eigenvalues of the normalized to the same values range and eliminate the correlation between different characteristics but can also improve the convergence speed and precision of the network; the figure shows the scope of training data characteristic value of each dimension is different will lead to most of the position of the gradient direction is not the best search direction, which requires many iterations to convergence, and for the normalized operation. Most of the position of the gradient direction of approximate loss function, the optimal search direction. The algorithm in the selected generation, in which every step is basic to the minimum, greatly improves the efficiency of the algorithm. The commonly used normalization method has Figure 5.17.

1. Scale normalization. Type, x_{max} and x_{min}, respectively, the maximum and the minimum of sample weight, normalization of the sample interval for [0, 1]. The expressions are shown in (5.14):

$$\bar{x} = \frac{x - x_{min}}{x_{max} - x_{min}} \tag{5.14}$$

2. Standard normalization. Standard normalization became z^{-} score normalization, characteristics of each component of the sample data processing to meet the standard of fall, N samples $\{x_i\}$, $i = 1, \ldots, N$, a feature for each component, the mean and standard deviation are shown in (5.15).

$$\bar{x} = \frac{x - x_{min}}{x_{max} - x_{min}} \tag{5.15}$$

wherein x_{max} and x_{min} refer to the max and min value of x. According to the formula to calculate the characteristics of each component of the mean and variance, as shown in (5.16):

$$\hat{x}^i = \frac{x^i - \mu}{\sigma} \tag{5.16}$$

Data samples after (5.16) treatment accord with the standard of fall, the mean to 0, the standard deviation of 1, μ for all the sample data of average, and σ for all sample data of standard deviation.

3. Nonlinear normalization.

Nonlinear normalization is to point to some mathematical function, the original value of the sample data to map, the method including exponential function and logarithmic function, tangent function, and so on; this method is often used in some data distribution larger scene, without detailed in this chapter.

In this section, the Python S_k learn modules are used for preprocessing; min and max scaler class completes the sample data of normalized operation, the principles after normalization are belonged [0, 1], and the foundation data lay into a subsequent sample data of neural network training.

When a neural network topology structure is determined, the number of input layer and output layer neurons network tends to be relatively fixed, the analysis of the ternary lithium-ion battery SOC cannot be obtained through direct measurement, but some parameters including voltage, current, and temperature are used. Some parameters in the process of charging and discharging are there is a strong nonlinear mapping relationship with the battery SOC, so in this chapter, the neural network input layer neuron number is three, respectively, corresponding to the battery voltage, current, and temperature, and the amount of SOC is the need to estimate, as the output of the network, so the number of output layer neuron network for one. The number of nodes in the hidden layer is a sign for the number, not the real number of the experimental model.

How to determine the number of hidden layer neurons, at present there is a scientific theory that is not yet universal, and a method for determining the number of hidden layer neurons to determine a more complex is also a very important problem, is one of the important factors that affect the network performance if the number of deficiencies can lead to the model used to solve the problem of information was obtained from the sample is too little, will cause less fitting; if the number is too much, it will not only increase the complexity of the network, the run time will be longer, and there is a fitting phenomenon. The experience formula is commonly used to determine the number of hidden layers as shown in (5.17):

$$\begin{cases} l = \sqrt{m+n} + a \\ l = \log_2 n \\ k < \sum_{i}^{n} C(m, n) \end{cases} \tag{5.17}$$

where l is the number of hidden layer neurons, m and n, respectively, are the number of input and output layer neuron networks, k is the sample size, a refers to the constant between [1, 10], i is a constant between [0, n].

This chapter combines the experience formula and trial-and-error method to determine the best neurons in hidden layer number. First of all, according to the empirical formula to determine the number of hidden layer neurons is roughly range, the experiment is reused to collect the battery charge and discharge data set in front of the neural network model of python programming training, record the corresponding model of the output MSE and training time, and then more instead of hidden layer neuron number corresponding is trained, until the output of the model of MSE reduce speed slowed or tends to be stable, the corresponding network training data will provide the basis for the determination of a number of hidden layer neurons, the number of hidden layers and the corresponding network output MSE and the corresponding time consuming as shown in Figure 5.18.

The adjustment shows the neural network model of the number of hidden layer neurons, and the corresponding model of the output MSE is changing. With the

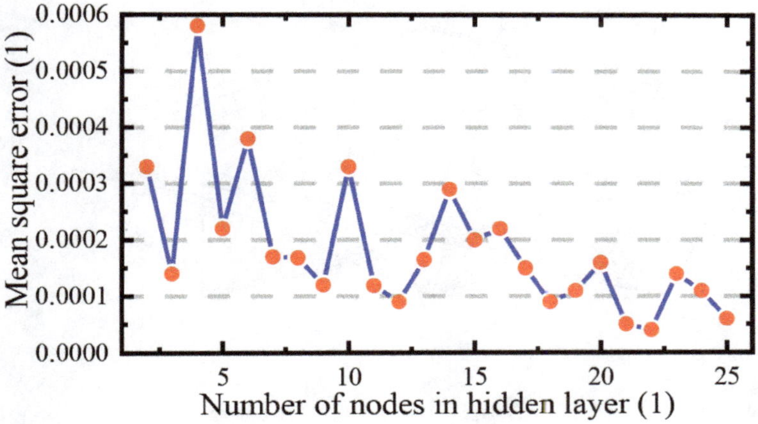

Figure 5.18 Hidden layer number and corresponding MSE diagram

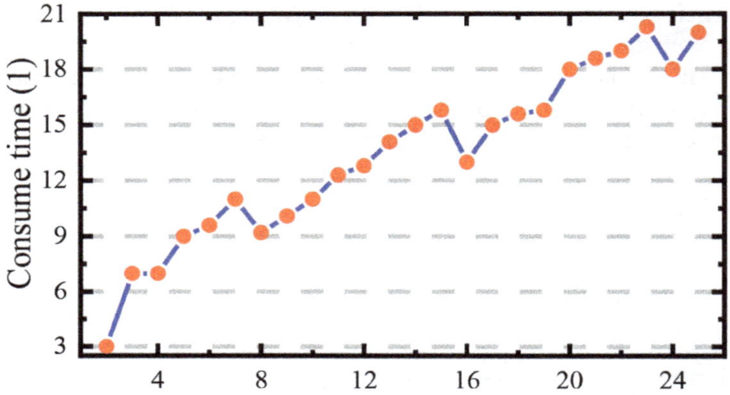

Figure 5.19 *The number of hidden layers and the corresponding running time*

increase in the number of hidden layers, the output of the model MSE is continuously decreasing trend, as shown in Figure 5.19.

Figure 5.19 shows that with the increasing number of hidden layers of the network, the model run time is increased. When the number of hidden layer networks is more than 18, the output of the model decrease of the MSE tends to slow down. The greater the number of hidden layer networks, model the longer run. The output MSE of the model and running time are integrated, and the research finally decided to set the number of hidden layer neuron network to 12. The network topology structure is shown in Figure 5.20.

According to Figure 5.20, the number of hidden layer neuron networks is set as 12. There are three input layers, including current, voltage, and temperature. The number of output layers is 1.

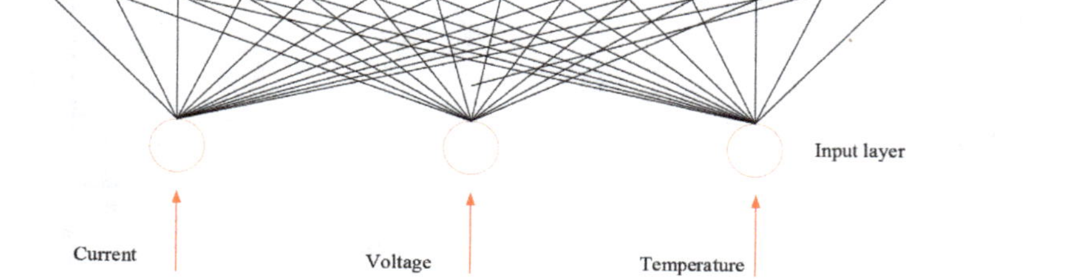

Figure 5.20 *Final network topology diagram*

5.3 Radial basis function-based SOC estimation

5.3.1 Radial basis function neural Network

RBFNN is a single-hidden layer, function approximation-based FFNN proposed in the late 1980s [175]. With the maturity of research, RBFNN has received great attention from researchers in various fields for its simple structure, strong nonlinear approximation capability, and good generalization ability and is widely used in many research fields such as pattern classification, function approximation, and data mining. RBFNN has the following advantages: (1) better approximation ability, classification ability, and learning speed than BP neural network; (2) simple structure, concise training, fast learning convergence, and ability to approximate arbitrary nonlinear functions; and (3) overcome the problem of local minima. The reason is that its parameter initialization has a certain method and is not random initialization.

RBFNN is a forward neural network with three layers, including an input layer, a hidden layer, and an output layer. The input layer consists of signal source nodes, which only act as data transmission and do not transform the input information in any way [176]. Unlike BP neural network, the hidden layer of RBFNN is composed of a radial basis function (usually Gaussian function), which is not connected with the weight of the input layer, and the input vector is directly mapped to the hidden layer. The center point of the radial basis function determines the mapping relationship between the input and hidden layers. The implicit layer to the output layer is linear mapping, and the output is a linear weighting of an implicit layer. The RBF neural network with *M* inputs, *N* hidden nodes, and one output is shown in Figure 5.21.

In Figure 5.21, $(x_1, x_2, \ldots, x_m)$ is the input layer, and the input when predicting SOC is the terminal voltage and load current of the lithium power battery, $(\theta_1,$

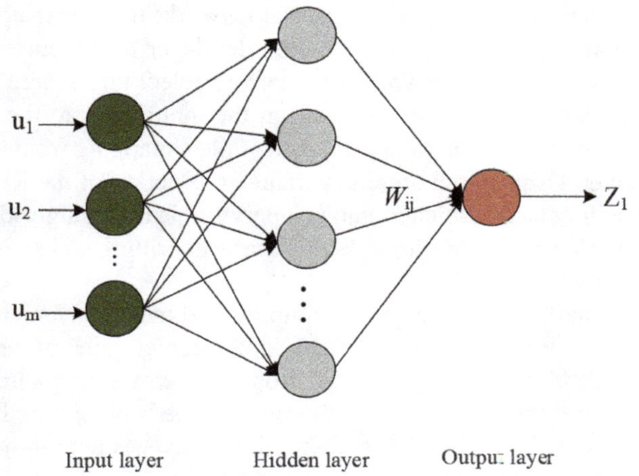

Figure 5.21 RBF neural network model

$\theta_2,\ldots\theta_n$) is the hidden layer, and the radial basis function θ_n is centered on c_i, which constitutes the hidden node in the hidden layer; y is the output layer, and the output result is the SOC of the battery. W_{ij} is the weight of the hidden layer and output layer.

There are many kinds of radial basis functions commonly used, but the Gaussian function with radial symmetry and concise expression is often selected, as shown in (5.18):

$$\varphi_i(x) = \exp\left(-\frac{x-c_i}{2\sigma_i^2}\right), \quad i = 1,2,\ldots\ldots,n \tag{5.18}$$

In (5.18), σ_i is the width of the basis function around the center point, c_i is the center value of the ith basis function, and $\varphi_i(x)$ is the output value of the ith implicit node. The output of the whole network can be obtained by linearly weighting the output results of the hidden layer, as shown in (5.19):

$$y_i = \sum_{k}^{j=1} w_{ij}\varphi_i(x), \quad j = 1,2,\ldots\ldots,k \tag{5.19}$$

In (5.19), W_{ij} is the connection weight between the ith basis function and y, and j is the number of output nodes. RBF neural network has the advantages of high speed, simple structure, and no local extremum, but the disadvantage is that the three parameters W, C, and δ_i cannot be determined [177]. At present, the trial-and-error methods such as computer selection, design, and retest are generally used, which have high blindness and low efficiency. The optimization strategy of RBF mainly focuses on the selection of its parameters.

When using RBFNN to predict SOC, the unknown parameters that need to be learned include the center value, expansion constant, and weight value of the basis function.

5.3.2 Neural network parameter learning method

According to the structural characteristics of the RBF neural network, there are three learning and solving parameters of this neural network: the center and variance of the basis function and the weight from the hidden layer to the output layer [178]. And training an RBF neural network is to learn to select these three most suitable parameters, and complete two mappings from the input layer to the hidden layer, from the hidden layer to the output layer, and complete mapping from the input layer to the output layer. Usually, the optimal parameter selection of the RBF neural network includes a direct calculation method, supervised learning algorithm in the data center, K-means clustering algorithm, least-square algorithm, and so on.

1. Direct calculation method

The direct calculation method is the simplest and most convenient of all methods, but it has some limitations. In this method, the center value of the hidden layer of the RBF network is randomly selected from a given training sample, which is determined and does not change. After the center value is determined, the mapping relationship between the input layer and the hidden layer is determined, the output value of each hidden layer unit can be obtained, and the weight between the hidden layer and the output layer can be solved by a series of linear equations. Therefore, this method is only applicable to the case where the distribution of sample data is obvious.

2. Supervised learning algorithm

Similar to the classical BP neural network, the most common situation is to train the center of radial basis function, the expansion constant, and the weight to the output layer in the hidden layer of the RBF neural network using a supervised learning algorithm, and all parameters are error-corrected and learned.

In the single-output RBF neural network, the squared error of the objective function is assumed to be shown in (5.20):

$$\begin{aligned} E &= \frac{1}{2}\sum_{P}^{i=1} e_i^2 \\ e_i &= d_i - F(X_i) = d_i - \sum_{M}^{j-1} w_j G(X_i - c_j) \end{aligned} \tag{5.20}$$

where p is the number of sample sets and e_i is the error of the input ith sample under the condition of ignoring the threshold. In order to obtain the best result, the objective function should be as small as possible, even if the correction value of each parameter is proportional to its negative gradient, such as (5.21):

$$\begin{cases} \Delta c_j = -\eta \frac{\delta E}{\delta c_j} \\ \Delta \delta_j = -\eta \frac{\delta E}{\delta \delta_j} \\ \Delta w_j = -\eta \frac{\delta E}{\delta w_j}, \quad \eta \in (0, 1) \end{cases} \tag{5.21}$$

The correction equation of each parameter can be derived, as shown in (5.22):

$$\begin{cases} \Delta c_j = \eta \frac{w_j}{\delta_j^2} eG(\| X_i - c_j \| (X_i - c_j)) \\ \Delta \delta_j = \eta \frac{w_j}{\delta_j^3} eG(\| X_i - c_j \| \| X_i - c_j \|^2) \\ \Delta w_j = \eta e_i G(X_i - c_j), \quad \eta \in (0, 1) \end{cases} \tag{5.22}$$

where η is the learning rate of the RBF neural network.

3. *K*-means clustering algorithm

K-means clustering algorithm is an unsupervised learning algorithm. Clustering means that objects with the same characteristics are clustered together as a cluster. The more similar the objects in the cluster, the better the clustering effect. *K* in *K*-means means that there are *k* clusters, which means that in each cluster, the average of all objects in the cluster is the center of the cluster, which is also called the center of mass. The centroid represents the characteristics of this cluster. In the clustering algorithm, first, *K* objects are randomly selected as the initial clustering centers in the sample set, the remaining samples are distributed to the nearest clustering centers, and the samples are integrated into *K* clusters, and then updated and redistributed until there is no change, that is, the clustering is completed.

In the *K*-means clustering algorithm, if the average value of the *i*th cluster Γ_i is *M*, *Y* is the sample data in the cluster, and there are *N* sample data in the cluster, the average value *M* and the error square *J* can be obtained as shown in (5.23):

$$\begin{cases} m = \frac{1}{N_i} \sum\limits_{y \in \Gamma_i} y \\ J = \sum\limits_{k}^{i=1} \sum\limits_{y \in \Gamma_i} |y - m|^2 \end{cases} \tag{5.23}$$

The smaller *j* is, the better the clustering result is.

Let the cluster centers of the *t*th cluster be $c_1(t)$, $c_2(t)$, ..., $c_k(t)$, and the corresponding clusters are $\Gamma_1(t)$, $\Gamma_2(t)$, ..., $\Gamma_k(t)$. When selecting the cluster center for the first time, $t = 1$, and selecting *S* different samples from the sample set, the distance between the selected sample and the cluster center is shown in (5.24):

$$d = X_j - c_i\left(t\right), i = 1, 2, \ldots, s, j = 1, 2, \ldots, N \tag{5.24}$$

After classifying according to the distance from small to large, the cluster center is recalculated, as shown in (5.25):

$$c_i\left(t+1\right) = \frac{1}{N} \sum_{x \in \Gamma_i(t)} x, i = 1, 2, \ldots, s \tag{5.25}$$

If $c_i(t+1) \neq c_i(t)$, start calculating the distance between the selected sample and the cluster center again and classification. If $c_i(t + 1) = c_i(t)$, the expansion constant is shown in (5.26):

$$\sigma_i = \lambda d_i, d_i = \min_i \left| c_j - c_i\left(t\right) \right| \tag{5.26}$$

where λ is the overlap coefficient, from which the output value of the hidden layer unit can be calculated.

4. Least squares algorithm

The orthogonal least square method comes from the linear regression model and is an important learning method in the neural network. After the center and expansion coefficient of the RBF kernel function are determined, the weight of the output unit can be calculated by the orthogonal least square method. In an RBF network with only one output neuron, its training sample is shown in (5.27):

$$\left\{X_n, d\left(n\right)\right\} \left(n = 1, 2, \ldots, N\right) \tag{5.27}$$

where X_n is the input data vector of the RBF neural network, *d*(*n*) is the expected output value, and *N* is the number of samples in the sample training set. The linear regression model is shown in (5.28):

$$\begin{gathered} d\left(n\right) = \sum_{M}^{i=1} p_i\left(n\right) w_i + e\left(n\right), M < N \\ p_i\left(n\right) = G\left(X_n - t_i\right), n = 1, 2, \ldots, N, i = 1, 2, \ldots, M \end{gathered} \tag{5.28}$$

where M is the number of hidden layer units, p_i is the regression operator, w_i is the model parameter, that is, the weight of the output layer connected to the hidden layer, and $e(n)$ is the residual error. The matrix form is shown in (5.29):

$$\mathrm{d} = \mathrm{PW} + e \tag{5.29}$$

where $\boldsymbol{P}$ is the regression matrix and the choice vector $\boldsymbol{P}_i$ is the solution to the regression equation. Once $\boldsymbol{P}_i$ is selected, the model parameter vector can be solved by linear equations. The orthogonal least square method is to learn and select the appropriate $\boldsymbol{P}_i$ and it's number M. The learning process is to orthogonalize $\boldsymbol{P}_i$, analyze the influence on reducing residual error, select the regression operator with significant residual error reduction, and determine its number. Orthogonal decomposition of $\boldsymbol{P}$ is shown in (5.30):

$$\mathrm{P} = \mathrm{UA} \tag{5.30}$$

where U is the N*M matrix with orthogonal columns and **A** is the upper triangular matrix. The least square solution is calculated, as shown in (5.31):

$$\hat{\mathbf{g}} = \mathbf{H}^{-1}\mathrm{U}^{\mathrm{T}}\mathbf{d} \tag{5.31}$$

The error compression ratio is calculated, as shown in (5.32):

$$\varepsilon_i = \frac{\mathbf{g}_i^2\mathbf{u}_i^{\mathrm{T}}\mathbf{u}_i}{\mathbf{d}^{\mathrm{T}}\mathbf{d}}, 1 \le i \le M \tag{5.32}$$

According to the value of ε_i, important regression operators can be selected. The upper triangular matrix **A** is calculated, as shown in (5.33):

$$\mathbf{AW} = \mathbf{G} \tag{5.33}$$

That is, the connection weight between the hidden layer and the output layer can be solved. According to the learning algorithm mentioned above, the center, base width, and weight error of the RBF network are corrected by clustering and the least square method. The flow chart of the RBF algorithm is shown in Figure 5.22.

The following is the process of realizing the RBF network algorithm: (1) Initialize RBF neural network parameters. As mentioned above, the center and expansion coefficient of the radial basis function can be initialized by the K-means clustering method, and the first K samples in the training sample set can be randomly selected as the centroid: the values between [0,1] can be randomly selected as the weights for the initialization of weights (in general, all weights are selected as 1); (2) giving the learning rate N and the precision requirement ε value until the iteration ends;

3. The input vector $\mathbf{X} = [x_1, x_2, ..., x_m]^{\mathrm{T}}$ and the output vector $\mathbf{R} = [Y_1, Y_2 ..., Y_k]^{\mathrm{T}}$ of the training set sample are determined, where n is the number of neurons in the input layer and m is the number of neurons in the output layer, and find the output vector $Q = [q_1, q_2 ..., q_n]^{\mathrm{T}}$ of the hidden layer unit.
4. The RMSE value of the output layer is calculated:

$$\varepsilon_i = \frac{\mathbf{g}_i^2\mathbf{u}_i^{\mathrm{T}}\mathbf{u}_i}{\mathbf{d}^{\mathrm{T}}\mathbf{d}}, 1 \le i \le M \tag{5.34}$$

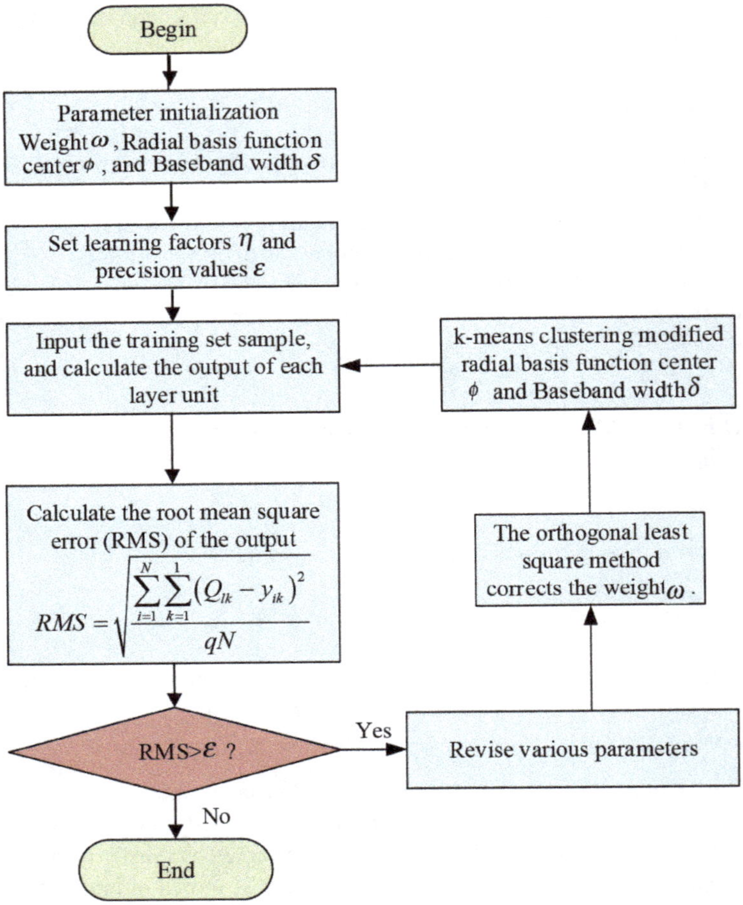

Figure 5.22 RBF algorithm flow chart

Comparison with the ε value to see if the required accuracy is achieved (it is generally not achieved after initialization of parameters). If the accuracy is not met, continue step (5), stop training when the required accuracy is achieved, and the algorithm is terminated.

5. The weight is corrected by the least square method introduced above, and the center and expansion coefficient of the radial basis kernel function are corrected by the *K*-means clustering algorithm.
6. Returning to step (4) until the end of the cycle.

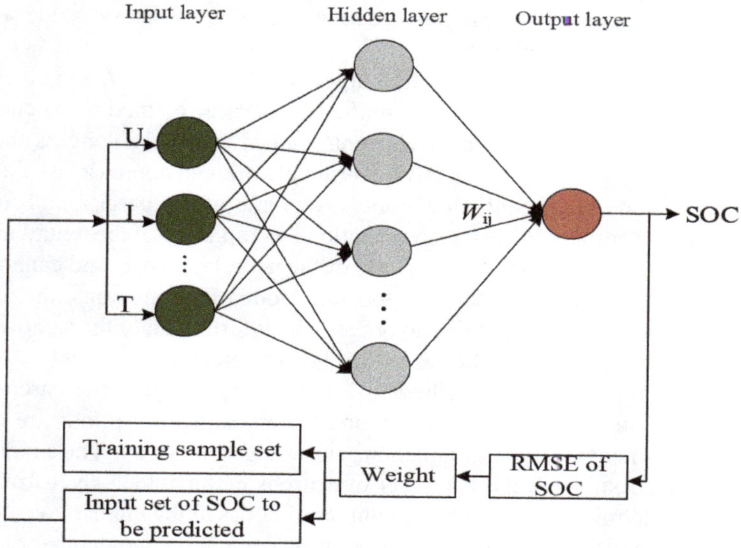

Figure 5.23 Flow chart of SOC estimation based on RBF algorithm

5.3.3 Improvement of RBFNN model for estimating battery SOC

In real life, the operating conditions of lithium-ion batteries for EVs are complex, and many external factors affect the working characteristics of lithium-ion batteries, thus limiting the estimation accuracy of battery SOC [179]. The main factors are working temperature, charging and discharging rate, cycle times, self-discharge, etc.

Because the self-discharge of the battery pack is uncontrollable, this section mainly characterizes the influence of temperature, discharge voltage, and discharge rate on battery SOC. Therefore, voltage, current, and other influencing factors are taken as three input parameters of the RBF neural network to predict SOC, and the SOC value is the only output of the neural network. Therefore, a basic three-layer neural network is constructed, and the network model proposed by A.N. Andrey Kolmogorov can approximate the nonlinear function with arbitrary precision in theory, as shown in Figure 5.23.

This chapter adopts the RBF neural network SOC prediction model shown in Figure 5.23. In this model, there are *M* neurons in the input layer, which are the voltage, current parameters, and other influencing parameters when the lithium battery discharges. The radial basis kernel function of the hidden layer adopts the Gaussian function, and the number of neurons is unknown, so it is necessary to find the appropriate number of neurons. There is only one neuron in the output layer, which represents the predicted result, that is, the SOC value of the lithium battery. In the process of training the sample set, by calculating the RMS value of the RMSE of the output layer, adjusting the connection weight by the orthogonal least square method, training the sample set again until the ideal error accuracy is reached, and inputting the test set after stopping the training, the SOC prediction value with the desired accuracy can be achieved.

Obviously, the input layer is the discharge voltage U, current I, and other influencing factors. At present, there is no unified theoretical method to calculate the appropriate number of hidden layer elements, but in theory, as long as there are enough hidden layer elements, the RBF neural network can approximate the linear nonlinear function of any continuous function with arbitrary precision. Too many hidden layer neurons will bring the computational complexity of the neural network to a sharp increase, reduce the learning rate of the neural network, and cannot guarantee real-time prediction. If there are too few hidden layer neurons, the network will not have enough learning ability to process the information of the sample set, so it is necessary to select the number of hidden layer neurons appropriately.

At present, there are many methods to determine the number of neurons in the hidden layer of the RBF neural network, and the mainstream methods are random selection experiments, empirical equation, clustering, and so on. The random test method is to randomly select the number of neurons in the hidden layer from small to large, and determine the appropriate number of nodes in the hidden layer through experimental prediction. According to the empirical equation, the number of hidden layer nodes M is calculated, as shown in (5.35):

$$\begin{gathered} m = \sqrt{n + l} + \alpha \\ m = \log_2 n \\ m = n = \sqrt{nl} \end{gathered} \tag{5.35}$$

where α is a random constant in the interval [1, 10], and N and L are the number of neurons in the input layer and the output layer, respectively.

RBF neural network is a forward neural network that can approximate nonlinear functions with arbitrary precision. Compared with the traditional BP neural network, it has global approximation ability and can avoid falling into local optimum. However, in an RBF network, the connection weight between the hidden layer and the output layer is a crucial parameter, which directly affects the reliability of the prediction results. Because the performance of the K-means clustering algorithm in optimizing the parameters of RBF kernel function is perfect enough for this subject, this paper improves the prediction accuracy by improving the algorithm of modifying the connection weights in the process of training the sample set. Subjectively, the least square method has regarded the mapping from hidden layer to output layer as a linear mapping and needs to calculate the inverse matrix, ignoring the assumption that the inverse matrix does not exist. When the sample vector is very large, it is time consuming to calculate the inverse matrix. The particle swarm optimization (PSO) algorithm is used to optimize the connection weight between the hidden layer and output layer of the RBF neural network, to improve the accuracy and learning rate of SOC prediction.

The PSO algorithm is one of the swarm evolutionary algorithms (EAs). In the PSO algorithm, the solution to the optimization problem is to search the particles in the space, and all particles have the fitness controlled by the optimization function.

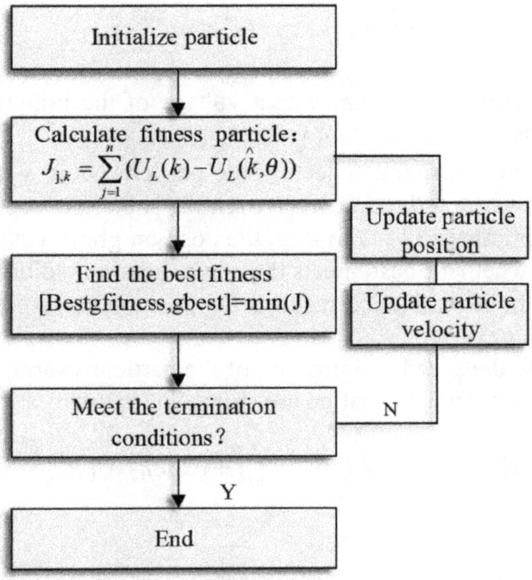

Figure 5.24 Flow chart of PSO algorithm the specific steps of the improved PSO algorithm are as follows

Each particle has a vector velocity that determines the direction and distance, and the particle follows the best particle search in the current space. The value of the objective function obtained at the best position of the particle is called the individual extreme value (*p*best), and the maximum or minimum value among all the individual extreme values of the particles in the global extreme value (*g*best), which corresponds to the position of the global optimal particle. Particles update themselves by tracking *p*best and *g*best in each iteration. The flow chart of the PSO algorithm is shown in Figure 5.24.

Step 1: Let $i = 0$ and initialize a group of particles with dimension m. Each particle represents a possible solution in the solution space. In order to have a faster convergence speed and calculation accuracy, it is necessary to set the scope of the solution space and the maximum search speed according to historical experience. Initialize a group of three-dimensional particle swarms according to the number of parameters to be identified, including the number, speed, and position setting of the particle swarm.

Step 2: Calculate the fitness of particle population at the time i. Calculate the fitness of each particle. The fitness function is a standard for judging the advantages and disadvantages of the particle position. In order to make the identified parameters accurately characterize the space state of the lithium battery, the absolute difference between the model output value corresponding to the particle position and the actual output value of the battery is selected as the fitness function j, as shown in (5.36):

$$J = \sum_{j=1}^{n} \left(\left| U_m(k) - U_m\left(\hat{k}, \theta\right) \right| \right) \tag{5.36}$$

where $U_m(k)$ represents the actual output voltage of the lithium-ion battery, and $U_m(k, \theta)$ represents the model output voltage.

Step 3: Selecting particles. Selecting the particle pbest_i with the best fitness from the particle swarm at the time $t = i$, even if j is the local optimal value. Then, from each local optimum in $t = 1$ to $t = i$, the position gbest_i with the best fitness is selected. Judging whether gbest_i meets the convergence condition, if so, the global search ends, and the identification result is the position corresponding to gbest_i. If not, proceed to the next step.

Step 4: Update the speed and position of the particle swarm. The speed update equation and position update equation are shown in (5.36):

$$\begin{cases} v_{id}^{k+1} = w * v_{id}^{k} + c1 * \eta_1 * \left(P_{id}^{k} - x_{id}^{k}\right) + c2 * \eta_2 * \left(P_{gd}^{k} - x_{id}^{k}\right) \\ x_{id}^{k+1} = x_{id}^{k} + \gamma v_{id}^{k+1} \end{cases} \tag{5.37}$$

Among them, v_{id}^{k+1} is the speed at the moment of $k + 1$ and ω is the inertia weight coefficient, which is generally taken as a value between 0.4 and 0.9. The larger the value of ω, the larger the range of particle search, but the convergence speed will also slow down correspondingly. $c1$ and $c2$ are self-cognitive weight coefficient and social cognitive weight coefficient, respectively, and their values determine the following situation of particle position with local optimum and global optimum. η_1 and η_2 are random numbers between (–1,1).

Steps 2–4 are carried out cyclically and iteratively and keep approaching the optimal value until the convergence condition is reached.

The PSO algorithm is easy to fall into the local optimum, so it is considered to improve the PSO algorithm with an adaptive adjustment strategy to improve the convergence speed and estimation accuracy of the algorithm as a whole. It can be seen from (5.37) that the search speed of particles changes with the value of inertia weight w, and a larger inertia factor is beneficial to jump out of the local minimum and facilitate global search, while a smaller inertia factor is conducive to an accurate local search of the current search area to facilitate algorithm convergence; However, if ω is too large, it will easily lead to premature convergence and oscillation near the global optimal solution in the later stage of the algorithm. In order to make a proper trade-off between search speed and accuracy of results, a PSO algorithm based on adaptive weight is established, and an improved PSO algorithm whose inertia weight ω dynamically changes with iteration times and particle position is established. The equation of dynamic change of inertia factor ω is shown in (5.38):

$$\begin{cases} w = w_{min} + \dfrac{\left(w_{max} - w_{min}\right) * \left(J - J_{min}\right)}{J_{avg} - J_{min}}, J < J_{avg} \\ w = w_{max}, J > J_{avg} \end{cases} \tag{5.38}$$

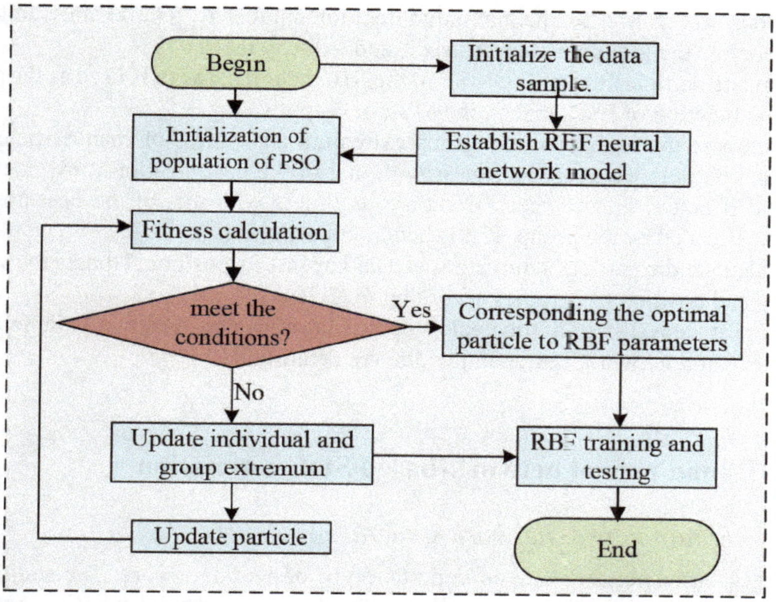

Figure 5.25 Flow chart of the PSO-RBF neural network

where J represents the current target function value of particles and J_{avg} represents the average target value of particles. According to the iteration times of the algorithm and fitness function of particles, the inertia weight ω will increase when the target values of each particle tend to be consistent or local optimum but will decrease when the target values of each particle are scattered. Meanwhile, for particles whose target function value is better than the average target value, the corresponding inertia weight factor is smaller, thus retaining the particle. On the contrary, for particles whose target function value is worse than the average target value, the corresponding inertia weight factor is larger, which makes the particle orientation better.

The PSO algorithm iterates a part of the previously adjusted weights to the next one and generates new weight changes according to the BP characteristics of the RBF network, which can quickly achieve the training accuracy required by the RBF network in several iterations. The flow chart of the PSO-RBF neural network is shown in Figure 5.25.

The detailed steps of the RBF neural network for the PSO algorithm are as follows.

1. Use a *k*-means clustering algorithm to cluster the samples and determine the number of basis function centers.
2. Initialization of particle swarm. Determine the particle swarm size, that is, the particle swarm dimension, and establish the mapping between the dimension space of PSO particles and the connection weights of neural networks. Set the

group size as $N = 20$, the maximum iteration number $T_{max} = 100$, and randomly generate the group position matrix X and velocity matrix V.

3. Update particle fitness. The MSE of the RBF neural network is used as the adaptive function of PSO, that is, the MSE of output voltage.
4. Compare the individual and group extremum and fitness of each particle. For each particle, compare its fitness with that of the best position it experienced, and if better, update *p*best. Compare its fitness with that of the best position experienced by the group. If it is better, update gbest.
5. Calculate the particle's moving speed and update its position. Update the velocity and position of particles according to (5.20).
6. Repeat steps 3–5 until the cycle stop condition is met. And two parameters of the neural network, center and width, are determined.

5.4 Elman neural network-based SOC estimation

5.4.1 Elman neural network structure

Feedback-based networks are an important type of neural network. The addition of inter- or intra-layer feedback links allows them to express the delay in time between the input and the output and therefore needs to be described by dynamic equations, whereas feedforward networks only implement nonlinear mappings [180]. Elman networks, a class of neural networks with local feedback, were originally proposed by Elman in 1990 based on the Jordan network, who proposed Modified Elman Networks, which are now mostly used as the standard Elman networks.

Currently, the most widely used in the field of lithium battery charge state measurement is the series Kalman algorithm, although there are also neural networks introduced into lithium battery charge state prediction [181]. The use of static feedforward networks to identify dynamic systems is changing the dynamic time modeling problem into a static modeling problem, which inevitably leads to many problems.

A network that can reflect the dynamic characteristics of the system more directly and vividly should be a dynamic neural network, that is, a regression neural network (recurrent neural network), and the Elman regression network is a typical dynamic neuronal network, which is based on the basic structure of the BP neural network and is equipped with the function of mapping dynamic features by storing the internal states, so that the system adapt to time-varying characteristics.

Its main structure is a feedforward connection, consisting of an input layer, an implicit layer, and an output layer, whose connection rights can be modified by learning; the feedback connection consists of a set of "structural" units, which are used to remember the output values of the previous moment, and whose connection rights are fixed. In this network, in addition to the normal implicit layer, there is a special implicit layer called the association layer (or linking unit layer),

This layer receives feedback signals from the implicit layer, and each implicit layer node has a corresponding association layer node connected to it. The role of

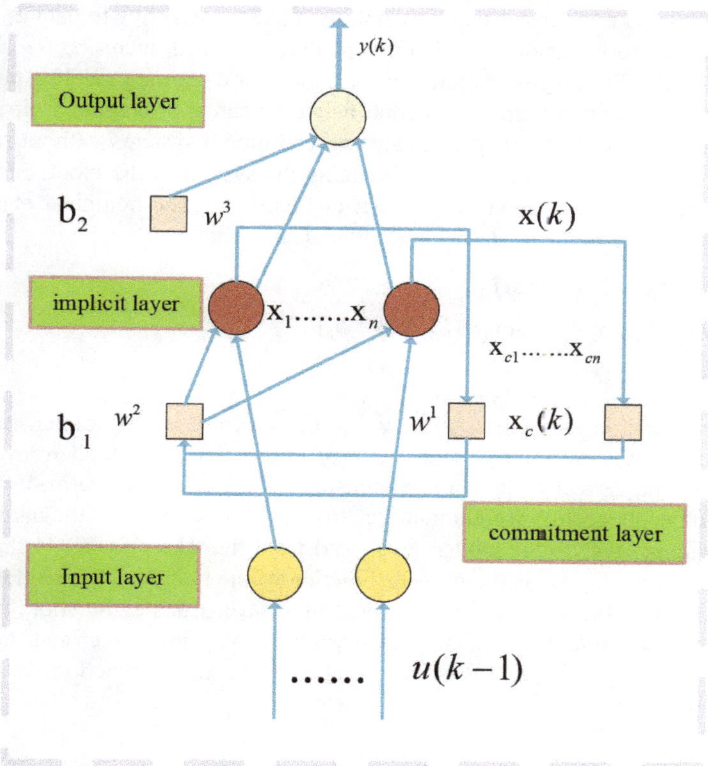

Figure 5.26 *Elman network structure*

the association layer is to use the state of the hidden layer at the previous time as input to the hidden layer, together with the network input at the current time, through linkage memory, which is equivalent to state feedback. The transfer function of the hidden layer is still some kind of nonlinear function, usually a sigmoid function, the output layer is a linear function, and the association layer is also a linear function.

Elman neural networks (ENNs) are typically dynamic neural networks with four layers: input layer, intermediate layer (implicit layer), commitment layer, and output layer, as shown in Figure 5.26.

The connection between the input, implicit, and output layers is similar to that of a feedforward network, with the units in the input layer acting only as signal transmission and the units in the output layer acting as linear weighting.

The transfer function of the cells in the implicit layer can be linear. It is also called the context layer or state layer, which is used to remember the output value of the implicit layer units at the previous moment and can be considered as a one-step delay operator.

The Elman regression neuron network is characterized by the output of the implicit layer being self-linked to the input of the implicit layer through the delay and storage of the take-up layer. This self-linking makes it sensitive to data on historical states, and the inclusion of an internal feedback network increases the ability of the network itself to process dynamic information for dynamic modeling purposes.

In addition, Elman regression neural networks can approximate arbitrary nonlinear mappings with arbitrary accuracy and can model systems without regard to the specific form of external noise influencing the system, if the input–output data pairs of the system are given. As shown in Figure 5.1, the nonlinear state space expression for the ENN, as shown in (5.39–5.41):

$$y(k) = g\left(w^3 x(k) + b_2\right) \tag{5.39}$$

$$x(k) = f\left(w^1 x_c(k) + w^2\left(u(k-1)\right) + b_1\right) \tag{5.40}$$

$$x_c(k) = x_c(k-1) \tag{5.41}$$

where k denotes the moment, y, x, u, x_c. Each denotes a one-dimensional output node vector, an m-dimensional vector of node cells in the hidden layer, an n-dimensional input vector, and an m-dimensional feedback state vector. w^1, w^2, and w^3 denote the connection weight matrices from the hidden layer to the output layer, from the input layer to the hidden layer, and from the take-up layer to the hidden layer, respectively. $f(x)$ is the transfer function of the neurons in the hidden layer, using the transit function, and $g(.)$ is the output layer transfer function, using the purely function. b_1 and b_2 are the closed values of the input layer and the hidden layer, respectively, where $f(x)$ is mostly taken as a sigmoid function, as shown in (5.42) and (5.43):

$$f(x) = \frac{1}{1 + e^{-x}} \tag{5.42}$$

$$y_k = w^3 x(k) \tag{5.43}$$

The ENN learning algorithm uses an optimized gradient-descent algorithm, the adaptive learning rate momentum gradient-descent BP algorithm, which both increases the training rate of the network and effectively inhibits the network from falling into local minima.

Learning aims to modify the weights and closures by the difference between the actual output value of the network and the output sample value so that the sum of squared errors in the output layer of the network is minimized. The objective is to minimize the sum of squares of the errors in the output layer. The objective is to minimize the sum of squares of the errors in the output layer. Let the actual output vector of the system at the kth step be $y_j(k)$, in the time period (0, T) define the error function, as shown in (5.44):

$$E = \frac{1}{2}\sum_{k=1}^{T}\left[y_d(k) - y(k)\right]^2 \tag{5.44}$$

Taking w^2 and w^3 as an example, by taking the partial derivative of E with respect to w^2 and w^3, respectively, we can obtain the weight correction formula as (5.44) and (5.46):

$$\Delta w_{1j}^3(k+1) = (1-mc)\,\eta\,(y_d(k)-y(k))\,.g'(.)\,x_j(k) + mc\Delta w_{1j}^3(k) \tag{5.45}$$

$$\text{j} = 1,2,3\ldots\ldots\ldots\ldots\ldots..m \tag{5.46}$$

$$\Delta w_{jq}^2(k+1) = (1-mc)\,\eta\,(y_d(k)-y(k))\,.f_j'(.)\,u_q(k-1) + mc\Delta w_{jq}^3(k) \tag{5.47}$$

$$j = 1,2,\ldots,m, q = 1,2,3,\ldots,n \tag{5.48}$$

where η is the learning rate and m is the momentum factor with a default value of 0.9. This reduces the sensitivity of the network performance to parameter adjustment by considering not only the current gradient direction but also the gradient direction of the previous moment when performing the update. This effectively suppresses the local minima.

In order to give the reader a deeper understanding of how Elman's algorithm works, the following section provides a detailed derivation of the algorithm. Define the error function as shown in (5.49)

$$E = \frac{1}{2}(y_d(k)-y(k))^T(y_d(k)-y(k)) \tag{5.49}$$

In (5.49), $y(k)$ represents the value of the output layer. Biasing E against the connection right W^3 from the hidden layer to the output layer gives (5.50):

$$\frac{\partial E}{\partial w_{ij}^3} = -(y_{d,i}(k)-y(k))\frac{\partial y_i(k)}{\partial w_{ij}^3} = -(y_{d,i}(k)-y(k))\,g_i'(.)\,x_j(k) \tag{5.50}$$

In (5.50), $y_{d,i}(k)$ represents the value of the output layer when different times and w_{ij} represents the connection weight between neuron i and neuron j in the network. Let $\delta_i^0 = (y_{d,i}(k)-y(k))\,g_i'(.)$ to get (5.51):

$$\frac{\partial E}{\partial w_{ij}^3} = -\delta_i^0 x_j(k)\; i = 1,2,\ldots...,m; j = 1,2,\ldots\ldots,n \tag{5.51}$$

In (5.51), w_{ij} is also represents the connection weight between neuron i and neuron j in the network. Bias E against the connection right W^2 from the input layer to the hidden layer, as shown in (5.52):

$$\frac{\partial E}{\partial w_{ij}^2} = \frac{\partial E}{\partial x_j(k)}\frac{\partial x_j(k)}{\partial w_{ij}^2} = \sum_{i=1}^{m} -\left(\delta_i^0 w_{ij}^3\right) f_j'(.)\,u_q(k-1) \tag{5.52}$$

In (5.52), δ_i indicates the actual output and expected output in the network. u_q indicates the control neuron. Let $\delta_j^h = \sum\left(\delta_i^0 w_{ij}^3\right) f_j'(.)$, (5.53) as flow:

$$\frac{\partial E}{\partial w_{ij}^3} = -\delta_j^h u_q(k-1)\, j = 1,2,\ldots,n; q = 1,2,\ldots,r \tag{5.53}$$

Similarly, the bias is applied to the connection right W^1 from the structural unit to the hidden layer. Without considering the dependence of $x_c(k)$ on the connection right w_{jl}^1, as shown in (5.54):

$$\frac{\partial E}{\partial w_{jl}^1} = \sum_{i=1}^{m} -\left(\delta_i^0 w_{ij}^3\right) \frac{\partial x_j(k)}{\partial w_{jl}^1} j = 1, 2, \ldots, n; \quad l = 1, 2, \ldots, n \tag{5.54}$$

$$\frac{\partial x_j(k)}{\partial w_{jl}^1} = f_j'(.)\, x_{C,l}(k)$$
$$f_j'(.)\, x_{C,l}(k) = f_j'(.)\, x_l(k-1) + \alpha . f_j'(.)\, x_{C,l}(k) \tag{5.55}$$

In (5.55), w_{jl} represents the connection weight between neuron j and neuron l in the network. α is the self-connecting feedback gain factor. $x_{C,l}(k)$ represents the lth structural unit:

$$\frac{\partial x_j(k)}{\partial w_{jl}^1} = f_j(.)\, x_l(k-1) + \alpha . \frac{\partial x_j(k-1)}{\partial w_{jl}^1} \tag{5.56}$$

In (5.56), $x_l(k-1)$ represents the output of the lth hidden layer unit. The improved Elman learning algorithm can be derived from $\Delta w = -\eta \frac{\partial E}{\partial w}$ as follows (5.57–5.59)

$$\Delta w_{ij}^3 = \eta \delta_i^0 x_j(k)\, i = 1, 2, \ldots, m; j = 1, 2, \ldots, n \tag{5.57}$$

$$\Delta w_{jq}^2 = \eta \delta_j^h u_q(k-1)\, j = 1, 2, \ldots, n; q = 1, 2, \ldots, r \tag{5.58}$$

$$\Delta w_{ij}^3 = \eta \sum_{i=1}^{m} \left(\delta_i^0 w_{ij}^3\right) \frac{\partial x_j(k)}{\partial w_{jl}^1} \delta_i^0 x_j(k) \tag{5.59}$$

$$j = 1, 2, \ldots, n; l = 1, 2, \ldots, n \tag{5.60}$$

Through the above derivation process, we can have a clear understanding of Elman algorithm and improve our understanding of the algorithm. For the above algorithm, it will also have some small shortcomings. It only considers the feedback of hidden layer nodes but does not consider the feedback of output layer nodes. The two network models given below consider the feedback of output nodes, which are, respectively, called output–input feedback Elman (OIF Elman) with output–input feedback mechanism, and the improved Elman network (OHF Elman) with output hidden feedback mechanism. The structural diagram is shown in Figure 5.27.

In Figure 5.27, w^1, w^2, and w^3 are learning steps. OIF Elman network adds the feedback of the output node, which we call structural unit 2, and puts it on the first layer as the input of the hidden layer node together with the input unit and structural unit.

The OHF Elman network only puts structural unit 2 on the second layer and serves as the input of the output layer together with the hidden layer nodes. Next, we give their mathematical models and learning formulas, respectively.

5.4.2 Adaptive learning process design

At present, there are many types of lithium-ion battery equivalent circuit models. The second-order RC model is one of the commonly used ones, which can well reflect the dynamic and static characteristics of the battery, with moderate calculations and fewer identification parameters [182]. Therefore, the second-order RC model is used as the battery equivalent circuit model, as shown in Figure 5.28.

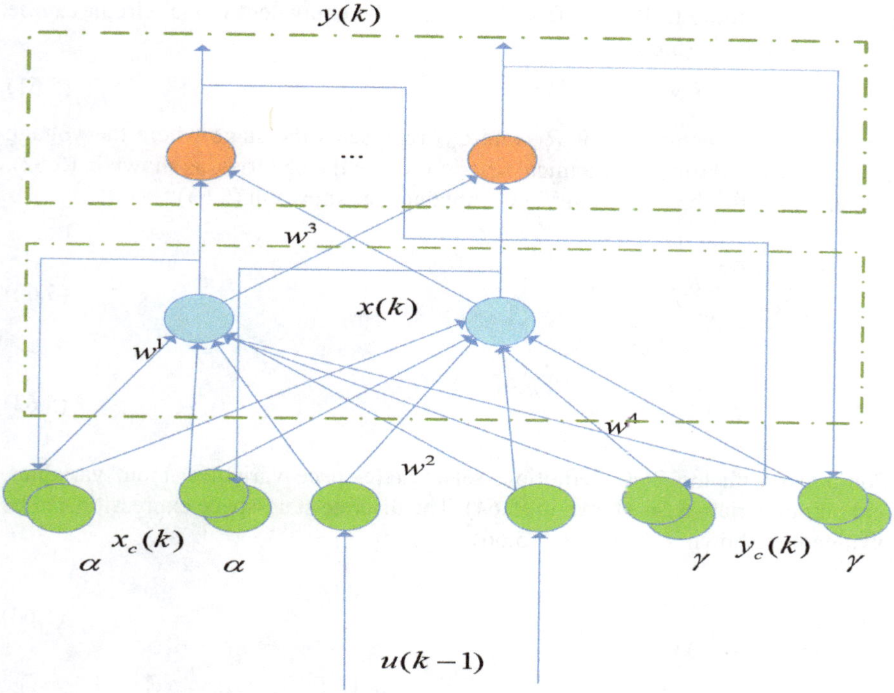

Figure 5.27 OIF Elman network

In Figure 5.28, U_O represents the OCV of the circuit; U_L represents the terminal voltage [183]. R_0 is the ohmic resistance and U_R is the ohmic voltage, which characterizes the battery voltage drop effect at the time point the battery is discharged and the end of the discharge. i_L is the battery charge–discharge current, and the discharge direction is set to positive.

The RC loop is composed of R_{p1} and C_{p2} which characterizes the internal chemical reaction of the battery. The stage where the voltage changes rapidly during the

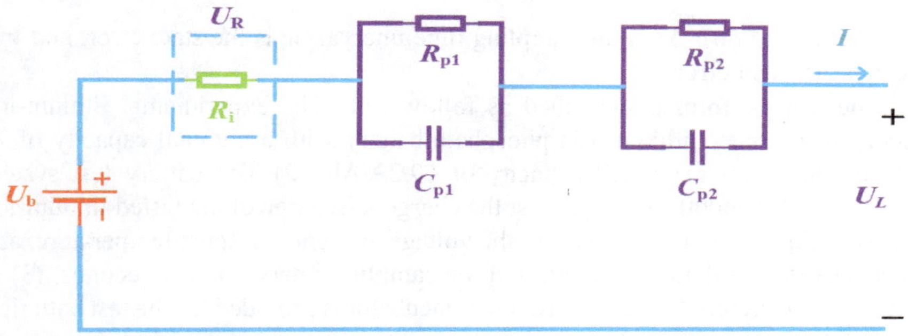

Figure 5.28 Second-order RC equivalent circuit model

process. According to Kirchhoff's circuit law, the equivalent model circuit can be listed, as shown in (5.61):

$$U_L = U_{oc} - U_R - U_{p1} - U_{p2} \tag{5.61}$$

The RC loop is composed of R_{p2} and C_{p2} represents the stage where the voltage changes slowly during the chemical reaction within the batteries, as shown in (5.62). The equation for the SOC definition is obtained, as shown in (5.63):

$$\begin{cases} \dfrac{dU_{p1}}{dt} = -\dfrac{U_{p1}}{R_{p1}C_{p1}} + \dfrac{i}{C_{p1}} \\ \dfrac{dU_{p2}}{dt} = -\dfrac{U_{p2}}{R_{p2}C_{p2}} + \dfrac{i}{C_{p2}} \end{cases} \tag{5.62}$$

$$SOC_t = SOC_{t_0} - \frac{\int_{t_0}^{t} I(t)\,\eta dt}{Q_0} \tag{5.63}$$

Combined with the SOC definition, select state-space variables, input variables, and output variables, as shown in (5.64). The discrete state-space expression can be obtained, as shown in (5.65) and (5.66):

$$\begin{cases} x_k = \left[SOC_k, U_{p1,k}, U_{p2,k}\right]^T \\ U_k = \left[I_k\right], y_k = \left[U_{L,k}\right] \end{cases} \tag{5.64}$$

$$\begin{cases} \begin{bmatrix} SOC_{k+1} \\ U_{p1,k+1} \\ U_{p2,k+1} \end{bmatrix} = \begin{bmatrix} 1 & 0 & 0 \\ 0 & e^{-\frac{\Delta t}{\tau_1}} & 0 \\ 0 & 0 & e^{-\frac{\Delta t}{\tau_2}} \end{bmatrix} \begin{bmatrix} SOC_k \\ U_{p1,k} \\ U_{p2,k} \end{bmatrix} + \begin{bmatrix} -\frac{\Delta t}{QN} \\ R_1\left(1 - e^{-\frac{T}{\tau_1}}\right) \\ R_2\left(1 - e^{-\frac{T}{\tau_2}}\right) \end{bmatrix} I_k + \begin{bmatrix} w_{1,k} \\ w_{2,k} \\ w_{3,k} \end{bmatrix} \\ \tau_1 = R_1C_1,\ \tau_2 = R_2C_2 \end{cases} \tag{5.65}$$

$$U_{L,k} = U_{oc,k} - R_{0,k}I_k + \begin{bmatrix} 0 \\ -1 \\ -1 \end{bmatrix}^T \begin{bmatrix} SOC_k \\ U_{p1,k} \\ U_{p2,k} \end{bmatrix} + v_k \tag{5.66}$$

In (5.65) and (5.66), Δt is the sampling time interval, w is the state error, and v is the measurement error.

The test platform is described as follows: (1) The experimental lithium-ion battery is a ternary lithium iron phosphate battery with a nominal capacity of 72 Ah and an actual calibrated capacity of 69.23 Ah. (2) The battery test system (NEWAREBTS-4000) is used to test the charge–discharge of the tested lithium-ion batteries. The equipment can record the voltage, current, capacity, temperature, and other data during the experiment, and the sampling interval is 0.1 seconds. (3) A constant temperature (25°C) environment incubator is provided for the test batteries.

HPPC is performed on the battery as an identification experiment. The battery model parameters are identified by analyzing the working characteristics of the battery during the

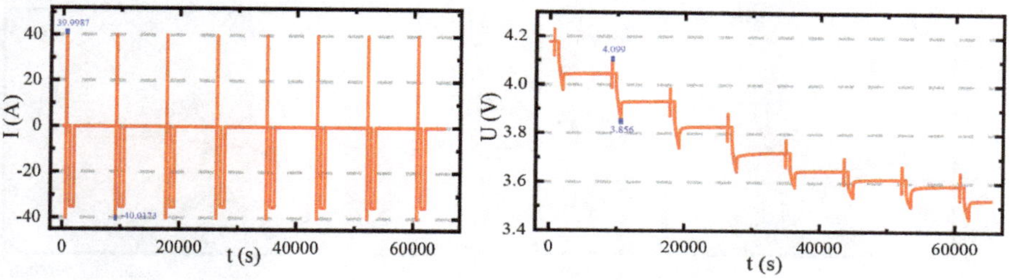

Figure 5.29 *HPPC voltage and current curve. (a) Current curve. (b) Voltage curve.*

experiment. The parameters that need to be identified are U_{OC}, R_0, R_{p1}, C_{p1}, R_{p2}, and C_{p2}. The steps of the HPPC experiment are described as follows.

1. Charging the battery with constant voltage and constant current to full capacity and let it shelve fully.
2. Pulse discharge with 1 C (69.23 A) current for 10 seconds; after discharge, let it rest for 40 seconds.
3. Using 1 C current, pulse charging lasts for 10 seconds and rests for 40 seconds after the end of the charge.
4. Discharging at a constant current of 1 C lasts for 6 minutes, so that the SOC value drops by 0.1, and rests for 30 minutes after the discharge ends.
5. Cycling the test steps (2)–(4) until the SOC value drops to 0.1. The HPPC experiment current curve and voltage response curve are shown in Figure 5.29.

Then, the single HPPC voltage and current change curve can be obtained, as shown in

The graph of the HPPC experimental voltage variation curve is shown, and the variation characteristics can be extracted to obtain the model parameters. Analyzing Figures 5.5–5.30, the following four stage characteristics can be obtained.

1. Discharge starts at t_1. The sudden drop of the voltage at the lithium-ion battery terminal from U_1 to U_2 is mainly due to the voltage change caused by the ohmic resistance of the lithium-ion batteries.
2. During the period from t_2 to t_3, the voltage of the lithium-ion battery terminal slowly drops from U_2 to U_3, which is due to the existence of the battery polarization effect. The process of charging the polarization capacitor by the discharge current is the zero-state response of the double RC series circuit.
3. During the period from t_3 to t_4, the voltage of the lithium-ion battery terminal rises suddenly from U_3 to U_4, which is also due to the voltage change caused by the ohmic resistance of the lithium-ion batteries.

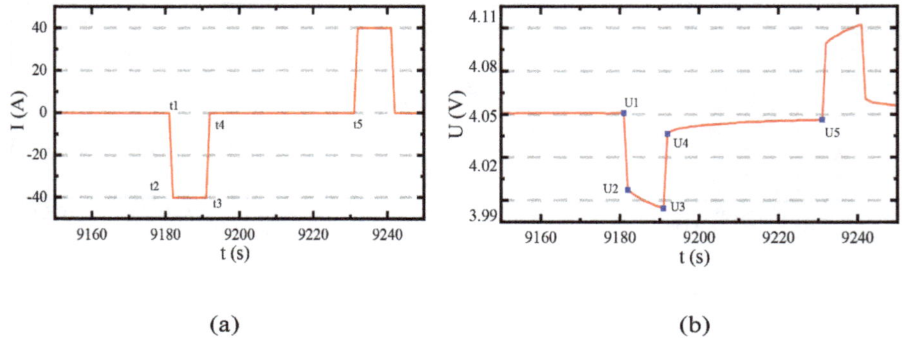

Figure 5.30 *Single HPPC voltage and current change curve. (a) Single HPPC current change curve. (b) Single HPPC voltage change curve.*

4. During the period from t_4 to t_5, the voltage of the lithium-ion battery terminal slowly rises from U_4 to U_5, which is the process of discharging the polarizing capacitor to the polarizing resistance, which is the zero-input response of the dual RC circuit.

Combining the second-order equivalent circuit model and its state-space in Figure 5.4, the adaptive EKF algorithm based on the Elman algorithm is proposed, so that the AEKF-Elman algorithm can be obtained. The flowchart for estimating SOC is shown in Figure 5.31:

$$\begin{cases} x_{k+1} = Ax_k + Bu_k + \Gamma w_k \\ y_k = Cx_k + v_k \end{cases} \tag{5.67}$$

In (5.67), k is the discrete time, x_k is the state of the system at time k, y_k is the observation value of the corresponding state, A is the state transition matrix, B is the system control matrix, and C is the observation matrix. Γ is the noise driving matrix. w_k is the system noise and v_k is the measurement noise, which, respectively, obey the following distribution, as shown in (5.68):

$$\begin{cases} w_k \sim N\left(q_k, Q_k\right) \\ v_k \sim N\left(r_k, R_k\right) \end{cases} \tag{5.68}$$

The specific steps of combining the adaptive algorithm based on the Kalman filter are described as follows: all steps include initialization, prediction stage, calculating the Kalman gain, update stage, and noise update stage.

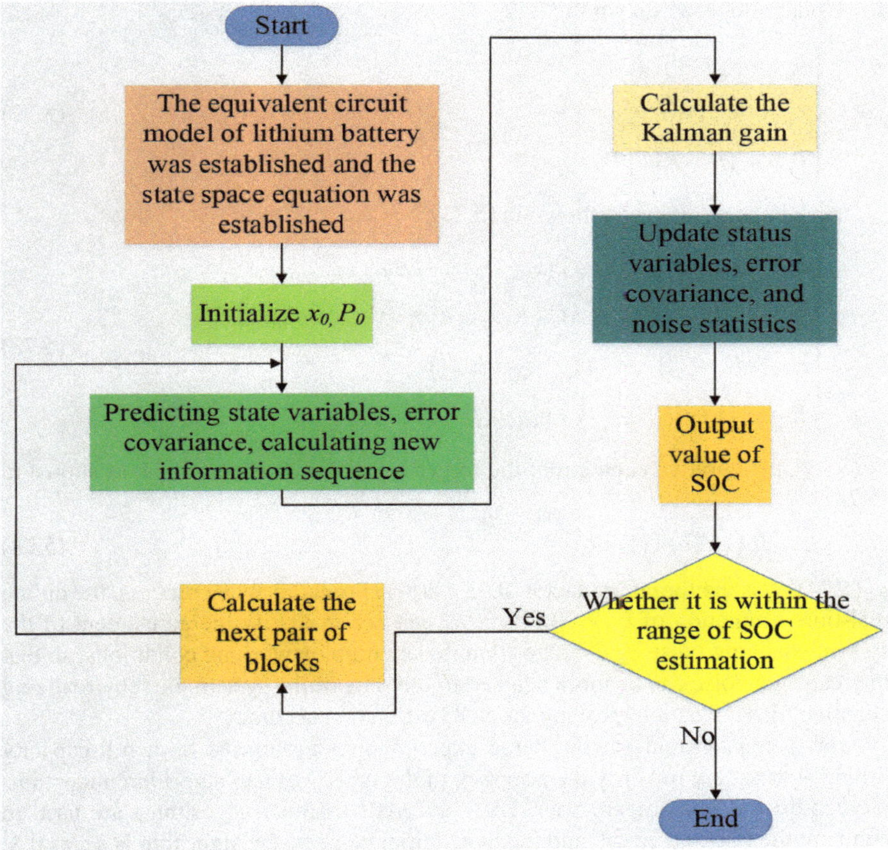

Figure 5.31 Flowchart of SOC estimation based on AEKF-Elman algorithm

1. Initialization, as shown in (5.69):

$$\begin{cases} \hat{x}_0 = E[x_0] \\ P_0 = E\left[(x_0 - \hat{x}_0)(x_0 - \hat{x}_0)^T\right] \end{cases} \tag{5.69}$$

2. Prediction stage, as shown in (5.70):

$$\begin{cases} \hat{x}_{k+1/k} = Ax_k + Bu_k + \Gamma w_k \\ P_{k+1/k} = AP_kA^T + \Gamma\hat{Q}_k\Gamma^T \end{cases} \tag{5.70}$$

3. Calculating the Kalman gain, as shown in (5.71):

$$K_k = P_{k+1/k}C^T\left(CP_{k+1/k}C^T + \hat{R}_k\right)^{-1} \tag{5.71}$$

4. Update stage, as shown in (5.72):

$$\begin{cases} \tilde{y}_{k+1} = y_{k+1} - \hat{y}_{k+1} \\ \hat{x}_{k+1} = \hat{x}_{k+1/k} + K\tilde{y}_{k+1} \\ P_{k+1} = (E - K_k C)\, P_{k+1/k} \end{cases} \tag{5.72}$$

5. Noise update stage, as shown in (5.73):

$$\begin{cases} \hat{q}_{k+1} = (1 - d_k)\,\hat{q}_k + d_k G\,(\hat{x}_{k+1} - A\hat{x}_k - Bu_k) \\ \hat{Q}_{k+1} = (1 - d_k)\,\hat{Q}_k + d_k G\left(K_k \tilde{y}_k \tilde{y}_k^T K_k^T + P_k - AP_{k+1/k}A^T\right) G^T \\ \hat{r}_{k+1} = (1 - d_k)\,\hat{r}_k + d_k\,(y_{k+1} - C\hat{x}_{k+1/k}) \\ \hat{R}_{k+1} = (1 - d_k)\,\hat{R}_k + d_k\left(\tilde{y}_{k+1}\tilde{y}_{k+1}^T - CP_{k+1/k}C^T\right) \end{cases} \tag{5.73}$$

In (5.73), the iterative equation of the forgetting factor d_k is described, as shown in (5.74):

$$d_{k-1} = (1 - b)\,/\,(1 - b^k) \tag{5.74}$$

In (5.74), b is the forgetting factor, $0.95 < b < 1$. From the above process, the online real-time estimation of $\hat{q}_k$, $\hat{r}_k$, $\hat{Q}_k$, and $\hat{R}_k$ can achieve real-time adjustment of the system changing noise by the algorithm under complex working conditions, so that the real-time noise can be more adapted to the state of the system, thereby realizing adaptive filtering, and improving the SOC estimation accuracy.

*A*1 *C* constant current discharge experiment is performed on a full-capacity lithium-ion battery to verify the accuracy of the AEKF-Elman algorithm under simple conditions. The ampere-hour, EKF, and AEKF-Elman algorithms are used to compare the tracking effect, and the SOC initial value of the algorithm is set as 0.8; the convergence effect of the algorithm under the unknown initial state value and the tracking effect in the discharge process are verified. The comparison chart of the algorithm operation is shown in Figure 5.32.

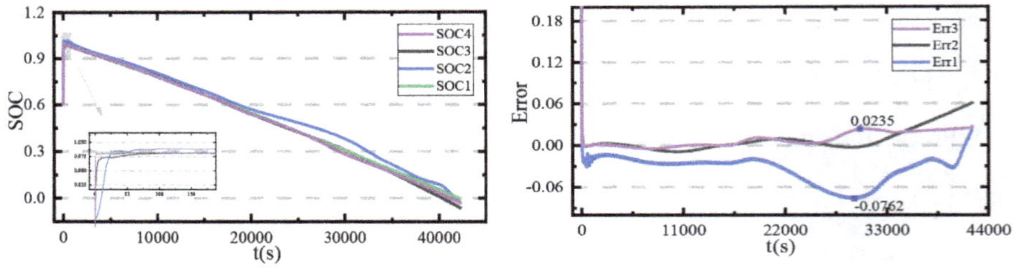

(a) (b)

Figure 5.32 Comparison chart of SOC estimation. (a) SOC estimated graph. (b) Error comparison chart.

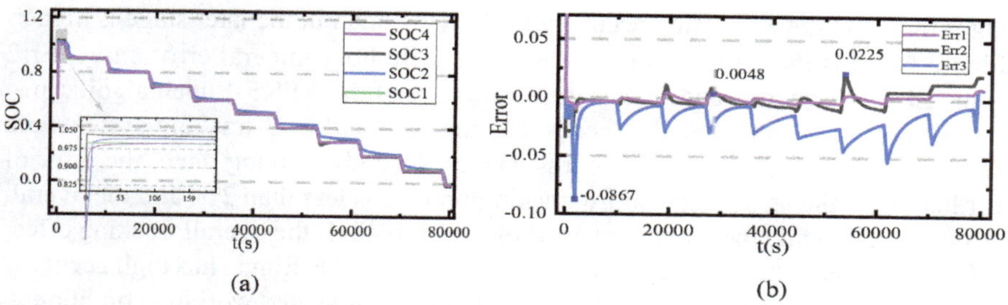

Figure 5.33 Comparison chart of SOC estimation under DST conditions. (a) SOC estimated graph. (b) Error comparison chart.

In Figure 5.32a, SOC1 refers to the reference SOC value, SOC2 is the SOC value calculated using the ampere-hour integration method, SOC3 is the SOC value estimated using the EKF algorithm, and SOC4 is the SOC value estimated using the AEKF-Elman algorithm. In the case of an inaccurate initial value of SOC = 0.6, the ampere-hour integration method cannot track the real state of the system; EKF can track the real state of the system faster.

The time required for the EKF algorithm to track the theoretical value is about 45 seconds, but the estimation accuracy is high enough. The tracking effect of the AEKF-Elman algorithm is almost the same as that of the EKF algorithm, and the later estimation accuracy of AEKF-Elman is higher than that of EKF. Figure 5.32b is the error curve.

Err1 is the ampere-hour integral error curve, Err2 is the EKF error curve, and Err3 is the error curve of the AEKF-Elman algorithm. The overall error of Err1 exceeds 7%, with a small error, the tracking speed of Err2 and Err3 is fast, and the overall tracking effect of Err3 is better, indicating that the AEKF-Elman algorithm has a better SOC estimation effect than the EKF algorithm.

To verify the estimation effect of the AEKF-Elman algorithm under complex working conditions, the HPPC algorithm is used to simulate the actual operating conditions.

After obtaining the battery discharge data, the ampere-hour integration method, EKF, AEKF, and AEKF-Elman algorithms are used to compare the tracking effect. The initial value of the algorithm is set as SOC = 0.8 to verify the convergence effect and tracking effect during the discharge process. The running effect diagram of the algorithm is shown in Figure 5.33.

In Figure 5.33a, SOC1 refers to the SOC value, SOC2 is the SOC value calculated using the ampere-hour integration method, SOC3 is the SOC value calculated using the EKF algorithm, and SOC4 is the AEKF algorithm. Under the condition of an inaccurate initial value of SOC = 0.6, the ampere-hour integration method cannot track the real state of the system. EKF can track the real state of the system faster.

The time required for the EKF algorithm to track the theoretical value is about 30 seconds, but the estimation accuracy is not high enough. The convergence effect of the AEKF-Elman algorithm in the early stage is almost the same as that of the EKF algorithm, and the estimation accuracy of AEKF-Elman in the later stage is higher. Figure 5.33b is the error curve, Err1 is the ampere-hour integral error curve, Err2 is the EKF error curve, and Err3 is the error curve of the AEKF-Elman algorithm.

The overall error of Err1 exceeds 8%, the Err2 and Err3 tracking systems are fast, and the convergence speed is less than 50 seconds. Among them, the error of Err2 is small, the error of the entire tracking process is less than 2%, and the overall error is small; while the maximum error of Err3 is 0.48%, the overall tracking effect is fluctuating and the accuracy is low. It shows that AEKF-Elman has high accuracy and strong applicability in estimating SOC under complicated working conditions.

5.4.3 Learning process optimization

Usually, in the original Elman algorithm, a combination of genetic algorithm (GA), PF, or EA is considered as a way to improve the accuracy of the algorithm in order to obtain a higher accuracy of the algorithm results [184]. In this subsection, basic information about these three improved algorithms is presented.

Evolutionary computation (EC), also known as EAs, uses simple coding techniques to represent a variety of complex structures and directs learning, and determines the direction of search through simple genetic operations and natural selection of winners and losers.

Since it uses a population (i.e., a set of representations) to organize the search, this makes it possible to search multiple regions within the solution space simultaneously and is particularly suitable for massively parallel computation [185]. EC is self-organizing, self-adaptive, and self-learning, is not constrained by the restrictive conditions of its search space (differentiable, single-peaked), and does not require other auxiliary information (derivatives). This makes EC not only obtain high efficiency but also simple and general in operation.

EAs all belong to iterative algorithms, i.e., the best solution or the next best solution is obtained from a given initial solution by iterating continuously and optimizing gradually. They all have a relatively simple structure and can be described generally as follows.

1. Randomly generate an initial population P with evolutionary generation $t = 1$.
2. Calculate the fitness of all individuals in the population $P(t)$.
3. Select a number of parents from $P(t)$ based on individual fitness values.
4. Apply various evolutionary operators to the selected parents to generate the next generation population $P(t + 1)$.
5. $t = t + 1$, if the termination condition is not satisfied turn (2).
6. Output the optimal solution and end.

The solution process of the EA can be depicted in Figure 5.1. The initial population is set randomly for the specific problem: the evaluation operation is based on a predefined evaluation function to calculate the "fitness" of each individual in the current population in the environment; the function of the generation mechanism is to select some evolutionary methods (such as selection, crossover, and mutation) from the current population. The function of the generation mechanism is to generate new individuals from the current population by selecting some evolutionary methods (e.g., selection, crossover, and mutation); the next operation is to obtain the next generation population from the current population and the newly generated individuals. The next operation is to obtain the next generation population from the current population and the newly generated individuals, which completes an evolution; the termination condition of the algorithm is that the solution satisfies the requirements or the number of iterations reaches a given value.

The algorithm terminates when the solution meets the requirements or the number of iterations reaches a given value.

Compared with traditional search algorithms, EAs have two distinctive features: one is the use of a population-based search mechanism and the other is the communication and information exchange among individuals through selection and/or recombination during the search process. Specifically, the superiority of EAs is mainly demonstrated by the following.

1. EAs use only information about the fitness values of individuals for searching, without using derivatives or other padding information.

   As long as a problem can be described as a certain optimization problem, EAs can be used to solve it without

   EAs can be used to solve a problem as long as it can be described as an optimization problem, without being limited by the mathematical nature of the problem, and thus can be used in many fields.
2. EAs use a probabilistic transfer mechanism instead of a deterministic transfer mechanism, which is less likely to fall into a local optimum in the search process.

   EAs use a probabilistic transfer mechanism, rather than a deterministic one, and are not prone to fall into a local optimum during the search process, and they can find the global optimal solution with a high probability even when the defined fitness function is discontinuous, non-regular, or noisy.
3. EAs are particularly suitable for parallel implementation due to their inherent parallelism.
4. EAs use evolutionary ideas to solve complex problems with very simple concepts and are able to solve more reliably very difficult problems.
5. EAs are a general problem-solving method and can be used when little knowledge of the problem domain is available. EAs are a general problem-solving method that can be used when little knowledge of the problem domain is available, and they are scalable and easy to combine with existing algorithms and models.

EAs have been widely used in the fields of optimization, ML, and parallel processing. The use of EAs is not guaranteed to find the global optimal solution, but they can generally find a satisfactory optimal solution, which is shown in Figure 5.34.

The GA is a parallel stochastic search optimization method that simulates natural genetic mechanisms and biological evolution. The elements of BP neural

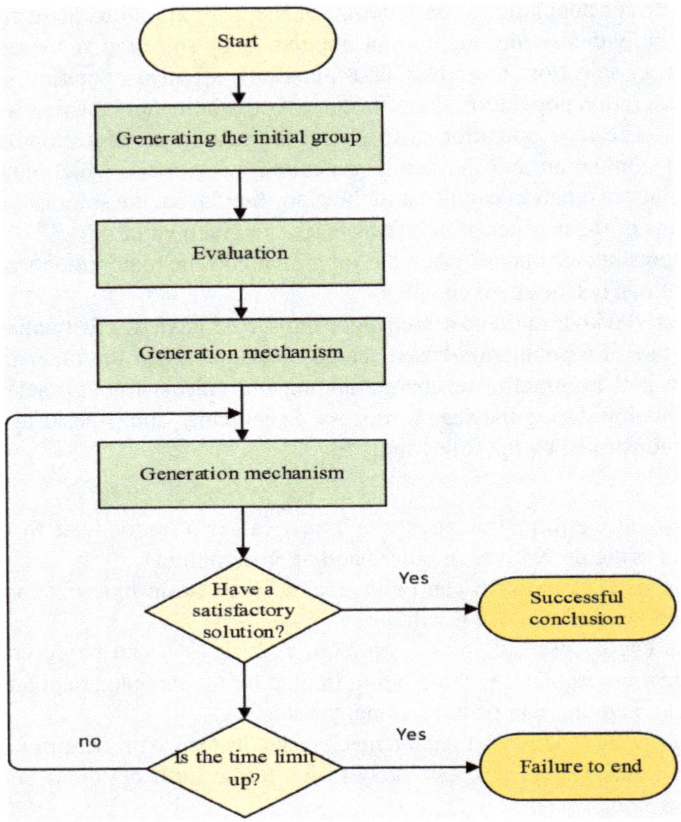

Figure 5.34 EA framework diagram

network optimization by the GA include population initialization, fitness function, selection operation, crossover operation, and mutation operation.

1. Population initialization

   The individual coding method is real number coding, and each individual is a real number string, which is composed of four parts: the connection weight between the input layer and hidden layer, the threshold value of the hidden layer, the connection weight between the hidden layer, output layer, and the threshold value of output layer. The individual contains all the weights and thresholds of the neural network. If the network structure is known, a neural network with its structure, weights, and thresholds can be constituted.

2. Fitness function

   According to the initial weights and thresholds of the BP neural network obtained by individuals, the BP neural network is trained with training data to predict the system output. The absolute error value between the predicted output and the expected output is taken as the individual fitness value F. The calculation equation of F is shown in (5.75):

$$F = k\left(\sum_{i=1}^{n} |y_i - o_i|\right) \tag{5.75}$$

In (5.75), n is the number of output nodes, y_i is the expected output of the ith node of the BP neural network, o_i is the predicted output of the ith node, and k is the coefficient.

3. Selection operation

The selection operation of GA includes roulette, tournament, etc. When the roulette method is used to calculate the selection probability p_i of each individual i, the calculation equation is shown in

$$\begin{cases} f_i = \frac{k}{F_i} \\ p_i = \frac{f_i}{\sum_{j=1}^{N} f_j} \end{cases} \tag{5.76}$$

In (5.76), F_i represents the fitness value of individual i. Since the smaller, the fitness value, the better, the inverse of the fitness value is calculated before individual selection. N represents the number of individuals in the population, and k represents the coefficient.

4. Crossover operation

The crossing operation of the kth gene a_k and the lth gene a_l at the j position is realized by using (5.77):

$$\begin{cases} a_{kj} = a_{kj}\left(1 - b\right) + a_{lj}b \\ a_{lj} = a_{lj}\left(1 - b\right) + a_{kj}b \end{cases} \tag{5.77}$$

5. Mutation operation

The jth gene of the ith individual factor is selected for mutation operation, as shown in

$$\begin{cases} f(g) = r_1\left(1 - \frac{g}{G_{max}}\right)^2 \\ a_{ij} = \begin{cases} a_{ij} + \left(a_{ij} - a_{max}\right) * f(g), r > 0.5 \\ a_{ij} + \left(a_{min} - a_{ij}\right) * f(g), r \le 0.5 \end{cases} \end{cases} \tag{5.78}$$

In (5.78), (5.77), a_{max} and a_{min} represent the upper and lower bounds of a_{ij},, r represents the random number, g represents the current iteration, and G_{max} represents the maximum number of evolutions.

To verify the feasibility of the GA-Elman algorithm, the lithium-ion battery SOC estimation model is established. SOC estimation based on GA-Elman is carried out under different working conditions such as constant current discharge and HPPC and the ampere-hour integration method under HPPC working conditions. The estimation accuracies of EKF and GA-Elman are compared to verify the superiority of the proposed algorithm. In the experiment, R_0, R_1, R_2, C_1, and C_2 are

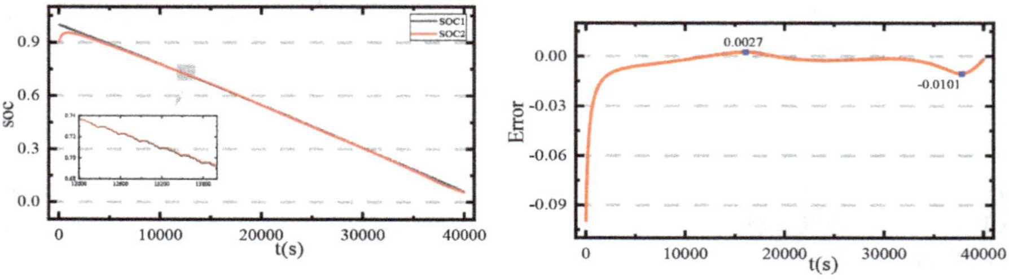

(a) (b)

Figure 5.35 SOC estimation result based on GA-Elman. (a) Comparison of SOC estimation results. (b) SOC estimation error.

substituted into the algorithm in the form of average values, and the test battery model is a ternary 50-Ah lithium-ion battery. The 2 C constant current discharge experiment is used to verify the effectiveness of the EKF algorithm in battery SOC estimation. The simulation effect is shown in Figure 5.35a, and the error is shown in Figure 5.35b.

In Figure 5.35a, the SOC1 is the actual SOC curve, and SOC2 is the GA-Elman estimated SOC curve. The experimental results in Figure 5.35b show that the maximum estimation error of SOC during 2 C constant current discharge is 0.0027, which is less than 0.27%, and it is within the allowable error range. It can be observed from the figure that as the number of filtering increases, the predicted value curve gradually approaches the true value curve and fluctuates near the true value curve. The error value also gradually decreases from the initial larger error, and finally fluctuates around 0, showing a trend of convergence.

After deriving and analyzing the principles and steps of the algorithm, the GA-Elman algorithm is used to estimate the battery SOC, and the results are compared and analyzed with the ampere-hour integral method and the EKF algorithm. The parameter design of the first-level EKF algorithm remains unchanged. In the s-level KF algorithm, the internal noise R to the model is 0.001, and the system output noise Q is obtained by the posterior estimated covariance of the EKF algorithm. The simulation results under HPPC working conditions are shown in Figure 5.36.

According to Figure 5.36a, SOC1 is the SOC curve estimated by GA-Elman, and SOC2 is the actual SOC curve. The experimental results in Figure 5.36b show that under HPPC conditions, the maximum estimation error of SOC is 0.0067, which is less than 0.67%.

The ampere integral method is a simple and widely used SOC estimation method. By integrating the current flowing through the battery within a certain working period, the charge–discharge capacity of the battery can be obtained. The simulation results of HPPC operating conditions based on Ah points are shown in Figure 5.37.

In Figure 5.37a, the SOC1 is the actual SOC curve, and SOC2 is the ampere-hour integral estimated SOC curve. The experimental results in Figure 5.37b show

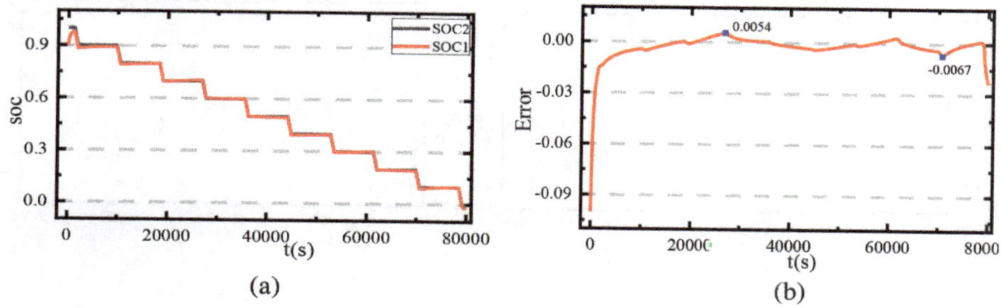

Figure 5.36 SOC estimation results based on GA-Elman under HPPC conditions. (a) Comparison of SOC estimation results. (b) SOC estimation error.

that under HPPC conditions, the maximum estimation error of SOC is 0.0068, which is less than 0.68%, and it is within the allowable error range. It can be observed from Figure 5.37 that the estimation error of the ampere-hour integration method continues to increase during the entire discharge process. This is closely related to factors such as current detection accuracy, fluctuations, battery capacity, and temperature. Under HPPC working conditions, the SOC estimation results based on the EKF algorithm are shown in Figure 5.38.

In Figure 5.38a, SOC1 is the SOC value estimated by the EKF algorithm, and SOC2 is the actual SOC value. Figure 5.38b shows that the maximum estimation error of SOC is 0.0528 under the HPPC condition, which is less than 5.28%.

It is worth noting that when the battery is at the end of discharge, the SOC estimation error fluctuates greatly, which is mainly caused by the nonlinearity of the battery itself. Through the above comparison, although the ampere-hour integration method is simple and easy to implement, the initial value cannot be obtained. The accuracy requirements are strict, and the lack of feedback correction links leads to a large accumulation of errors. The EKF algorithm has good robustness, but the GA-Elman estimation accuracy is high; it is not easily interfered with by the accuracy of the battery model with strong stability and convergence.

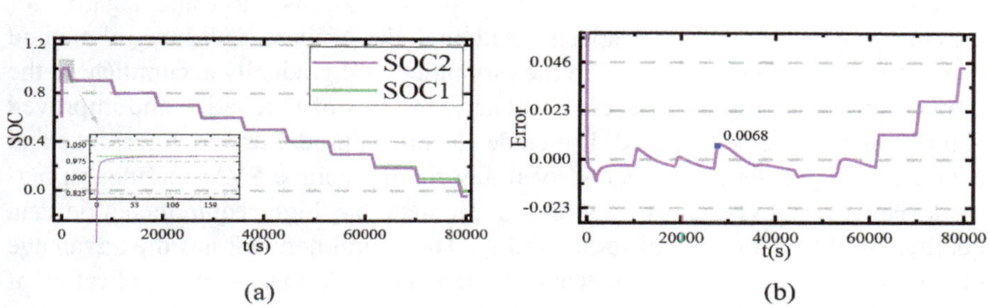

Figure 5.37 SOC estimation under HPPC working conditions. (a) Comparison of SOC estimation results. (b) SOC estimation error.

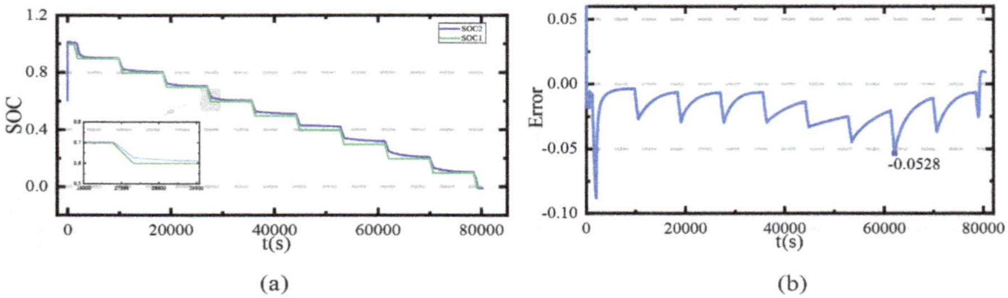

Figure 5.38 *SOC estimation results based on EKF under HPPC conditions. (a) Comparison of SOC estimation results. (b) SOC estimation error.*

5.5 Nonlinear autoregressive neural network-based SOC estimation

5.5.1 Nonlinear autoregressive neural network structure

SOC is an important index to evaluate the remaining effective capacity of lithium batteries. It is crucial to accurately estimate the SOC of lithium battery to ensure the safe operation of lithium battery, prolong the service life of lithium battery, and avoid over-charge or over-discharge of the battery pack [186]. However, the chemical reaction inside lithium batteries is extraordinarily complex in the normal operation process. With uncertain temperature changes, lithium battery SOC changes are nonlinear, so it is difficult to accurately estimate SOC.

It is known that the SOC estimation of lithium batteries is in a time-series system. Traditional SOC estimation is based on ampere-hour integration, as shown in (5.79):

$$SOC\left(t\right) = SOC_0 - \frac{\int \eta I\left(t\right) dt}{Q} \tag{5.79}$$

In the model, $SOC\left(t\right)$ is the value of SOC at time t; SOC_0 is the initial value. η is the coulomb efficiency, which is regarded as one in the discharge case. $I(t)$ is the current flowing through the battery at time t, positive for discharge and negative for the charge; Q is the reference capacity. Although the Ampere-hour integral method computes the gradient, the errors in the estimation will gradually accumulate in the timing system and affect the accuracy. Therefore, lots of innovative and improved algorithms have been proposed. It includes technical model-driven algorithms and data-driven algorithms. The data-driven algorithms include SVM and neural network. However, SVM relies on large sample data, has high requirements on data quality, and has poor generalization ability. The neural network has the advantage of not relying on the battery model and mathematical relations, so the application of the neural networks to the estimation of lithium batteries has a more efficient ability.

The neural network algorithm is an algorithm that can be used for nonlinear regression, but its core is "linear," and autoregression is an algorithm represented by

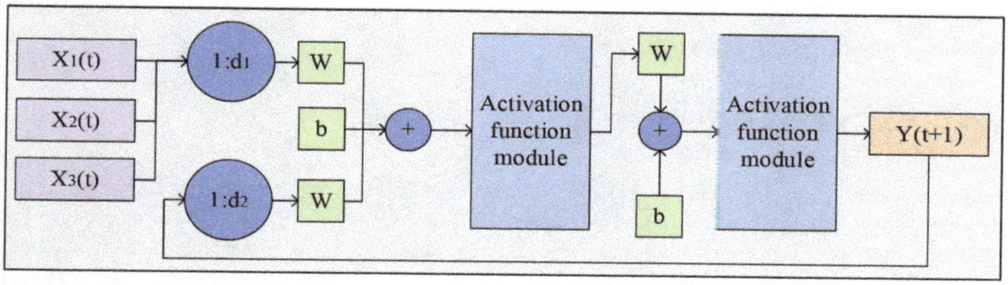

Figure 5.39 Open-loop NARX neural network

a linear model. To apply the autoregressive algorithm to the current and voltage data with unstable time series, the neural network is trained to gain and the minimum error is taken as the weight to obtain the final result. Nonlinear autoregressive neural network is a recurrent neural network (RNN) based on static MLP which is often used in time-series model by introducing delay element and feedback and is very suitable for lithium battery SOC estimation system.

Nonlinear autoregressive with exogenous input (NARX) neural network is a nonlinear autoregressive neural network based on a linear autoregressive neural network with external input and considering the temporal correlation of input and output. By adding the external excitation of the past moment in the function, comprehensively considering the influence of the external excitation on the current system output, as shown in (5.80):

$$\hat{y}_i = F\left(u,y\right) = F\left\{u_{i-1}, u_{i-2} \ldots, u_{i-d1}, u\left(t\right), y_{u-1}, y_{u-2}, \ldots, y_{u-d2}\right\} \tag{5.80}$$

In the model, u_i and y_i are the input and output of the network at the time i, respectively; y_i represents estimates of the system's output at the time i; d_1 and d_2, respectively, represent the delay order of input and output; so the $u_{i-1}, u_{i-2} \ldots, u_{i-d1}$ is the historical input relative to time i; $y_{u-1}, y_{u-2}, \ldots, y_{u-d2}$ is the historical output relative to time i; F is the nonlinear function obtained by network fitting. The models are shown in Figure 5.39.

The open-loop and close-loop neural networks are used in NARX neural network. The open-loop neural network feeds the output to the input when the output is shown in Figure 5.40.

The close-loop neural network uses its own generated estimates in the calculation, which can highlight the dynamic characteristics of the system in multi-step and long-sequence calculations. For outputs that cannot be measured directly, using the open-loop neural network to train the model, the initial estimate can be used for the close-loop neural network, and finally get the final estimate by using the close-loop neural network.

First, network parameters were identified based on the open-loop NARX network topology, and the obtained initial SOC estimate was used as the output of the close-loop neural network. On this basis, the lithium battery SOC estimation model was constructed based on the close-loop NARX neural network topology. The neural network is shown in Figure 5.41.

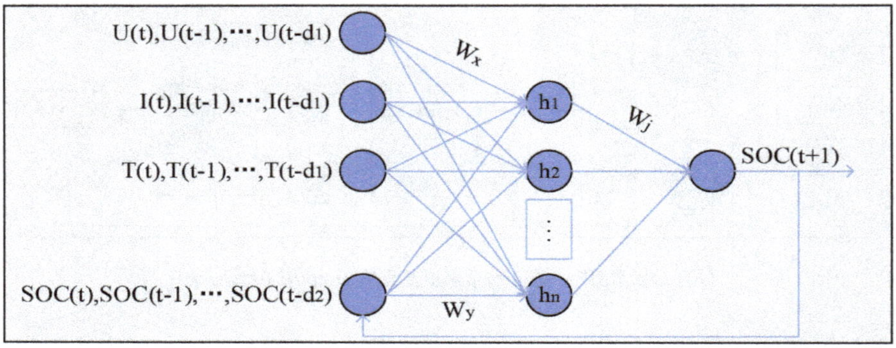

Figure 5.40 Close-loop NARX neural network

In Figure 5.41, $U(t)$, $I(t)$, $T(t)$, and $SOC(t)$ are, respectively, the voltage, current, temperature measured value, and SOC estimated value at time t. W_x, W_y, and W_j are the connection weights of hidden layer neurons, respectively. h_n is the output value of the neuron of the n hidden layer, and its value is as shown in

$$\begin{cases} h_n(t+1) = f_n\left[b_n + \sum W_x \cdot x + \sum W_y \cdot y\right] \\ x = \left[x(t), x(t-1), \ldots, x(t-d_1)\right] \\ y = \left[y(t), y(t-1), \ldots, y(t-d_2)\right] \end{cases} \tag{5.81}$$

where f_n is the activation function corresponding to the n neuron in the hidden layer; b_n is the threshold of the n neuron in the hidden layer. x and y are inputs, respectively, time series of states U, I, T, and output state SOC.

The output value of neurons in the output layer can be expressed as shown in (5.82):

$$\begin{cases} y(t+1) = f_g\left[b_g + \sum w_j \cdot h\right] \\ h = \left[h_1, h_2, \ldots, h_n\right] \end{cases} \tag{5.82}$$

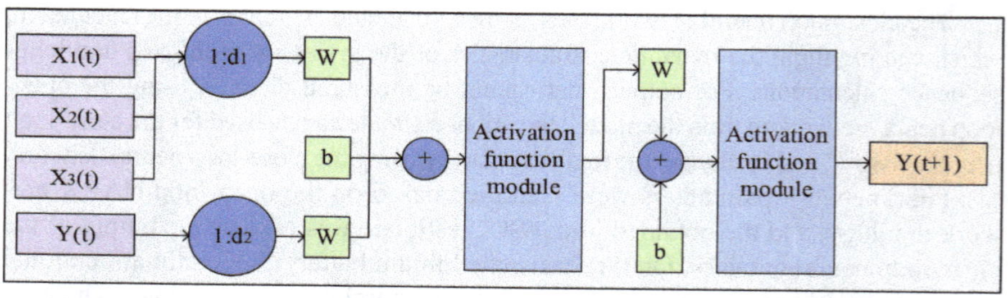

Figure 5.41 Nonlinear autoregressive with external inputs neural networks

where f_g is the activation function corresponding to neurons in the output layer; b_g is the threshold value of neurons in the output layer; and h is the hidden layer output set.

The NARX natural network can realize the function of feedback and historical data memory. Therefore, the NARX natural network can greatly reduce the amount of data needed for model training. However, with the increases in model training time, NARX regression neural network models are prone to gradient vanishing and gradient explosion, and the model complexity and model training time are also increasing rapidly. Simultaneously, NARX neural network usually shows gradient disappearance and gradient explosion when the voltage and current change periodically. To solve this problem, a moving window method is used based on the traditional finite element principle. The size of the moving window can be changed with the change in the discharge rate. Then, the model expands the window at an appropriate time to retain historical information proportional to the current state and reduces the size of the window at certain times to reduce the historical information that interferes with the SOC estimation of the current state, thus avoiding the problem of gradient explosion and the gradient vanishing, and the accuracy and modeling the speed of the obtained model are also improved.

5.5.2 Network creation and training

The NARX RNN is a dynamic regression neural network that can realize the function of feedback and historical data memory. With feedback and memory, the NARX can retain the data of the previous moment and add it to the calculation of the next moment, which makes the neural network not only dynamic but also more complete because of the retained system information.

The Levenberg–Marquardt (LM) algorithm is used in the parameter calculation in the NARX model [187]. The LM algorithm combines the advantages of the gradient-descent method and the Gauss–Newton method. The second-order derivative method of LM, which has fast convergence speed and strong stability, is used in the calculation, and the algorithm prevents the neural network from falling into the local optimum. The iterative model is shown in (5.83):

$$\begin{cases} x_{k+1} = x_k + \Delta x \\ \Delta x = -\left(A_k^T A_k \mu I\right)^{-1} A_k^T e_k \end{cases} \tag{5.83}$$

where x_k is the input vector at the kth iteration; x_{k+1} is the input vector of the $k+1$th iteration; and Δx is the vector of the change amount at two iterations. A_k is the Jacobian matrix, I is the identity matrix, and μ is a parameter used to adjust the calculation accuracy and speed as a constant.

First, the training data are inputted for model training. The input and output of the current calculation are retained by NARX and are inputted to the next input data. Second, the moving window method ends the model training of the current stage and starts the model training of the next stage when the problems of gradient explosion and gradient vanishing are about to happen. Finally, the SOC estimation model proposed is obtained.

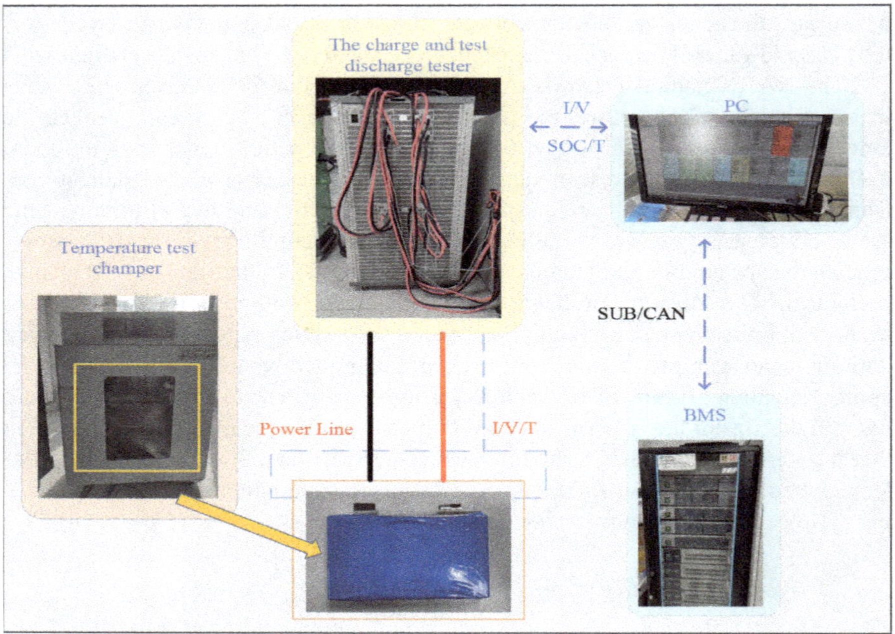

Figure 5.42 Battery test bench

The SOC estimation of lithium-ion batteries starts from the establishment of the experimental platform followed by data collection. As shown in Figure 5.42, a battery test bench, which included the incubator, battery test system, and a host computer, was used to collect the NMC battery data. In addition, to obtain the experimental data, an NMC battery was used to charge and discharge, and the constant current and constant voltage (CCCV) and the dynamic stress test (DST) were set up by the battery test software at the host computer to control the battery test system.

The incubator was set up at 20°C in the thermal chamber. The upper and lower limits of the battery voltage were 4.2 and 3.4 V, and the interval time of data sampling was 1 second. After the initial cycles, the NMC battery was tested under different working conditions at 20°C and 3 000 samples were chosen as model training samples, which is shown in Figure 5.42.

The SOC estimation process of the proposed method is shown in Figure 5.5, which mainly includes three stages. The first stage: the experimental data of the NMC battery are collected under CCCV and DST. The collected data are analyzed with its discharging rates, and a reasonable window size is determined using the experimental data under CCCV working conditions. A reasonable window size directly affects the accuracy of the model and can avoid the gradient disappearance and gradient explosion phenomena. The second stage: the training data are inputted into the proposed model for SOC estimation model training, and the battery model for SOC estimation is obtained. In the third stage: the proposed method is

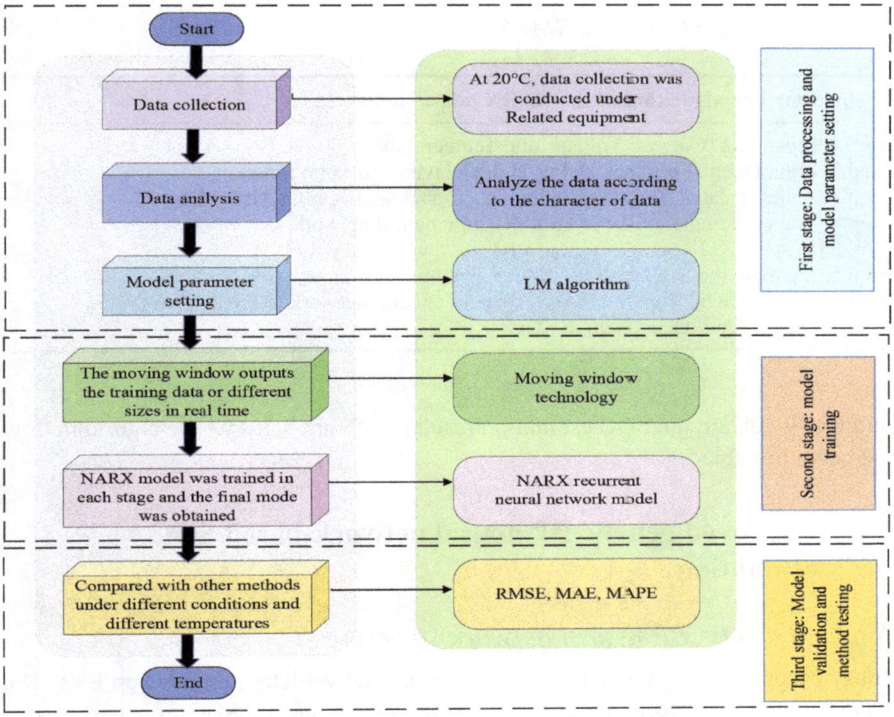

Figure 5.43 Schematic diagram of the proposed model for SOC estimation

tested by using the experimental data under different working conditions and different temperatures.

The mathematical expressions of RMSE, MAE, and MAPE are shown in (5.84):

$$\begin{cases} RMSE = \sqrt{\frac{1}{n}\sum_{i=1}^{n}\left(T_e - T_a\right)^2} \\ MAE = \frac{1}{n}\sum_{i=1}^{n}\left(T_e - T_a\right) \\ MAPE = \frac{1}{n}\sum_{i=1}^{n}\frac{|T_e - T_a|}{T_a} \end{cases} \tag{5.84}$$

where n represents the number of test samples, T_e represents the estimated value of SOC, and T_a represents the actual value of SOC. A schematic diagram of the proposed model for SOC estimation is shown in Figure 5.43

The neural network algorithm combined with nonlinear autoregression was used to train and simulate the current and voltage of lithium battery in time series,

Table 5.3 The algorithm of NARX

Algorithm: The algorithm of the NARX neural network
Step 1. Input data (Current, Voltage, and Temperature)
Step 2. Input Delay, Feedback Delay, Hidden layer, and Train function
Step 3. Training date (70%), Validation data (15%), and Testing data (15%)
Step 4. Train the model of the NARX dynamic neural network $\hat{y}_i = F(u, y) = F\{u_{i-1}, u_{i-2} \ldots, u_{i-d1}, u(t), y_{u-1}, y_{u-2}, \ldots, y_{u-d2}\}$
Step 5. Estimate the SOC with the NARX dynamic neural network
Step 6. Add the SOC from the NARX dynamic neural network to the database
Step 7. Output the final estimation SOC

and finally, obtain the SOC estimation result of lithium battery. The algorithm flow is shown in Table 5.3.

5.6 Improved genetic BP neural network-based SOC estimation

5.6.1 Genetic BP neural network structure

It has become an inevitable trend to phase out fuel vehicles and develop EVs. The power battery is used as an energy storage device, and its SOC affects the driving range and driving performance of the vehicle, so it is necessary to accurately estimate the SOC of the battery [188]. But due to the complex internal reaction of the battery and the nonlinear relationship between SOC and various electrical parameters, it is impossible to measure directly and accurately, which increases the difficulty of accurately predicting the SOC of the battery.

At present, the commonly used methods are: the ampere-hour integration method requires an initial SOC value to give an accurate estimation result. Because of the error in the current detection, the estimated SOC will eventually deviate; the OCV method records the voltage and SOC data through the discharge experiment and predicts the SOC value according to the size of the voltage data, but this method does not support dynamic online detection; Kalman filtering algorithm needs to obtain an accurate model, and the parameters in the model are updated by the algorithm to predict the value of SOC. The battery model, to be sure, is difficult to build accurately and is not very versatile [189].

The BP neural network algorithm can effectively overcome the shortcomings of the above three estimation methods. It has a strong nonlinear approximation ability and can be used to estimate the SOC of the battery. However, the relationship between internal resistance and SOC is rarely studied in the existing literature. The measurement experiment found that the internal resistance of the lithium battery has an important influence on its SOC. In this chapter, the voltage, current, internal resistance, and temperature data of the battery during discharge are collected by the

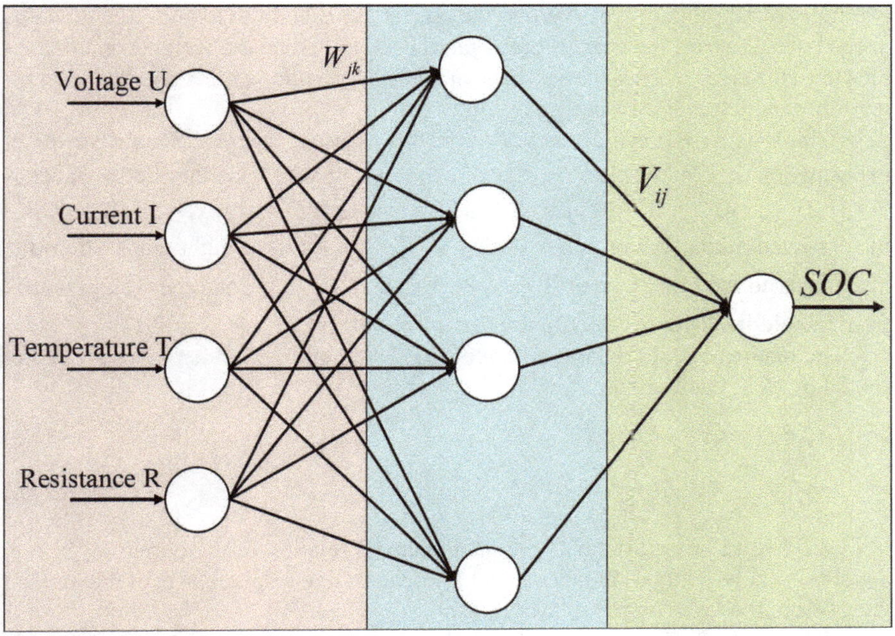

Figure 5.44 *BP neural network predicts battery SOC structure diagram*

battery charge and discharge tester and transmitted to the input layer, and the output layer outputs the result of SOC and trains the network model to achieve the purpose of online accurate estimation of SOC. The SOC prediction method established in this chapter can be applied to different types of power batteries.

The neural network method has a high degree of nonlinearity and strong self-learning ability. Through the learning of existing training samples, the neural network can map the internal law between the input and output well. The advantages of strong generalization ability, high fault tolerance, and excellent robustness are very suitable for SOC estimation of lithium-ion batteries.

Considering that the GA has the characteristics of global optimality, it can overcome the local optimal defect of the BP algorithm, and it can optimize the initial weights and thresholds of the BP neural network to improve the convergence of the network and shorten the occupied time.

During the training of the BP neural network, the first step is to calculate the error by the forward transmission of the signal, and the second step is to correct the weight and threshold by the reverse transmission of the calculated error. Selecting the parameter that has a high correlation with the battery SOC value as the input. Therefore, the network structure shown in Figure 5.1 is used to estimate the SOC of the battery. Through the experiment, the collected current, voltage, internal resistance, and temperature data are transmitted to the input layer, and then processed by the middle layer, and finally, the signal is transmitted to the output layer, which is a forward learning process. When the output value and the expected SOC value exceed the set error, the error is propagated back to update the weights and thresholds until the error reaches the set range or stops by reaching the set number of learning times, which is shown in Figure 5.44.

A three-layer BP neural network structure is used, where the input vector of the network is $X = (x_1, x_2, ..., x_i, ..., x_n)^T$; the output vector of the hidden layer is $Y = (y_1, y_2, ..., y_j, ..., y_n)^T$; the vector of the output layer is $O = (O_1, O_2, ..., O_k, ..., O_l)^T$; the expected output vector is $d = (d_1, d_2, ..., d_k, ..., d_l)^T$; the weight matrix from the input layer to the hidden layer is $V = (V_1, V_2, ..., V_j, ..., V_m)^T$; and the weight matrix from the hidden layer to the output layer is $W = (W_1, W_2, ..., W_k, ..., W_l)^T$.

The mathematical relations of the output layer and the hidden layer are represented by (5.85) and (5.86):

$$o_k = f(net_k), \quad k = 1, 2, ..., l \tag{5.85}$$

$$net_k = \sum_{j=0}^{m} w_{jk} y_j, \quad k = 1, 2, ..., l \tag{5.86}$$

In (5.85) and (5.86), o_k denotes the mathematical relation of the output layer, net_k denotes the network layer at layer k, and W_{jk} denotes the weight matrix of the implied layers from the layer j input to the layer k.

Mathematical relationship of the hidden layer is shown in (5.87) and (5.88):

$$y_i = f(net_j), \; j = 1, 2, ..., m \tag{5.87}$$

$$net_j = \sum_{i=0}^{n} v_{ij} x_i, \; j = 1, 2, ..., m \tag{5.88}$$

In (5.87) and (5.88) are unipolar sigmoid functions. $f(x)$ has a certain continuity and the inverse is also a continuous first derivative, and the network output error is defined as shown in (5.89):

$$f(x) = \frac{1}{1 + e^{-x}} \tag{5.89}$$

In (5.89), $f(x)$ is a unipolar sigmoid function, and $f(x)$ has continuity. Its derivative is also continuous.

The reciprocal is also continuous, and the first derivative is shown in (5.90):

$$f'(x) = f(x)\,[1 - f(x)] \tag{5.90}$$

In (5.90), $f'(x)$ is the first-order derivative of $f(x)$, which also has continuity like $f(x)$.

The network output error is defined as:

$$E = \tfrac{1}{2}(d - o)^2 = \tfrac{1}{2}\sum_{k=1}^{l}(d_k - o_k)^2 \tag{5.91}$$

Substituting (5.90) and (5.91) into the relationship between the parameters of the hidden layer and the error, and (5.92) and (5.93) are substituted into the above equation, the relationship between the parameters of the input layer and the error are shown in

$$E = \frac{1}{2}\sum_{k=1}^{l}\left[d_k - f(net_k)\right]^2 = \frac{1}{2}\sum_{k=1}^{l}\left[d_k - f\left(\sum_{j=0}^{m} w_{jk}y_j\right)\right]^2 \tag{5.92}$$

$$E = \frac{1}{2}\sum_{k=1}^{l}\left\{d_k - f\left[\sum_{j=0}^{m} w_{jk}f(net_j)\right]\right\}^2 = \frac{1}{2}\sum_{k=1}^{l}\left\{d_k - f\left[\sum_{j=0}^{m} w_{jk}f\sum_{j=0}^{m} v_{ij}x_i\right]\right\}^2 \tag{5.93}$$

According to (5.93), it can be found that the network output error can be represented by the function of the input layer weight and the hidden layer weight, and the weight update direction is the direction of gradient descent, as shown in (5.94) and (5.95):

$$\Delta w_{jk} = -\eta\frac{\partial E}{\partial w_{jk}},\ j = 0, 1, 2, ..., m;\ k = 1, 2, ..., l \tag{5.94}$$

$$\Delta v_{jk} = -\eta\frac{\partial E}{\partial v_{jk}},\ i = 0, 1, 2, ..., m;\ j = 1, 2, ..., m \tag{5.95}$$

The negative signs in (5.94) and (5.95) are the direction of the weight update direction and gradient-descent direction, which is the learning efficiency of the network. The above equations reflect the process of network weight update and error reduction.

For the output, (5.95) can be expanded to get, for the hidden, (5.96) can be expanded to get, and the error signals of the output layer and the hidden layer are, respectively defined as δ_k^0 and δ_j^y:

$$\Delta w_{jk} = -\eta\frac{\partial E}{\partial w_{jk}} = -\eta\frac{\partial E}{\partial net_k}\frac{\partial net_k}{\partial w_{jk}} \tag{5.96}$$

In (5.96), W_{jk} denotes the weight matrix of the implied layers from the layer j input to the layer k, net_k denotes the network layer at layer k, and η denotes network learning efficiency.

$$\Delta v_{ij} = -\eta\frac{\partial E}{\partial v_{ij}} = -\eta\frac{\partial E}{\partial net_k}\frac{\partial net_k}{\partial v_{ij}} \tag{5.97}$$

In (5.97), V_{jk} denotes the weight matrix of the input layers from the layer j input to the layer k, net_k denotes the network layer at layer k, and η denotes network learning efficiency.

$$\delta_k^0 = -\frac{\partial E}{\partial net_k} \tag{5.98}$$

$$\delta_j^y = -\frac{\partial E}{\partial net_j} \tag{5.99}$$

Then, substituting (5.98) and (5.99) into (5.96) and (5.97), respectively, for the output layer expansion and hidden layer expansion, we can get:

$$\Delta w_{jk} = \eta \delta_k^0 y_j \tag{5.100}$$

$$\Delta v_{ij} = \eta \delta_j^y x_i \tag{5.101}$$

In (5.100), W_{jk} denotes the weight matrix of the implied layers from the layer j input to the layer k and V_{jk} denotes the weight matrix of the input layers from the layer j input to the layer k

$$\delta_k^0 = -\frac{\partial E}{\partial net_k} = -\frac{\partial E}{\partial o_k}\frac{\partial o_k}{\partial net_k} = -\frac{\partial E}{\partial o_k} f'(net_k) \tag{5.102}$$

For the hidden layer expansion, finding the inverse of the network error for each output layer, for the hidden layer, using (5.102) to substitute the above results into (5.103) and obtain the first derivative of the unipolar sigmoid function which is shown in (5.103).

$$\delta_j^y = -\frac{\partial E}{\partial net_j} = -\frac{\partial E}{\partial y_j}\frac{\partial y_i}{\partial net_j} = -\frac{\partial E}{\partial y_j} f'(net_j) \tag{5.103}$$

In (5.104) δ_j^y denotes the error signal of the output layer, y_i denotes the mathematical relation of the implied layer, y_j denotes the error signal of the output layer.

$$\frac{\partial E}{\partial o_k} = -(d_k - o_k) \tag{5.104}$$

In (5.104), o_k denotes the mathematical relation of the output layer, d_k denotes the desired output vector at the kth level, and E denotes network output error.

$$\frac{\partial E}{\partial y_i} = -\sum_{k=1}^{l} (d_k - o_k) f'(net_k) w_{jk} \tag{5.105}$$

In (5.105), E denotes the network output error, W_{jk} denotes the weight matrix of the implied layers from the layer j input to the layer k, and y_i denotes the mathematical relation of the implied layer.

$$\delta_k^o = (d_k - o_k) o_k (1 - o_k) \tag{5.106}$$

$$\delta_j^y = \left[\sum_{k=1}^{l} (d_k - o_k) f'(net_k) w_{jk}\right] f'(net_j) = \left(\sum_{k=1}^{l} \delta_k^o w_{jk}\right) y_i (1 - y_i) \tag{5.107}$$

The above is the derivation process of the error signals δ_k^o and δ_j^k. Calculation expression for weight update of each layer:

$$\Delta w_{jk} = \eta \delta_k^0 y_j = \eta (d_k - o_k) o_k (1 - o_k) y_j \tag{5.108}$$

$$\Delta v_{ij} = \eta \delta_j^y x_i = \eta \left(\sum_{k=1}^{l} \delta_k^o w_{jk}\right) y_i (1 - y_i) x_i \tag{5.109}$$

Proposing a three-layer network model structure is selected to predict the SOC of the battery. In the three-layer network, a structure with four parameters input and one parameter output is used. Figure 5.45 is the flow chart of BP neural network.

A reasonable selection of experimental data can improve the accuracy of SOC estimation. The relationship between four sets of data and SOC is recorded and analyzed through discharge experiments. These four parameters are highly correlated

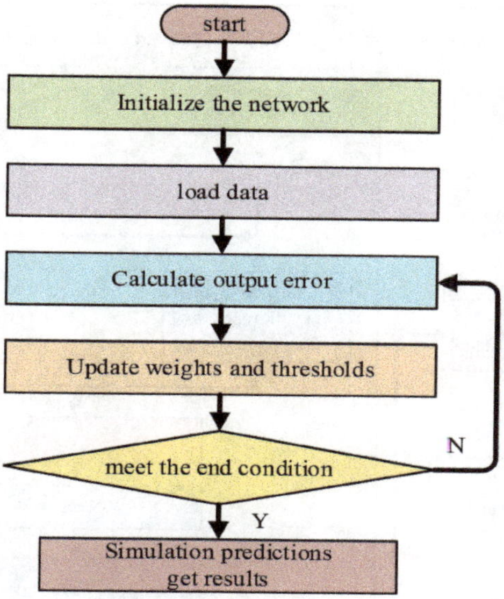

Figure 5.45 BP neural network flow chart

with SOC. The choice of experimental data coverage can also make the SOC estimation results more accurate. Therefore, a wide coverage of data is required for experiments. To reduce the experimental simulation running time and improve the convergence speed, the data are normalized. It can be processed in (5.110):

$$Y = \frac{X - X_{min}}{X_{max} - X_{min}} \tag{5.110}$$

The accuracy of estimating SOC can be increased by optimizing the structure of the setup network. The structure model with four inputs and one output is adopted in this chapter. According to (5.111), it can be known that the number of nodes in the hidden layer of the network is set to 9.

$$N = 2I + O \tag{5.111}$$

Among them, N is the number of hidden layer nodes, I represents the number of input parameters, and O is the number of output parameters.

5.6.2 BP calculation procedure

GA is an adaptive global optimization search algorithm formed by simulating the biological genetic and evolutionary processes in the natural environment. The genetic search algorithm can process multiple individuals in the group simultaneously and quickly optimize the initial weights and thresholds of the BP neural network so that the optimized BP neural network can better predict the function output [190]. The

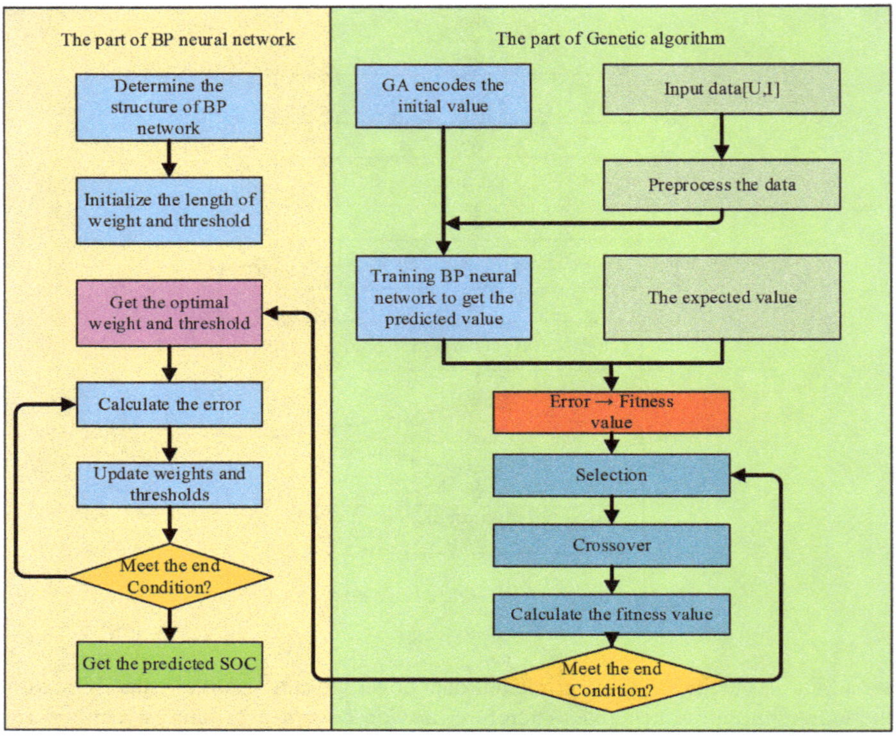

Figure 5.46   The flowchart of the GA-BP algorithm

elements of GA optimization of BP neural network include population initialization, fitness function, selection operation, crossover operation, and mutation operation.

GA-BP neural network uses a GA to optimize the initial weights and thresholds of the BP neural network so that the optimized neural network can better predict the function output. The flowchart of the GA-BP algorithm is shown in Figure 5.46.

The optimization process can be divided into three parts.

1. The network structure is determined according to the functional requirements.
2. The parameter values of the BP neural network structure are optimized by a GA. All initial weights and thresholds in the network are genetically operated to obtain the initial optimal weights and thresholds.
3. Maintain the topological structure of the BP neural network, assigning the initial optimal weight and threshold to the network, and then using the BP neural network for training, testing, and verification.

The BP neural network is a multi-layer network that generalizes the W-H learning rules and performs weight training on nonlinear differentiable functions.

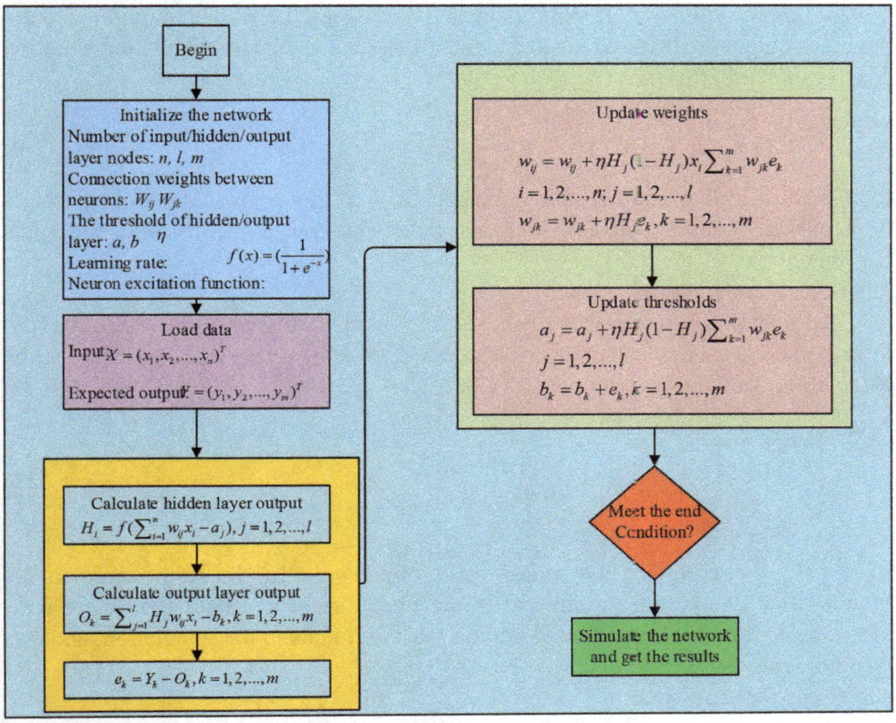

Figure 5.47 The flowchart of the BP algorithm

BP network is mainly used for function approximation, pattern recognition, classification, and data compression. The flowchart of the BP algorithm is shown in Figure 5.47.

The GA is a parallel stochastic search optimization method that simulates natural genetic mechanisms and biological evolution. The elements of BP neural network optimization by the GA include population initialization, fitness function, selection operation, crossover operation, and mutation operation.

1. Population initialization

   The individual encoding method is a real number encoding, where each individual is a real number string consisting of four components: the connection weight between the input layer and hidden layer, the threshold value of the hidden layer, the connection weight between the hidden layer, and output layer, and the threshold value of output layer. The individual contains all the weights and thresholds of the neural network. If the network structure is known, a neural network with its structure, weights, and thresholds can be constituted.

2. Fitness function

   According to the initial weights and thresholds of the BP neural network obtained by individuals, the BP neural network is trained with training data to predict the system output. The absolute error value and E between the predicted output and the expected output are taken as the individual fitness value F. The calculation equation of F is shown in (5.112):

$$F = k\left(\sum_{i=1}^{n} |y_i - o_i|\right) \tag{5.112}$$

In (5.112), n is the number of output nodes, y_i is the expected output of the ith node of the BP neural network, o_i is the predicted output of the ith node, and k is a coefficient.

3. Selection operation

The selection operation of GA includes roulette, tournament, etc. When the roulette method is used to calculate the selection probability p_i of each individual i, the calculation equation is shown in (5.113)

$$\begin{cases} f_i = \dfrac{k}{F_i} \\ p_i = \dfrac{f_i}{\sum_{j=1}^{N} f_j} \end{cases} \tag{5.113}$$

In (5.113), F_i represents the fitness value of individual i. Since the smaller, the fitness value, the better, the inverse of the fitness value is calculated before individual selection. N represents the number of individuals in the population, and k is a coefficient.

4. Crossover operation

The crossing operation of the kth gene a_k and the lth gene a_l at the j position are realized by using (5.114). In (5.114), b represents a random number in [0,1]:

$$\begin{cases} a_{kj} = a_{kj}(1 - b) + a_{lj}b \\ a_{lj} = a_{lj}(1 - b) + a_{kj}b \end{cases} \tag{5.114}$$

5. Mutation operation

The jth gene of the ith individual is selected for mutation operation, as shown in (5.115):

$$\begin{cases} f(g) = r_1\left(1 - \frac{g}{G_{max}}\right)^2 \\ a_{ij} = \begin{cases} a_{ij} + (a_{ij} - a_{max}) * f(g), r > 0.5 \\ a_{ij} + (a_{min} - a_{ij}) * f(g), r \le 0.5 \end{cases} \end{cases} \tag{5.115}$$

In (5.115), a_{max} and a_{min} represent the upper and lower bounds of a_{ij}, r represents the random number, g represents the current iteration, and G_{max} represents the maximum number of evolutions.

5.6.3 GA fitness correction

GA is a kind of EA whose basic principle is to imitate the evolutionary law of "natural selection and survival of the fittest" in biology [191]. GA is to encode the problem parameters as chromosomes in the population by iterative selection and crossover and mutation operations and finally generate chromosomes in line with the optimization goal.

In GAs, chromosomes correspond to data or arrays, which are usually represented by one-dimensional string structure data, and the values of genes correspond to each position on the string. Clusters of genes are called chromosomes or genotypes. A certain number of individuals form a group. The number of individuals in a population size, also known as population size. The degree to which each individual adapts to the environment is called fitness. The basic steps of GA are as follows.

1. Code

GA first represents the solutions of the solution spaces as genotype string structure data of the genetic space before searching, and the different combinations of these string structures constitute different points.

2. Generation of initial population

N initial string structure data are randomly generated. Each string structure data is called an individual, and N individuals constitute a group. GA starts evolution with these N string structure data as the initial point.

3. Fitness assessment

Fitness indicates the advantages and disadvantages of an individual or solution. The definition of fitness function is different for different problems. Therefore, there are different solutions to different problems, and it needs to start from all aspects.

4. Selection

The purpose of selection is to select excellent individuals from the current population, so that they can have the opportunity to reproduce offspring as parents for the next generation. GA embodies this idea through the selection process. The principle of selection is that individuals with strong adaptability have a high probability of contributing one or more offspring to the next generation. The selection embodies Darwin's survival of the fittest principle.

5. Crossover

Crossover operation is the most important genetic operation in GA. Through crossover operation, we can get a new generation of individuals, which combines the characteristics of their parents. Crossover reflects the idea of information exchange.

6. Mutation

Mutation first randomly selects an individual in the population, and the value of a string in the string structure data is randomly changed with a certain probability for the selected individual. As in the biological world, the probability of variation in GA is very low, and the value is usually very small.

For example, GA and nonlinear programming are used to solve the minimum of the following functions, where x_1, x_2, x_3, x_4, x_5 are real numbers between 0 and 0.9π. The minimum value of this function is –2, and the minimum position is $\frac{\pi}{2}, \frac{\pi}{2}, \frac{\pi}{2}, \frac{\pi}{2}, \frac{\pi}{2}$.

$$f(x) = -5 \sin x_1 \sin x_2 \sin x_3 \sin x_4 \sin x_5 - \sin 5x_1 5x_2 5x_3 5x_4 5x_5 + 8 \tag{5.116}$$

The population initialization module initializes the population according to the solving problem, and the fitness value calculation module calculates the fitness function. Selection, crossover, and variation are the search factors of the GA, and N is a fixed value. When the number of times is multiple of N, the nonlinear optimization method is adopted to accelerate the evolution. Nonlinear optimization using the current chromosome values using the function "fmincon" to find the local optimal value of the problem. The steps of GA are as follows.

1. Population initialization

GA cannot deal with the parameters of problem space directly. Therefore, the feasible solution of the required problem must be expressed as chromosomes or individuals in the genetic space by coding, Gray coding, real number coding (floating point coding), multi-level parameter coding, ordered string coding, structural coding, and so on.

Real coding does not need numerical conversion, and GA operation can be performed directly on the phenotype of the solution. Therefore, this case adopts this coding method, and each chromosome is a real vector.

2. Fitness function

Fitness function is used to distinguish the good or bad individuals in a group and is the only basis for natural selection, which is generally obtained by the transformation of the objective function. This case is to find the minimum value of the function, the reciprocal of the function value as the individual fitness value. The smaller the function value, the larger the fitness value, and the better the individual. The fitness calculation function is shown in (5.117):

$$\mathrm{F} = f(x) \tag{5.117}$$

3. Selecting operation

The selecting operation selects good individuals from the old population with a certain probability to form a new population, in order to reproduce the next generation of individuals. The probability of an individual being selected is related to the fitness value. The higher the fitness value of an individual is, the higher the probability of being selected. The selection operation of GA includes roulette method, tournament method, and other methods. In this case, roulette method is selected, that is, the selection strategy based on fitness ratio. The probability of individual i being selected is (5.118), where F_i is the fitness value of individual i and N is the number of individuals in the population:

$$p_i = \frac{F_i}{\sum_{j=1}^{N} F_j} \tag{5.118}$$

4. Crossover operation150567

The crossover operation refers to the random selection of two individuals from the population, through the exchange combination of two chromosomes, the excellent characteristics of the parent string to the child string, so as to produce new excellent individuals. As individuals use real number encoding, real number crossing method is adopted for crossover operation. The crossover operation method of a_k on the chromosome k and a_l on the chromosome l at j position is shown in (5.119), where b is a random number between $[0, 1]$:

$$a_{kj} = a_{ij}\left(1 - b\right) + a_{lj}b \tag{5.119}$$

$$a_{lj} = a_{lj}\left(1 - b\right) + a_{kj}b \tag{5.120}$$

5. Mutation operation

The main purpose of mutation operation is to maintain population diversity. Mutation operation selects a random individual from the population and selects a point of the individual to mutate to produce a better individual. The operation method for mutation of the j gene of the i individual is shown in (5.121):

$$a_{ij} = \begin{cases} a_{ij} + \left(a_{ij} - a_{max}\right) * f(g), & r \geq 0.5 \\ a_{ij} + \left(a_{min} - a_{ij}\right) * f(g), & r < 0.5 \end{cases} \tag{5.121}$$

where a_{max} is the upper bound of gene a_{ij}, a_{min} is the lower bound of gene a_{ij}, $f(g) = r_2\left(1 - \frac{g}{G_{max}}\right)$, r_2 is a random number, g is the current iteration number, G_{max} is the maximum evolution number, and r is the random number in the interval of $[0, 1]$.

6. Nonlinear optimization

The GA was iterated once. The result of the current calculation of the GA was taken as the initial value, the linear programming function "fmincon" in the MATLAB optimization toolbox was used for local optimization, and the local optimal value was used as the new individual chromosome to continue to evolve.

According to the GA and nonlinear programming theory, MATLAB software programming is based on GA and nonlinear programming function optimization algorithm. The program realizes the following process.

1. Fitness function

The individual fitness value is the reciprocal of the fitness function value, and the fitness function is also called the evaluation function. The reciprocal is the individual fitness value.

2. Selecting operation

The selection operation adopts roulette method to select individuals with good fitness from the population to form a new population. Roulette is a commonly used random algorithm, which is widely used in computer simulation random applications. Its purpose is to make the number of individuals selected follow the fitness and selection probability of individuals on the whole.

3. Crossover operation

The crossover operation is to select two individuals from the population and get new individuals according to a certain probability. Multipoint crossover, or generalized crossover, refers to the random setting of multiple intersections in individual chromosomes and then gene exchange. The operation process is similar to that of single-point crossing and two-point crossing. If only one intersection is selected for multipoint intersection becomes single-point intersection.

4. Mutation operation

The mutation operation is to randomly select an individual from the population and obtain a new individual according to a certain probability of mutation. The basic content of mutation operator is to change the gene value on some loci of individual string in the population. According to the different representation methods of individual coding, there can be the following algorithms: (i) real value variation and (ii) binary variation. Generally, the basic steps of mutation operator operation are as follows: (i) Judge whether to mutate by the preset mutation probability for all individuals in the group. (ii) The mutation sites were randomly selected for mutation.

5. Algorithm principle function

The main function flow of genetic function is as follows: (i) randomly initialize the population; (ii) the fitness value of the population was calculated to find the optimal individual; (iii) selecting operation; (iv) crossover operation; (v) mutation operation; (vi) nonlinear optimization; and (vii) judge whether evolution is over, if not, return to step (ii).

6. Nonlinear optimization and interpretation of results

Nonlinear optimization, that is to call MATLAB optimization toolbox to process the function. According to the theory of GA, it is programmed in MATLAB to find the optimal solution of function by using the basic GA, so as to improve the search performance of GA.

5.7 Experimental verification for large-capacity batteries

5.7.1 Experimental design and data set establishment

In many applications, experimental design has become a more and more concerned field [192]. Initially, experimental design found application in agriculture, biology, and other areas of hard science. It has spread through the engineering arenas to the social sciences of economics and behavioral analysis. Since the mid-1960s, experimental design seems to have been used in traditional commercial and management applications. Recently, the experimental design method has been applied to nonprofit and government departments.

There are many reasons for this progression; the training of managers and professionals in the latter field in statistical and quantitative methods has increased, followed by the wide use of quantitative decision-making technology. This trend was further encouraged by the total quality management (TQM) movement originating in the mid-1980s and continuing today. On March 11, 1996, Forbes magazine, a highly respected media, mentioned in an article entitled “new mantra”: The article of MVT shows that the advantages of experimental design are widely accepted. MVT stands for multivariable testing, which is a term proposed together with experimental design in this article to distinguish factor design from experiments one factor at a time.

The following are the six steps of experimental design.

1. Plan the experiment.
2. Design the experiment.
3. Experiment.
4. Analyze the data from the experiment.
5. Confirm the results of the experiment.
6. Evaluate the conclusions of the experiment.

The planning process is critical to the success of the experiment. In this process, it is necessary to understand the close cooperation between researchers (process experts) and people (designers) with experience in experimental design methods. Efforts at this stage can greatly increase the opportunity to make appropriate and meaningful assumptions; ignoring this stage may waste time and resources in everything that follows.

The planning stage itself consists of five steps

1. Identify the dependent, or output, variable(s).
2. Translate output variables to measurable quantities.
3. Determine the factors (independent, or input, variables) that potentially affect the output variables that are to be studied.
4. Determine the number of levels or values for each factor and what those levels are.
5. Identify potential synergy between different factors.

The dependent variable to be studied should be chosen carefully. Sometimes, the choice is obvious, but other times, it is not. The general guideline is to ensure that the right performance measure is chosen – typically, the quantity that is the important one; maybe it's someone who can tell you how good a product is, how well it is done, or how it is accepted in the market. For example, in the experiment of helping to develop better covers for professional catalogs, the wise choice of dependent variable is unlikely to be the brightness of the cover. One way to choose a dependent variable is to make sure that all vested interests are represented—those of the user, producer, marketing department, and relevant others—perhaps in a formal brainstorming session. If the goal is to produce popcorn that sells more, is it obvious to the consumer what the key qualities should be? Fluffiness? Color? Taste? Texture? Percent popped? Once the dependent variable is selected, it must usually be transformed into a quantitative measure to make it useful.

Many variables are subjective. What is the measure of taste or appearance? What about the units of measurement? The choice of units is rarely important if those under consideration are linearly related to one another (such as inches versus feet, or dollars versus yen). However, how about the choice between circumference, area, and volume to measure the size of a spherical product? Two known measures determine one of the other measures, but if the value of one measure changes, the values of the other measures do not change proportionally. So, although one measure may vary linearly with the level of a factor, another measure may vary nonlinearly. As noted later, linear variation and nonlinear variation may be an important problem. Choosing the factors to study is sometimes quite straightforward; however, other times it is not as easy as it might seem.

In our consulting experience, the process experts sometimes propose an unworkably large number of possible factors for study. An effective way to propose all candidate input variables is to use cause and effect diagrams during formal brainstorming meetings. This is a technology developed in the context of quality control to help identify factors affecting product quality. This technology is applicable to any situation where you want to determine the factors affecting dependent variables. This approach generally yields a nearly exhaustive list of choices in categories such as people factors, equipment factors, environmental factors, methods factors, and materials factors. Pareto analysis can reduce the initial list to its main points. In this context, Pareto analysis classically states the now well-known concept that 80% of the identifiable impact will be related to 20% of the factors, although the exact values of "80" and "20" are unlikely to be realized accurately. Of course, you should first reduce the list in any obvious way.

Recalling one case in which (in simplistic terms) the factor "temperature greater than 212 For not" was identified along with another factor, "presence or absence of steam" (with the experiment run at sea level in an unpressurized container). One is usually motivated to minimize, to the degree possible, the number of factors under study. In general, everything else being equal, a higher number of factors under study is associated with an increased size of an experiment; This in turn increases the cost and time of the experiment. The relationship between the number of factors in the experiment and the scale and efficiency of the experiment is discussed from many angles in this chapter. How many different levels should a factor appear in an experiment? This is not always easily answered. A quick-and-dirty answer is that if the response of the dependent variable to the factor is linear (i.e., the change in the dependent variable per unit change in the level of the factor is constant), two levels of the factor will suffice, but if the response is nonlinear (not a straight line), one needs more than two. In many nonlinear cases, if the factor is measured on a numerical scale (representing some unit of measurement, not simply categories), three levels will suffice.

However, this answer has the major flaw of circular reasoning: the answer is clear if the response function (the true relationship between dependent and independent variables) is known, the reason for conducting experiments is usually to find the response function. Naturally, if the number of factor levels is high, the total number of factor level combinations increases. Eight factors, each at two levels, have a total of 256 different factor level combinations, while eight factors, each at three levels, have 6,561 different factor level combinations—this is a big difference! In the whole text, multi-level problems have been solved in different environments. The last planning step mentioned earlier is to determine the synergy (or anti synergy) between factors.

The more formal word used for synergy is interaction. Interaction is the combined effect of factor levels that is above and beyond the sum of the individual effects of the factors considered separately. That is, the total is greater or lower than (i.e., not equal to) the sum of the parts. For example, if adding a certain amount of shelf space for a product adds 10 units of sales, and adding a certain amount of money toward promoting the product adds 8 units of sales, what happens if the store increases shelf space and promotion expenses at the same time? Did the store get sales of 18 units (the sum of 10 and 8 units)? If so, then these two factors, the six steps of experimental design and five shelf space and promotion, are considered not to interact. If the revenue exceeds 18 sales units, it indicates that there is positive interaction or positive synergy; if the gain is less than 18 units, this situation is called negative interaction or negative synergy. The text expounds this concept in depth; in fact, the number of possible interactions between the factors in the study has a significant impact on the selected experimental design.

Designing the Experiment Having completed the planning of the experiment, design the experiment, completing the plan of the experiment, and then starting the design stage of the experimental design, which is the primary theme of this section. First, design type selection is still needed. A basic principle of this chapter is factorial design, in which the experiment includes different levels of factor combinations, which is usually much better than one-at-a-time design. The level of each factor is different, but there is only one factor at a time. This is a major theme of the Forbes article mentioned earlier; In fact, this directly leads to the discussion of the article called "multivariate testing."

Having chosen the design type, the specific design still needs to be made. A critical decision is to determine how much "fractionating" is appropriate. When there are a large number of different levels of factor combinations, it is inevitable that only a small part is running. Determining which fraction is appropriate and the best subset of combinations of levels of factors to make up this fraction is a large part of the skill in designing experiments. The degree of fractionating can be dramatic. For example, if 13 different factors are studied and each factor has three levels, the result will be 1 594 323 different factor level combinations. However, if it is assumed that none of these factors interact with other factors, it is necessary to carefully select a subset of only 27 of these combinations, but not unique, and may have moderate replication to obtain an effective estimate of the impact of each of the 13 factors.

The accuracy of estimation depends on the complexity of the experimental design, that is, the number of data values obtained under the same experimental conditions. Another key factor in designing experiments is to consider blocking, which is the control factor. These factors may not be the main concern, but if not controlled, it will increase the variability of data and may mask the real impact of the factors of real concern.

For example, suppose this experiment wants to study the effect of hair color on the ability of a particular brand of shampoo to reduce the number of forked hair. Further suppose that independent of hair color, male and female hair react differently to the shampoo. Then, if the experiment wants to take sex as the second factor (although some texts will not mark it as a factor, but just a block to distinguish it from the main factor, namely, hair color). By being controlled, or accounted for, the variability associated with the factor (or block) of sex could be calculated and extracted so that it does not count against and obscure differences due to hair color. The discussion in this section must be somewhat superficial. Designing experiments does not always follow such a set of independent sub-steps that are easy to describe.

Other considerations must be taken into account, at least tangentially if not more directly. For example, perhaps the experiment under consideration is to be one of a series of experiments; in this case, expecting to design a slightly different experiment—an experiment that does not include all the factors that may be required for an independent experiment. When selecting the design, there may also be the problem of combining the data of one experiment with the data of another experiment.

Perform the experiment. It goes without saying that once the experiment has been designed, it must be performed ("run") to provide the data that are to be analyzed. Although not spending much time discussing the operation of the experiment, it does not mean that this is a trivial step. The experiment that was designed must be the experiment that is run. In addition, the combination order of operation factors should be random. The randomization prevents the introduction of unexpected effects or bias in the experiment. For instance, in an experiment to determine the effectiveness of a new drug in comparison to a standard option, the patients could be randomly allocated to the two treatments. Indeed, some statistical software programs that use information decided during the planning stage to provide designs for the user also provide a worksheet in which the order of the combinations has been randomly generated.

Sometimes, the conclusions drawn from the analysis of experimental data seem obvious. However, that can be deceptive. Often the results are not clear-cut, even when they appear that way. It is important to tell whether an observed difference is indicating a real difference or is simply caused by fluctuating levels of background noise. To make this distinction, it should go through a statistical analysis process called hypothesis testing. The statistical analysis in the six steps of experimental design cannot prove that a certain factor will produce difference, but it can indeed provide a measure of result consistency on the premise that the factor does not affect the dependent variable. If the results are relatively inconsistent with this premise (a term that gets quantified), experiments usually conclude that this factor does have an impact on the dependent variable and consider the nature of the impact. The statistical analysis also provides a measure of how likely it is that a given conclusion would be in error. Somewhat formidable mathematics is required to derive the methods of hypothesis testing that are appropriate for all but the simplest of experiments.

However, the good news is that, for the most part, it is the application of the methodology, not its derivation, that is a core requirement for the proper analysis of the experiment. Most texts provide illustrations of the application. In addition, numerous statistical software packages do virtually all calculations for the user. Indeed, far more important than the mechanics of the calculations is the ability to understand and interpret the results. The ability to understand and interpret results is enhanced by the ability to "calculate numbers" manually (as mentioned earlier, using this phrase includes using a calculator). The principal statistical method used for the analysis of the data in experimental designs is called analysis of variance (ANOVA), a method developed by Sir Ronald Fisher. The primary question ANOVA addresses is whether the level of a factor (or interaction of factors) influences the value of the output variable. Other statistical analyses augment ANOVA to provide more detailed inquiries into the data's message.

Once practical conclusions are drawn from the experimental analysis, the results of the experiment can be confirmed, and it is usually a good idea to try to verify these conclusions. What factors affect the variables and to what extent? Which factor level combination can provide the best value of the dependent variable. "Practical safety statistics" shows that at the combined level of these factors, the result is indeed the expected result. Note that if the experiment is intended for research purposes, some scientific journals will not accept manuscripts without a verification step. Why? Well, it is very likely that while running only a fraction of the total number of combinations of levels of factors. In this process, there is no ideal effect; Or it does run, but the process is not comprehensive.

The selected design wisdom may be based in part on the assumption that some interactions exist and do not exist, and the results derived from the analysis surely assumed that no results were mis-recorded during the performance of the experiment and that no unusual conditions which would harm the generalizability of the results occurred during the experiment. Thus, why not perform a confirmation test (with several replicates) that verifies our conclusions? If there is a discrepancy, better to identify it now rather than later!

After any experiment, it is best to know whether the test results are meaningful by evaluating the experimental conclusions. If it doesn't make sense, further thinking is needed to determine what to do next. As the sixth step of the experimental design process, the special evaluation considered is the economic evaluation of the results.

Not all situations lend themselves to an explicit utilization of this step. However, according to experience, a significant proportion of experiments applied in the management areas do lend themselves to, indeed mandate, a cost and benefit analysis to determine whether the solutions suggested by the results of the experiment are economically viable.

Today, it is clear that companies generally cannot, and should not, embrace quality for quality's sake alone. Quality improvements need to be economically justified. A fruitful application of designed experiments is in the area of product configuration, which involves running experiments to examine which combination of levels of factors yields the highest purchase intent or sales. However, the combination of levels of factors that has the highest purchase intent (let's put aside the issue of if and how intent translates to sales) may be a big money loser. Three scoops of sugar-cone ice cream for nickel would surely yield very high revenue, but not for long—the company would soon go out of business.

As another example, suppose that a combination of certain levels of ingredients that has a variable cost higher than the current combination would result in the same average quality indicator value (such as battery life), but with a lower amount of variability from product to product. It is a consensus that lower variability is desirable, but, for example, does a 20% decrease in this variability mean a 30% increase in variable costs? The answer lies with an economic evaluation or cost/

benefit analysis. It may or may not be easy to do such an evaluation; however, it is difficult to conclude without one.

Data set establishment

Power is the basis of the continuous development of modern science and technology, is the important guarantee of improving human life quality, and is the lifeblood of the national economy and security. The traditional power supply system adopts centralized power generation and then uses the transmission network to transmit power to users. Electric power development has experienced two important milestones: first, in 1882, the United States built the world's first DC generator driven by the steam engine to transmit the power system formed by DC, marking the beginning of the power era; second, Germany in 1891 established the first by the AC generator power generation, and then through the length of 179 km, the voltage level of 12-kV transmission lines transmit power transmission system, a symbol of the modern AC supply system the establishment of the basic patterns.

For the research results of new energy power generation scattered at home and abroad, these results are mainly limited to unilateral research. For example, research power prediction only considers forecasting precision and has no effect on the capacity of the energy storage device. The energy storage device only considers how to select energy storage device and no late combined with energy storage devices in the network behavior. In addition, the new energy research hotspot focused on its implementation in forms, such as the topology of the inverter, harmonic elimination circuit design, and the energy of fine management. A reasonable system structure is designed with load fluctuation curve and new energy power generation curve, and select a reasonable optimization algorithm.

The model is based on IEEE 30, School of electrical engineering, University of Washington. The power structure is based on research. The elements included in it include fossil fuel power generation, whose power can be between the upper limit of power generation and the lower limit of power fluctuation rate. It represents the peak shaving power plant or the rotating standby unit in the system, and its capacity is a fixed value, which represents the basic load power generation part of the system; for wind power generation, the power changes with the size of wind force; for solar power generation, the power changes with the intensity of sunlight change; it also has wave load, and the power demand changes with time, representing the dynamic load in the system. The rest of the power demand is a constant, representing the base load part of the system. At the same time, in order to simulate the influence and power characteristics of the energy storage power station in the system, the whole system is equipped with a certain amount of energy storage devices. The load data used in this chapter are shown in Table 5.4.

The universal model contains the type structure of transmission line, the series impedance for this; a parallel structure of capacitive reactance, the capacitive reactance to it; an ideal phase-shifting transformer, the impedance of 0, only phase shift effect, there is no loss of energy. Ideal phase shift change than the value of, phase shifting angle for shift.

Start injecting current in the plural form of I and end; each variable in the model of solving solution can be obtained using the method of node voltage, and

Table 5.4 Node loads

Node	Load capacity	Node	Load capacity	Node	Load capacity
1	0	11	0	21	17.5
2	21.7	12	11.2	22	0
3	2.4	13	0	23	3.2
4	7.6	14	6.2	24	8.7
5	94.6	15	8.2	25	0
6	0	16	3.5	26	3.5
7	Change	17	9	27	0
8	30	18	3.2	28	0
9	0	19	9.5	29	2.4
10	5.8	20	2.2	30	10.6

the computational expressions can be represented with a two-order matrix. Equation (5.122) is for the node admittance matrix, and the pattern of (5.123) is for the use of the node voltage method to solve the related variables used in the process of equations:

$$Y_{br} = \begin{pmatrix} \left(y_s + j\frac{b_c}{2}\right)\frac{1}{\tau^2} & -y_s\frac{1}{\tau e^{j\theta_{shift}}} \\ -y_s\frac{1}{\tau e^{j\theta_{shift}}} & y_s + j\frac{b_c}{2} \end{pmatrix} \tag{5.122}$$

$$\begin{pmatrix} i_f \\ i_t \end{pmatrix} = Y_{br}\begin{pmatrix} v_f \\ v_t \end{pmatrix} \tag{5.123}$$

The electric power system in the network users' demand for energy is to change to stimulate new energy access to the system after the electric network needs to join a dynamic load in the electric grid. In the power system, load change conforms to specific rules, although it will not conflict with the actual physical meaning; for example, it generally does not appear, such as a large pulse load fluctuation trend, but also exist obvious relevant features, such as lamp of peaks and troughs. To ensure that the dynamic load curve and actual basic are consistent, the design of the load curve should contain peaks and troughs and the main characteristics of curve error. The design principle is adopted, as shown in (5.124).

$$f_{load_v}(t) = g\left(t, \delta_{peak}, \delta_{valley}, P_{fixedmean}, P_{vmean}, \delta_{vibration}, per\right) \tag{5.124}$$

$$per = \frac{\delta_{peak/valley}}{P_{fixed\_mean} + P_{v\_mean}} \times 100 \tag{5.125}$$

$$\delta_{peak} = \frac{per \cdot P_{fixed\_mean} + per \cdot P_{v\_mean}}{100} \tag{5.126}$$

To simplify the analysis, the sine wave function is used to simulate the dynamic load curve and set curve peaks, and troughs of volatility percentage are 30%. Equation (5.126) shows the average of the sine function, the function of the peak and average difference model, and the average value and function in the valley of the differential

Table 5.5 *Dynamic load data*

Time point	Amount of volatility	Time point	Amount of volatility	Time point	Amount of volatility
1	0.665355	11	72.05607	21	192.8479
2	2.65413	12	83.92482	22	202.7278
3	5.944536	13	96.20478	23	2117174
4	10.20052	14	108.7614	24	219.718
5	16.27217	15	121.4571	25	226.6421
6	23.19625	16	134.1529	26	232.4138
7	31.1969	17	146.7095	27	236.4138
8	40.18645	18	158.9895	28	240.2602
9	50.06643	19	170.8582	29	242.2489
10	60.72857	20	182.1857	30	242.9143

mode. The above value generation in (5.121) easy solution for type is shown in (5.127)

$$f(t) = A sin\left(t - T\right) + A \tag{5.127}$$

$$A = \frac{per \cdot P_{fixed\_mean}}{100 - per} \tag{5.128}$$

The parameter values of the dynamic load fluctuation data are shown in Table 5.5.

New energy curve design includes wind power and photovoltaic power generation curve. Also, adopt the method of summary on the related data for the reasonable design of the curve; the difference is needed combined with the actual situation to modify the related curve. Photovoltaic power generation at night, for example, should be revised to 0, wind power output volatility and volatility of photovoltaic power output are not the same, and so on. The initial volatility of photovoltaic power generation is set up as 10%, the average for the system as a whole load of 0.3 times; wind power fluctuation was 20%, and the average for the whole load of 0.4 times. Need of special note is that there is only an initial curve, the design of the curve is not the end of the simulation curve, and the related parameters can combine the actual conditions of the specific problems in the application or the emphasis of the research problem appropriately modified.

New energy grid runtime through some optimization algorithm controls the size of the capacity of the new energy power generation access networks, to make the new energy power generation in the power supply system to achieve the best balance between the stability of the economy and safety. Looking for system optimal equilibrium is the essential connotation of the system of optimal power flow. The concept of optimal power flow was first put forward by John Casteen Pentair French engineer in the early 1960s, its core content is in the guarantee system of various constraint conditions under the condition of established, and many of the variables in the control system by some means value target of the system to achieve the desired minimum or maximum.

As a result of the optimal power flow including content is numerous, widely involved economic optimal control and scheduling of the system operation safety, power grid planning. Optimal power flow, therefore, has been proposed since got rapid development, many excellent algorithms have been proposed. The original dual method belongs to an important branch of the interior-point method; it has strong practicability and appeal. The basis of the algorithm framework is established by a meiddo, interior, and others, then promotion by Kojima, to the complementary problem. The original-dual algorithm can easily be applied to nonlinear convex programming, nonlinear complementarity, variational inequality, semidefinite optimization, and second-order cone optimization problems such as field, which has a similar approach in the best polynomial complex boundaries. This chapter will adopt this kind of method, which is the core of the system optimization research method, algorithm detailed steps as follows: extreme-value proposition of the objective function, set up the constraints in the problem domain, which is shown in (5.126):

$$\begin{aligned} &min f(X) \\ &G(X = 0) \\ &H(X) \le 0 \end{aligned} \tag{5.129}$$

The inequality constraint is transformed into equality constraint, and the objective is modified. The core idea of this step is to put into Lagrange extremum problem with constraint condition equation form optimization inequality forms of extreme value problem, which is shown in (5.130):

$$\begin{aligned} min &\left[f(X) - \gamma \sum_{m=1}^{n} ln(Z_m) \right] \\ G(X) &= 0 \\ H(X) + Z &= 0 \\ Z &> 0 \end{aligned} \tag{5.130}$$

First, the optimality condition. The PDE in (5.130) is appropriately transformed to obtain the first-order optimality condition equation of 0.

The p-ply structure is based on the structure of the IEEE 30 power supply network. According to the inherent characteristics of photovoltaic and wind power generation, the simulation curve of photovoltaic power generation is designed, wind power generation simulation curve, and simulated load curve; using universal transmission line structure simplifies the structure of the power supply system. The advantages and disadvantages of various optimization algorithms are compared and analyzed. The final choice of the primal-dual optimization algorithm is optimizing the kernel algorithm.

5.7.2 BP-based SOC estimation analysis

To verify the feasibility and accuracy of the BP neural network algorithm, an experimental platform is established to test the performance of lithium-ion batteries. The experimental platform consists of a host, battery testing equipment

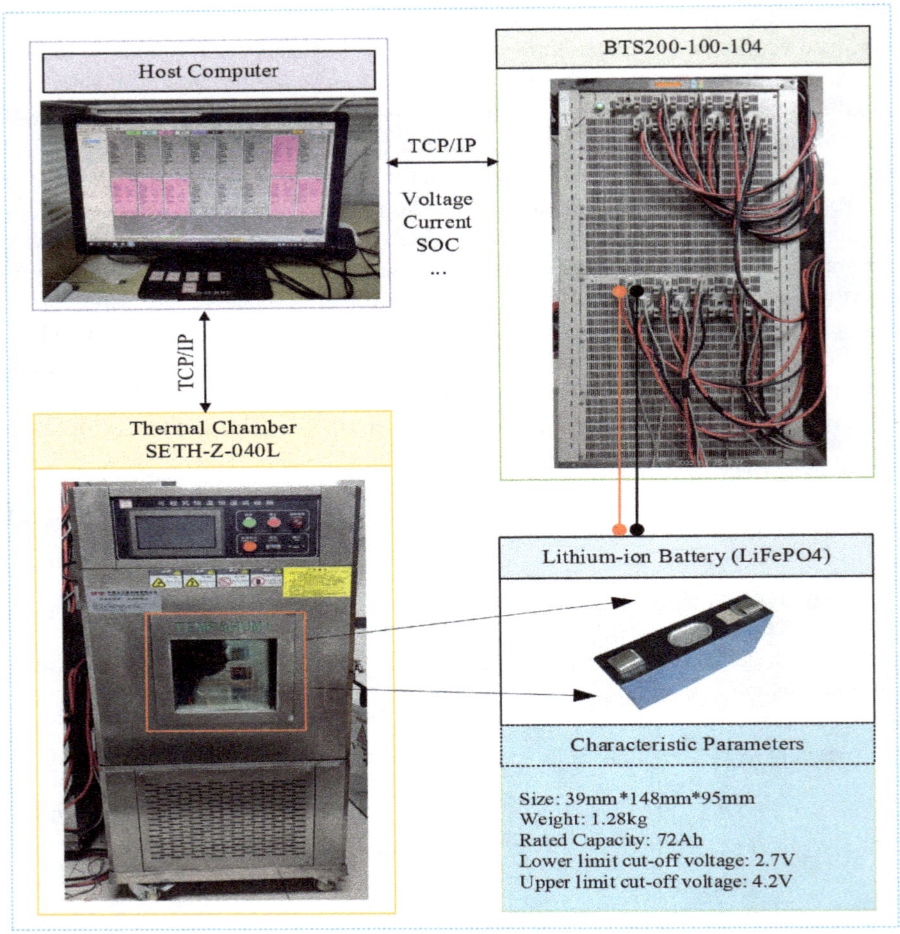

Figure 5.48 Experimental platform

(BTS200-100-104), and a temperature chamber (SETH-Z-040L). The experimental platform is shown in Figure 5.48.

Battery test equipment is used as a programmable power source and electronic load to simulate the fluctuating driving conditions of EVs. The temperature chamber is used to provide a constant ambient temperature for the battery under test. Finally, the host computer controls the temperature and the charging and discharging process and collects data from the battery test system and the temperature chamber through TCP/IP.

The experimental battery was a ternary lithium-ion battery with a rated capacity of 72Ah, and the DST test was performed at 20°C. The training samples are shown in Figures 5.49 and 5.50.

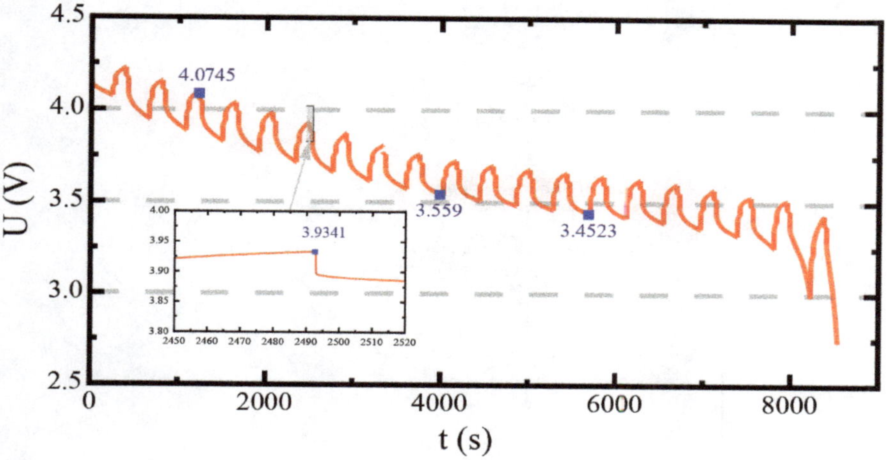

Figure 5.49 The voltage of the train sets

Figures 5.49 and 5.50 are the training samples and test samples under the DST condition of the BP neural network simulation experiment.

Import the normalized sample data into the BP neural network model for training, and the corresponding flowchart is shown in Figure 5.51.

In the process of Figure 5.51, first initialize the entire model, select a random decimal not greater than one to initialize the weight, and then input the corresponding training sample points, which include information such as voltage, current, and SOC. Then solve for the outputs of the hidden and output layers. Solve the corresponding errors and judge whether they meet the requirements. If it is satisfied, the training is over; if it is not satisfied, the hidden layer unit error is calculated in turn,

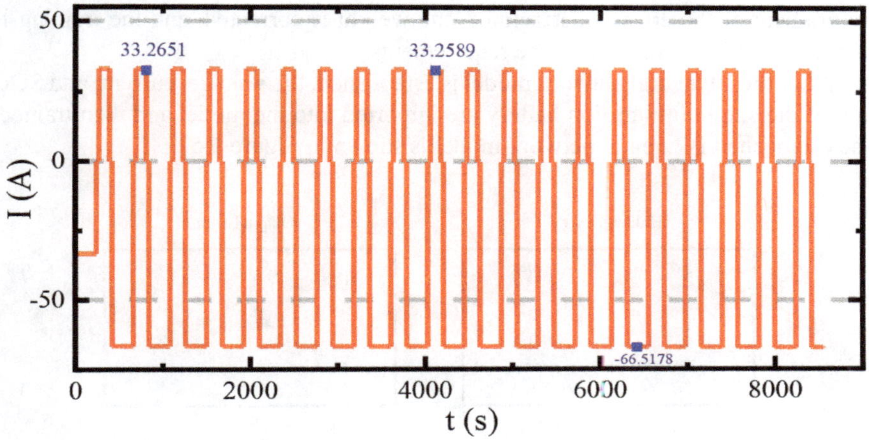

Figure 5.50 The current of the train sets

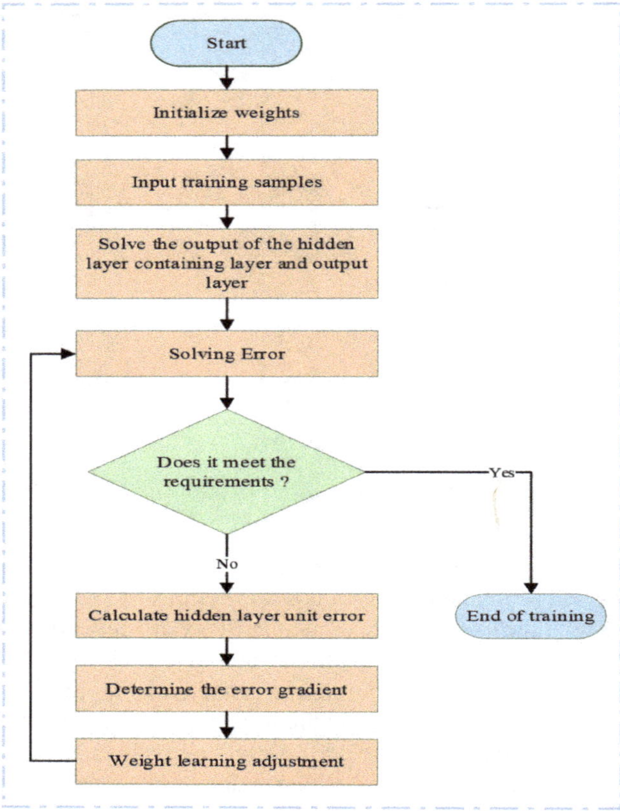

Figure 5.51 BP neural network training flow chart

the gradient of the error is determined, the weight is corrected, and the training is re-trained until the error meets the requirements.

After the BP neural network model is established, the voltage, current, and SOC data of the single lithium-ion battery are imported into the model and then trained. The established BP neural network model is shown in Figure 5.52

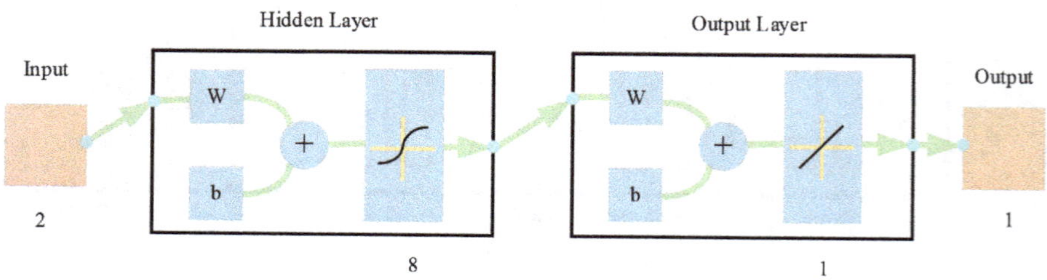

Figure 5.52 BP neural network model

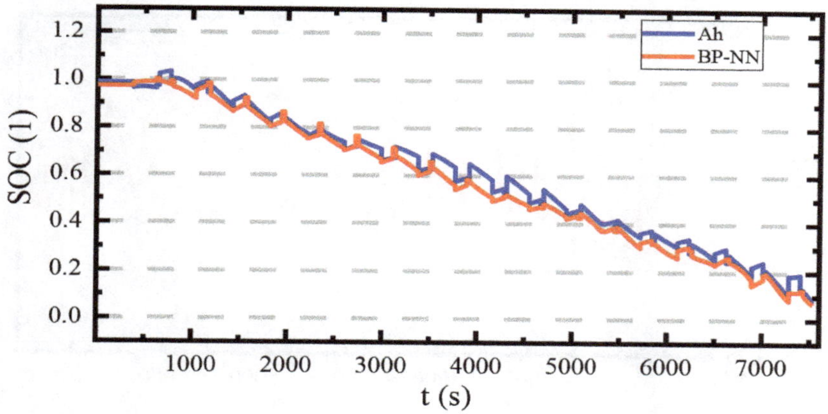

Figure 5.53 SOC estimation curve

The number of hidden layers of the BP neural network model is set to eight layers, and the number of hidden layers is modified according to the training results.

After repeated debugging of the set parameters of the BP neural network, the final number of iterations is 200, the learning rate is 0.01, and the training target error is controlled at 0.001. After 93 steps of training, the network error can meet the design requirements, and the error converges to 0.0021567.

The results of training, validation, and testing of the neural network can be seen from the regression curve. For a good fit, the data should be along a 45° angle, indicating that the output is equal to the expectation. The R values of the fitting output results are all greater than 0.99, indicating that the fitting degree is theoretically better, that is, the established neural network model is better.

Based on the trained neural network model, input test samples, and test the effectiveness and accuracy of the algorithm by comparing the actual value with the predicted output value of the neural network. The simulation results are shown in Figures 5.53 and 5.54.

The actual value of the SOC of the lithium-ion battery is obtained by the ampere-hour integration method and is used as the reference object for the BP neural network to predict the SOC. It can be seen from the simulation results that the predicted numerical curve of SOC changes along the actual numerical curve of SOC. In the actual test data, the SOC value changes in real time, and the SOC changes slightly every second. However, affected by the acquisition accuracy, the actual SOC curve is not smooth but is stepped. It can be seen from Figure 5.53 that the trained BP neural network can accurately predict the SOC of the lithium-ion battery, showing good followability. The SOC estimation error curve in Figure 5.54 further verifies the accuracy of the BP neural network. The average error of the algorithm is 0.12%, which meets the accuracy requirements of lithium-ion battery SOC.

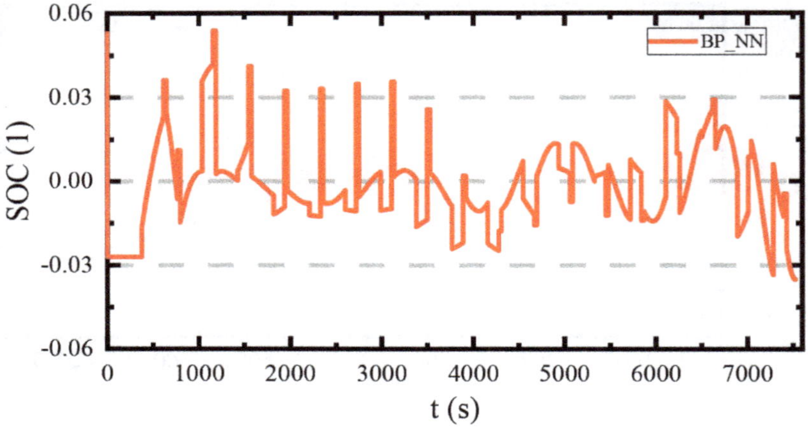

Figure 5.54 *SOC estimation error*

5.7.3 Radial basis function-based SOC estimation analysis

A lithium iron phosphate battery with a nominal capacity of 72 Ah is taken as the research object, and the charge and discharge tests are carried out at a constant temperature of 20°C. Tested at different charge and discharge rates, including 0.5, 1, 2, and 4°C. The test data are shown in Table 5.6.

Compared with the traditional BP neural network, the RBF neural network has the advantages of simple structure and fast convergence speed and is more suitable for estimating the battery SOC. The process of modeling lithium-ion battery voltage and SOC is mainly shown in Figure 5.55.

To ensure the accuracy of network modeling, the calculation procedure selects $\varepsilon = 10^{-10}$, that is, when the error index is less than 10^{-10}, the network training ends,

Table 5.6 *Partial data of test samples*

U/V	**SOC/C**			
	0.5	**1**	**2**	**4**
3.25	0.816	0.905	0.953	0.983
3.17	0.786	0.834	0.927	0.976
3.12	0.73	0.755	0.855	0.945
3.09	0.56	0.506	0.745	0.837
3.04	0.351	0.401	0.508	0.688
2.91	0.227	0.314	0.435	0.475
2.78	0.182	0.237	0.324	0.334
2.65	0.143	0.188	0.272	0.276
2.52	0.115	0.164	0.227	0.203
2.39	0.098	0.136	0.202	0.161
2.26	0.089	0.112	0.182	0.135

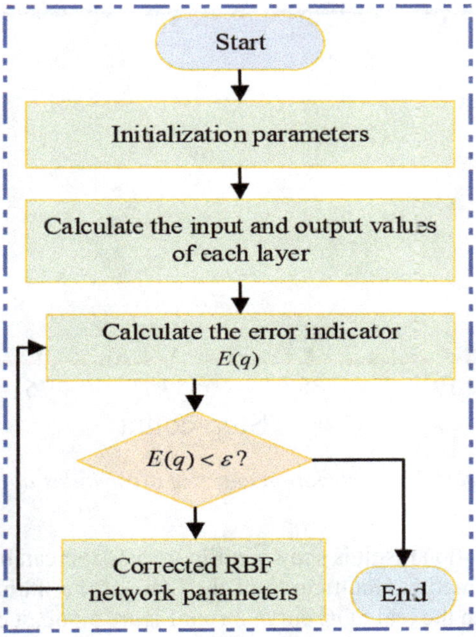

Figure 5.55 Flow chart of RBF neural network modeling

to obtain accurate weights, Gaussian function width, and center vector parameter value, to implement the modeling process. In the simulation process, the estimation accuracy is higher when $\eta = 0.1$ and $\alpha = 0.05$.

Through the above analysis, this experiment first selects the charging data at 0.5 C for modeling, and the SOC estimation result of the lithium-ion battery is shown in Figure 5.56.

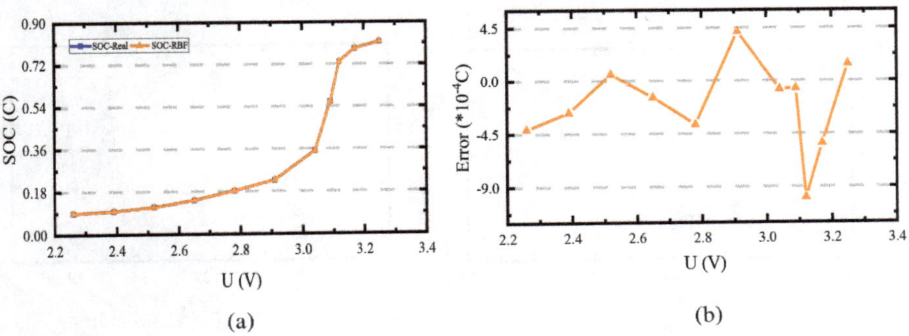

Figure 5.56 SOC estimation result of neural RBF intraneural neural network at 0.5 C

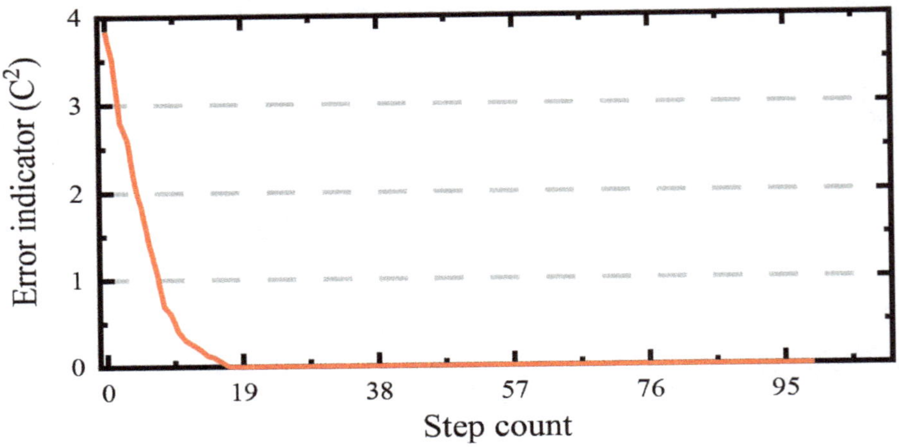

Figure 5.57 Variation curve of error index at 0.5 C

The SOC estimation result is shown in Figure 5.56a; it can be seen that the RBF neural network estimation result curve of the SOC of the lithium-ion battery is in good agreement with the real SOC curve. From Figure 5.56b, it can be analyzed that the error value between the estimated output value of the battery SOC of the RBF neural network and the actual SOC data at 0.5 C is less than 10^{-4}, which indicates that the RBF neural network has a high estimation accuracy.

Among them, the change process of the battery SOC estimation error index of the RBF neural network when the rate is 0.5, as shown in Figure 5.57.

It can be analyzed from Figure 5.57 that the convergence speed of the RBF neural network is fast, and the SOC of the lithium-ion battery can be quickly estimated.

Similarly, the same method was used to model the relationship between the battery voltage and the response SOC value at other rate states. Among them, the RBF neural network battery SOC estimation results at 1, 2, and 4 C are shown in Figures 5.58–5.60.

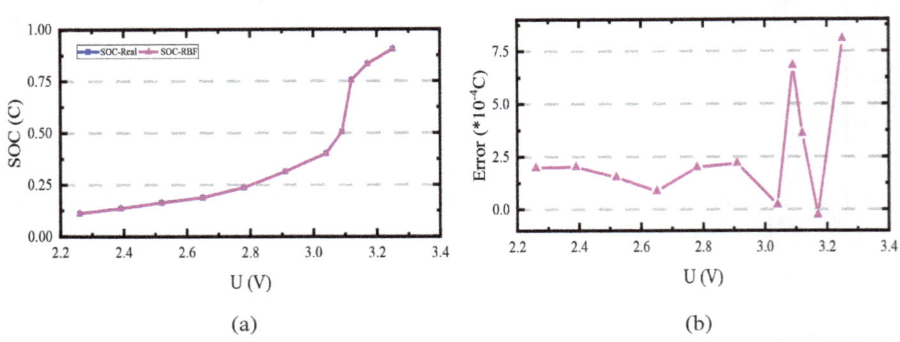

Figure 5.58 SOC estimation result of RBF neural network at 1 C

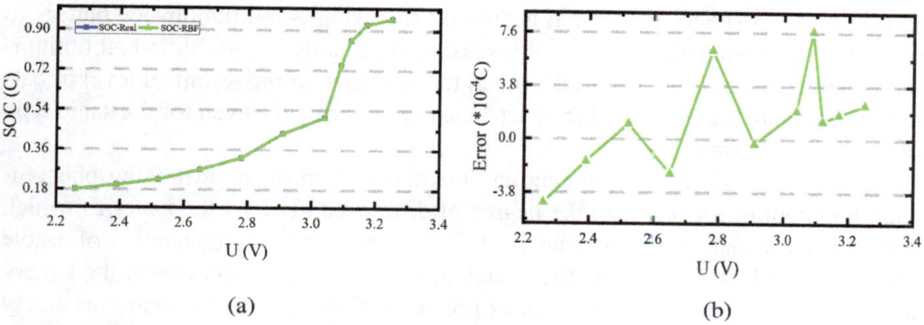

Figure 5.59 SOC estimation result of RBF neural network at 2 C

From the above results, it can be analyzed that the error value between the estimated output value of the battery SOC of the RBF network and the actual SOC data is less than 10^{-4}, and the algorithm has a fast convergence speed, which can meet the accuracy requirements of the lithium-ion battery SOC. The accuracy of the simulation results proves the rationality and effectiveness of the RBF algorithm applied to the SOC estimation of lithium-ion batteries.

5.7.4 Elman neural network-based SOC estimation analysis

1. Introduction

Lithium-ion batteries have been widely used in EVs. Accurate estimation of the SOC of lithium-ion batteries can prevent the damage of lithium-ion batteries due to overcharging and over-discharge and further prolong the service life of lithium-ion batteries. However, SOC does not allow direct measurement and can only be calculated indirectly by other methods [193]. There is a nonlinear relationship between the SOC of the battery and the signals such as current, voltage, temperature, and so on. It will be a very effective method to estimate SOC by studying the relationship between them. In practical applications, especially in the field of clean energy

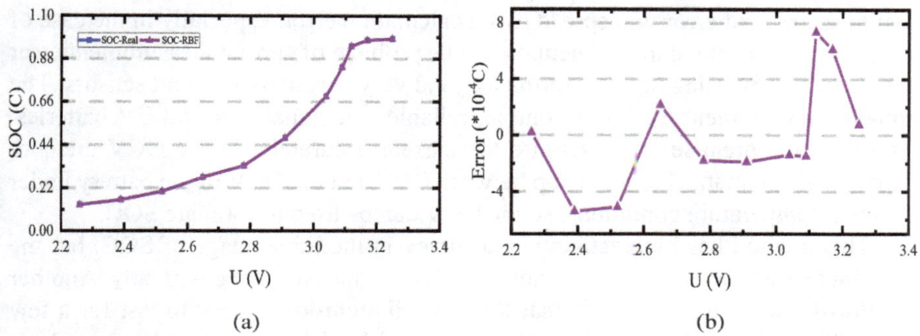

Figure 5.60 SOC estimation result of RBF neural network at 4 C. (a) SOC estimation result. (b) SOC estimation error.

storage systems (ESS) and EVs, it is difficult for a single-cell lithium-ion battery to meet the high-power demand. For this reason, we usually use multiple-cell lithium-ion batteries in series and parallel. Due to the influence of the technical level of battery production, there will be inconsistencies in performance even for the same type of battery in the same process.

The change of the environment and the aggravation of its own aging phenomenon will continue to increase the degree of difference between the batteries, which will eventually make the performance of the battery worse and the number of usable times less and less. In addition, the overcharge and discharge behavior of the battery will also cause safety problems such as fire. Therefore, in order to ensure the safety and durability of the battery, a BMS must be established to monitor and control the battery in real time. The BMS can monitor the status of each single cell in the battery pack in real time, analyze the power of each battery, realize the balance of the power of the single cells in the battery pack, and ultimately prevent the battery from overcharging and over-discharging, and ensure the safe operation of the battery. Based on the above functional requirements, it is essential to accurately estimate the lithium-ion SOC method. In the process of using a lithium-ion battery, the SOC of the battery is the most important parameter. SOC is generally defined as the ratio of the remaining power of the battery at the current moment to its total capacity when it is fully charged. As a characteristic parameter of the battery, the SOC of a lithium-ion battery is affected by various factors such as current, voltage, temperature, and aging degree.

The relationship expression between the measured parameters is estimated. Due to the nonlinear relationship between SOC and its related influencing factors, the research that leads to accurate online estimation faces challenges, especially in complex working conditions. It is very important to study higher accuracy SOC estimation algorithms for the performance of BMSs. That is to say, the SOC of the battery indicates the degree of use of the battery power. The accurate SOC can predict the use time of the battery, provide a basis for reasonable energy distribution, and make more effective use of limited energy.

Current methods to estimate SOC include ampere measurement method, OCV, impedance analysis, Kalman filter, and neural network. There are some problems in the application of the commonly used ampere measurement method. The current measurement error will spread to the calculation error, especially in the case of high temperature and current fluctuation. The problem of current measurement error can be solved by using high-performance (and very expensive) current sensors. The ampere measurement method is simple, reliable, and suitable for all EV batteries. However, the premise is to measure the current accurately. In the OCV method, there is a mathematical relationship between SOC and the OCV of the battery under different temperature conditions, so the OCV can be used to estimate SOC.

This method has high estimation accuracy in the initial stage of SOC, but the error increases when the current changes rapidly or the load changes greatly. Another big drawback of this method is that it takes a lithium-ion battery to last for a few hours. The OCV method is only applicable to parking EVs. The impedance analysis method estimates the SOC by studying the relationship between battery resistance and SOC. This relationship can form the electrochemical impedance spectrum of lithium-ion battery. Therefore, SOC can be obtained by analyzing the impedance spectrum. However, the impedance spectrum is very sensitive to temperature. When the temperature varies greatly, it is difficult to estimate the SOC accurately by using the impedance spectrum.

The Kalman filter method regards the battery as a power system, and SOC is the internal state parameter of the system. The principle of estimating SOC by Kalman filter is to minimize the MSE. Two main disadvantages affect this method. First of all, this method is highly dependent on the lithium-ion battery model. Second, the method contains a large number of matrix operations, resulting in a large amount of computation to be performed. To avoid the shortcomings of the above SOC estimation methods, the neural network method is proposed as another method to estimate SOC. The SOC estimation using this method does not take into account the internal state and characteristics of the battery. It only needs to collect the input and output data needed by the lithium-ion battery neural network. SOC can be accurately estimated through adaptive training within the network.

In practical applications, many optimization algorithms have been proposed based on the above basic methods. Some scholars have proposed a learning system based on fuzzy logic for battery SOC estimation. Fuzzy logic is used in the learning system. During the learning process, the system will automatically generate a new fuzzy rule base and replace the old one; this method can continuously reduce the estimate of SOC.

In order to overcome the shortcomings of IEKF in estimating SOC, the LM algorithm was used to optimize the iterative process of IEKF. It is shown that the LM-IEKF algorithm has better accuracy in SOC estimation; some scholars propose to use the UKF to estimate the battery SOC and recursively calculate the system state variables through the nonlinear mapping process, thereby effectively suppressing the EKF. The error is caused by the Mann filter (EKF); the experimental results show that the UKF method is better than the EKF method; some scholars combine the neural network with the EKF, use the data collected during the charging and discharging process to train the network, and use the trained network to determine the EKF. The corresponding SOC value can be obtained through the state variables of the state space equation model. This method is relatively difficult, but the estimation accuracy is high. Some scholars have proposed an adaptive unscented EKF algorithm to estimate the SOC value.

By improving the model at the same time, the parameter accuracy and observation equation accuracy are analyzed and studied under the unknown interference noise. The variation curve and error relationship of the SOC estimation value are analyzed. Compared with the SOC estimation error, the SOC estimation error is reduced, and it has better robustness; some scholars have established a BP neural network based on GA to estimate the SOC model of lithium-ion batteries. Based on the BP neural network, which has advantages in nonlinear problems, and the GA, the traditional BP neural network is easy to fall into the minimum value, and the estimation accuracy is effectively improved. At the same time, the GA and immune algorithm are used to improve the weight threshold of the neural network.

The results show that the performance of the network improved by the hybrid algorithm is significantly improved, that is, the learning time is short and the prediction accuracy is high; some scholars combine the neural network and fuzzy control. In combination, an adaptive fuzzy neural network is proposed to estimate SOC, which uses a subtractive clustering method to generate an initial fuzzy inference system for training. Through the training and learning of test samples, the real system can be approximated to the greatest extent, while the accuracy is guaranteed, the training times are reduced, and the convergence speed is improved.

ENN is a kind of FFNN that is especially suitable for time-series prediction. It is a typical representative of the neural network method. Moreover, ENN has a better prediction performance because it has a load-bearing layer that other neural networks do not have. Compared with the simple FFNN, it has higher SOC estimation accuracy. By analyzing the disadvantages of existing SOC estimation algorithms, neural network is selected as the method for estimating SOC. An SOC prediction method based on ENN is proposed.

2. Principle introduction

ENN is a dynamic RNN with local memory units and local feedback connections. On the basis of the commonly used BP neural network, a succession layer is added, which can memorize the data output by the hidden layer at the previous moment, so that the entire network structure has the ability to process dynamic information. ENN is divided into four layers, namely, input layer, hidden layer, successor layer, and output layer. The connection and function of the input layer, hidden layer, and output layer are similar to those of the commonly used FFNN. The newly added successor layer, as a one-step delay operator, takes a delay operation on the memorized output data of the hidden layer, and then feeds the data back to the hidden layer, which can enhance the sensitivity of the system to historical data and improve the estimate of the precision of the data. ENN has the ability to adapt to time series due to the existence of the successor layer and can be well applied to dynamic modeling problems. Its structure is shown in Figure 5.61.

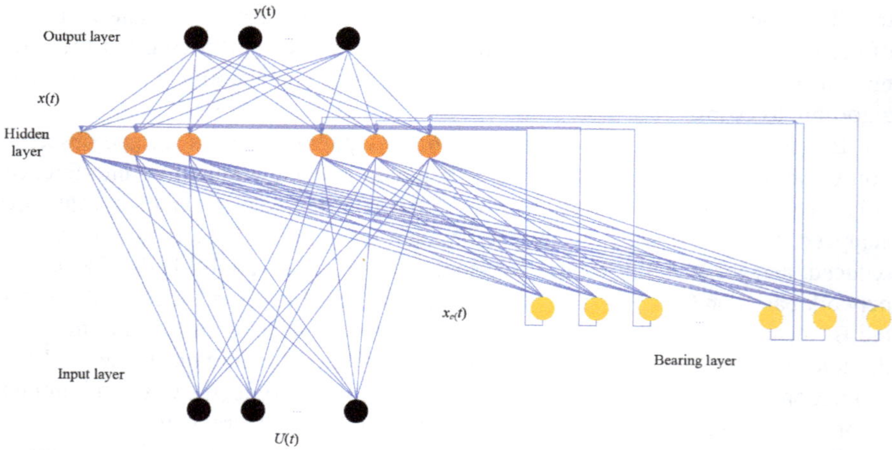

Figure 5.61 ENN structure

The Elman network does not need to store all the input information, but it can reflect the influence of all historical signals on the current system in the network and can approximate any nonlinear mapping with arbitrary precision. It is mainly used for dynamic prediction in engineering. Let the Elman network input vector be $u(t)$, the output vector be $y(t)$, the hidden layer output vector be $x(t)$, the successor layer output vector be $x_c(t)$, the connection weight from the successor layer to the hidden layer be w^1, the connection weight from the input layer to the hidden layer be w^2 and the connection weight from the hidden layer to the output layer is w^3, then the input–output relationship of the Elman network, as shown in (5.130–5.133):

$$y(t) = g\left(w^3 x(t) + b_2\right) \tag{5.131}$$

$$x(t) = f\left(w^1 x_c(t) + w^2 u(t) + b_1\right) \tag{5.132}$$

$$x(t) = x(t-1) \tag{5.133}$$

where f is the transfer function of the neurons in the hidden layer, g is the transfer function of the neurons in the output layer, b_1 is the threshold of the hidden layer, and b_2 is the threshold of the output layer. Here, the sigmoid function is chosen as the transfer function between the hidden layer and the output layer.

The learning process of the ENN is to learn and train through the training set data, obtain the dynamic characteristics between the input and output parameters, and finally obtain stable network parameters. The ENN adopts the error correction learning rule. Calculate the error between the actual output result and the expected output result of the output layer, and judge the size of the error. If the error exceeds the set threshold, BP is performed, that is, the error signal is reversed in some way. It is propagated to the neurons of each layer to change the weights and threshold matrices of neurons based on this. Suppose the actual output of the output layer of the kth step of the ENN training process is $y(k)$, the expected output is $d(k)$, and W is the weight matrix, then define the error function for judging whether the algorithm ends, as shown in (5.134):

$$E(W) = \sum_{k=1}^{n} \left[y(k) - d(k)\right]^2 \tag{5.134}$$

The purpose of network learning is to find a weight matrix that minimizes the error function, which can transform error correction learning into an optimization problem. The gradient-descent algorithm is the most commonly used optimization algorithm. The traditional ENN, like the BP neural network, uses the gradient descent method to update the weight matrix. Equations (5.135) and (5.136) are the update equations of the weight matrix. Since the gradient-descent method needs to input all the training data every time the weight matrix is updated, if the amount of training data is very large, the learning process of the entire network will take a lot of time. This method not only leads to long network training time but also misidentifies the local extreme value as the optimal estimation result.

$$W = W + \Delta W \tag{5.135}$$

$$\Delta W = -\eta \frac{\partial E}{\partial W} \tag{5.136}$$

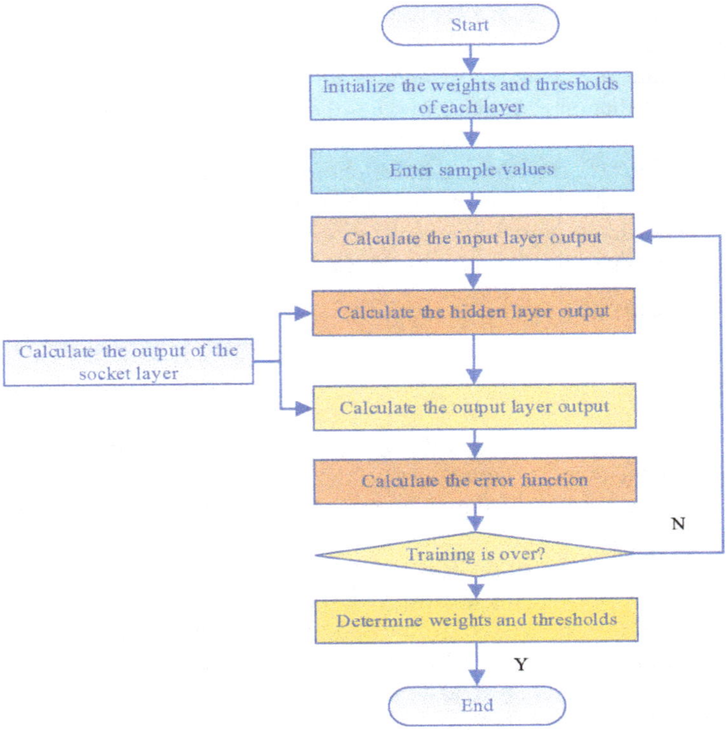

Figure 5.62 ENN training flow chart

where E represents the error function and η represents the learning efficiency. The training process of ENN is shown in Figure 5.62.

First, initialize the weights and thresholds of the network; then divide the data into training data and test data, and use the training data for network training. During the training process, the network weights and thresholds are continuously adjusted according to the errors obtained during each iteration. When the network is trained and the number of iterations reaches the set maximum value or the error of the network training result reaches the target error, the current network training can be stopped; in this way, we obtain the trained ENN, that is, the optimal weights and thresholds are fixed, and finally enter the test. The data can be predicted to obtain the ideal target value.

3. Experiment analysis

In order to verify the SOC estimation accuracy of the Elman model under actual working conditions, DST working conditions are used to simulate actual working conditions. The specific working steps are set as follows. (1) Charge the experimental lithium battery to full capacity with CCCV. (2) Set aside for 30 minutes. (3) Perform constant current discharge for 4 minutes at a rate of 0.5 C, and leave it for 30 seconds after stopping the discharge. (4) Constant current charging at 0.5 C rate

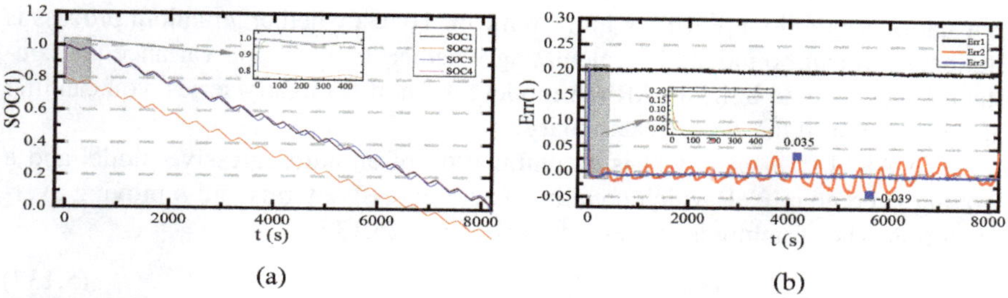

Figure 5.63 Comparison chart of SOC estimation under DST conditions. (a) SOC estimated graph. (b)Error comparison chart.

for 4 minutes, stop charging, and set aside for 30 seconds. (5) Cycle steps (3), (4) until the battery discharge ends. After obtaining the discharge data of the battery, use the ampere-hour integration method, the EKF algorithm, and the Elman model to estimate the SOC, and verify the convergence effect and tracking effect of the algorithm. The initial value of the algorithm is set as SOC = 0.8 to verify the convergence effect and tracking effect during the discharge process. The SOC estimation curve and estimation error are shown in Figure 5.63.

In Figure 5.63, SOC1 refers to the SOC value, SOC2 is the SOC value calculated using the ampere-hour integration method, SOC3 is the SOC value calculated using the EKF algorithm, and SOC4 is the Elman model. Under the condition of an inaccurate initial value of SOC = 0.8, the ampere-hour integration method cannot track the real state of the system. EKF can track the real state of the system faster. The time required for the EKF algorithm to track the theoretical value is about 30 seconds, but the estimation accuracy is not high enough. The convergence effect of the Elman model in the early stage is almost the same as that of the EKF algorithm, and the estimation accuracy of the Elman model in the later stage is higher. Figure 5.63b is the error curve, Err1 is the ampere-hour integral error curve, Err2 is the EKF error curve, and Err3 is the error curve of the Elman model. The overall error of Err1 exceeds 20%, the Err2 and Err3 tracking systems are fast, and the convergence speed is less than 50 seconds. Among them, the error of Err2 is small, the error of the entire tracking process is less than 2%, and the overall error is small; while the maximum error of Err3 is 3.5%, the overall tracking effect is fluctuating and the accuracy is low. It shows that the Elman model has high accuracy and strong applicability in estimating SOC under complicated working conditions.

5.7.5 Nonlinear autoregressive-based SOC estimation analysis

1. Autoregressive and moving average model

Statisticians G. E. P. Box and G. M. Jenkins proposed the autoregressive moving average model in the 1970s, which is a statistical model for describing time series. As a data-driven approach, ARMA models can predict the future trends of a series by studying the interconnectedness of the system through historical data [194]. A necessary condition for establishing an ARMA model is that the sequence is a stationary random process with zero mean. To test whether a random process is stationary, it can be judged by calculating whether its mean and variance are constant. If you want to build an ARMA model for a non-stationary series, you can first difference it until the series is stationary.

The ARMA (p, q) model is a combination of an autoregressive model and a moving average model, which contains p autoregressive terms and q moving average terms. The structure is expressed, as shown in (5.137):

$$X_t = e_t + \sum_{j=1}^{p} \varphi_i x_{t-i} + \sum_{j=1}^{q} \theta_j \varepsilon_{t-j} \tag{5.137}$$

wherein X_t is the time series, e_t is white noise, p is the AR model order, q is the MA model order, and θ and φ are the coefficients of the AR model and the MA model, respectively. Both AR(p) and MA(q) are special forms of the ARMA (p, q) model. When the order p is 0, the ARMA (p, q) model degenerates into the MA(q) model. When the order q is 0, the ARMA (p, q) model degenerates into the AR(p) model. As the primary determinant of the ARMA model, the order p, q directly affects the accuracy of the prediction result. There are three traditional order determination methods.

1. Box-Jenkins (B-J) order method

Calculate the autocorrelation function (ACF) and partial autocorrelation function (PACF) of the sequence, respectively, and then observe the ACF diagram and PACF. Figure to get the truncation and trailing numbers of ACF and PACF for order determination. The disadvantage of this method is that there are too many candidate order combinations, which need to be further screened in combination with other criteria.

2. Minimum information criterion (Akaike information criterion, AIC) order method.

Use maximum-likelihood estimation to determine the order. AIC is defined, as shown in (5.138):

$$AIC = -2L(\beta) + 2k \tag{5.138}$$

where k is the number of independent variables; β is the maximum likelihood estimate, and L is the likelihood function. The AIC criterion effectively avoids the subjectivity factor in the B-J order determination method, but in this method, the order must be obtained through a continuous iterative process, which increases the computational complexity.

3. Bayesian information criterion (BIC) order method

In view of the shortcomings of the AIC method, BIC has improved it and is defined, as shown in (5.139):

$$BIC = Nlg\sigma^2 + p(lgN)/N \tag{5.139}$$

Among them, N is the number of sampling points, and σ^2 is the residual. Although BIC solves some of the defects of AIC, it still has shortcomings, such as complex calculation process and the appearance of local minima.

4. Autoregressive model

As part of the ARMA model, the AR model can be defined, as shown in (5.140):

$$X_t - \varphi_1 X_{t-1} - \varphi_2 X_{t-2} - \cdots - \varphi_p X_{t-p} = \varepsilon_t \tag{5.140}$$

where is $\{X_t, t = 0, \pm 1, \pm 2, \cdots\}$ the stationary time series with zero mean, p is the order of the AR model, φ_i is a parameter of the AR model, and $\varepsilon_t t$ is the stationary white noise with zero mean. Introducing the backshift operator B, the AR model can be simplified, as shown in (5.141) and (5.142):

$$\varphi(B) X_t = \varepsilon_t \tag{5.141}$$

$$\varphi(B) = 1 - \varphi_1 B - \varphi_2 B^2 - \cdots - \varphi_p B^p \tag{5.142}$$

To determine the specific form of the AR model, it is divided into two parts: determining the order and parameters of the AR model. The methods for determining the order and parameters are described below.

1. Determine the order of the AR model. In the previous section, we introduced the traditional method of determining the order of the ARMA model. For the specific research object, this section adopts an adaptive order-determining method and selects the RMSE as the fitness function, which is defined as shown in (5.143):

$$RMSE = \sqrt{\frac{1}{N} \sum_{i=1}^{N} (x_i - y_i)^2} \tag{5.143}$$

where N is the sample length, x_i is the actual data, and y_i is the predicted data. In order to obtain higher prediction accuracy and get the smallest RMSE, the particle swarm algorithm is used to calculate the order.

2. Determine the parameters of the AR model.
The AR model can be rewritten, as shown in (5.144–5.148):

$$X_t = \varphi_1 X_{t-1} + \varphi_2 X_{t-2} + \cdots \varphi_p X_{t-p} + \varepsilon_t \tag{5.144}$$

$$X = \begin{pmatrix} X_p & X_{p-1} & \cdots & X_1 \\ X_{p+1} & X_p & \cdots & X_2 \\ \vdots & \vdots & & \vdots \\ X_{N-1} & X_{N-2} & \cdots & X_{N-P} \end{pmatrix} \tag{5.145}$$

$$Y = \begin{bmatrix} X_{p+1} & X_{p+2} & \cdots & X_N \end{bmatrix}^T \tag{5.146}$$

$$\varepsilon = \begin{bmatrix} \varepsilon_{p+1} & \varepsilon_{p+2} & \cdots & \varepsilon_N \end{bmatrix}^T \tag{5.147}$$

$$\beta = \begin{bmatrix} \beta_{p+1} & \beta_{p+2} & \cdots & \beta_N \end{bmatrix}^T \tag{5.148}$$

where φ_i is the parameter of the AR model. Let $\varepsilon_1 = \varepsilon_2 = \cdots = \varepsilon_p = 0$, can be written as shown in (5.149):

$$Y = X\beta + \varepsilon \tag{5.149}$$

where X is the training data, Y and ε are random variables, and X_{t-1} is a random variable. The ACF of the least squares estimate is defined as shown in (5.150):

$$R(m-n) = \frac{1}{N}\sum_{t=1}^{N-p} X_{t+p-n}X_{t+p-m} \tag{5.150}$$

When the time is $t-1$, it can be considered that the random variables at this time are all fixed values, and the AR model is regarded as a linear regression model, so the least squares estimation method commonly used in linear regression analysis methods is used to obtain the parameters of the AR model.

2. Nonlinear autoregressive model.

Compared with the other battery technologies, lithium-ion batteries have achieved dominance in terms of applicability in ESS such as electric mobility and stationary applications. This relies on their promising characteristics, such as superior energy and power density, high efficiency, and long lifetime. Nevertheless, lithium-ion batteries fail gradually to supply sufficient energy and power for the application in which they operate. This happens as a result of several operating conditions, mechanical stress, and chemical reactions. Therefore, ensuring the reliable performance of lithium-ion batteries is of great importance. To this end, over the past decades, many efforts have been conducted to diagnose lithium-ion batteries' SOC.

The SOC is a criterion for describing the rate of capacity or power fade in a battery. A broad range of SOC estimation methods has been developed in recent years. There exists a considerable amount of literature that reviews the existing SOC estimation methods. For instance, some scholars present a review on different health diagnosis methods used in hybrid and battery EVs. The authors provide an overview of existing methods for the onboard state of available power prediction of lithium-ion batteries. Similarly, some scholars present a critical review on different techniques for state monitoring of lithium-ion batteries and outline the strength and weakness of each model. A wide range of SOC estimation methods focus on the characterization of battery parameters and modeling of battery behavior. These methods are known as model-based methods. Model-based methods apply a variety of observers and adaptive filters to model parameterization and optimization. Sliding mode observer, Kalman filters, H-infinity filters, and PFs are examples of adaptive filters that are applied in model-based SOC estimation methods.

On the other hand, data-driven methods, particularly ML-based techniques have been gaining much attention to cope with the complexity of model-based methods. A variety of ML algorithms have been developed to estimate battery SOC. SVM, ANNs, Gaussian process regression (GPR), relevant vector machine, as well as ensemble methods such as random forest regression and gradient, boosted trees are examples of ML techniques that are applied for SOC estimation. Among the aforementioned methods, different implementations of ANN are widely proposed to estimate SOC due to their superior accuracy, adaptation ability, good generalization to untrained data, and robust performance on nonlinear data.

For instance, a recent study by Duo *et al.* employed BP neural networks for SOC estimation. The extreme learning machine is another type of NN that is used by many researchers. NARX is one of the most convenient types of ANN for the

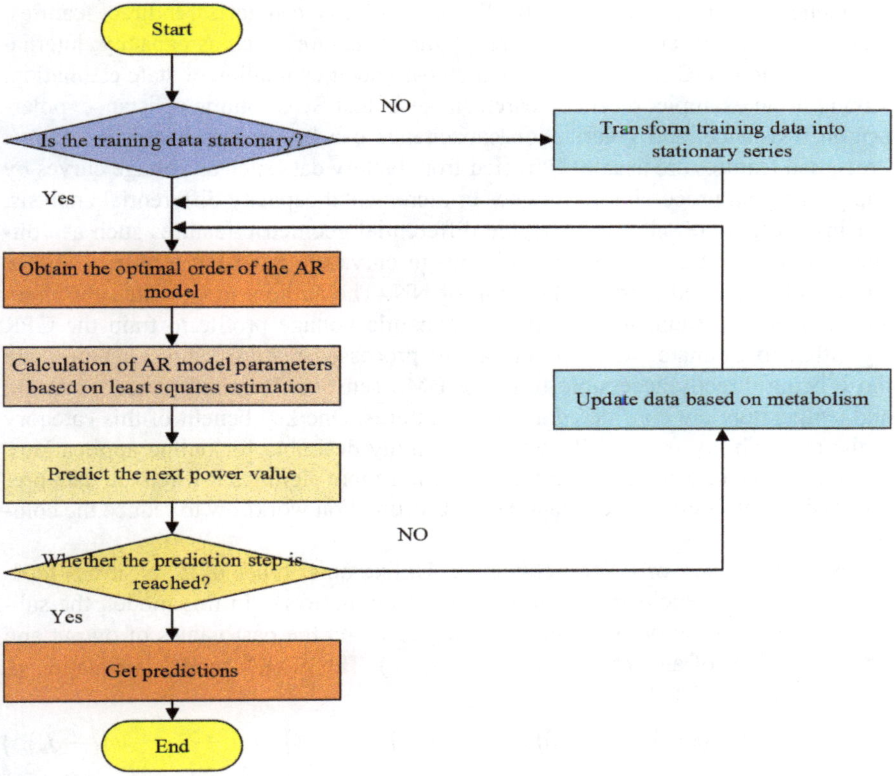

Figure 5.64 The AR model prediction flowchart

estimation purpose in nonlinear dynamic systems. This method has the ability to keep the past values of the input data and output data and to update the current value of output accordingly. This ability is appropriate to track the correlations of capacity data through battery degradation. The AR model prediction flowchart is shown in Figure 5.64.

Despite the promising capabilities of this method, only a few researchers deployed this method for SOC estimation. For instance, some scholars applied this method to estimate SOC estimation in two lithium iron phosphate ($LiFePO_4$) and lithium titanate cells. Nevertheless, a deeper study of the NARX method in different battery chemistries, as well as in different operating conditions is demanded. To fill this gap, this chapter attempts to apply the NARX method for SOC estimation of 21 nickel manganese cobalt oxide (NMC) battery cells, which are experimented under various operating conditions. It is notable that approximately all the aforementioned ML methods need to be trained using the battery health indicators. Health indicators are features of the battery which are able to represent a predictable trend during battery trajectory.

Generally, these features can be divided into three categories: indirect features, processed features, and direct features. Indirect features such as capacity, internal resistance, and SOC should be attained by parameter estimation or state estimation. Taking as an example, recent research suggests that SOC, ohmic resistance, polarization resistance, and polarization capacitance can be used as health indicators. Processed features are usually extracted from battery data such as voltage curves by applying signal processing techniques or incremental capacity/differential analysis. For instance, some scholars extracted differential geometric features such as voltage velocity, voltage arclength, and voltage curvature from the terminal voltage curve and employed them as the input of NN. The authors in extracted the time- and frequency-domain features from a dynamic voltage profile to train the GPR algorithm to estimate SOC. In contrast to processed features, direct features are those being directly accessible from the BMS sensors. Terminal voltage, current, and temperature are considered as direct features. One key benefit of this category is the accessibility from BMS; thus, it is highly desirable for online applications. The advantage of using these features becomes more significant when the data preprocessing step is eliminated from the SOC estimation workflow to reduce the computation costs.

NARX is a sort of RNN architecture. Unlike other types of RNN, it has feedback connections enclosing several layers of the network. In this model, the subsequent value of the output data $y(t)$ is regressed on the past values of output and previous values of an exogenous input data (u). The NARX model is equationed, according to the (5.151):

$$y(t) = f(y(t-1), y(t-2), \cdots y(t-d_y), u(t-1), u(t-2), \cdots u(t-d_u),) \tag{5.151}$$

where $f()$ is the function of nonlinear mapping, and dy and du are the time delays for the output and input, respectively. The detailed diagram of the NARX resulting network is shown in Figure 5.65.

The network constitutes an input, a two-layer feedforward network including the hidden and output layers, and output delay. It also includes the bias (b), the input weight (IW), and the layer weight (LW). The f^1 defines the activation function

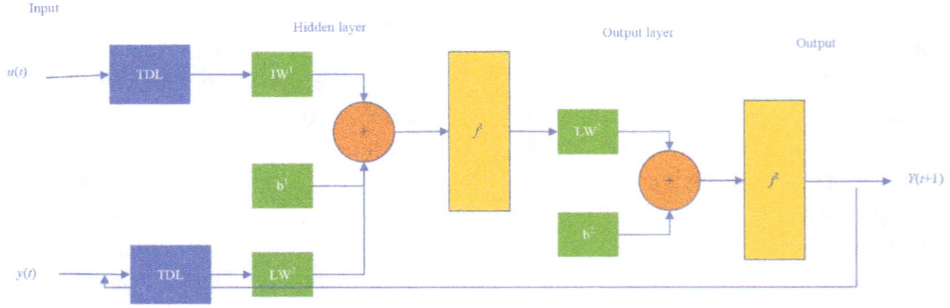

Figure 5.65 The detailed diagram of NARX

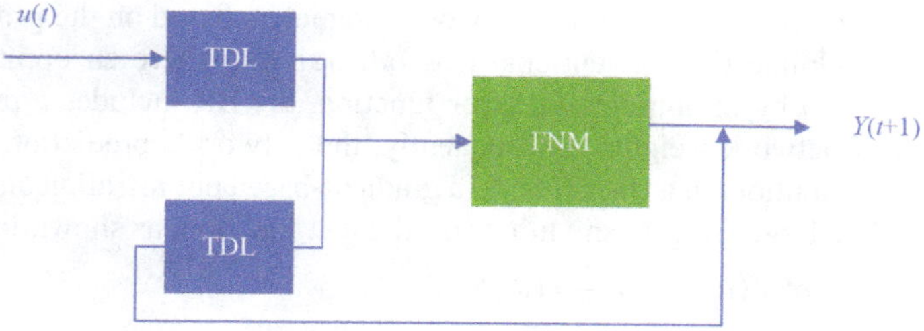

(a) Parallel architecture

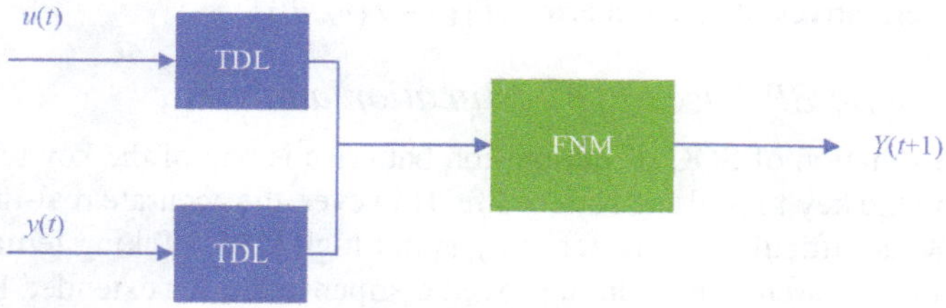

(b) Series-Parallel architecture

Figure 5.66 *NARX design. (a) Parallel architecture. (b) Series–parallel architecture.*

of the hidden layer and, $f^2$1 is the output layer activation function. Before training the NARX, the following parameters should be specified: input and feedback time delays, the size of the hidden layer, the feedback mode, and the training function. In the NARX algorithm, the input values of *if*(*t*) include time-delay lines (TDLs) that are presented in (5.152):

$$x(t) = \begin{bmatrix} u(t-1) & u(t-2) & \cdots & u(t-d_u) \\ y(t-1) & y(t-2) & \cdots & u(t-d_y) \end{bmatrix}^T \tag{5.152}$$

Regarding the feedback mode, NARX includes two modes: close or series and open or series–parallel. In the close mode design of NARX, the estimated target feedbacks to TDLs of FFNN. The schematic representation of the close mode architecture of NARX is depicted in Figure 5.66a. However, due to the availability of a real target, the standard design can be converted to an open or series–parallel architecture (Figure 5.66b) in which the real target is replaced with the output estimated by FNNN. The series–parallel design is superior to the former design because of two reasons. First, it avoids the feedback loop inside the model; thus, the computational cost is reduced. Second, since the resulting network has a pure feedforward design, the static BP can be used for training. Therefore, the series–parallel architecture of NARX is adopted in this work.

There are various functions to train the NARX network. LM and Bayesian regularization (BR) are the most convenient and common functions. The latter is selected for this research. The BR defines the network parameters based on the probabilistic interpretation. Unlike the conventional network training where an optimal set of weights is chosen by minimizing an error function, the BR includes a probability distribution of network weights. Consequently, the network's prediction will be a probability distribution. In a BR network, a gradient-based optimization algorithm is used to penalize large weights and minimize the lost function, as shown in (5.153):

$$\theta_{t+1} = \theta_t + a\nabla f(u_t, \theta_t)\left[y_t - f(u_t, \theta_t)\right] \tag{5.153}$$

where θ is the set of weight and biases, α is the learning rate, and $\nabla f(u_t, \theta_t)$ is the gradient vector of f concerning θ. The phrase $\nabla f(u_t, \theta_t)\left[y_t - f(u_t, \theta_t)\right]$ defines the first-order derivatives of squared error of $\left[y_t - f(u_t, \theta_t)\right]^2$.

5.7.6 Genetic BP-based SOC estimation analysis

Accurate estimation of SOC of lithium-ion batteries is one of the key technologies of BMS and the key to prolong battery life. However, the accurate real-time estimation of SOC is difficult and the accuracy is not high [195]. Taking ternary lithium battery as the research object, an improved compensation for extended Kalman filter (CEKF) algorithm was proposed to solve the nonlinear errors caused by THE linearization of EKF. In this algorithm, GA-optimized BP neural network is used to predict the nonlinear errors of EKF at *k* time, so as to compensate for the nonlinear errors of extended Kalman *K* time, and adaptive FFRLS is used to identify the parameters of the model. Experimental results with DST and BBDST show that the accuracy range of the proposed algorithm is about 2%, and the maximum error and average error are much smaller than EKF, which can track the theoretical value of SOC more effectively. Moreover, the estimated SOC stability of the proposed algorithm is also more stable than EKF.

1. The equivalent model

When selecting the equivalent model of the battery, the model containing RC circuit can be selected. RC circuit can be used to simulate the polarization reaction of the battery, but it should not be too many RC circuits, as too many will lead to more and more complicated calculation. In this chapter, the second-order RC equivalent model is constructed, and a set of RC circuits are added on the basis of the common Thevenin equivalent model, which can more accurately simulate the chemical reaction inside the battery. The order of RC is not high, and the calculation degree is not complicated while the estimation accuracy is improved. The second-order RC equivalent model is shown in Figure 5.67.

In Figure 5.67, U_{oc} represents the OCV. U_L indicates the terminal voltage. R_0 is the ohmic internal resistance. C_1 and C_2 represent the polarization capacitance, while R_1 and R_2 represent the polarization resistance. According to Kirchowski's law, the constructed Davinan equivalent circuit model is analyzed, and the voltage and current expression of the equivalent circuit can be obtained, as shown in (5.154):

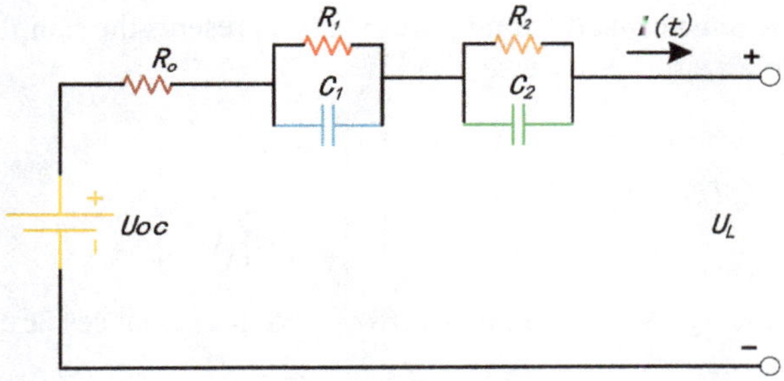

Figure 5.67 *The second-order RC*

$$\begin{cases} U_L = U_{oc} - U_o - U_1 - U_2 \\ \dfrac{dU_1}{dt} = -\dfrac{U_1}{R_1C_1} + \dfrac{I}{C_1} \\ \dfrac{dU_2}{dt} = -\dfrac{U_2}{R_2C_2} + \dfrac{I}{C_2} \end{cases} \tag{5.154}$$

For the second-order *RC* equivalent model shown in Figure 5.1, the OCV can be expressed by *SOC*. By discretizing the equivalent circuit model and combining *SOC* definition (5.155), selecting state space variables, input variables, and output variables, discrete state space equations can be obtained:

$$SOC = SOC_0 - \frac{1}{C}\int \eta I dt \tag{5.155}$$

SOC_0 represents the initial *SOC* of the battery. *SOC* represents the *SOC* of the battery at the current time:

$$\begin{bmatrix} SOC(k+1) \\ U_1(k+1) \\ U_2(k+1) \end{bmatrix} = \begin{bmatrix} 1 & 0 & 0 \\ 0 & e^{-\frac{\Delta t}{\tau_1}} & 0 \\ 0 & 0 & e^{-\frac{\Delta t}{\tau_2}} \end{bmatrix} \begin{bmatrix} SOC(k) \\ U_1(k) \\ U_2(k) \end{bmatrix} + \begin{bmatrix} -\dfrac{\eta \Delta t}{Q_N} \\ R_1\left(1 - e^{-\frac{T}{\tau_1}}\right) \\ R_2\left(1 - e^{-\frac{T}{\tau_2}}\right) \end{bmatrix} I_k + w_k \tag{5.156}$$

Q_N indicates the rated capacity of the battery. Δt represents the sampling interval. R indicates the polarization resistance.

$$U_L(k) = U_{oc}(k) - R_0 I(k) + \begin{bmatrix} 0 \\ -1 \\ -1 \end{bmatrix}^T \begin{bmatrix} SOC(k) \\ U_1(k) \\ U_2(k) \end{bmatrix} + v_k \tag{5.157}$$

w_k is state error. v_k is measurement error, and its variances are Qk and Rk, respectively.

2. Online parameter identification based on adaptive FFRLS

Forgetting factor is introduced on the basis of RLS, which can avoid data saturation and weaken the influence of old data and strengthen the influence of new data. If the forgetting factor is too large, the ability to weaken old data will be very small; if the forgetting factor is too small, the tracking ability will be enhanced, but the ability to suppress noise will be greatly reduced. Therefore, the FFRLS of the adaptive forgetting factor is proposed.

$$\begin{cases} \hat{\theta}_{k+1} = \hat{\theta}_k + K_{k+1}\left(y_{k+1} - \phi_{k+1}^T \theta_k\right) \\ K_{k+1} = P_k \phi_{k+1}\left(\lambda + \phi_{k+1}^T P_k \phi_{k+1}\right)^{-1} \\ P_{k+1} = \frac{P_k}{\lambda} - \frac{K_{k+1}}{\lambda} \phi_{k+1}^T P \end{cases} \tag{5.158}$$

where K is the recursive gain, P is the covariance, and λ is the forgetting factor. Adaptive forgetting factor:

$$\begin{cases} \varepsilon_k = y_{k+1} - \phi_{k+1}^T \theta_k \\ d_k = \frac{1}{l} \sum_{i=k-l+1}^{k} \varepsilon_i^T \varepsilon_i \\ \lambda = \lambda_0 & d_k < \rho \\ \lambda = \lambda_L + \frac{\lambda_0 - \lambda_L}{n} & d_k \geq \rho \qquad k = 0, 1, \cdots, n \end{cases} \tag{5.159}$$

represents the forgetting factor. Change the ability to forget old data by changing the size of the forgetting factor in real time.

3. Improved GA-optimized BP neural network

Although BP neural network is widely used, due to the failure to accurately obtain the optimal initial parameters and thresholds, the network training has many iterations and long training time, and the results of each training are different, with large errors and so on. Therefore, in order to solve these problems, a GA is proposed to optimize BP neural network and find the optimal initial weight and threshold value of BP neural network.

The general BP neural network algorithm uses the gradient-descent method for iterative training, but the gradient-descent method only uses the first-order reciprocal of the objective function for training. The number of iterations is more, the iteration time is longer, and the final error is not ideal. Therefore, on the basis of the gradient-descent method, a numerical optimization method is proposed to train,

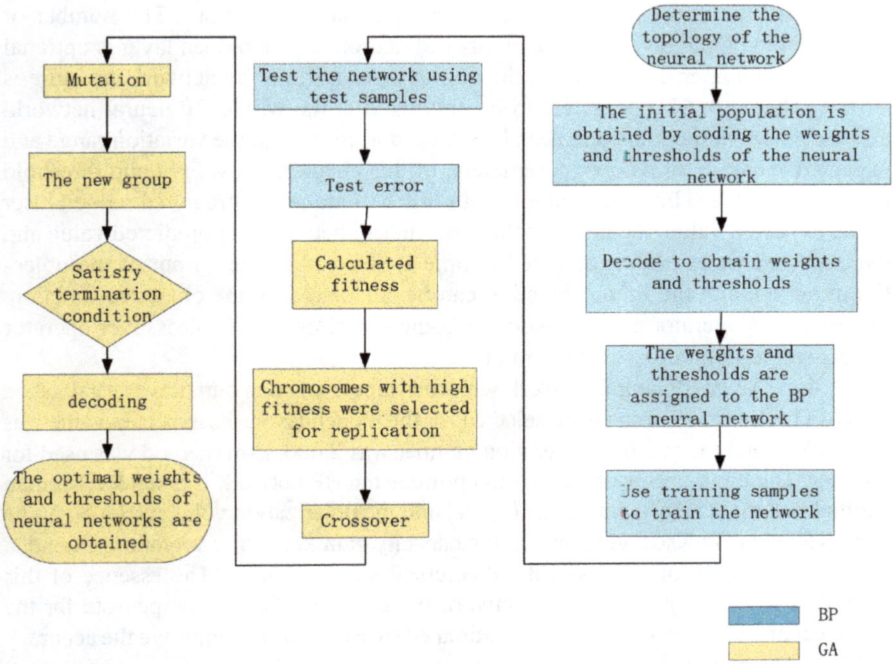

Figure 5.68 The overall flow chart

which not only makes use of the first-order reciprocal of the objective function but also makes use of the second-order reciprocal information of the objective function. The specific description is shown in (5.160):

$$\begin{cases} f\left(X^{(k+1)}\right) = \min f\left(X^{(k)} + \eta^{(k)} S\left(X^{(k)}\right)\right) \\ X^{(k+1)} = X^{(k)^{\eta}} + \eta^{(k)} S\left(X^{(k)}\right) \end{cases} \tag{5.160}$$

where $X^{(k)}$ is the vector composed of network ownership value and bias value; $S\left(X^{(k)}\right)$ is the search direction in the vector space composed of components of X; and $\eta^{(k)}$ is in the direction of $S\left(X^{(k)}\right)$, so that the minimum step size. The optimization of network weight can be divided into two steps: first determine the optimal search direction of the current iteration, and then seek the optimal iteration step in this direction. The search direction is shown in (5.161):

$$S\left(X^{(k)}\right) = -\left(H^{(k)} + \lambda^{(k)} I\right)^{-1} \nabla f\left(X^{(k)}\right) \tag{5.161}$$

Optimization of BP neural network by GA includes determination of BP neural network structure, optimization weight and threshold value of GA, and training and prediction of BP neural network. Among them, BP neural network generally adopts a hidden layer, and the overall flow chart of BP neural network optimized by GA is shown in Figure 5.68.

The BP network part consists of four inputs and one output. The number of neurons in the input layer is 4, the number of neurons in the hidden layer is optional 10, and the number of neurons in the output layer is 1, so the network structure is 4-10-1. There are 61 parameters to be optimized in the whole BP neural network, and the binary number encoded can be adjusted according to the variation range and expected accuracy of its own parameters. In this chapter, the weight and threshold encoding choose 6-bit binary number. In order to make the predicted value closer to the expected value, the norm of the error matrix between the predicted value and the expected value of the predicted sample is selected as the output of the objective function, and the fitness function can be calculated by the objective function. The selection operator adopts random ergodic sampling and the crossover operator selects single-point crossover operator.

A three-layer BP neural network with four inputs and one output was established on MATLAB. Ten neurons were selected for the hidden layer, the expected value was set to 10–5, and the maximum iteration number was 2,000. L-m method was used for training. The input vector of the GA to optimize the BP network is extended Kalman k initial battery SOC estimation $S|k\left(k-1\right)$ and the polarization voltage $U(k)\,|k-1$, the SOC Kalman gain Ksoc and polarization capacity gain Kc, output vector are extended Kalman estimation of SOC and the theoretical value of *SOC*. The essence of this algorithm is to use the BP neural network optimized by GA to compensate for the nonlinear and equal error of the *SOC* estimated by EKF, so as to improve the accuracy of *SOC*. The overall flow chart of the improved CEKF is shown in Figure 5.69.

4. Experimental verification and result analysis

   DST working condition experiment

In practical application, the current of lithium battery is complex and changeable, and the dynamic performance of lithium battery is strict, which also brings difficulties to the estimation of lithium battery *SOC* under complicated working conditions. In order to verify the reliability of the algorithm proposed in this chapter,

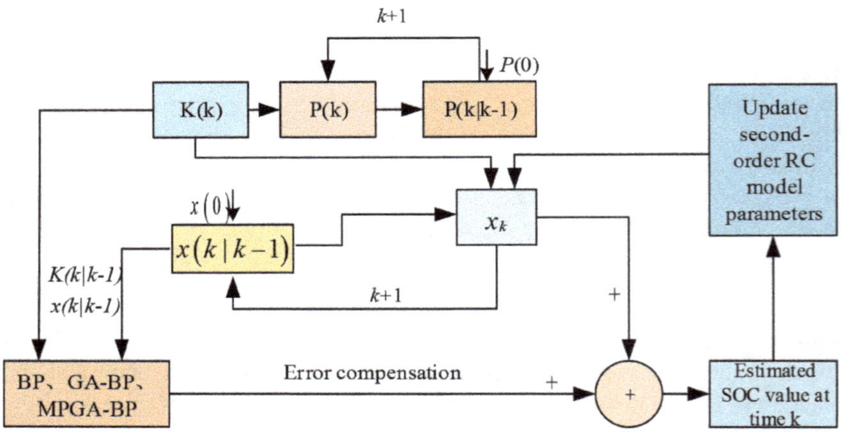

Figure 5.69 The overall flow chart

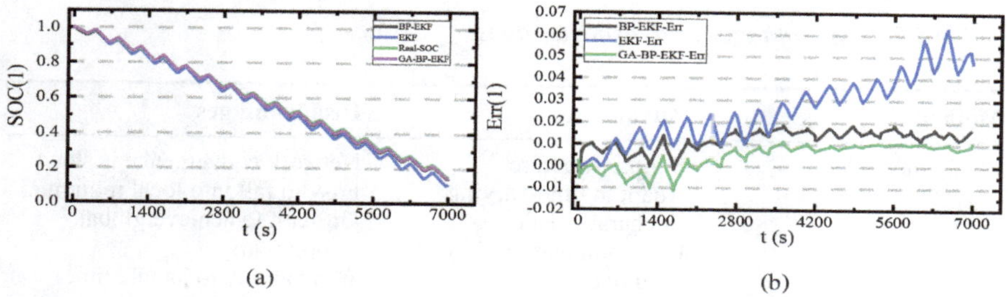

Figure 5.70 Experimental result. (a) Results of various SOC estimates. (b) Error curves of each algorithm.

DST conditions are used to verify and make a comparative analysis. The experimental results are shown in Figure 5.70.

According to Figure 5.70, the maximum error of improved CEKF is 1.1695%, 0.7152% less than CEKF and 5.1405% less than EKF. The average error was 0.6836%, 0.5401% less than CEKF and 1.9154% less than EKF. The RMSE was 0.7559%, 0.5403% less than CEKF and 2.2489% less than EKF. According to the error diagram in Figure 5.70b and Table 5.1, the error of the improved CEKF is less than EKF and CEKF in all aspects, and the maximum error is within 2%. The accuracy is higher than EKF and CEKF algorithms, and the error stability is better than EKF and CEKF algorithms. It can better track the theoretical value of SOC and is more suitable for SOC estimation.

Many neural network methods have been proposed for SOC estimation of lithium-ion batteries. Although they are designed for various applications, they are promising to use battery management in EVs. For a more comprehensive understanding of the existing neural network methods.

Table 5.7 summarizes the advantages and disadvantages of the existing neural network methods.

The structure of FFNNs is simple, with a relatively small number of neurons per layer. The FFNN converges slowly, and the neural network may fall into local minima of the nonlinear function during operation. In addition, FFNNs only have ENNs, and time-delay neural networks can have the input of processing time information. But only simple delay processing can be done. Other FFNNs have almost no ability to handle temporal data at all.

Deep learning methods can extract complex feature information from the input data in layers by building a deep neural network with multiple layers of nonlinear transformations. Some deep learning methods can handle time-series data considering the time dependence of SOC estimation. Training data and training methods have a significant impact on the accuracy of *SOC* prediction. Further research is needed in practical applications, as its algorithm is too complex and requires much computation.

The model-based method allows for relatively easy online measurement of battery SOC. There are fewer components for the equivalent circuit model, so the

Table 5.7 Advantages and disadvantages

Method	Advantages	Disadvantages
FFNN method	Simple structure Adapt to time-varying characteristics Fast computational cost for online phase	Network convergence is slow Easy to fall into local minimum Difficult to achieve global optimum Weak ability to handle time-series data Requires data to be stable and complete
Deep learning method	Easier expansion Better generalization capability Handle more complex problems High ability to handle time-series data	Connection between visible and hidden layers is more complicated Computationally intensive for deep networks Require a large amount of reliable data Require large storage devices
Hybrid method	Preserve the advantages of each Handle various ambient temperatures Online Reduce the effect of noise	Computationally intensive Difficult to train High computational cost Limitations in embedded device applications

relationship between the parameters is easier to derive. However, the model-based method tends to ignore the battery aging and diffusion phenomena in combining with neural networks, which has a particular impact on the *SOC* estimation results. In practical applications, the model is required to perfectly match the state changes of the battery, which causes great difficulties in model building and increases the computational effort of the BMS.

The hybrid Kalman filter-based method can fully play the advantages of both algorithms. On the one hand, it can deal with the battery as a nonlinear system with the help of a neural network. On the other hand, it can use the self-correcting property of the Kalman filter to deal with the problem of drastic changes of current in the input data. At the same time, Kalman filtering can reduce the influence of noise on the SOC estimation results. Of course, the application of the Kalman filter has its limitations, such as the need for highly complex mathematical calculations, high computational costs, and possible linearization errors.

Gaussian process and vector autoregressive moving average algorithm can make the SOC of the last time as the input of the next time, which makes the SOC have time-series characteristics—in essence, dealing with interpolation or linear relationships, which requires the data to be stable and complete. When the number of characteristic parameters is too large, its effectiveness is also significantly reduced. More detailed battery states can be obtained with electrochemical impedance spectroscopy

(EIS). If the EV has the capability of EIS measurements, it can participate in estimating the battery status as a part of the BMS.

As each method has its advantages and disadvantages, developing a new generation of BMSs is imminent. To this end, this review provides some research directions and recommendations to obtain accurate and robust SOC estimates in solving the following existing problems.

The studied neural network method can be applied to other types of lithium-ion batteries by transfer learning and other methods to improve the generalization ability and estimation efficiency of the method.

The hybrid approach may increase the computational complexity, but the computational power of the onboard BMS is limited. In the future, the complexity of the model can be reduced by selecting suitable optimization algorithms or network compression technology to make it more suitable for EV systems.

To cope with the effect of battery aging on SOC prediction, SOH and RUL can be used as input parameters of the network. Also, updating the network with a specific time interval can reduce the effect of aging.

Existing works have studied *SOC* estimation at above-zero temperatures, but this is not the actual environment. Therefore, it is necessary to improve the network structure to enhance its estimation performance at low environmental temperatures.

Although it is challenging to measure EIS in real time, enabling the EV with EIS measurement capability allows the *SOC* to be measured by inputting raw EIS data into the neural network.

These recommendations make a remarkable contribution to the future *SOC* and point to developing intelligent BMSs. The realization of these recommendations, therefore, is crucial for the future of EVs.

The neural network method can perform the lithium-ion battery *SOC* estimation task excellently based on the excellent nonlinear mapping and self-learning capability. This study first outlines the definition of *SOC* for lithium-ion batteries and shows that there is a non-negligible link between *SOC* and battery aging. Then existing neural network methods are critically reviewed and classified in the context of specific studies. Finally, future directions are given based on the existing problems.

The FFNN method is characterized by slow convergence and poor ability to handle time-series data, although their relatively simple structure is also a unique advantage. While deep learning methods can handle time-series data, they also suffer from the problem of overly complex algorithms that are difficult to apply. The hybrid method can fully exploit the advantages of multiple methods, gradually gaining attention from researchers. Considering battery aging, actual temperature, and operating condition, the application of neural network methods to practice will be the future trend of *SOC* estimation.

Overall, *SOC* estimation methods based on neural networks for lithium-ion batteries still hold great promise for development. Researchers can find an overview of various existing neural network methods and recommendations from this review, significantly contributing to future research on SOC estimation methods.

5.8 Conclusion

Neural networks have been widely used for image recognition and result prediction based on input data. Compared with other *SOC* estimation methods, the neural network method does not need to accurately consider the electrochemical state inside the battery to estimate the *SOC* by self-learning capability. With the continuous development of artificial intelligence, neural network methods have received increasing attention from researchers in the battery field. However, few articles summarize neural network methods individually. This review examines recent articles that use neural network methods alone to estimate the SOC of lithium-ion batteries, dividing the methods into FFNN method, deep learning method, and hybrid method. Then, the neural network method is summarized and discussed from the aspects of network structure, algorithm principle, appropriate environment, advantages, disadvantages, and estimation errors. The challenges to neural network use in the EV field are discussed briefly, and the prospects and opportunities for neural networks are examined in closing.

Chapter 6
AI-based battery parameter estimation

6.1 AI-based battery state of health estimation

6.1.1 Definition of state of health

With the application of lithium-ion batteries (LIBs) in aerospace, new energy vehicles, civil power generation and other fields, ensuring the safety and reliability of LIBs in complex working conditions has become a hot issue in the field of LIBs research [196]. Researchers use the state of health (SOH) of LIBs to indirectly represent the current performance of batteries, so it is very important to accurately estimate the SOH of lithium batteries and evaluate the health status of lithium batteries. To extend the life of the battery while ensuring reliability throughout its life, it is crucial to accurately diagnose the SOH of the battery in real time [197]. Battery SOH represents the capacity of the current battery to store electric energy relative to the new battery and represents the state of the battery from the beginning to the end of its life in percentage form, which is used to quantitatively describe the current battery performance state.

The significance of the study on the health status of lithium batteries lies in the difficulty and slow progress of SOH research, but SOH research is of great value to the use, maintenance and evaluation of batteries, and can provide basis and reference for planning, policy and industrial development, which is of great significance.

- Implications for battery management system

1. Implications of battery use and maintenance
   SOH study is helpful to understand the influencing factors of battery aging and provide theoretical guidance for the use and maintenance of batteries. For the use and maintenance of batteries, understanding the factors that affect the aging of batteries can reduce high and low temperature and overcharge and overdischarge, which can damage the use of batteries. Knowing the health status of the battery helps you determine potential risks and service life of the battery, and provides reference for battery maintenance and replacement. Therefore, the study of the health status of LIBs plays an important role in battery use and maintenance.

2. Significance of battery economy evaluation

   Different application scenarios, use modes and maintenance means of lithium batteries lead to the difference in battery life, resulting in difference in the use cost, economic benefit and other economic evaluation of batteries. The establishment of a battery aging model through SOH research will provide data support for the analysis of battery economy and will provide effective assistance for enterprise investment decision-making, government policymaking and industrial development planning. Therefore, the accurate evaluation of SOH is of great significance to the economic evaluation of batteries.

   There are many performance indicators of batteries. There are so many definitions of SOH at home and abroad that the concept is not unified. At present, THE definition of SOH is mainly reflected in capacity, electric quantity, internal resistance, cycle times and peak power. The automobile industry standard "QC/T743-2006" issued by China in 2006 clearly stipulates the conditions for the end of LIB life: the available capacity attenuates to 80% of the standard capacity. Currently, SOH is generally recognized in the industry as shown in (6.1).

$$SOH = \frac{Q_n}{Q_0} \times 100\% \tag{6.1}$$

   In (5.11), Q_0 is the rated capacity or nominal capacity of the new battery before it leaves the factory, and Q_n is the actual available capacity of the battery after it is put into use. Its value will decrease with the increase in the number of charge and discharge cycles.

   In addition, there are two other ways to define SOH.

3. Internal resistance defines SOH

$$SOH = \frac{R_{EOL} - R_c}{R_{EOL} - R_{new}} \times 100\% \tag{6.2}$$

   In (6.11), R_{EOL} is the internal resistance at the end of the battery life, R_c is the internal resistance of the current battery and R_{new} is the internal resistance of the new battery.

4. Residual cycle number defines SOH

$$SOH = \frac{Cnt_{\mathrm{remain}}}{Cnt_{\mathrm{total}}} \times 100\% \tag{6.3}$$

   In (5.11), Cnt_{remain} is the number of remaining battery cycles and Cnt_{total} is the total number of battery cycles.

   Among the SOH definitions of the above three batteries, the capacity definition is highly operable, while the internal resistance and residual number are not. The internal resistance is related to state of charge (SOC) and temperature, and it is not easy to measure, and the remaining cycle number and total cycle number cannot be accurately predicted.

   SOC estimation focuses on LIBs' residual capacity and the ratio of the current available capacity; the electric car users will use SOC as gasoline vehicles "oil" according to the current estimates to make travel planning and expected charging time; if the SOC estimation is not accurate, power failure will cause driving not to complete or frequent charging wastes owners' time. Therefore, SOC focuses on the single charge and discharge process of LIBs, which is the data parameter under micro timescale in the whole life cycle of LIBs. SOH is similar to the fuel tank capacity of an automobile, i.e., the available capacity of a battery in its whole life cycle. It is a parameter in a macro timescale. Different from the fuel tank of a gasoline car, the SOH value changes with time and presents an overall downward trend. If the SOH estimation is inaccurate, especially online estimation, it cannot only guarantee the safety and reliability of power battery in the use process but also seriously affect the SOC accuracy.

There are various definitions of SOH at home and abroad, mainly reflected in capacity, electric quantity, internal resistance, cycle number and peak power, etc. The generally accepted definition is the percentage value of the electric capacity released by the battery from the full state to the cut-off voltage at a certain rate under standard conditions and its corresponding initial capacity. SOH represents the degree of battery aging and deterioration. When SOH reaches a certain threshold, the vehicle should send an alarm signal that the battery needs to be replaced. It is generally believed that LIBs can no longer be used as a power source for electric vehicles when the SOH drops up to 80%. In recent decades, extensive studies have been carried out on capacity estimation and SOH prediction estimation of LIBs, which can be divided into three types: mechanism model, equivalent model and data model, which can be divided into micro timescale and macro time scale according to the time perspective. The micro timescale focuses on the state of single charge and discharge of LIBs. The macro timescale focuses on data over the life of LIBs.

5. Mechanism modeling

The mechanism model mainly starts from the electrochemical reaction mechanism of the battery itself and estimates the SOH of the battery by studying the electrochemical law of the decline of the battery capacity. This kind of method mainly builds the mechanism model of the battery based on the porous electrode theory and concentrated solution theory, and describes the physical and chemical reaction process between the electrode and the active material surface. The model parameters related to degradation were identified by measurement or system identification method to estimate the SOH of LIBs. Commonly used degradation mechanism models include quasi-two-dimensional mechanism model, single particle model and extended single particle model.

Compared with other models, the mechanism model can better evaluate and simulate the key performance indicators of the power battery, but there are many parameters related to the internal materials of the battery, so it is difficult to update regularly in practical application calibration; at the same time, the model calculation is large, generally only used for battery performance analysis. Although the simplified mechanism model can reduce the computational cost of the model, the dynamic simulation accuracy of the model is poor because the nonlinear characteristics of the battery and the variation of the characteristic parameters with SOC are ignored.

6. Equivalent modeling

   An equivalent model is a research model and method that replaces the actual and complex physical and chemical working process of the battery with an equivalent, simple and easily researched physical or mathematical process under a certain premise. Therefore, the accuracy of the equivalent model mainly depends on the adaptability of the model to the working process of the battery. The equivalent circuit model is mainly from the perspective of electrical engineering to equivalent lithium battery into a circuit model, and the corresponding relationship between the open-circuit voltage and SOC is established. The dynamic characteristics and capacity decline characteristics of the battery are described by using the measured operating voltage, current and other external characteristics. The core technology includes the accurate measurement of battery internal resistance, which describes battery characteristics, and estimation of the functional relationship between the open-circuit voltage and SOC. The main research methods are particle filter and Kalman filter.

7. Data-driven modeling

   Although an equivalent model can effectively estimate capacity under certain conditions, the equivalent model first to offline simulation model of the battery impedance, the battery internal resistance measurement results cannot be guaranteed under the condition of traceability; the accuracy of such simulation results is not reliable, and in addition, these technologies require a lot of matrices and high demand for computing resources, hampering their real-time application. The data model can directly analyze the hidden SOH information and its evolution rule from the external characteristic parameters of LIBs. The battery data mostly come from the use of process data and laboratory data, which contains the influence of various working states on the battery, and can also reflect individual differences to a certain extent. To overcome the problem of poor adaptability of equivalent model, at present, such methods mainly include neural networks, support vector machines (SVMs) and autoregressive models.

   At present, the research on the health status of lithium batteries faces the following difficulties:

8. The research period is long and the experimental conditions are strictly controlled

   LIBs' cycle life is long and the battery aging experiment cycle is long; in the test process, temperature, charge and discharge current and charge and discharge cut-off voltage need to be strictly controlled every certain time to evaluate the aging of the battery.

9. It is difficult to monitor and analyze the internal status of batteries

   SOH research of LIBs involves internal state variables, such as internal temperature, electrolyte concentration and internal resistance in the electrochemical model. However, it is very difficult to accurately monitor the internal state of batteries, and quantitative analysis of these state variables is also required, which makes SOH research of batteries very difficult to solve.

Table 6.1 Comparison of advantages and disadvantages of health status estimation models

Model type	Methods	Advantages	Disadvantages
Mechanism model	P2D model	High accuracy	Large amount of calculation
Equivalent model	Particle filtering	Strong nonlinear data processing ability	Complex calculation
	Kalman filtering	Good dynamic characteristics and easy calculation	State following cannot be performed
Equivalent model	The neural network	Strong ability to handle data	Weak anti-interference ability
	SVM	The generalization ability is strong	Difficult to train large-scale samples
	Autoregressive (AR) model	Strong reliability and easy realization	The test workload is large, long time and less robustness

10. Various influencing factors are coupled

    Battery operating temperature, charge–discharge ratio and discharge depth are all factors that affect battery aging and life span, and these factors work together. The study of battery SOH requires the decoupling of various influencing factors. However, these factors are interrelated and the decoupling conditions are difficult to control, so it is difficult to conduct decoupling analysis at present.

    LIB SOH prediction refers to the estimation and prediction of battery performance in the case of a decline in the whole life cycle of a LIB It is an important part of LIB health management. Inaccurate SOH prediction may lead to the operation of electric vehicles under unsafe systems. To further understand the accuracy of each model and compare the advantages and disadvantages of the health status estimation model (see Table 6.1).
11. The health status of LIBs is an important part of battery management. The battery management system estimates the state of charge and the remaining power and capacity of the battery. If the battery management system can master the aging law and health status of the battery, it will be helpful for the battery management of the battery life cycle.

The electrochemical process of P2D model is described comprehensively and has high accuracy; particle filter has strong processing ability for nonlinear data; Kalman filter has good iterative ability, good dynamic characteristics and moderate amount of calculation; neural network has a strong ability to process big data; SVM is a reconstruction risk minimization algorithm, which mainly aims at small sample training problems and has a strong generalization ability. The autoregressive model is not highly dependent on the model. The real-time capacity is obtained by looking up the table and fitting the function. The algorithm has strong reliability and is easy to implement in engineering.

6.1.2 Main affecting factor analysis

SOH refers to the health status of the battery, which is one of the most important internal states of the power battery [198]. It is used to measure how long the battery has left to use and whether the battery is safe. In the process of using the power battery, some factors will affect the health status of the power battery and even shorten the health status of the battery [199]. However, we can also use the relationship between some influencing factors and battery SOH to quickly obtain the health status of the battery.

1. Influence of temperature on battery SOH

   For LIBs, performance is greatly affected by the ambient temperature. The ambient temperature of the battery will have an important influence on the available capacity of the battery, the internal resistance of the battery and the rate of chemical reaction in the battery.

   When the ambient temperature of the battery is high, the rate of chemical reaction in the battery increases, the available capacity of the battery increases, and the internal resistance of the battery decrease. When the ambient temperature is low, the rate of chemical reaction inside the battery will decrease, the available capacity of the battery will decrease, and the internal resistance will increase. According to the definition of battery SOH, when the available capacity decreases, the SOH of the battery will decrease [200]. In addition, the ambient temperature will affect the internal resistance of the battery, and the internal resistance is also the characterization parameter of the battery SOH [201]. It can be seen that the ambient temperature of the battery will indirectly affect the SOH of the battery by affecting the available capacity and internal resistance of the battery. When the battery works in an unsuitable temperature environment for a long time, its service life will be greatly reduced, and the SOH of the battery will be rapidly reduced. Therefore, it is very important to ensure that the power battery works at the right temperature.
2. Influence of charge–discharge current ratio on battery SOH.

   The charge–discharge ratio will affect the battery life. The 300 cycle experiment of SONY 18650 battery with three different discharge rates showed that the battery capacity decreased by 9.5%, 13.2% and 16.9%, respectively, and the battery internal resistance increased by 12.4%, 18.3% and 27.7%, respectively. Meanwhile, the high discharge rate will generate more heat inside the battery. The solid-electrolyte interphase (SEI) film on the electrode surface of high-rate battery discharge is thicker than that of low-rate battery discharge under electron microscope.

3. Influence of discharge depth on battery SOH

   The depth of charge and discharge of a battery has an impact on the health and aging of the battery [202]. It is believed that the battery has accumulated total transferred energy, and the capacity attenuation and aging of the battery are analyzed based on the total transferred energy. Gao Fei *et al.* analyzed the relationship between the cumulative transferred energy of the battery and the attenuation of the battery capacity through cyclic testing of LIBs at different discharge depths, and concluded that before the battery capacity attenuates to 85%, the cumulative transferred energy of the battery is basically the same in the two modes of deep charge and deep discharge and shallow charge and shallow discharge. When the battery capacity attenuates to 85~75% in terms of cumulative transferred energy and energy efficiency, deep charge and deep discharge mode are superior to shallow charge and shallow discharge mode.

4. Influence of cycle interval on battery SOH

   The charge–discharge cycle range also affects the battery aging process. The internal resistance of the charge–discharge battery varies according to the cycle range. Therefore, the battery heating and reaction are slightly different during the cycle, which affects the battery health and aging in the long run. Therefore, some experts recommend a battery SOC range of 20–80% , which is conducive to battery health and cycle life.

5. Influence of charge–discharge cut-off voltage on battery SOH.

   Overcharge and overdischarge of a battery affect the battery health. Improper upper and lower voltages affect the battery health. The lower the discharge cut-off voltage, the greater the internal resistance of the battery, resulting in internal heating of the battery, while causing an increase in side reactions, the reduction of battery active substances and negative graphite layer collapse, accelerated aging and capacity attenuation of the battery. High charging cut-off voltage causes the increase of internal resistance of the battery, the increase of internal heat of the battery, overcharging causes the phenomenon of "lithium evolution" in the negative electrode and the increase of the corresponding side reaction, which affects the capacity and aging of the battery.

6. Influence of internal resistance on battery SOH

   In LIBs, the internal resistance mainly includes ohmic internal resistance and polarization internal resistance, among which the ohmic internal resistance is much higher than that of polarization internal resistance. The ohmic resistance inside the battery is formed by the resistance of the material inside the battery. The sum of the resistance of the material inside the battery is the ohmic resistance of the battery. The ambient temperature of the battery has a great influence on the value of ohmic internal resistance. When the temperature rises, the ohmic internal resistance decreases, and when the temperature drops, the ohmic internal resistance increases. When the battery is charged and discharged, the chemical substances inside the power battery will participate in the chemical reaction, thus producing the polarization internal resistance. Different from the ohmic internal resistance, the polarization internal resistance is less

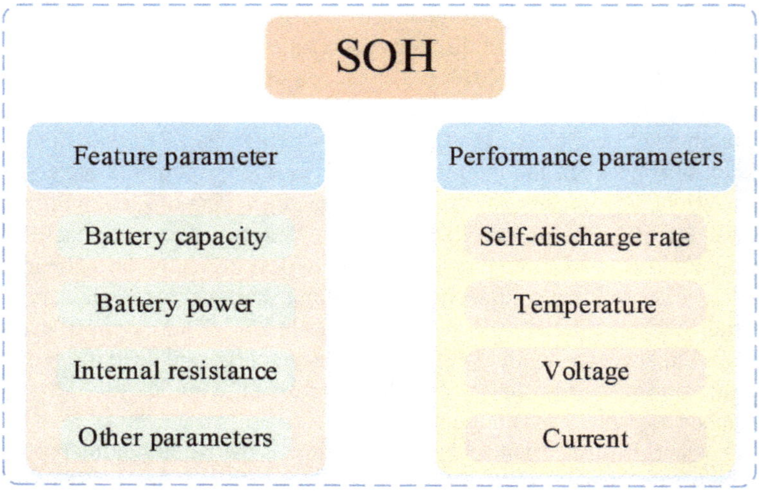

Figure 6.1 The main affecting factors of SOH

affected by the temperature of the battery. When the temperature increases and decreases, the polarization internal resistance will not have significant changes.

7. Influence of inconsistencies on battery SOH

   When electric vehicles are in normal use, the voltage provided by the power battery will reach hundreds of volts and the current will reach hundreds of amps, while a ternary LIB has a maximum voltage of 4.2 V and a discharge current of several amps. Such voltage and current are far from the use requirements of electric vehicles [203]. Therefore, in actual use, a certain number of LIBs are connected in series to form battery modules, and then multiple battery modules are connected in series to form battery packs for electric vehicles.

   However, when battery manufacturers produce lithium-ion cells, they cannot guarantee that the production process of each cell is exactly the same. Therefore, this leads to the inconsistency of the battery, and the health status of each cell is not consistent. When these monomer batteries are used in series and parallel into battery modules, the inconsistency of these batteries will be aggravated and the performance of these batteries will be reduced. The battery lasts longer than the battery serial-parallel modules [204]. That is, when the SOH of the battery module is lower than 80%, the charge and discharge times of the battery module are smaller than when the SOH of the battery module is lower than 80%. It can be seen that the inconsistency of the battery accelerates the aging of the battery. Therefore, improving the consistency of the battery module will delay its aging degree, which is shown in Figure 6.1.

To sum up, the operating temperature, charge–discharge ratio, discharge depth, cycle interval and a charge–discharge cut-off voltage of the battery all affect the health status and life of the battery [205]. At present, the research on the influencing factors of battery health status is in the stage of qualitative research. The quantitative analysis of these influencing factors on battery aging and the coupling relationship between each factor are the difficulties of the research, and also the hot spot of the research on battery health and life in the future.

6.1.3 Summary of AI-based SOH estimation

In the above section, several artificial intelligence algorithms were introduced to estimate the SOC of Li-battery with good results, and in this chapter, we will estimate the SOH based on the long- and short-term memory network and the gated cyclic unit, respectively.

The following are current LIB SOH estimation methods.

6.1.3.1 Characteristic method

Based on the evolution of the characteristic parameters shown during the aging process of the battery, a mapping between the characteristic quantities and the SOH of the battery is established, and the SOH is then estimated.

1. Internal resistance analysis

   The internal resistance method uses the internal resistance of the battery as the main indicator of battery life [206]. As the battery ages and its capacity decreases, the internal resistance of the battery will gradually increase. The difficulty in SOH estimation by the internal resistance method lies in the extraction of the mapping relationship between SOH and internal resistance, especially when SOC, temperature and multiplicity are taken into account. In addition, the extracted characteristic relations are only applicable to a certain brand and model of battery, which is not very general.
2. Electrochemical impedance spectroscopy EIS analysis

   The impedance spectrum is first measured at different stages of battery aging, then the EIS curve is related to the equivalent circuit model parameters of the battery, and then the SOH is located according to the relationship between the model parameters (e.g., series resistance, charge transfer resistance, Warburg coefficient) and the SOH.

   However, the EIS measurement is complex and requires special instruments, and can only be applied offline; in addition, similar to the internal resistance method, the EIS method is less versatile and is only applicable to the battery production, design and process improvement process, and is not suitable for application in practical power systems.
3. Differential analysis method: (a) incremental capacity IC or incremental voltage IV method

   For example, the peak potential, peak sharpness, asymmetry and peak area of the DC curve are directly related to the thickness of the SEI film, the stress state between the electrochemical reaction centers, the degree of reversibility of the electrode reaction and the content of active substances [207]. The aging characteristics of the cell can be extracted from the set of DC curves under different aging conditions, which enables the estimation of SOH.

4. Other methods

   The discharge curve of the battery is used directly as a basis for measuring the health of the battery. Based on this phenomenon, the relationship between the characteristic values of the battery discharge curve and SOH has been established to estimate SOH [208]. However, in practice, the battery discharge mechanism is random, so this method cannot estimate the SOH of the battery online but can only be an offline tool.

6.1.4 Modeling methods

Ageing mechanism model analyzes the physical and electrochemical processes within the battery at a microscopic level, focusing mainly on the battery aging process.

1. Destructive

   Dismantling of lithium batteries with different levels of aging to obtain samples of the internal materials of the batteries and analyzing the relationship between the aging parameters and the residual capacity of the batteries, extracting the aging mechanism from the large amount of data obtained and combining the data to establish an objective and realistic model of the aging of lithium batteries.

2. Nondestructive

   Find the appropriate aging degree characterization parameters and establish the correspondence between these parameters and the aging degree of the battery, and then obtain the aging mechanism model [209]. This method is similar to the characteristic method, but its SOH estimation accuracy is less than that of the destructive method, and the workload is smaller; so it is one of the mainstream development directions at present, and the technical difficulty is the location and acquisition of characteristic parameters.

   1. Arrhenius model: it focuses more on the influence of temperature on battery durability
   2. Inverse power-law model: electrical stress (e.g., voltage, current, power)
   3. Others: mechanical stress (product of SOC increment and SOC mean)
   4. Probabilistic modeling method

Combine the battery equivalent circuit model with probabilistic analysis methods (e.g., Bayesian regression and classification algorithms) to describe the aging and capacity decay process of the battery, and validate the model through experiments.

A probabilistic monotone echo state network (MONESN) algorithm is used to track the nonlinear battery aging process based on the time intervals corresponding to the voltage drop sequence in the battery capacity and discharge process.

The probabilistic model has two advantages: (a) it only requires partial charge/discharge testing of the battery for SOH estimation [210]; (b) it is relatively simple and easy to implement, but the extraction of the feature mapping requires consideration of various factors and the workload is high.

6.1.5 Data-driven methods

This method takes the battery test data as the original sample and uses it to extract the evolution of the battery performance during the battery aging process through some mechanism, and then uses this law for SOH estimation.

These methods include autoregressive AR models, neural networks, SVM and Gaussian process regression (GPR).

The advantage of AR is that it is less computationally intensive and less complex, but the disadvantage is that the accuracy of the results deteriorates over time

NN, SVM and fuzzy methods: NN has the disadvantage of requiring a large number of comprehensive sample data for training and the estimation error is largely influenced by the training data; in addition, NN is inherently difficult to train online due to its open-loop structure. The SVM also requires a large number of samples for training and is sensitive to missing data. The fuzzy rule base requires a large amount of empirical knowledge and data, and the control rules, once established, cannot be adjusted online and are not very versatile.

6.1.6 Statistical law method

This is also known as the empirical method, which is based on the empirical knowledge accumulated during the use of the battery and is used to estimate the SOH of the battery by mathematical and statistical means [211]. These methods are used to estimate SOH for specific applications where the empirical knowledge of the battery is sufficient.

In this prediction estimation of SOH of a power battery, we will train and validate the long- and short-term memory network based on the number of cycles and SOH values obtained from the cyclic charging and discharging experiments [212]. This experiment is also conducted for multiple batteries simultaneously to ensure sufficient training and validation data for the long- and short-term memory network model [213]. For this network, we set the input vector as a one-dimensional vector, which is the number of cycles of the charge/discharge cycle test, and the output vector as a one-dimensional vector, which is the SOH value obtained at each charge/

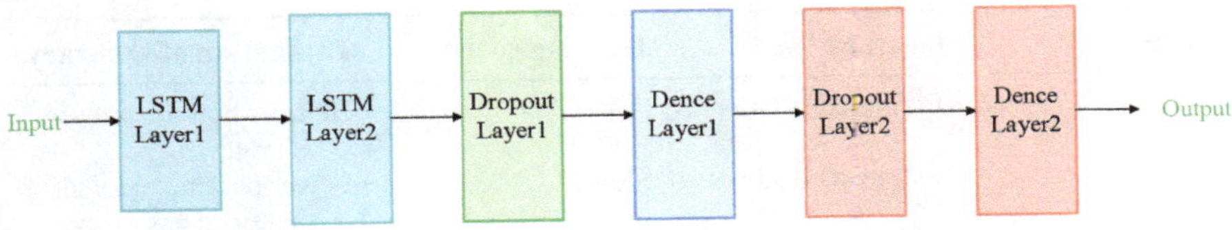

Figure 6.2 The estimation model for SOH based on LSTM

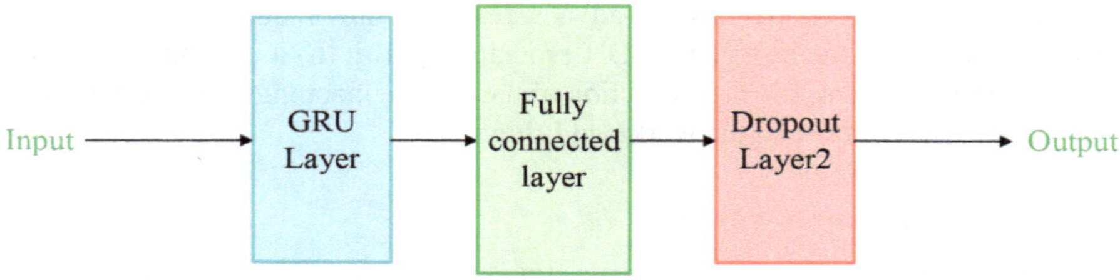

Figure 6.3 The estimation model for SOH based on GRU

discharge completion. Figure 6.2 shows the structure of the SOH prediction model for LIBs based on long- and short-term memory (LSTM).

While predicting the estimated battery SOH, the input vector dimension of the long- and short-term memory network is 1, which is the number of cycles of the charge–discharge cycle experiment, and the output vector dimension is 1, which is the SOH value of the battery; therefore, the mathematical expressions of the input and output vectors can be obtained as shown in (6.4) and (6.5), respectively.

$$X_t' = (N_t) \tag{6.4}$$

$$y_t' = f_{lstm}(N_t) \tag{6.5}$$

To verify the effectiveness of the gated loop unit for battery SOH estimation further, in this section, we apply the gated loop unit to power cell SOH estimation and verify its estimation accuracy by statistical errors. Figure 6.3 shows the structure of the SOH prediction model for lithium-ion batteries based on Gated Recurrent Unit (GRU).

To verify the effectiveness of the gated loop unit for battery SOH estimation, this section configures the gated loop unit used according to the LSTM configuration described in this chapter. Table 6.2 shows a comparison of the advantages and disadvantages of LSTM and GRU.

At present, with the increasingly serious problems of global environmental pollution and energy crisis, new energy vehicles are gradually receiving attention from various countries. As the core component and power source of electric vehicles, lithium-ion batteries have received widespread attention due to their many advantages [214]. Therefore, in the application field of lithium-ion battery, it is necessary

Table 6.2 Comparison of the advantages and disadvantages of LSTM and GRU

Model	Structure	Training speed	Estimation of accuracy
LSTM	Missing cell status	Faster	Better when the data set is large
GRU	Contains cell status	Slower	Fewer parameters and therefore easier to converge

to study the modeling of lithium-ion battery and its parameter identification and internal state estimation. In this chapter, we have conducted research in the field of improving the parameter identification of the second-order Resistance capacitance (RC) equivalent circuit equivalent model of lithium-ion battery and the internal SOC and SOH state estimation of lithium-ion battery with lithium-ion power battery as the research object, and achieved certain results. Its main work can be divided into the following parts:

The main innovation points are the following two aspects:

1. The output response reconstruction method is used to reconstruct the battery voltage and improve the modeling accuracy. Due to the noise in the battery test data, the accuracy of the battery model estimation is reduced, and the traditional noise filtering method, low-pass filter method, has the problem that the optimal cut-off frequency is not easy to choose and the filtered waveform is distorted, so the output response reconstruction method is proposed [215]. This method is based on the principle of convolution and correlation function and other mathematical methods to reconstruct the battery terminal voltage; compared with the low-pass filter method, this method can effectively suppress the noise in the battery voltage and use this terminal voltage for battery model parameter identification, which greatly improves the identification accuracy.
2. The long- and short-term memory network algorithm and the gated cyclic unit algorithm are applied to the field of battery internal state estimation, and it is experimentally verified that the gated cyclic unit is more suitable for the short timescale state like SOC, and the long- and short-term memory networks are better suited for long timescale states like SOH.

For the estimation of the battery SOH in this chapter, LSTM network and gated cyclic unit are used to estimate the battery SOH for such a long timescale state, respectively [216]. Finally, the estimation accuracies of both are verified experimentally. Therefore, when estimating the SOH, which is a long timescale state, we can give priority to guarantee the estimation accuracy and choose the LSTM network for its estimation.

At present, we have achieved good results in the identification of the parameters of the second-order RC equivalent circuit model and the estimation of the internal state of the battery by applying the output response reconstruction method and the long- and short-term memory network method, respectively, but some work has not yet been carried out due to time and experimental constraints [217]. Future research should be continued in the following directions to achieve better results:

1. Although the output response reconstruction method has achieved good results in the parameter identification of the second-order RC equivalent circuit model, it is currently only used in the offline identification field and has not yet been studied in the online identification field, thus limiting its application prospects [218]. Therefore, further research should be carried out to apply the output response reconstruction method to the online identification field, so as to lay

the foundation for the practical application of the vehicle battery management system.

2. For the estimation of the internal state of the battery SOC and SOH, long- and short-term memory networks and gated cyclic cell algorithms were used to estimate both, and algorithms were judged to be suitable for the estimation of SOC and SOH, respectively. However, in real-world in-vehicle battery management systems, the joint estimation of multiple variables is often required, and the joint estimation of internal battery states has not yet been investigated in this topic. Therefore, future research can be carried out in the direction of joint estimation of multiple internal states of lithium-ion batteries to enrich the research results.

6.2 AI-based battery state of power determination

6.2.1 Definition of state of power

The state of power (SOP) of the lithium-ion battery is an important parameter for electric vehicles to ensure safety performance and improve the braking energy recovery rate [219]. On the premise of the normal operation of the battery system, the dynamic performance of the electric vehicle is optimized to better meet the functional requirements of climbing, acceleration and braking energy recovery.

The level of SOP greatly affects the safety and efficiency of the battery. If the estimated value of the SOP is too high, it will lead to overcharge and overdischarge of the battery, which will affect the life of the battery, and even cause the danger of bulging and fire. If the estimated value of SOP is too low, it will reduce the efficiency of the battery and reduce the dynamic performance of the electric vehicle [220]. Therefore, SOP estimation can restrict the charging and discharging process of the battery, prevent the occurrence of overcharge and overdischarge, and play an important guiding role in the rational use and life extension of lithium-ion batteries.

In the existing research, the description forms of the lithium-ion battery SOP are mainly divided into two types: one is to directly use the current SOP that the battery can provide to reflect the power state of the battery [221]. The maximum power that can be absorbed or released is a physical quantity with dimension *W*; the other is the ratio of the current SOP that the battery can provide to the rated power of the battery as the power state of the battery. Compared with the first description, the second description can not only provide the motor controller and the vehicle controller with battery SOP information but also reflect the attenuation of battery power; at the same time, because of the percentage expression, it maintains a better unity in form with the two states of battery SOC and SOH. Therefore, this research chooses to express the power state of the battery, as shown in (6.6).

$$SOP = \frac{P_{\max}}{P_{\text{rate}}} \tag{6.6}$$

Table 6.3 Hybrid pulse power characterization test method

The experimental discharge current is 4/3 C, and the charging current is 1 C.
1. put the battery under test in a 25°C incubator for 1 hour;
2. fully charge the battery according to the charging specification specified by the battery;
3. adjust the SOC to 90% with 0.5 C, and let it stand for 1 hour;
4. 1 C constant current charging for 10 seconds, standing for 40 seconds;
5. 4/3 C constant current discharge for 10 seconds, standing for 40 seconds;
6. discharge 10% SOC at 0.5 C discharge rate, switch to 1/3 C to charge for 10 seconds, and let it stand for 1 hour;
7. 1 C constant current charge for 10 seconds, let it stand for 40 seconds;
8. 4/3 C constant current discharge for 10 seconds, let it stand for 40 seconds;
9. repeat steps (6)–(8) for seven times.

In the formula, P_{max} represents the SOP of the battery at the current stage and P_{rate} represents the rated power of the battery. Therefore, obtaining the battery power percentage representation of the lithium-ion battery at different stages through analysis can reflect the safety and use efficiency of the lithium-ion battery, and construct the relationship expression with different state quantities.

The mutual verification of experiments and algorithms is an important means of SOP estimation [222]. The SOP test method obtained from the experiment is an important extension of the definition and verification of SOP. At present, the SOP test methods of power batteries mainly include hybrid pulse power characterization (HPPC) test method, Japan Electric Vehicle Society (JEVS) test method and China Constant Power test method.

1. HPPC test method

   The HPPC test method is proposed by the "FreedomCAR Battery Experiment Manual." Two pulse experiments are carried out at every 10% SOC point, one charging pulse lasting 10 seconds and one discharging pulse lasting 10 seconds. The size of the discharge pulse is 4/3 times the charge pulse, the equivalent charge and discharge internal resistance are calculated according to the charge and discharge pulse, and then the SOP is calculated. The specific test method is shown in Table 6.3.

   The test current of HPPC is fixed and single, which cannot reflect the relationship between impedance and discharge rate, and the test result has a large error, which is likely to cause overcharge or overdischarge of the battery. This test method also does not take into account the safety boundary of the current. If the current is too large, it will affect the life of the battery.
2. JEVS test method

   The JEVS test method is to alternately charge and discharge with different pulses for 10 seconds at different SOC points, and select 1C, 2C, 5C, and 10C for the pulse rate, to obtain the relationship between different pulse rates and the battery terminal voltage at the end of 10 seconds of charge and discharge.

Table 6.4 Japan Electric Vehicle Society test method

Taking 10% SOC as an example, the specific test steps of the Japanese JEVS test battery 10 seconds peak current are as follows:
1. put the battery under test in a 25°C incubator for 2 hours;
2. discharge with 1C constant current to the cut-off voltage, and let it stand for 2 hours;
3. adjust the SOC to 10% at 0.5C and let it stand for 2 hours;
4. 1 C constant current charging for 10 seconds and let it stand for 1 hour;
5. 1 C constant current discharge for 10 seconds and let it stand for 1 hour;
6. 2 C constant current charging for 10 seconds and let it stand for 1 hour;
7. 2 C constant current discharge for 10 seconds and let it stand for 1 hour;
8. 5 C constant current charging for 10 seconds and let it stand for 1 hour;
9. 5 C constant current discharge for 10 seconds, and let it stand for 1 hour;
10. 10 C constant current charging for 10 seconds and let it stand for 1 hour;
11. 10 C constant current discharge for 10 seconds and let it stand for 1 hour.
The voltage value corresponding to the current at the end of each charge and discharge for 10 seconds was recorded, and the current value and the voltage value were fitted.

The current value corresponding to the cut-off voltage of charge and discharge is obtained by linear fitting, which is the SOP of charge and discharge of the battery. The specific test method is shown in Table 6.4.

Since the battery cannot be charged at a high rate at a high SOC point, the Japanese JEVS test method is not suitable for the high SOC test, resulting in a large error in the high SOC.

3. Constant power test method

The constant power test method is proposed in the performance test specification of high-power lithium-ion power battery for Hybrid Electric Vehicle (HEV) in China's "863 Program" energy-saving and new energy vehicle project. Carry out at least five high-power charge–discharge tests and record the charge-discharge time when the battery reaches the charge–discharge cut-off voltage or the maximum rate specified by the battery manufacturer as the termination condition. By fitting the relationship between power and time, the charging and discharging power corresponding to 10 seconds is obtained as the peak charging and discharging power of the battery. The specific test method is shown in Table 6.5.

The power value obtained by this test method can theoretically be regarded as the true value of SOP, but due to the lag in the PI adjustment of the test instrument, the power cannot be kept constant, resulting in errors.

To sum up, different lithium-ion battery test verification methods have different test advantages and disadvantages. With the rapidly increasing demand for high-power and high-energy power batteries, peak power capability over long-time scales and wide temperature ranges has received increasing attention.

6.2.2 *Main affecting factor analysis*

SOP is an important parameter in the operation of lithium-ion batteries, which is used to measure the energy conversion intensity and efficiency of batteries over a

Table 6.5 Constant power test method

Taking the 50% SOC discharge direction as an example, the specific process of the constant power test is as follows: 1. put the battery under test in a 25°C incubator and let it stand for 2 hours; 2. 0.5C constant current charge to a cut-off voltage of 4.2 V, constant voltage charge to currently less than 0.01C and let it stand for 1 hour; 3. adjust to 50% SoC with 0.5C and let it stand for 1 hour; 4. apply constant power discharge P1 to the battery under test, record the time t1 used to discharge to the cut-off voltage and record the discharge capacity C1, and let it stand for 1 hour; 5. fill the capacity C1 with 0.5C and let it stand for 1 hour; 6. apply Constant Power P2 to the battery under test (if t1 is less than 10 seconds, then P2 is less than P1. If t1 is greater than 10 seconds, then P2 is greater than P1) to discharge, record the time t from the discharge to the cut-off voltage, and record the discharge capacity C2, static set for 1 hour; 7. repeat steps 5–6 at least three times; 8. fit the constant power P obtained each time with the discharge time t, and the power value corresponding to 10 seconds is the SOP.

certain period during charging and discharging [223]. In the state detection of lithium batteries, SOP generally refers to the continuous peak power of the battery over a certain period [224]. Many parameters affect the battery SOP; the main factors are temperature, SOC, current, internal resistance and voltage. The characteristics of the main influencing factors of SOP and their mode of action will be introduced separately.

1. Influence of current and voltage

   Generally, in the process of charging and discharging the battery, its main energy transfer method is in the form of power, and the calculation method of power is the product of the terminal voltage and current of the battery at a specific time node, and the algorithm method is shown in (6.7).

$$P = I * V \tag{6.7}$$

As shown in (8.19), the calculation of power is directly related to the voltage and current at the battery terminal and has a positive correlation with both factors [225]. For the battery, the voltage and current have upper and lower limits, and both factors will change due to the limitations of the working state and load requirements of the battery. Generally, the current of the battery is relatively stable relative to the terminal voltage, which is determined by the artificial magnification setting, while the voltage, as a dependent variable, is in a state of nonlinear change. The voltage variation range of common batteries is generally between 2.8 and 4.2 V [226]. Under the condition of not exceeding the upper limit current of the battery, the charge–discharge magnification (c) of the battery can be set arbitrarily according to the situation.

According to the power calculation method and the basic characteristics of battery current and voltage, the current ratio has a great influence on SOP, but it mainly affects the peak power in the process of battery charge and discharge. According to the law of energy conservation, the influence of voltage and current has little influence on the continuous charge and discharge power of the battery.

2. Influence of battery internal resistance

During the operation of the battery, the open-circuit voltage of the battery is greater than its terminal voltage because its internal composition and structure form an obstacle. In the circuit model, this obstacle is called internal resistance. The partial voltage effect of internal resistance will consume the energy of the battery and mainly dissipate this part of energy in the form of thermal energy to form invalid power, resulting in energy waste. Therefore, the SOP of the battery will be affected by the internal resistance of the battery. Excessive internal resistance will reduce the energy supply efficiency of the battery. The power consumed by the internal resistance is calculated in (6.8).

$$P_R = I^2 * R \approx (E - U_L) * I \tag{6.8}$$

In (6.8), I represents the working current of the battery and R^0 represents the equivalent internal resistance of the battery. In the battery equivalent circuit model, the internal resistance effect is the main cause of battery energy waste, so the energy consumed by internal resistance is approximately equal to the difference between battery electromotive force E and terminal voltage U_L multiplied by battery working current.

3. Influence of temperature and battery SOC

The reason why the SOC of the battery is combined with the working temperature as a comprehensive influencing factor is that the common method of measuring SOP is the look-up table method. The ideal SOP table is a three-dimensional data table of {minimum temperature, maximum temperature} {0% SOC, 100% SOC} (temperature value accuracy 1°C, SOC accuracy 1%). However, for such a table, assuming that the temperature range is −30–60 °C, there must be 9,000K data points to be tested, which is basically impossible to complete. Therefore, in the actual application process, the accuracy is generally relaxed. For example, a point is tested at 10°C and 10% SOC and finally the look-up table is obtained by interpolation method, or these characteristic points are directly processed by battery management system (BMS) to obtain the allowable charge–discharge current value under specific temperature and specific SOC. HPPC test is a commonly used method for the determination of these data.

In recent years, the control of SOP by various research institutions is mostly divided into the following: discharge power within 10 seconds, discharge power within the 30 seconds, charging power within 10 seconds and charging rate within 30 s. The power limit of discharge is well understood, and the power limit of charging is mainly the constraint on charging at high and low temperatures, as well as the energy feedback system. The way of control is the lookup table method, and the lookup table method and the equivalent circuit model combined with the prediction method; at this stage most manufacturers use the temperature—SOC determined power MAP two-dimensional lookup table [227]. This power MAP is also a new electric vehicle or a new set of battery packs of the nascent attributes; this power MAP needs to be combined with the dynamic parameters of the vehicle to generate power demand, such as 100 km acceleration requirements, the highest speed requirements, ordinary working conditions, and track conditions and other performance; MAP also combined with the battery pack life management needs, the current OEM for the battery pack warranty policy is often 8 years/150,000 this standard, i.e., the power MAP also needs to consider the life dimension when proposing. Figures 6.4 and 6.5 show the power limit of a lithium battery system.

It can be seen from the power limit diagram of lithium batteries in the 10 seconds and 30 seconds that the effects of SOC and temperature on the SOP of lithium batteries have a combined effect, and it is more meaningful to study the comprehensive effect of the two factors on SOP.

Based on the look-up table method, the SOP of a lithium battery can be deduced and calculated. First, set the peak discharge power as X_P (10 seconds), the constant discharge power as X_C (the 60 seconds), and the current actual discharge power of the whole vehicle as X_F. Calculate the integral value Y_1 of the actual power exceeding the constant power as of t_0, as shown in (6.9).

$$Y_1 = max\left(\int_0^{t_0} \left(X_f - X_c\right) dt, 0\right) \tag{6.9}$$

Then, the value Y_2 of the peak power of the battery exceeding the constant power is calculated by the integral, as shown in (6.10).

$$Y_2 = \int_0^{10s} \left(X_p - X_c\right) dt \tag{6.10}$$

After the calculated values of Y_1 and Y_2 are completed, a coefficient f is defined to indicate the proximity between them. The calculation method is shown in (6.11).

$$f = max\left(0, 1 - Y_1/Y_2\right) \tag{6.11}$$

The power calculation is analyzed from the perspective of energy (the integration of power and time). The closer the value of f is to 1, the more available energy and the greater the available power in the next second; the closer f is to 0, the less available energy is. When $f = 0$, it is generally considered that the available power of the cell is constant; combined with F, the real-time power of the next second can be estimated, as shown in (6.12).

$$X_{next} = X_c + \left(X_P - X_c\right) \tag{6.12}$$

According to the calculation method of (8.19), X_{next} must be a value between X_c and X_p, and conforms to the law of power use, the larger the f, the closer the available power at the next moment to X_p.

The SOP estimation method using power map and look-up table method has the following limitations:

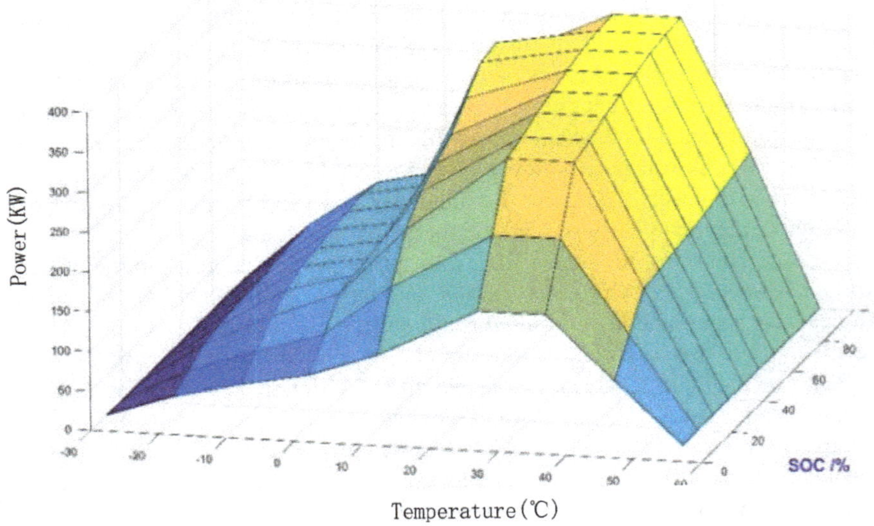

Figure 6.4 Ten-second discharge three-dimensional diagram

1. Life effect. The limitation of the power map is only the parameters of the new battery system. With the gradual aging of the battery, the available capacity gradually decreases and the internal resistance gradually increases. The problem

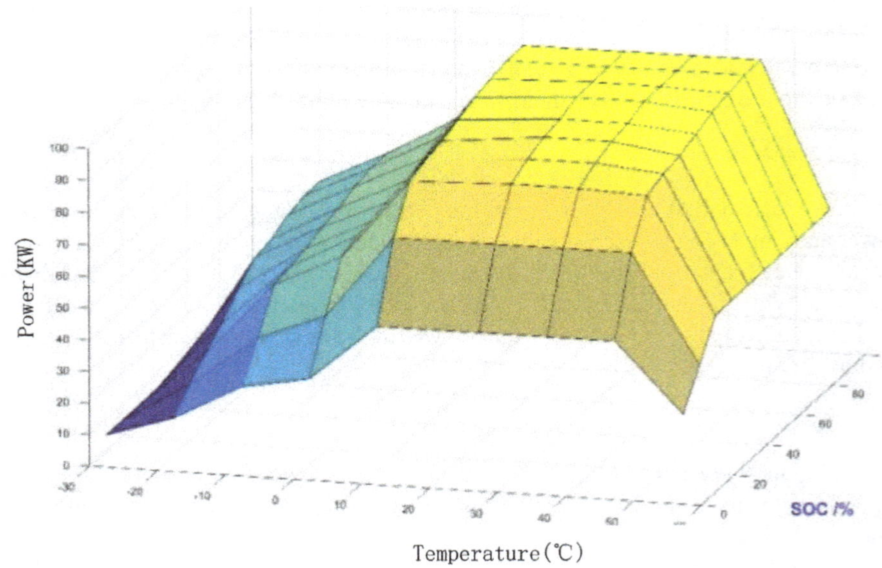

Figure 6.5 Thirty-second discharge three-dimensional diagram

will be that the discharge battery voltage will reach the cut-off voltage faster, and there will be no power output capacity of the new battery.

2. Model accuracy. The core idea of SOP using an equivalent circuit model is to predict terminal voltage and battery temperature. The accuracy of the battery thermoelectric coupling model will be the key to the accuracy of SOP prediction.
3. Switching between various limited powers. Ten-second discharge power and 30-second discharge power will be calculated at the same time, and when to switch the 10 seconds, 30 seconds, or even 60 seconds power limit will be judged from many aspects of calculation logic and control strategy.

6.2.3 Summary of AI-based SOP estimation

At present, with the development of artificial intelligence technology, the lithium battery SOP estimation method based on the AI algorithm is also called the nonparametric model estimation method [228]. It regards the battery as a "black box," does not consider the internal reaction mechanism and characteristics of the battery, uses the methods of data analysis and machine learning, takes the SOP to be estimated as the output of the model and the influencing factors as the input, tests a large amount of data, and uses the model for learning and training. The estimation of battery SOP can be realized.

1. BP neural network method

BP neural network is a multilayer feedforward network trained by an error back-propagation algorithm, which is composed of an input layer, hidden layer and output layer [229]. The input layer is the basic quantity of the input battery, including voltage, temperature, SOC, etc., and the output layer is the SOP of the battery. The output layer reversely calculates the error according to the original line and adjusts the weight and threshold of each layer node by the gradient descent method until the output layer obtains the desired output result. The structure of the BP neural network is shown in Figure 6.6.

BP neural network selects the appropriate number of input layer, hidden layer and output layer according to the modeling requirements, uses the hidden layer to connect the input and output layer, and establishes the node weight between layers [230]. Then the normalized training samples are sent to the output layer to train the network. Finally, adjust the network weight according to the error until the error reaches the allowable range. The construction steps of the battery peak power estimation model are shown in Figure 6.7.

2. Adaptive neuro-fuzzy inference system (ANFIS) model

A fuzzy inference system is a new type of neural network, which combines fuzzy inference with a neural network and automatically extracts fuzzy rules from a fuzzy inference system. The ANFIS network structure of the first-order T-S fuzzy inference system is shown in Figure 6.8.

The overall prediction process of the system is data acquisition, model training and simulation verification. Using the experimental method, the peak power

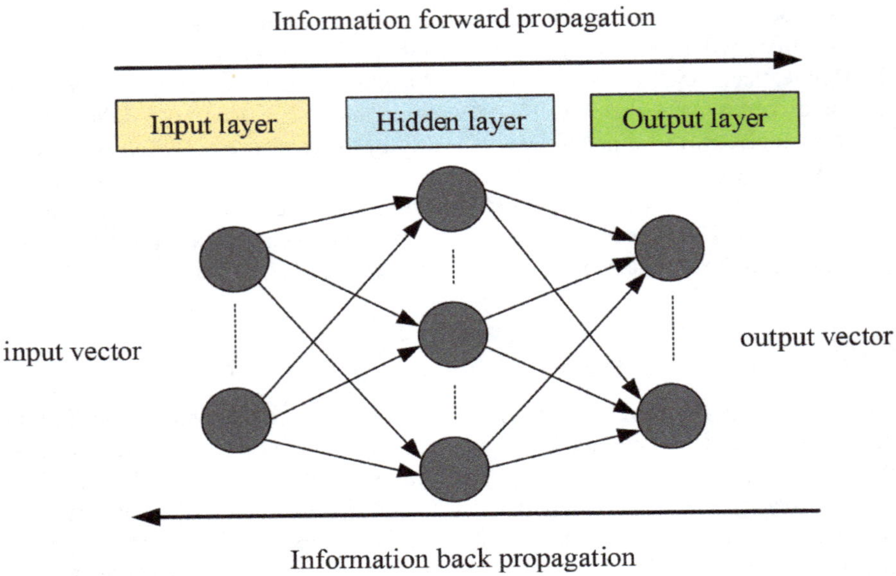

Figure 6.6 BP network structure diagram

and battery internal resistance under different SOC at different temperatures are first obtained. In the model training stage, the fuzzy interval of variables can be divided by the mesh generation method and subtractive clustering method, and the model training can be carried out by a single BP neural network and comprehensive training method, which can reduce the complexity and improve the accuracy [231]. Finally, the simulation is verified, and the method solves the local optimal problem of a single BP neural network, which has high prediction accuracy.

3. Support vector machine method

The SVM method is a machine learning algorithm developed based on statistical theory, which has strong learning ability, is especially suitable for the identification and estimation of small sample data, and has good applicability and robustness. The method is divided into three layers: the input layer, the intermediate layer and the output layer. The input layer stores data such as voltage, current, temperature, SOC, etc., and the middle layer selects the appropriate kernel function before data training, and finally constructs the regression estimation function, i.e., the output layer [232]. By introducing the method of grid search GS to obtain global optimal values, the calculation efficiency of SVM is improved and the search for global optimal values is guaranteed.

The data-driven nonparametric model estimation method does not consider the internal reaction mechanism of the battery and can estimate the SOP of any battery, which is widely applicable, but it needs to be premised on a large amount of experimental data, and the sample selected during training will have a great impact on the

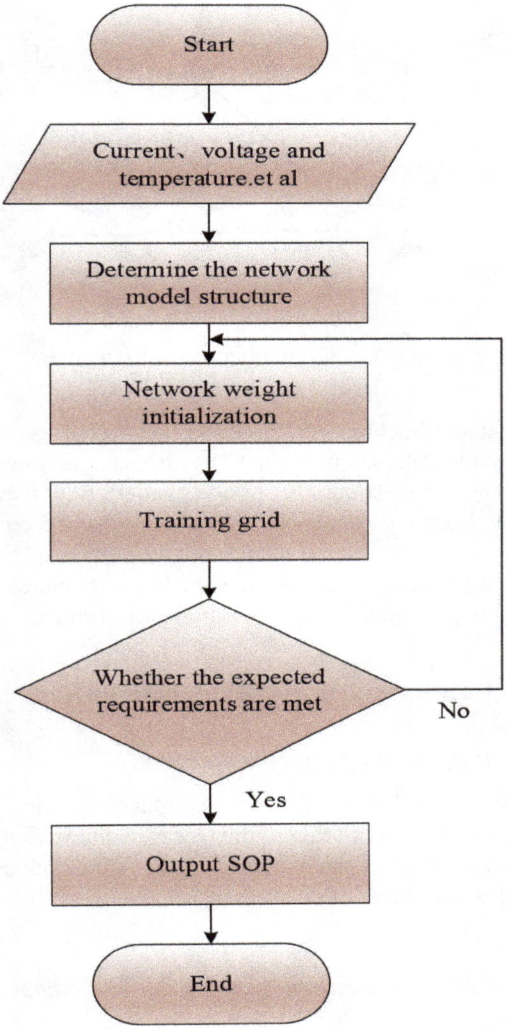

Figure 6.7 SOP estimation flow chart of neural network method

results. Compared with other methods, the data-driven method is computationally complex and hardware-demanding, and the upfront training takes a long time.

The above is the current commonly used and typical lithium battery SOP estimation method based on AI algorithms. The advantages and disadvantages of several methods are summarized in Table 6.6.

To improve the estimation accuracy of SOP, it is necessary to establish a more accurate power battery model to more accurately reflect the dynamic and static changes of the battery. In addition, it is necessary to pass a large number

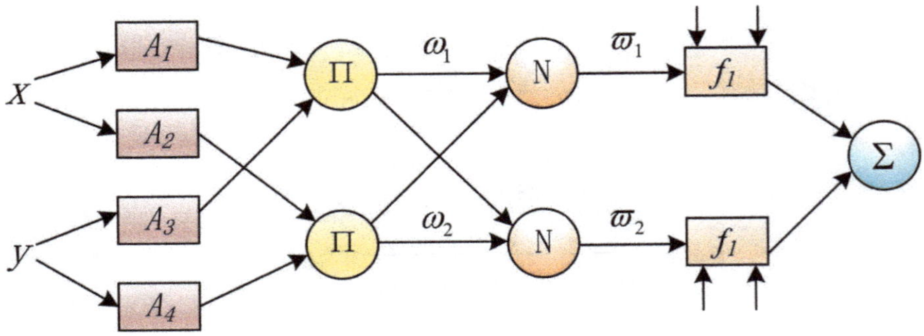

Figure 6.8 Fuzzy inference system model

of experimental tests and actual working conditions tests, establish a more complete database, provide data samples for SOP estimates and develop towards the era of big data. In short, the estimation of power battery SOP requires the synthesis of various methods, learning from each other's strengths and complementing each other's weaknesses, realizing accurate and rapid estimation of power battery SOP, providing an accurate reference for electric vehicle power matching, and providing guarantee for the safe and reliable operation of electric vehicles.

6.3 AI-based battery state of energy calculation

6.3.1 *Definition of state of energy*

The state of energy (SOE) of the power battery, as the proportional parameter of the residual energy on the w • h unit scale of the electric vehicle, is a direct description of the energy supply capacity of the battery [233]. It is more advantageous to take it as the charging index parameter of the electric vehicle user.

Table 6.6 Advantages and disadvantages of AI-based methods

Estimation methods	Advantages	Disadvantages
BP neural network	There is no need for an accurate battery model; there is a self-learning ability	It is easy to appear local optimal, high computational requirements and requires a lot of experiments
An adaptive fuzzy neural inference system	Strong learning ability, overcome the local optimal problem and high estimation accuracy	It takes a lot of experiments and the calculations are complex
SVM method	It has good robustness	It takes a lot of experiments and the calculations are complex

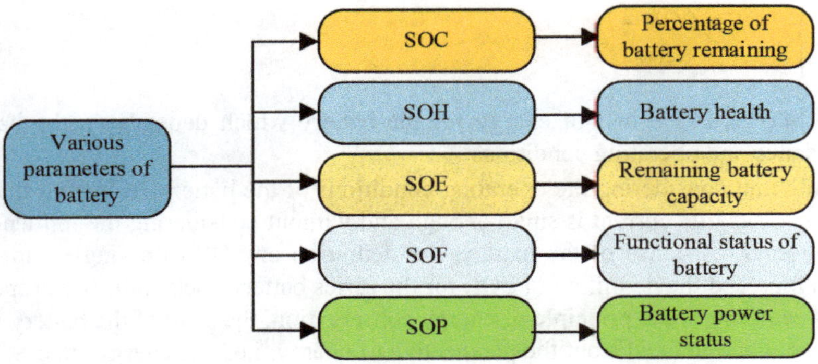

Figure 6.9 Various parameters of the battery

The abbreviations and meanings of various parameters of the battery are shown in Figure 6.9.

At present, the battery SOE estimation method is usually the application expansion of the SOC estimation algorithm [234]. The battery SOE is estimated by using the mapping relationship between SOC and SOE. However, the SOE obtained by SOC conversion is not accurate. This is because, in practical application, the power battery will cause different degrees of energy loss with the change in vehicle driving conditions, resulting in the change of the actual supply energy of the power battery.

The battery model is the basis of battery state estimation [235]. It can be divided into the electrochemical mechanism model, empirical model, and equivalent circuit model. Among them, because the equivalent circuit model has the advantages of easy analysis and application and good universality, it has been widely used.

The SOE estimation algorithm is studied based on the battery equivalent circuit model [236]. The model-based battery Stete of Energy (SOE) estimation algorithm, the driving condition identification algorithm based on information entropy theory, and the driving condition prediction method based on Markov chain theory are proposed.

The battery SOR is the ratio of the remaining energy of the battery to the carrying energy when the battery is fully charged under the condition of constant current and constant voltage charging, and the expression is shown in (6.13).

$$SOR = \frac{E_{res}}{E_N} = \frac{E_{res}}{C_N \int_0^1 U_{oc}\left(SOC\right) dSOC} \tag{6.13}$$

In (6.1), E_{res} is battery residual energy, E_N is battery full load energy, $U_{oc}\left(SOC\right)$ is battery open circuit voltage and is a function of SOC, and C_N is battery capacity.

SOE is defined as the difference between the battery SOR and the energy loss rate $H_\int$ caused by the battery charging, discharging to the charging, and discharging cut-off voltage under a certain working condition, and the expression is shown in (6.14).

$$\begin{cases} SOE = SOR - H_{int} \\ H_{int} = \frac{E_h}{E_N} \end{cases} \tag{6.14}$$

In (6.14), E_h is the loss of energy for the battery which depends on the battery impedance and operating conditions.

Without considering the operating conditions of the battery, assuming that the charge–discharge current is small enough and without considering the influence of the internal resistance of the battery, the definition of SOE of a single battery is given first, and the definition of SOE for the series battery pack is further proposed.

According to the principle of energy conservation, the state of the battery is re-expressed and defined from the perspective of energy, i.e., the energy state SOE of the battery, which is expressed by the ratio of the residual energy of the battery to the rated total energy of the battery [237]. Similar to the definition of SOC, the full energy state of the battery corresponds as shown in (6.15).

$$SOE = 1 - \frac{W_y}{W_z} \times 100\% = \frac{W_s}{W_z} \times 100\% \tag{6.15}$$

In (6.15), W_y is the energy consumed by the battery, W_s is remaining energy of battery and W_z is rated total energy of the battery.

The principle of energy conservation is that energy is transformed from one form to another, or transferred from one object to another, and the total energy remains unchanged in the whole process [238]. The total energy of the battery in the working process is certain, so the remaining energy of the battery can be obtained only by considering the change of battery energy.

It is difficult to obtain SOE directly, which can only be obtained by calculating the energy consumed by the power battery. Taking the battery discharge as an example, the environmental state of the battery and its state change in real-time during the whole process of battery energy consumption, the accuracy of battery energy calculation in the whole process cannot be guaranteed [239]. Therefore, the original definition of battery SOE is improved to calculate the SOE of the next time t+1 through the SOE of the previous time t and the energy change between the two times. In this way, the iterative calculation of SOE is carried out in turn, as shown in the formula in (6.16)

$$SOE(t+1) = SOE(t) + \frac{W_b}{W_z} \tag{6.16}$$

In (6.16), W_b is a change of power battery energy (during t and t+1, charging is positive and discharging is negative).

After the introduction of research, the definition method of SOE at this stage is mainly defined by the ratio of battery residual energy EREM to battery maximum available energy E_{max}, and E_{max} is also specified as rated energy in some literature. The specific expression for the definition of SOE in the discharge direction is given in (6.17). For the definition of SOE, the reference value is taken as the maximum available energy. Refer to the manufacturer's data manual. The specific working conditions corresponding to the reference value are a temperature of 25°C and a charge–discharge ratio of 0.5 C.

This working condition is defined as the reference working condition, and all the energy released by the battery under this working condition is defined as the reference energy. Using $E_{n(25^{\circ}C,0.5C)}$ indicates that, to make a unified analysis and comparison, this chapter puts forward the idea of SOE normalization, i.e., the maximum available energy at any temperature magnification is based on the energy released by the battery at 25°C and 0.5 C magnification. The internal resistance of lithium iron phosphate batteries is low, so the heat energy lost in the research process is ignored. At this time, the residual energy in the discharge direction can be expressed as the difference between the total energy and the released energy, which is defined as SOE. The general expression of normalization is shown in (6.17).

It can be seen that the research of SOE at this time focuses on the molecular part of the ratio, i.e., the residual energy EREM, which represents SOE. Because the specified energy reference value is unified, it can be seen from (6.17) that the key to the research goal of the energy state in this chapter is the released energy of the battery under any working condition. The energy mentioned later in this chapter refers to the energy of the battery in the discharge direction. At the same time, it is also the energy value that people care about for the practical application process. Therefore, the follow-up study of the energy state of the battery takes the energy change in the process of battery discharge as the main research goal.

$$SOE = \frac{E_{\text{rem}}}{E_{\text{max}}} = \frac{E_{\text{max}} - E_d}{E_{\text{max}}} \tag{6.17}$$

In (6.17), E_n is the total energy released by the battery and E_d is the accumulated energy released by the battery.

$$SOE = \frac{E_{max}}{E_{n(25^{\circ}C,0.5C)}} = \frac{E_{n(25^{\circ}C,0.5C)} - E_{d(T,i_{\text{rate},j})}}{E_{n(25^{\circ}C,0.5C)}} = 1 - \frac{E_{d(T,i_{\text{rate},j})}}{E_{n(25^{\circ}C,0.5C)}} \tag{6.18}$$

In (6.18), (T, i_{rate}) is working conditions at temperature T °C and different current ratios J.

To measure the energy state of the battery, the concept of SOE is introduced. Because E_R and E_P are closely related to the working condition of the battery, if the external voltage is used to determine whether it is fully charged or discharged, and the residual energy is defined, the charge and discharge energy of the battery will change with the change in working condition. It is difficult to determine the current and working model of the battery for electric vehicles in the future operation, and this will bring a great difficulty to the estimation of battery SOE. Therefore, this chapter uses the battery open-circuit voltage to reach the end voltage of charge and discharge as the judgment basis of full charge and discharge.

The maximum available energy (E_{max}) of the battery is defined as the energy released during the process of discharging the battery from the fully charged state with a small enough current to the fully discharged state. The residual energy (E_{rem}) is defined as the energy released during the process of discharging the battery from the current state with a small enough current to the fully discharged state. The SOE of the battery is defined as the percentage of E_{max} and E_{rem}.

When the charge–discharge current is small enough, U_r and U_p can be ignored. At this time, $U_O = U_{OCV}$, and the charge–discharge energy efficiency reaches the maximum value of 1 and this part is shown in (6–19).

$$\begin{cases} E_{max} = Q_{max} \cdot U_{OCV} \left| SOC = [0\%, 100\%] \right. \\ E_{rem} = Q_{max} \cdot SOC_0 \cdot U_{OCV} \left| SOC = [SOC_0, 0\%] \right. \end{cases} \tag{6.19}$$

In (6–19), Q_{max} is the maximum usable capacity of the battery, U_{OCV} is the average value of open-circuit voltage corresponding to the whole capacity interval, Q_{rem} is remaining battery capacity and SOC_0 is the current state of charge of the battery.

Besides, $U_{OCV} \left| SOC = [SOC_0, 0\%] \right.$ is the average value of open-circuit voltage from SOC_0 to discharge. So, it can be shown in (6–20).

$$SOE = SOC_0 \times \frac{U_{ocv} \left| SOC = [0\%, 100\%] \right.}{U_{ocv} \left| SOC = [SOC_0, 0\%] \right.} \tag{6.20}$$

It can be seen that for the same battery, OCV-SOC is fixed, so E_{max} and E_{rem} are only related to Q_{max} and SOC, and is decoupled from the operating conditions. Therefore, the SOE of the battery is decoupled from the operating conditions.

The initial energy state is SOE_0 and the charge state is SOC_1. When a certain energy ΔE is released, the energy state becomes SOE_1 and the charge state is SOC_1, which can be shown in (6.21).

$$SOE_1 = \frac{E_{rem_0} - \Delta E}{E_{max}} \times 100\% = SOE_0 - \frac{\Delta SOC \cdot U_{OCV} \left| SOC = [SOC_1, SOC_0] \right.}{U_{OCV} \left| SOC = [0\%, 100\%] \right.} \tag{6.21}$$

This makes the battery SOE recursive.

6.3.2 Main affecting factor analysis

In the SOE expression, W is the total internal energy of the battery, which represents the maximum storable energy of the battery in the current state. The accuracy of the W value has a great influence on the calculation accuracy of the SOE. By analyzing the working characteristics of the battery, it can be known that the ambient temperature, discharge rate, cycle life, self-discharge, etc., have an impact on the total internal energy W of the battery [240]. To make the SOE calculation result more accurate, it is necessary to determine the relationship between the influencing factors and W. The monthly self-discharge rate of lithium batteries is low (<10%) and many factors affect self-discharge. In this research, the influence of self-discharge on the estimation accuracy is ignored in the SOE calculation, and the environmental temperature, charge and discharge current and battery health status are mainly analyzed. The effect on the total energy inside the battery and the calculation of SOE are corrected.

Table 6.7 Temperature influence coefficient at different temperatures

Ambient temperature	0°C	10°C	20°C	30°C	40°C
Total discharge energy	7,209.9	7,436.44	9,077.96	9,820.07	9,442.01
Temperature influence coefficient η_T	0.734	0.757	0.924	1	0.962

1. Ambient temperature influence analysis

Through the experimental analysis of the battery temperature characteristics, the difference in ambient temperature will affect the chargeable and dischargeable energy of the battery [241]. The discharge energy of the battery is a necessary parameter in SOE estimation, so the difference in ambient temperature will inevitably bring SOE estimation error. Based on the total discharge energy at 30°C, the influence of other ambient temperatures on the total discharge energy was analyzed, and the temperature influence coefficient η_T was used to describe the effect of temperature on the discharge energy. The calculation formula is shown in (6–22).

$$\eta_T = \frac{\textit{Total discharge energy at different temperatures}}{30^{\circ}C\ \textit{total discharge energy}} \tag{6.22}$$

Through the calculation formula of the influence of temperature on the discharge energy, the temperature influence coefficients under different ambient temperatures are calculated as shown in Table 6.7, and the fitted curve is shown in Figure 6.10.

The ambient temperature is time-varying during the working process of the power battery for pure electric vehicles. The temperature and temperature influence coefficients in Table 6.7 are made into a two-dimensional table, as shown in Figure 6.10.

According to the corresponding relationship between temperature and temperature influence coefficient, the current η_T value is obtained by looking up the table according to the measured temperature when the battery is working, and the total discharge energy at this temperature is converted into the total discharge energy at 30°C, which provides a correction basis for SOE calculation.

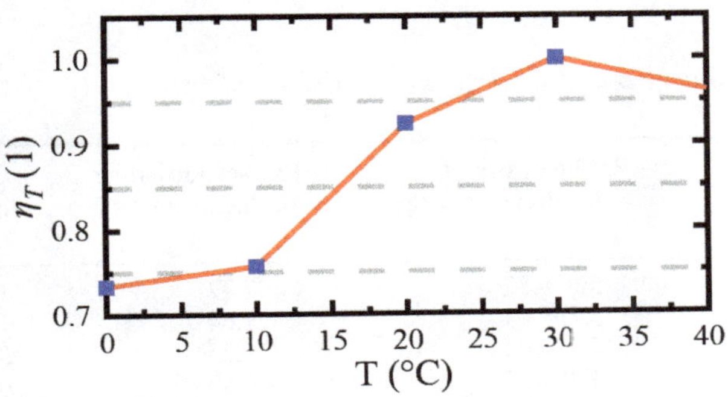

Figure 6.10 Temperature influence coefficient curve

Table 6.8 Two-stage discharge test method

	Constant current-constant voltage charging	First-stage discharge		Shelved	Second-stage discharge
Model Step	**Recharging current**	**Discharge current**	**Discharge time**	**Time**	**Discharge current**
Step 1	C/3	C/5	2.5 h	1 h	C/3
Step 2	C/3	C/2	1 h	1 h	C/3
Step 3	C/3	C	0.5 h	1 h	C/3

2. Working on current impact analysis

Through the influence of the battery rate on the charge and discharge performances, it is known that the energy that the battery can release under different discharge currents is very different, and the total energy *W* the battery can release under different working conditions will be different, which brings errors to the SOE estimation, so the SOE. The effect of the discharge current on the *W* value needs to be considered in the estimation [242]. The influence of different discharge currents on the total discharge energy is analyzed based on C/3 (110A), and the influence of the discharge current on the discharge energy is described by the current influence coefficient η_I. The calculation method is shown in (6–23).

$$\eta_T = \frac{A}{B - C} \tag{6.23}$$

In the formula, *A* is the total energy of the first-stage discharge of the battery at different rates, *B* is the total energy of discharge at 30°C and *C* is the total energy of the second-stage discharge at different rates.

Since the polarization produced by the current influences the determination of the relationship between the discharge current and the discharge energy, a two-stage discharge test method is adopted. The test scheme is shown in Table 6.8.

Table 6.9 Charge and discharge test results at different rates

Discharge rate	The first-stage discharge energy	The second-stage discharge energy	Current influence factor η_I
C/5	5,606.87	4,402.72	1.035
C/3	5,408.43	4,411.64	1
C/2	5,170.04	4,421.25	0.958
C	4,881.05	4,461.59	0.911

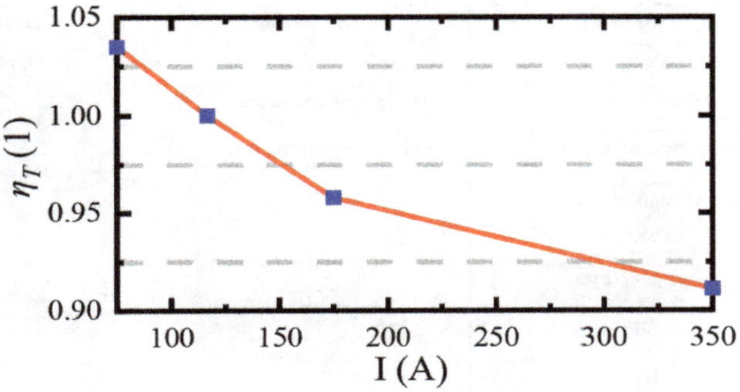

Figure 6.11 The relationship between discharge current and current influence coefficient

According to the second-order discharge experimental scheme adopted in Table 6.8, by carrying out specific experimental operations, the relevant data of the experimental process can be obtained, as shown in Table 6.9.

The power battery for pure electric vehicles often works under variable current. The current and current influence coefficients in 0 are made into a two-dimensional table, and the relationship between discharge current and current influence coefficient obtained is shown in Figure 6.11.

During the battery operation process, the current η_I value is obtained by looking up the table according to the working current, and the total discharge energy is converted to C/3. The total discharge energy can be used to provide a correction basis for SOE calculation.

3. Health status impact analysis

The total internal energy W of the battery represents the maximum releasable energy in the current state, which is an important parameter for SOE estimation. Through the analysis of the battery cycle characteristics, it can be known that the maximum releasable energy of the battery decreases with the increase of the number of cycles. To obtain this data, based on the related research on the SOH of the battery in the early stage, the adaptive voltage curve fitting method is used to obtain the SOH of the battery, and then the current most dischargeable capacity and energy are calculated.

6.3.3 Summary of AI-based SOE estimation

Owing to the consideration of nonlinear battery equilibrium potential, seeking accurate SOE estimates seems to be more complicated and challenging, in comparison with SOC estimation tasks [243]. Until now, extensive attempts have been being made to continually improve the effectiveness of SOE estimation. The main SOE estimation methods at home and abroad and their advantages and disadvantages are shown in Figure 6.12.

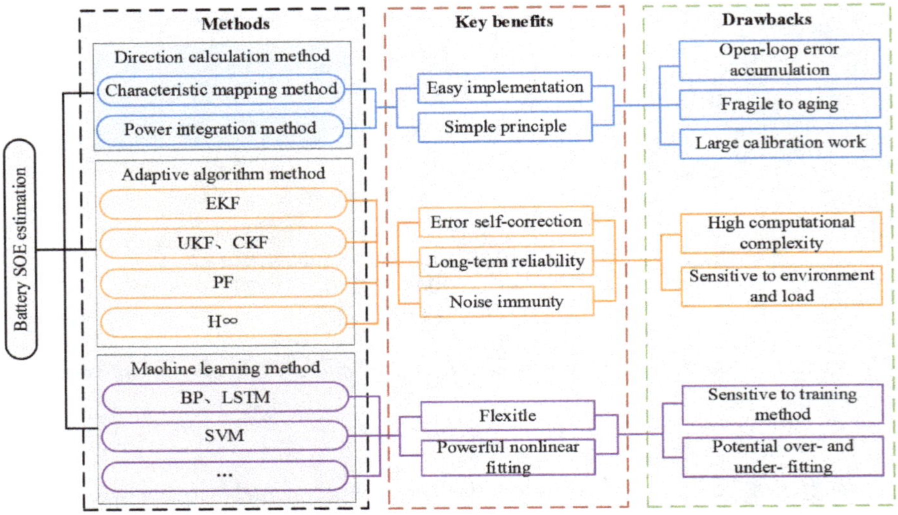

Figure 6.12 SOE estimation methods in terms of key benefits and drawbacks

As can be seen from Figure 6.12, the SOE estimation method commonly used at home and abroad can be roughly divided into three categories: power integration method, data-driven estimation method and state-space model-based method.

1. Power integration method

The power integration method is simple to calculate and easy to implement and is often used to estimate the SOE of a battery, but this method has obvious drawbacks: First, it is difficult to obtain the initial value of the battery SOE accurately. In the power integration method, an inaccurate initial value used to calculate SOE is easy but causes the accumulation of errors, resulting in the continuous increase of errors [244]. Second, this method is sensitive to the accuracy of the current sensor. The working conditions of lithium-ion batteries are complex, and the sensor is often affected by uncontrollable factors such as external temperature and noise, resulting in accumulated errors in the integration process. Finally, the algorithm has high requirements on the initial battery energy accuracy of the battery, and the aging of the battery will lead to energy attenuation, which will reduce the reference energy, resulting in a decrease in the estimation accuracy of SOE. The expression of the power integration method is shown in (6.24).

$$\begin{cases} SOE(t) = SOE(t_0) + \dfrac{1}{En}\int_{t_0}^{t} uIdt \\ SOE(t) = SOE(t_0) + \dfrac{uI\Delta t}{En} \end{cases} \tag{6.24}$$

(6–24)Equation (6.24) shows the continuous and discrete forms of the power integral expression, where *SOE*(*t*) is the energy state value at the current moment, and

$SOE(t_0)$ is the initial energy state value. *En* is the rated energy of the battery. *I* represents the charging and discharging current of the battery under actual operating conditions. Because of the shortcomings of the power integration method, scholars at home and abroad have proposed a series of improvements, such as solving the SOE value by constructing the mapping relationship between the discharge power, UOC, residual energy and SOE. Compared with the power integration method, this kind of method has a certain improvement but still needs a lot of calibration and characteristic testing.

2. Data-driven SOE estimation method

With the rise in machine learning techniques, data-driven methods are widely used in battery SOE estimation. For example, a back-propagation neural network (NN) is used to capture the nonlinear and coupled properties of batteries considering irreversible energy losses due to Joule heating, electrochemical reactions and phase shifts. Similarly, considering the effects of temperature and discharge rate, a wavelet NN model was also applied to simulate battery electrodynamics. In addition, algorithms such as long short-term memory networks and SVMs are also used for SOE estimation. The estimation effect of this type of method depends to a large extent on the quantity and quality of training data and the choice of fitting algorithm, which usually requires a large amount of calculation and requires high sensor accuracy.

3. Data-driven SOE estimation method

To overcome the shortcomings of the power integration method, scholars at home and abroad have proposed many SOE estimation methods based on the state-space model. This kind of method first establishes the equivalent model of the battery, such as the Thevenin model, the second-order RC model, the composite equivalent model, etc., and then selects the appropriate observer or filter for SOE estimation, such as EKF, UKF, CKF, PF, H_∞ and other algorithms and corresponding optimization improvements.

1. Extended Kalman filter algorithm

The EKF algorithm is an improvement from the traditional KF algorithm. By expanding the linear system Taylor and ignoring higher-order terms, the algorithm can be applied to nonlinear systems such as lithium-ion batteries. The algorithm iterates the state vector at the current moment according to the estimated value of the state vector at the previous moment and the output observation value at the next moment. At the same time, the error covariance is updated, and the state vector is gradually corrected by the coefficient gain so that the result is close to the true value. The structure of the EKF algorithm is simple and easy to understand, and it is one of the most widely used SOE estimation methods. Taking the second-order RC model shown in Figure 6.13 as an example, taking SOE, U_1, and U_2 as state variables, and the terminal voltage U_L as the observation variable, the state space expression of the system is shown in (6.25).

In Figure 6.13, U_{OC} represents the open-circuit voltage of the battery, U_L indicates the terminal voltage of the battery, R_0 is the ohmic internal resistance of the battery, which is used to characterize the instantaneous voltage drop caused by the

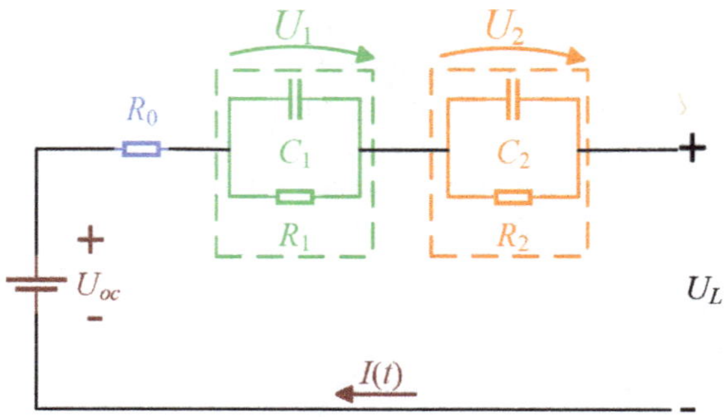

Figure 6.13 Second-order RC equivalent circuit model

battery current, R_1 and R_2 are polarization resistors, and C_1 and C_2 are polarization capacitors.

$$\begin{cases} \begin{bmatrix} SOE_k \\ U_{1,k} \\ U_{2,k} \end{bmatrix} = \begin{bmatrix} 1 & 0 & 0 \\ 0 & e^{\frac{\Delta t}{\tau_1}} & 0 \\ 0 & 0 & e^{\frac{\Delta t}{\tau_2}} \end{bmatrix} \begin{bmatrix} SOE_{k-1} \\ U_{1,k-1} \\ U_{2,k-1} \end{bmatrix} + \begin{bmatrix} \frac{-\eta U_{L,k-1}\Delta t}{E_N} \\ R_1\left(1 - e^{\frac{\Delta t}{\tau_1}}\right) \\ R_2\left(1 - e^{\frac{\Delta t}{\tau_2}}\right) \end{bmatrix} I_{k-1} \\ U_{L,k+1} = g\left(SOE_k\right) - U_{1,k} - U_{2,k} - I\left(k\right) R_0 + v_k \end{cases} \tag{6.25}$$

In (6.25), η is the Coulomb efficiency, which is generally 1, τ is the time constant, w_k is the process noise, v_k is the measurement noise, U_L represents the terminal voltage of the battery equivalent model and $g(SOE_k)$ is the functional relationship between the open-circuit voltage U_{OC} and SOE. The SOE estimation process based on the EKF algorithm can be divided into two parts: the prediction stage and the correction stage.

a. The expression of the prediction stage is shown in (6.26).

$$\begin{cases} \hat{X}_{\bar{k}} = A\hat{X}_{k-1} + BI_{L,k-1} + w_k \\ \hat{P}_{\bar{k}} = A\hat{P}_{k-1}A^T + Q_w \end{cases} \tag{6.26}$$

In (6.27), A is the state transition matrix, B is the input matrix and P is the error prior covariance.

b. The expression of the correction stage is shown in (6.27).

$$\begin{cases} K_k = \hat{P}_k^- C_k^T (C_k \hat{P}_k^- C_k^T + R_k)^{-1} \\ \hat{X}_k = \hat{X}_k^- + K_k[Z_k - h(X_k^-)] \\ \hat{P}_k = (1 - K_k C_k)\hat{P}_k^- \end{cases} \tag{6.27}$$

The correction stage includes Kalman gain calculation, state variable update and error covariance matrix update. In (6.27), $Z_k - h(*)$ represents the innovation at the current moment, the difference between the actual terminal voltage and the analog terminal voltage in the SOE estimation process, I represents the unit matrix, and C represents the system output matrix.

The SOE estimation process based on the EKF algorithm is calculated in a recursive way of continuous "prediction-correction." First, the predicted value is calculated, and then the predicted value is revised according to the innovation and Kalman gain obtained from the observed value. Prediction can be obtained from the filtering value, and filtering can be obtained from the prediction [245]. The filtering and prediction interact, do not require any observation data to be stored and can be processed in real time.

The EKF algorithm is less complex but has significant disadvantages. When the EKF algorithm is used in a nonlinear system such as a lithium-ion battery, the nonlinear system needs to be linearized. The specific implementation method is to perform Taylor expansion of the nonlinear equation, discard the second-order and above terms, and then use the linear Kalman method. The filter recursion equation is calculated. However, when the high-order terms of the Taylor expansion of the nonlinear equation contain a large amount of system information, blindly discarding the high-order terms will introduce a large linearization error, which will reduce the accuracy of the algorithm to estimate the SOC, and even lead to filter divergence.

2. Volumetric Kalman filtering algorithm

The CKF algorithm is based on the third-order spherical radial volume criterion and uses a set of volume points to approximate the state mean and covariance of nonlinear systems with additional Gaussian noise. This method is theoretically the closest approximation algorithm to Bayesian filtering and can effectively solve the problem of nonlinear system state estimation. The algorithm does not need to ignore high-order terms through Taylor expansion and linearizes the nonlinear system, and theoretically, the SOE estimation accuracy is higher than that of the EKF algorithm.

As a deterministic sampling filtering algorithm, CKF uses the mean and covariance of the prior probability density distribution of the system state to generate point sets according to a certain sampling strategy when dealing with nonlinear equations. Then, each sampling point in the point set is directly propagated nonlinearly, and finally, the mean and covariance of the posterior probability density distribution of the system state are calculated by weighted summation. The whole process does not need to linearize the nonlinear equation, which eliminates the linearization error, and does not need to calculate the Jacobian matrix during the iteration process of the filtering algorithm, which is easier to use in practice.

Compared with the UKF algorithm, although both UKF and CKF do not need to linearize the nonlinear equation, the linearization error is eliminated, and the accuracy is greatly improved than that of the EKF. However, the principle of avoiding linearization is different between UKF and CKF. UKF transforms the linearization method of omitting higher-order terms from Taylor's expansion of nonlinear equations into an approximation of the eigenvalues of probability and statistics of nonlinear equations. There is no strict mathematical theory as the basis. The CKF is based on the numerical integration theory, using the third-order spherical radial volume criterion to approximate the Gaussian integral, and has a strict and complete theory as the basis. Therefore, from the perspective of a theoretical basis, CKF is more rigorous and stable than UKF. In addition, for the same sampling filtering algorithm, UKF and CKF need to generate a point set through a certain sampling method in each iteration process. The point set size of UKF is $2n+1$, while the volume point set of CKF is only $2n$, i.e., each iteration is one sampling point less than UKF, so theoretically, the CKF algorithm has higher execution efficiency and better real-time performance.

In summary, CKF is superior to EKF and UKF inapplicability, theoretical basis, time complexity, and reliability, and is very suitable for SOE estimation of lithium-ion batteries.

3. Particle filtering algorithm

The particle filter obtains the posterior probability density distribution of the state vector of the system, approximates the probability density function by finding a set of random samples propagating in the state space, and replaces the integral operation with the sample mean. After the state update and time update, the weight and position of the particles are continuously adjusted according to the system observation, and the initial conditional distribution is corrected through the adjusted particle information. When the number of particles is large enough, the probability density distribution function of the corrected particles gradually approximates the probability density function of the state, and the mean value of the particles approximates the estimated value of the state vector.

In the prediction of the remaining life of the battery cell, due to the uncertainty of the remaining life prediction, the particle filter method can reduce the influence of the uncertainty on the estimation error. And in the real-time system, the requirements for the system model of the particle filter are lower than that of the Kalman filter. The particle filter does not require that the model used must be a linear model and has a wide range of applications in solving nonlinear problems.

In terms of SOC and SOE estimation, compared with other Bayesian filters, the main advantage of particle filter-based techniques is that even if there is uncertainty in non-Gaussian measurements, the approximate probability density function can be obtained through the obtained approximate probability density function estimated value. And in the real-time system, the requirements for the system model of the particle filter are lower than that of the Kalman filter. The particle filter does not require that the model used must be a linear model and has a wide range of applications in solving nonlinear problems. However, the PF algorithm requires a large amount of calculation and is prone to particle degradation, so it needs to be further optimized when estimating the state of lithium batteries.

4. Multistate collaborative estimation

Lithium-ion battery states are coupled and interacted with each other. One state quantity is estimated separately and other states are ignored. Satisfactory results can be obtained only when certain constraints are met. Considering the multiple coupling of the electrothermal aging mechanical conditions inside the battery, the multistate joint estimation and prediction can obtain more accurate results. In addition, the BMS monitors various states of the battery, such as SOC, SOE, SOH, etc., so the multistate collaborative estimation is more in line with the actual needs.

At present, domestic and foreign researchers have done a lot of research on multistate joint estimation, and the collaborative estimation of SOC and SOH is dominant in the existing literature. This is mainly because with the continuous use of the battery, the battery capacity and internal resistance will change, and the rated capacity and internal resistance of the battery have a great influence on the SOC, and the regular update of the SOH (capacity or resistance) can improve the accuracy of SOC estimation. Sexuality plays a vital role.

In addition to the joint estimation of SOC and SOH, some researchers also deal with other two-state estimations, such as joint estimation of SOC and SOP, joint estimation of SOC and SOT, and joint estimation of SOE and SOP, etc. Compared to estimation with only one state application, multistate estimation scenarios will require more computational effort, so designing simpler and more accurate equivalent models and more advanced state estimation methods such as multi-timescale estimators will be future research's important direction.

6.4 AI-based battery remaining useful life prediction

6.4.1 Definition of remaining useful life

Lithium-ion batteries have the advantages of high specific energy, low self-discharge rate, long life and environmental friendliness. They are widely used in special robots, aerospace and new energy vehicles [246]. The performance of lithium-ion batteries will continue to degrade with cyclic charge and discharge. If they cannot be replaced in time, the performance of electrical equipment may be degraded or even safety accidents may occur [247]. Therefore, how to improve the safety and reliability of lithium-ion batteries has become an urgent problem to be solved in engineering applications.

Remaining useful life (RUL) prediction of lithium-ion batteries is one of the important technologies for establishing a high-performance BMS, and it can timely remind users to maintain the battery, and improve the efficiency and safety performance of the battery [248].

RUL represents the length of time from the current time to the end of life (EOL). It is generally considered that when the available capacity of the battery drops to 70% or 80% of the rated capacity, the battery life reaches the end, and this capacity value is called the battery life failure threshold [249]. Lithium-ion battery life usually has four definitions, namely, calendar life, storage shelf life, standard cycle life and operating cycle life, respectively. Calendar life is the whole process of battery from production to scrapping, including all environments such as shelving, testing and use [250]. Storage shelf life refers to the time it takes for the battery capacity to drop to the capacity failure threshold under shelving conditions, and the entire process is affected by environmental factors. The standard cycle life refers to the time taken for the battery capacity to drop to the capacity failure threshold under specific charge and discharge conditions [251]. This process is suitable for laboratories and easy to analyze. The cycle life under working conditions refers to the cycle test of the lithium-ion battery according to the actual working conditions, i.e., the service life of the lithium-ion battery under the actual working conditions.

Lithium-ion battery RUL prediction methods are generally classified into three categories: model-based methods, data-driven methods and fusion-based methods, as shown in Figure 6.14.

1. Model-based method

The model-based method establishes the SOH degradation model of the lithium-ion battery by analyzing the internal mechanism of the lithium-ion battery [252], which can be divided into three categories: electrochemical model, equivalent circuit model and empirical model.

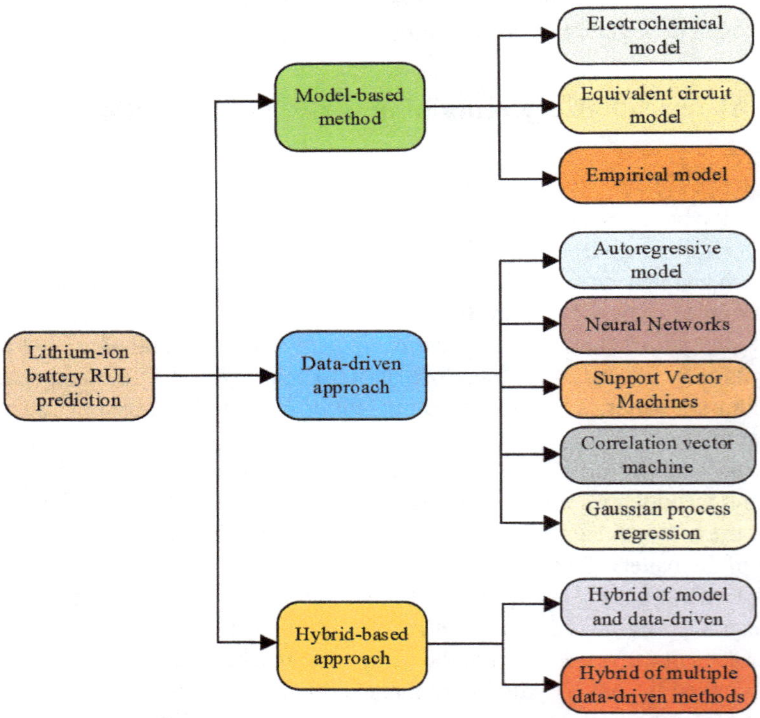

Figure 6.14 Lithium-ion battery RUL prediction method

The electrochemical model establishes the degradation model by analyzing the electrochemical properties of the lithium-ion battery and realizes the RUL prediction by analyzing the operation mechanism and electrochemical reaction of the lithium-ion battery [4]. This method can give the actual physical and chemical meaning of the model, but it is difficult to predict online due to factors such as the environment and equipment accuracy. Meanwhile, due to the different physical and chemical properties of different batteries, the robustness of the model is poor; for different batteries, it needs to be remodeled and the workload is larger.

The equivalent circuit model approximates the dynamic characteristics of a lithium-ion battery by analyzing a large amount of data and converting a complex lithium-ion battery into a simplified circuit model [253]. At present, the commonly used equivalent circuit models include the Rint model, RC model, Thevenin model and The Partnership for a New Generation of Vehicles (PNGV) model. Among them, the RC model is widely used because of its simple structure and high accuracy. Considering the battery polarization effect, hysteresis voltage characteristics and dynamically adjusting the order of the RC model in the traditional RC model can improve the model's ability to predict the dynamic characteristics [254]. The RUL prediction method based on the equivalent circuit model is simpler than the electrochemical model.

However, the equivalent circuit model is a simplified lithium-ion battery model and cannot fully reflect the electrochemical reaction characteristics inside the lithium-ion batteries; therefore, it cannot fully reflect the dynamic characteristics of lithium-ion batteries.

The empirical model is based on the data of the lithium-ion battery aging test, and various forms of polynomial models are obtained by fitting, which has a wider scope of application than the electrochemical model and the equivalent circuit model [255]. The RUL prediction method based on the empirical model constructs the degradation model by fitting the historical degradation data of the lithium-ion battery with the empirical model, uses the filtering method to update the model parameters and finally realizes the RUL prediction of the battery.

Commonly used empirical models include impedance linear variable parameter model, impedance exponential growth model, exponential capacity decay model, and battery capacity regeneration empirical model [256]. This method mainly realizes the prediction of battery RUL by combining it with a filtering algorithm. The main filtering methods are Kalman filter (KF), particle filter (PF) and their improved algorithms. Since the empirical model ignores the complex working mechanism inside the battery, this method is easier to implement and has wider applicability than other models.

2. Data-driven method

Compared with the model-based method, the data-driven method does not need to analyze the internal characteristics of the lithium-ion battery, and it obtains the implicit relationship between the input and output through a large amount of experimental data. This method does not need to establish a white-box model and directly obtains the prediction model from the data through machine learning and statistical analysis methods, but this method is too dependent on data, not only needs a large amount of data to support but also needs to ensure that the data used have sufficient accuracy. The accuracy of the data directly affects the accuracy of the method to predict the battery RUL [257]. Data-driven methods include the AR model, artificial neural networks (ANN), SVM, relevance vector machine (RVM) and Gaussian process regression (GPR).

The AR model is a time series model, which predicts the state of the system at a certain time in the future based on the linear combination of the observed values at several times in the past [258]. It has the advantages of simple model parameters and a small amount of calculation, but the AR model is a linear prediction model, which is less accurate in long-term RUL prediction, the robustness is not high and it is difficult to achieve long-term RUL prediction.

The ANN is based on the knowledge of network topology. Through the abstraction of the structure and operation mechanism of the human brain, it simplifies and simulates the processing of complex data information. It has a strong self-learning ability and can handle complex nonlinear problems well. It has been widely used in the field of RUL prediction of batteries. This method generally consists of an input layer, hidden layer and output layer, and is a typical operation model based on a nonlinear method. At present, extreme learning machine (ELM), LSTM, Elman neural networks and deep neural networks (DNN) are widely used in RUL prediction [259]. However, the initial weights and hidden layer thresholds of ANN in the prediction process are randomly generated, which is easy to cause the problem of local minimum. To solve the above problems, the genetic algorithm (GA), particle swarm optimization (PSO) algorithm, simulate anneal (SA) algorithm and other optimization algorithms are widely used in the parameter optimization of ANN and have achieved good results.

The method based on ANN can accurately predict the battery RUL, but it relies too much on experimental data and requires a large amount of data [260]. Accurate experimental data are required to ensure the algorithm has a high prediction accuracy, and compared with the filtering algorithm, the estimation interval of ANN cannot be achieved.

SVM is a machine learning method that can solve classification problems and nonlinear problems. SVM has low sensitivity to singular values, which helps to eliminate useless data, reduce the amount of calculation and enhance the robustness of the algorithm. Based on these advantages, SVM has been widely used in battery RUL prediction in recent years [261]. However, the problem that SVM hyperparameters are difficult to determine always exists and seriously affects its prediction performance. Although the optimization algorithm has improved this problem to a certain extent, the problems of the optimization algorithm itself and the high complexity of the improved optimization algorithm all affect the prediction effect of the RUL prediction method based on how the optimized SVM is presented. In addition, SVM also has the problem that the kernel function must satisfy the Mercer condition and the sparsity is limited.

RVM is a learning algorithm developed based on the sparse Bayesian theory of probability learning, which was first proposed by Tipping. Compared with SVM, RVM has the advantages of small amount of calculation, the kernel function does not need to meet the Mercer condition and can give the probability distribution of the prediction results. It is a machine learning method that is very suitable for the RUL prediction of lithium-ion batteries (LIBs) [262]. Compared with SVM, RVM requires less computation, but its correlation vector is also sparser. At the same time, the actual degradation data of LIBs have large fluctuations, which leads to the low stability of the prediction results of the RUL prediction method based on RVM. In addition, the problem of poor prediction performance of the RVM algorithm in the long-term RUL prediction of LIBs has always existed.

GPR is a nonlinear probabilistic regression method based on statistical theory and the Bayesian framework. It provides confidence intervals for the prediction results by calculating covariance and means, and has good results in dealing with complex problems such as high-dimensional, nonlinear and small samples. Compared with ANN and SVM, GPR has the advantages of easy implementation, flexible nonparametric inference and probability distribution of output. However, the calculation process of the GPR model involves the inversion of the matrix, resulting in a relatively large amount of calculation for the GPR-based RUL prediction method. In addition, the online update ability of the GPR model is weak, and these defects make it difficult to apply the GPR model online.

3. Fusion-based method

The fusion-based method integrates a variety of prediction methods, which makes up for the shortcomings of a single model prediction, and achieves higher prediction accuracy by giving full play to the advantages of different methods, which has become a current research hotspot.

The RUL prediction method based on the model- and data-driven fusion combines the advantages of high accuracy of the model-based RUL prediction method and strong robustness of the data-driven method and has higher prediction accuracy [263]. For example, in response to the low long-term prediction accuracy of RVM, some scholars have proposed an iteratively updated RUL prediction method, which improves the accuracy of multistep RUL prediction. In addition, some scholars have proposed an RUL prediction method that fuses the double exponential empirical model and RVM. The experimental results show that the prediction performance of the proposed method is better than the basic RVM and BP neural network.

The RUL prediction method based on the fusion of multiple data-driven methods can improve the robustness of the algorithm and obtain higher prediction accuracy. Fusion-based methods can greatly improve prediction accuracy and generalization ability [264]. However, with the increase in fusion algorithms, the complexity of the algorithm will increase sharply. In addition, the fusion algorithm has many parameters and the difficulty of parameter identification is also an urgent problem to be solved.

6.4.2 *Main affecting factor analysis*

Lithium-ion batteries play an indispensable role, from portable electronic devices to electric vehicles and home storage systems [265]. Even though they are characterized by superior performance to most other storage technologies, their lifetime is not unlimited and has to be predicted to ensure the economic viability of the battery application. Furthermore, to ensure the optimal battery system operation, the RUL prediction has become an essential feature of modern BMSs [266]. Thus, the prediction of RUL of lithium-ion batteries has become a hot topic for both industry and academia.

1. Analysis of battery aging mechanism

With the development of the electrification era, the vigorous advancement of new energy vehicles, and the Internet of things, the importance of energy storage system performance has become prominent. Lithium-ion batteries stand out among various energy storage solutions due to their high energy density, high power capability and low self-discharge rate [267]. At the same time, this also puts forward higher requirements and challenges for the development of battery management technology. A comprehensive BMS should include the following functions: battery data collection, battery status determination and prediction, charge and discharge control, safety protection, thermal management, balance control and communication.

LIBs have attracted increasing interest and are bound to assume a predominant role in the various electrochemical energy storage systems. Preferred fields of application are the ones requiring storage systems with high energy density, such as the automotive and portable-electronics sectors.

However, the lifetime of everything is limited, and LIBs are no exception. The price and aging of lithium-ion batteries are the two main factors that hinder their acceptance in a wider range of applications. The performance of LIBs will decrease with calendar aging and cycle aging, due to various aging phenomena. Battery aging will increase operating costs, reduce the equipment service life, and affect the safe operation of equipment [268]. Generally, when the capacity drops to 80% of the initial value, the battery reaches the end of its service life. The RUL is defined as the amount of operation time in a certain application until the battery reaches the predefined end-of-life criterion; it represents the period from the present observation to the EOL.

The aging mechanism of the battery is affected by the battery material and the internal chemical reaction during charging and discharging. Battery aging is a complex process of change and presents a highly nonlinear trend [269]. In actual operation, factors such as operating temperature, charge/discharge rate and environmental stability will affect the aging process, thereby affecting the battery life.

While having excellent performances and cycling life, LIBs are affected by severe safety issues, mostly related to the high flammability of the organic electrolyte. One of the major issues regards the loss in performance at low temperatures, which is also related to the deposition (plating) of metallic lithium during charging, especially if the charge is performed at high rates. Lithium plating results in a loss of capacity and the growth of lithium dendrites. These can, in turn, pierce the separator and cause short circuits [270]. At low temperatures, the electrolyte's ionic conductivity and the lithium-ions diffusion coefficient in graphite decrease. Furthermore, the interface resistance between electrode and electrolyte (charge transfer and SEI) increases. The general worsening of the kinetic and diffusion parameters induces an increase in the cell polarization, with the consequent risk of the anode to slip into the plating potential region. A detailed understanding of the dynamics in battery materials is therefore required, which includes the determination of parameters such as the charge transfer resistances and the lithium diffusion coefficients in the electrode materials. While charge transfer resistances are usually obtained by electrical impedance spectroscopy (EIS), the diffusion coefficients can be estimated by a series of electroanalytical techniques, such as cyclic voltammetry (CV), EIS or by incremental titration techniques, i.e., the potentiostat (PITT) and the galvanostatic (GITT) intermittent titration techniques. Due to its higher accuracy, GITT has become one of the most widespread techniques for this purpose. Nonetheless, the results are still affected by severe scattering due, especially, by an incorrect estimation of the diffusion area and length.

GITT is becoming increasingly interesting also for the characterization of full cell performances and the parametrization before full cell modeling. However, to determine the diffusion coefficients of single electrodes, a three-electrode configuration is needed. Also in a half-cell configuration, the use of a reference electrode is recommended, to avoid the interference of the counter electrode. As far as the technique is used to characterize custom electrodes, this is not a problem. However, if one wants to determine the diffusion coefficients in a commercial cell, this may become a serious technical hurdle [271]. Indeed, the usual procedure implies disassembling the cell, cutting a sample of the electrode of interest and re-mounting it in a three-electrode laboratory cell. Alternatively, one could insert a reference electrode in the commercial cell. Both operations require working under an inert atmosphere (glove box) or, at least, under the exclusion of humidity (dry room), let alone having skilled personnel being able to perform it.

For health considerations, we need to constrain the difference between and throughout a charging process. The potential difference, arising when the charging begins, is tied to health detriment [272]. The larger it is, the more potential harm charging will bring. This conclusion is due to the resemblance that bears to the gradient of the concentration of ions within an electrode. Accordingly, there is an approximate mathematical equivalence between the RC model and the SPM, and the potential difference approximates the lithium-ion concentration gradient in the SPM. Created during charging, the concentration gradient drives the diffusion of ions. However, too large a gradient value will cause internal stress build-up, heating, SEI formation and other negative side effects. Mechanical degradation in the electrode and capacity fade can happen or accelerate when the magnitude of the gradient is too large. Thus, the uneven ion concentration should be suppressed during charging [273]. In addition, this restriction should be implemented more strictly as the SoC increases, when the battery becomes more vulnerable to the adverse effects caused by a large concentration difference. It has been noticed that the pulse charging was indeed intended to mitigate the build-up of the concentration gradient because the short rest period of a pulse cycle will allow for the ions to diffuse and distribute more evenly throughout the battery electrodes.

The SEI is a passivation layer on most NE surfaces, having the properties of a solid electrolyte and formed when the liquid electrolyte comes into contact with the electron-conductive surface of the NE [274]. This is usually operating at voltages below the electrochemical stability window of the electrolyte, accelerating redox processes irreversibly breaking down the electrolyte, leading to electrolyte loss. Li metal electrodes develop this SEI layer as well as graphite. A variety of compounds have been observed within the SEI, e.g., lithium fluoride (LiF), lithium carbonate (Li_2CO_3), lithium methyl carbonate ($LiOCO_2CH_3$), lithium ethylene dicarbonate ($LiOCO_2CH_2$) and lithium oxide (Li_2O). The SEI layer forms initially on the first cycle of the cell, resulting in ca. 10% reduction in capacity but then serves to stop the further reaction of the electrolyte at the NE. However, the thickness of the SEI layer increases (predominantly on the graphite NE) as the cell ages. The growth could be due to various reasons, including diffusion of solvent molecules through existing SEI, newly exposed electrode surfaces which result from cracking and deposition of side reaction products, such as plated Li and TM ions dissolved from the PE, which react with the electrolyte to form SEI. The SEI growth rate approximately correlates with the square root of time; as the SEI thickness increases, the rate of solvent molecule diffusion slows down.

The SEI begins to form as soon as the NE is lithiated and exposed to the electrolyte and will grow even if the battery is not then used. However, high temperatures increase diffusion rates and hence also the SEI growth rate. High currents also lead to particle cracking and new SEI formation. Under normal conditions, LTO anodes do not form an SEI layer, due to LTO being within the stability limit of most organic electrolytes; however, an SEI can form at potentials below 1 V. Capacity is irreversibly lost due to otherwise cyclable lithium being trapped within the SEI. In addition, the SEI layer is less permeable to Li + ions than the electrolyte, restricts electrolyte flow through pore blocking and consumes the electrolyte solvent [275]. All of these effects increase the overall impedance of cells, leading to power fade. While the SEI layer itself is not thought to cause catastrophic failure, at elevated temperatures, it can decompose and contribute to a thermal runaway event. SEI growth consumes the electrolyte solvents, reducing both the amount and the conductivity of the electrolyte.

Li plating is a side reaction where metallic Li forms on the surface of the NE instead of intercalating into it. This can be caused by the NE surface becoming fully lithiated, in which the Li has nowhere else to go (thermodynamic plating), or by fast charging, where the high electrolyte potential increases the rate of the side reaction relative to the main intercalation reaction (kinetic plating). Even at moderate charge rates, below-freezing temperatures slow down the main intercalation reaction enough to cause plating. As with any electroplating reaction, the Li metal can be recovered through the inverse reaction, known as stripping. Low temperatures, high SoC, high (charge) current, high cell voltage and insufficient NE mass or electrochemically active surface area can all cause lithium plating. It is standard practice to put 10–20% spare capacity in the NE to prevent overcharge. Local overcharge can also occur at the edges of the electrode, so spare surface area or "overhang" is included to minimize this. However, these mitigating strategies are not enough to stop plating during fast charging, or at below-freezing temperatures [276]. Local defects in the separator or NE are also known to cause plating; these can arise from either manufacturing or during use. Calendar aging is slow at low temperatures, implying that Li plating does not occur when the battery is at equilibrium. However, rest periods immediately after fast charging will favor the reaction of the plated lithium with the electrolyte rather than its removal by stripping.

The plated metallic Li quickly undergoes further side reactions with the electrolyte to form SEI. This SEI growth can electrically isolate the remaining Li, forming "dead lithium" that cannot be recovered. Li plating, therefore, has both reversible and irreversible components. Both the additional SEI growth and dead lithium manifest as loss of lithium inventory and reduced conductivity through pore-clogging. The plating of metallic lithium can lead to dendrite growth, which can puncture the separator and cause an internal short circuit.

The accuracy of state estimation is an important criterion for evaluating the performance of BMS [277]. A high-performance BMS can make the energy storage system operate reliably and extend the battery lifetime. The optimized BMS should provide multitask processing capabilities, which can make various tasks work together. At the same time, a real-time operating system is introduced to monitor system parameters and status in real time so that the system can be adjusted in time.

Therefore, how to accurately predict the RUL under complex working conditions is an important issue to be solved urgently. In the context of industry, RUL predicts that it can reduce investment costs and improve profitability [278]. In the academic world, improving the prediction accuracy of RUL can make predictive adjustments to the use of the energy storage system in advance, so that the system can operate safely and stably, and extend the service life of the battery.

RUL prediction methods can generally be divided into three categories: model-based methods, data-driven methods and hybrid methods. Among them, model-based methods usually include physical models, electrochemical models and so on. The electrochemical model uses detailed mathematical models to accurately express the internal chemical reaction process of the battery [279]. Therefore, these methods will also bring about problems such as high complexity and high calculation cost while achieving high estimation accuracy. In addition, to complete the parameterization process of the electrochemical model, it is usually necessary to disassemble the battery, which will bring great inconvenience in the actual application process. A lithium-ion battery is a highly complex electrochemical power system. When performing RUL prediction for lithium-ion batteries, model-based methods are usually complicated to implement, while data-driven methods are more suitable for estimating the RUL of lithium batteries, especially in the case of a large number of historical data applications. Therefore, data-driven prediction methods have attracted much attention.

Among the data-driven methods, ML integrates human knowledge into a machine to achieve previously unavailable functions and performance, and it promotes the interaction between humans and ML systems, making ML decisions understandable to humans [280]. When ML technology is applied to predict the RUL of Lithium-ion batteries, it also has the potential for high prediction accuracy and high calculation efficiency.

1. Machine learning for RUL prediction
2. SVM

SVMs have received widespread attention due to their strong advantages in processing small training data sets. SVM is a kernel-based nonparametric ML technology. When the size of the training data set increases, the number of support vectors will increase accordingly. In a complex system, this method can be modeled according to the characteristics of the system and can provide sufficient data support, so it has the characteristics of high flexibility.

SVM uses two parallel hyperplanes to classify linearly separable data sets. Equation (6.28) is the decision boundary between two parallel boundaries, where w is the weight and b is the deviation parameter vector. The distance between the decision boundary and each hyperplane is suitable for the standardized data set. SVM introduces the hinge loss function based on the hyperplane to reduce the classification error on the linear inseparable data set.

$$w^T x + b = 0 \tag{6.28}$$

$$y(x) = \frac{\lambda}{N} \sum_{n=1}^{N} w^T k(x_i, x_j) + b \tag{6.29}$$

$$w^T \varphi(x) + b = 0 \tag{6.30}$$

$$k(x_i, x_j) = \varphi(x_i)^T \varphi(x_j) \tag{6.31}$$

The problem becomes the function $y(x)$ in the minimization where n represents the number of samples and λ represents the regularization parameters. Specifically, SVM can also use the kernel function to transform the input low-dimensional vector into a high-dimensional feature space, and then use the hyperplane to separate the data, thereby applying the kernel method (the decision-making boundary element has (6.30) and (6.31) as shown above, where φ is the mapping function). The structure of the SVM algorithm is shown in Figure 6.15.

When SVM is used as a continuous value regression tool, it is called support vector regression (SVR). SVR needs to find a hyperplane, similar to SVM. The difference is that SVM needs to find a hyperplane with the largest gap. In SVR, a threshold ε needs to be defined, and only the loss of the data points in the strip area as shown in (6.32) is calculated, and then the points outside the area are regressed.

$$|y(x) - y| > \varepsilon \tag{6.32}$$

Regression is achieved by searching for the smallest marginal fit. SVR is one of the most commonly used regression methods at this stage. In the regression model, the convexity solution of SVR can be obtained by constructing the Lagrangian loss function. At the same time, the mapping function can be used to convert a low-dimensional nonlinear input space into a high-dimensional linear feature space and a nonlinear regression problem into a linear problem. The SVR algorithm proposes a nonparametric regression method, which can be updated through model retraining. Because of its ability to describe the nonlinear correlation of input and output data, SVR is suitable for health diagnosis and prediction tasks.

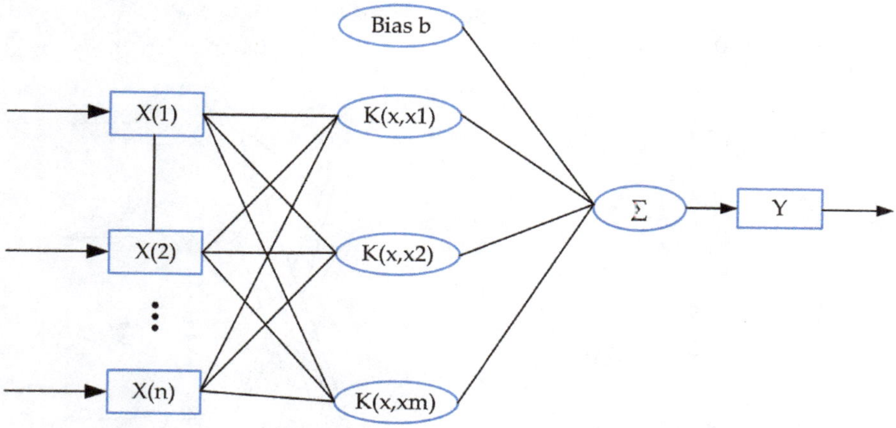

Figure 6.15 The structure of the support vector machine algorithm

When the SVM method performs RUL prediction, the activation function is usually selected as the radial-based kernel, the training algorithm is the logistic regression function margin and the hyperparameter adjustment methods are the regularization factor, the SVM type regression and the kernel parameter.

2. Gaussian process regression

The prediction part is added to the prior knowledge based on the Bayesian framework. Such a kernel-based ML method is GPR [281]. GPR uses the average forecast variance to describe the associated uncertainty. The structure of the GPR algorithm is shown in Figure 6.2. The GPR model is flexible, nonparametric and probabilistic. It can be updated through online retraining and has been widely used in the prognostic analysis. GPR was used in the initial stage to predict the decay trend of battery internal resistance, and the decay of battery internal resistance is the main factor in the decrease of battery capacity. Therefore, GPR is gradually being applied to the prediction of battery RUL. The attenuation of capacity is a very complicated nonlinear process. It is affected by many uncertain environmental factors and working conditions. The improper operation will also lead to the rapid attenuation of capacity [282]. Thus, a single covariance function will lead to unreliable predictions for nonlinear mappings with multidimensional input variables. Therefore, an anisotropic kernel with a high-level structure should be constructed, and the training part of GPR should be started by obtaining the training data set and then initializing the hyperparameters. GPR uses the conjugate gradient method to determine the optimal value of the hyperparameters, which will lead to a decrease in the negative marginal log-likelihood function. Finally, RUL is estimated in (6.33) and (6.34), which can be expressed as

$$\mu = k^T \left(K + \sigma_n^2 I\right)^{-1} y \tag{6.33}$$

$$\sum = \sigma_n^2 + k_{**} - k_*^T \left(K + \sigma_n^2 I\right)^{-1} k_* \tag{6.34}$$

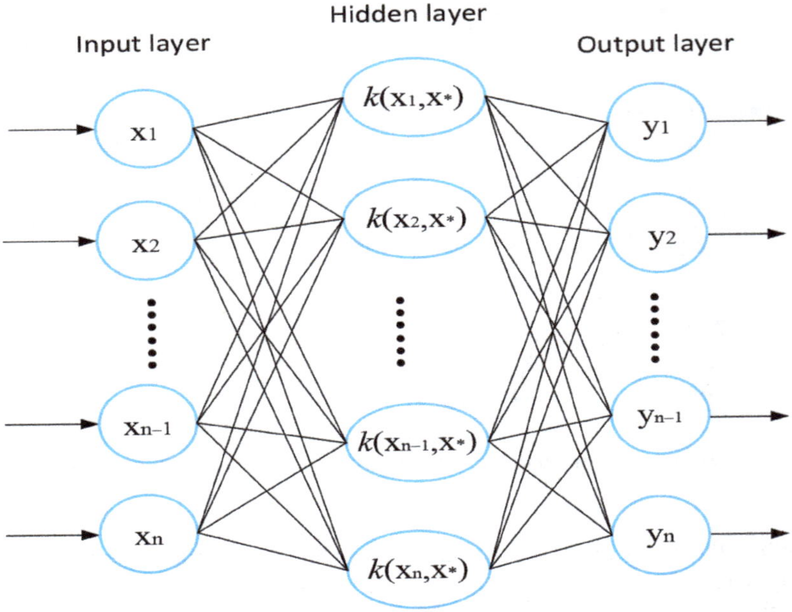

Figure 6.16 GPR algorithm structure diagram

where μ is RUL estimation; $k_* = \left[k\left(x_1, x_*\right), \ldots, k\left(x_N, x_*\right)\right]^T$; $k_* = k_s\left(x_*, x_*\right)$; the kernel matrix is denoted as K, the output of the trained data set is yy, I is the identity matrix, the inverse matrix is determined by the marginal log-likelihood function, and its gradient is $K+\sigma_n^2 I$, which is shown in Figure 6.16.

Although GPR can predict nonlinear systems, as the complexity of prediction increases, the accuracy of GPR will drop rapidly. To solve this problem, GPR introduced a new covariance function and a mean function based on the zero mean function and the diagonal covariance squared exponential function. The performance of GPR is highly sensitive to the covariance function, so proper kernel selection and hyperparameter optimization can avoid the problem of excessive sensitivity. To improve ground-penetrating radar, one method is to minimize the negative impact of logarithmic marginal probability. It is usually necessary to inverse the covariance matrix to train GPR, which will increase the calculation time and complexity of the algorithm and increase the memory requirements [283]. To solve the problems of long calculation time and high complexity, various sparse methods based on the use of a subset of the training sample size have been developed.

When selecting the GPR method for RUL prediction, the activation function is usually a kernel function, the training algorithm is a squared exponential kernel or marginal log-likelihood function, and the hyperparameter adjustment methods are the input dimension length scale and latent function values.

3. Extreme learning machine

With the development of battery RUL prediction technology, extreme learning machines have also been applied in this field. ELM can randomly select hidden layer unit settings. When the single hidden layer feedforward neural network (SLFNN) is determined, ELM can analyze the output weight of SLFN. Because of its fast-learning speed and high prediction accuracy, ELM has been widely used in single-step and multistep prediction algorithms. In the process of state estimation for nonlinear complex systems, ELM has strong flexibility, scalability and high learning performance, which can quickly approach the real value. ELM is a member of ML, and its structure is usually divided into three layers, namely, the input layer, hidden layer and output layer, as shown in Figure 6.3.

During the data input process, ELM randomly assigns the input weight [284]. When the data are transmitted between the input layer and the hidden layer, the hidden layer deviation is also set randomly, and the input weight and the hidden layer deviation do not need to be adjusted after setting. When the data pass through the hidden layer and enter the output layer, the connection weight will be determined by solving the equation once. Since the connection weight does not need to be adjusted iteratively, ELM can perform fast convergence. In (6.35), x_i represents the input layer, and y_i represents the output layer. The mathematical expression output by the hidden layer is represented by the following equation:

$$\sum_{i=1}^{\widetilde{N}} \beta_i f_i(x_i) = \sum_{i=1}^{\widetilde{N}} \beta_i f(a_i x_i + b_j) = t_j, j = 1, 2, \ldots, N \tag{6.35}$$

$$f(a_i \cdot x_j + b_j) = \frac{1}{1 + e^{-(a_j \cdot x_j + b_j)}}, i = 1, \ldots, L, j = 1, \ldots, N \tag{6.36}$$

$x = [x_{i1}, x_{i2}, \cdots, x_{iN}]^T$ is the input weight vector, b_i is the hidden layer deviation, $\widetilde{N}$ is the hidden neuron, the weight vector of the ith hidden node and the input node, and the output weight of the output layer neuron is denoted as $a_i = [a_{i1}, a_{i2}, \ldots, a]^T$ and $B_i = [B_{i1}, \beta_{i2}, \ldots, \beta]^T$, respectively. The activation function is $f(x)$ (e.g., sigmoid, tanh, linear), which is shown in Figure 6.17.

When ELM is used as the RUL prediction method, the activation function is usually selected as sigmoid, the training algorithm is a linear system function and the hyperparameters are adjusted through hidden neurons.

4. Deep neural network

Unlike the SLFNN structure of the standard ANN model, the DNN model contains multiple hidden layers. In the DNN algorithm, a functional relationship is established between the input vector and the output vector through nonlinear calculations. In the calculation process, the function parameters are calculated by a certain method. The DNN contains multiple hidden layers, as shown in (6.37).

$$y = f(x) = f_a\left(W_0^T f_a\left(W_0^T x + b_0\right) + b_1\right) \tag{6.37}$$

$$f(x) = f_a\left(W_i^T f_a\left(W_{i-1}^T \cdots f_a\left(W_0^T x + b_0\right)\right) + b_i\right) \tag{6.38}$$

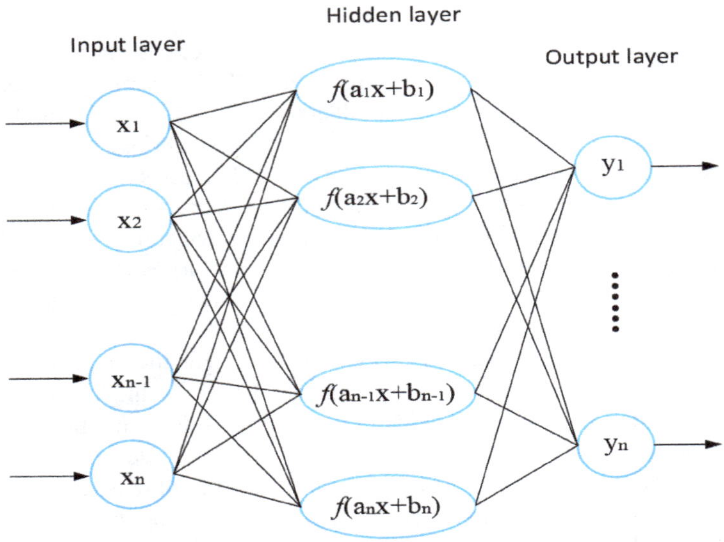

Figure 6.17 Extreme learning machine algorithm structure diagram

SLFNN is shown in (6.38), where the input data are x, the output data are y, the activation function is denoted by f_a, and the weight and deviation are W and b, respectively. In the training process, the real value is approached by continuous iterative updating of W. The He method is usually used for initialization, which is shown in Figure 6.18.

When DNN is used as an RUL prediction method, the activation function is usually selected as sigmoid or ReLU, the training algorithm is gradient descent, backpropagation through time and the hyperparameters are adjusted through hidden layers.

5. Recurrent neural network

Recurrent neural networks are widely used to process time-series data because of their time series memory. RNN is an SLFNN, with a classic three-layer model structure. According to time changes, the time series variables at each moment are used as the input of RNN. Through training, RNN can predict the changing trend of input variables. Among the many RNN architectures, the LSTM algorithm is the most representative. LSTM has a forget gate that can filter low-correlation inputs and enhance strong-correlation inputs. In this way, the problem of vanishing and exploding gradients can be solved. Compared with the traditional RNN algorithm [285], LSTM is more suitable for scenarios that require long-term prediction and has better robustness and accuracy.

Figure 6.19 shows the typical structure of an RNN. For an input sample $x = [x_1, x_2, \ldots, x_N,]$ where N denotes the sequence length, RNN calculates the hidden state vector sequence $h = [h_1, h_2, \ldots, h_N,]$ and outputs the sequence $y = [y_1, y_2, \ldots, y_N,]$ through iteration of the (6.39) and (6.40), from $t = 1$ to N.

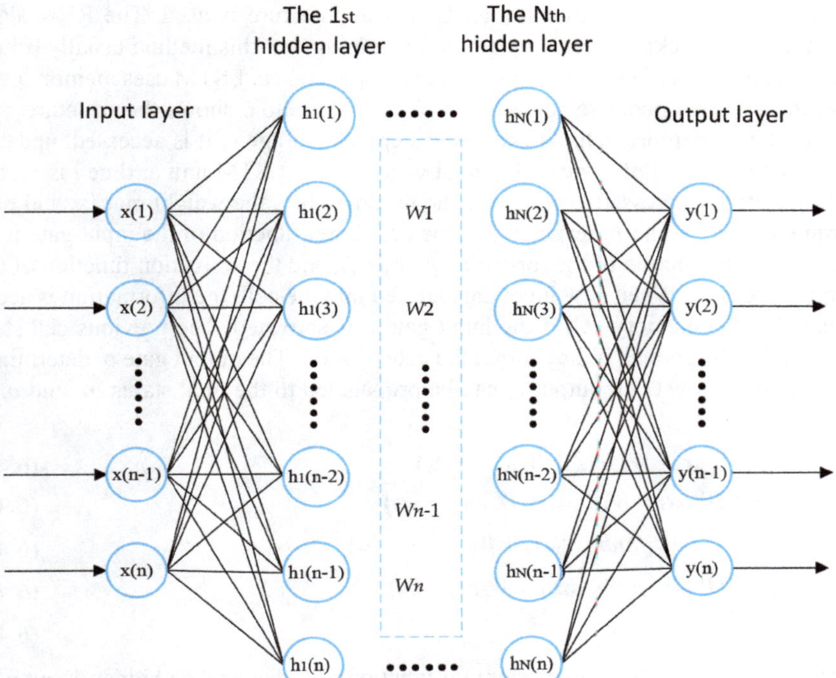

Figure 6.18 Deep neural network algorithm structure diagram

$$h_t = H\left(W_{xh}x_t + W_{hh}h_{t-1} + b_h\right) \tag{6.39}$$

$$y_t = W_{hy}h_t + b_y \tag{6.40}$$

where the weight and bias are W and b, respectively. The weight before the input layer and the hidden layer is represented by W_{xh}, and the bias vector and the non-linear activation function of the hidden layer are b_h and H, respectively, which is shown in Figure 6.19.

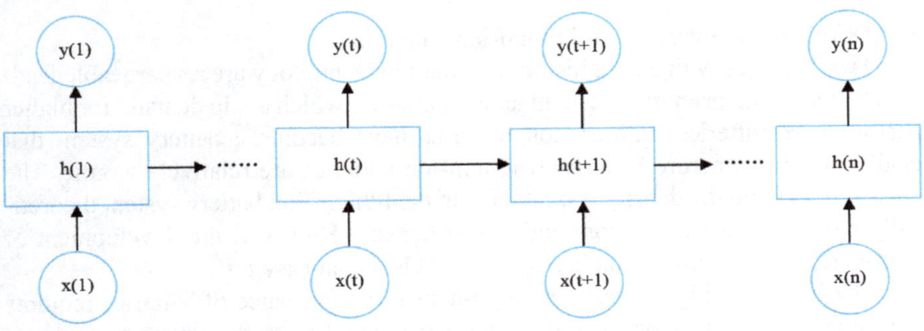

Figure 6.19 Recurrent neural network algorithm structure diagram

Based on the RNN structure, an LSTM architecture is used. The RNN algorithm uses the backpropagation method for training, but this method usually brings about the problem of gradient explosion or disappearance. LSTM uses memory cells instead of hidden nodes to solve this problem. Figure 6.6 shows the structure of a single LSTM memory cell. At each time step, the storage unit is accessed, updated and cleared by multiple gates. The input vector of the LSTM unit at time t is x_t, the hidden state is expressed as h_t, c_t are the unit memory, the weight matrix, and bias parameters are W and b, respectively, the activation function of the input gate is i_t, the activation function of the forgetting gate is f_t, and the activation function of the output gate is o_t. When new input data are fed into the cell, the information is accumulated in the memory cell if the input gate i_t is activated. The previous cell state c_{t-1} can be "forgotten" if they forget the gate f_t is on. The output gate o_t determines whether the newest cell output c_t can be propagated to the final status h_t, and o_t is implemented as

$$i_t = \sigma(W_{xi}x_t + W_{hi}h_{t-1} + W_{ci}c_{t-1} + b_i) \tag{6.41}$$

$$f_t = \sigma\left(W_{xf}x_t + W_{hf}h_{t-1} + W_{cf}c_{t-1} + b_f\right) \tag{6.42}$$

$$c_t = f_t c_{t-1} + i_t tanh\left(W_{xc}x_t + W_{hc}h_{t-1} + b_c\right) \tag{6.43}$$

$$o_t = \sigma\left(W_{xo}x_t + W_{ho}h_{t-1} + W_{co}c_t + b_o\right) \tag{6.44}$$

$$h_t = o_t tanh\left(c_t\right) \tag{6.45}$$

where σ represents the logistic sigmoid function, W_{hi} denotes the hidden-input gate matrix and W_{xo} is the input/output gate matrix. The LSTM algorithm structure is shown in Figure 6.20.

When RNN performs RUL prediction, the activation function is usually selected as sigmoid, the training algorithm is gradient descent, backpropagation through time and potential overfitting problems are solved through hyperparameter adjustment. Among them, the activation function of the LSTM algorithm is usually selected as sigmoid and tanh, the training algorithm is gradient descent, backpropagation through time and the hyperparameters are adjusted through hidden neurons.

6.4.3 Summary of AI-based RUL prediction

1. Development overview of lithium-ion batteries

Two batteries with a long history and mature technology are rechargeable lead-acid batteries and primary zinc-manganese batteries, which are in demand for higher performance batteries. Lithium-ion batteries have become a battery system that challenges these mature batteries. Lithium-ion batteries are relatively atomic. The mass is 6.941 and the density is 0.53/cm^3, so the lithium-ion battery system theoretically has higher battery energy and power density. However, the development of high-performance lithium-ion battery systems is not an easy task.

The improved high-energy and high-density performance of batteries requires the establishment of a systematic model and method to study novel positive and negative electrodes and nonaqueous electrolytes.

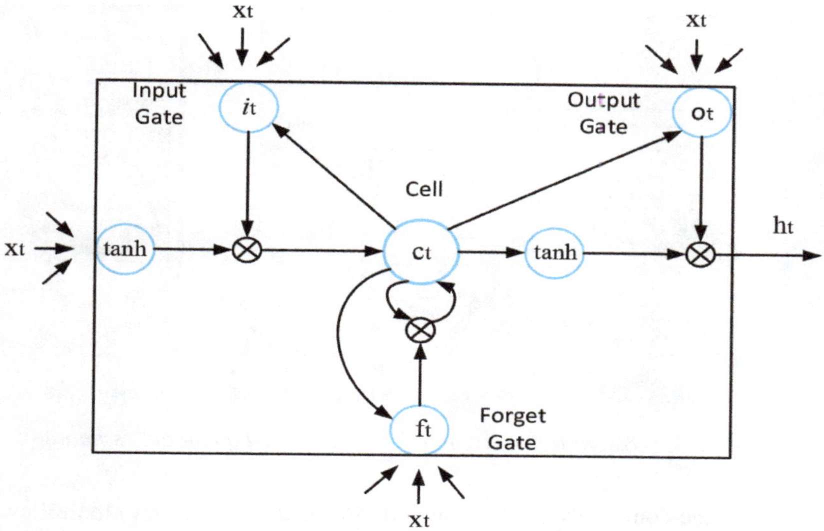

Figure 6.20 LSTM algorithm structure diagram

The research on lithium atomic batteries began around 1960. At that time, due to the oil crisis, people had to find a new alternative energy source. Among many metals, scientists found that lithium metal has the lightest mass, the lowest redox potential and energy density. In 1958, a graduate student from the University of California first proposed the idea of using active metals such as lithium and sodium as the negative electrode of the battery, and then began research work in this area; in 1962, the U.S. military proposed the idea of a "lithium nonhydrolyzable hydrolyzate system"; in 1970, scholars from Exxon used titanium sulfide as the positive electrode material and metal lithium as the negative electrode material to make the first LIB; in 1980, Goodenough J. proposed to use cobalt oxide as the cathode material for LIBs, which opened the prototype of LIB research; until 1991, Sony released the first commercial LIB; by 1995, BYD set up a small battery workshop, less than three. In 2018, the LIB produced instantly entered the top three rankings in the battery industry, becoming the world's second-largest battery supplier after Sanyo, occupying 15% of the global market. So far, LIBs have begun to be commercialized on a large scale, which has also led to the innovation of electronic products.

2. Model-based approaches

The model method is to analyze the physical and chemical principles and data of the battery, build a degradation model, use the parameter estimation algorithm to infer the optimal parameters of the model, and finally estimate the initial battery capacity and the degradation trend of the final battery capacity of the model to be predicted according to the parameters. To achieve the purpose of predicting the RUL of LIBs, before the prediction, it is necessary to build a model and analyze the parameters of the model online, which is shown in Figure 6.21.

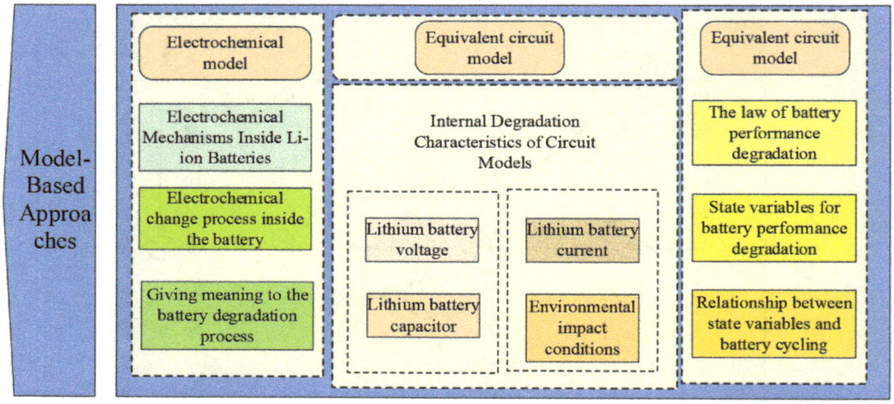

Figure 6.21 Estimation of lithium batteries based on model estimation

Usually, the commonly used prediction models are divided into the following three types: the first is the electrochemical model, which first by discussing the electrochemical mechanism inside the LIB, the internal electrochemical change process of the lithium-ion battery is explained through the mechanism, to better give the physical and chemical meaning of the battery degradation process; the second is the equivalent circuit model. The model mainly builds a circuit model through circuit components such as voltage sources, resistors and capacitors to characterize the internal degradation characteristics of the equivalent circuit model, to achieve the prediction effect; the third is an empirical degradation model. To reflect the law of battery performance degradation, by finding state variables that characterize battery performance degradation, such as battery capacity and internal impedance, establish a relationship between state variables and battery cycle time.

1. Electrochemical model

Since the degradation of battery performance is the result of the interaction of many factors when constructing a battery, it is not only necessary to deeply study the specific physical and chemical phenomena during battery operation but also to fully consider the factors that lead to the degradation of battery performance, to ensure the prediction model accuracy and reliability. For example, Reference [29] used electrochemical impedance spectroscopy to parameterize the circuit model of a 3.4 Ah lithium-ion battery to obtain a wide range of temperature and SOC levels. After parameterization, under the conditions of different temperatures and load currents, verify different parameterized models and build a model to estimate the voltage of the LIB in the discharged state. Reference [33] established a capacity decay model of the battery based on the kinetic reaction and mass transfer process of the SEI layer, combined with the thermal process and conducted experiments on multiple lithium-ion batteries with different electrode thicknesses. With the increase, the capacitance will decay and the total heat value will increase.

2. Equivalent circuit model

The circuit model is constructed by circuit components such as voltage sources, resistors and capacitors to characterize the internal degradation characteristics of the equivalent circuit model, to achieve the prediction effect. Compared with the electrochemical model, the equivalent circuit model is widely used in battery models because of its simple and intuitive structure and uncomplicated parameter identification: RC model and Thevenin model. The literature uses the second-order Thevenin model to analyze the internal parameters and polarization voltage characteristics of lithium-ion batteries, and finally, proposes a lithium-ion battery charging strategy based on the negative pulse charging theory. The literature analyzes the influence of battery internal temperature, discharge speed and battery voltage on battery capacity, and establishes a lithium-ion battery life prediction model. The experimental results show that the model can effectively predict the RUL of lithium-ion batteries.

Although the equivalent circuit model has been widely used, it also needs to consider other factors that affect the battery performance, such as current, voltage, temperature and power, etc. If the battery performance is affected by these factors, the prediction accuracy will be poor and the desired effect is not achieved.

3. Empirical degradation model

Empirical degradation models describe the health of a battery by analyzing the changes in battery internal parameters over time. The output variable of the empirical degradation model is the capacity of the battery, and the commonly used input variables are the number of charge–discharge cycles, temperature, current, SOC, depth of discharge, etc. The first and second methods need to consider the internal characteristics of the battery when used, which makes the modeling process relatively complicated. Empirical models look for state variables that can characterize battery performance degradation to reflect the law of battery degradation. The empirical degradation model is mainly to establish an empirical model, which can accurately reflect the changing trend of battery performance state parameters (capacity, equivalent internal resistance, etc.) over time, or the recursive relationship between the two-time state variables before and after the system, to realize the battery RUL predictions.

To sum up, the model-based method needs to deeply explore the aging mechanism inside the battery and establish the corresponding battery aging model. However, due to the complex structure of the battery, the degradation process of the battery has a high degree of uncertainty, and it is difficult to fully consider the influence of the environmental and load characteristics on the battery life in the mechanism model. Therefore, it is difficult to accurately model the battery system.

4. Data-driven approach

The data-driven method is the opposite of the model method. This method avoids the analysis of the internal electrochemical reaction and failure mechanism of the lithium-ion battery. It can overcome the problems of low dynamic accuracy and poor generalization ability of the model-based method to a certain extent. It has been widely used in many fields for its flexibility and usability. The data-driven prediction method occupies a high position in the failure prediction method because

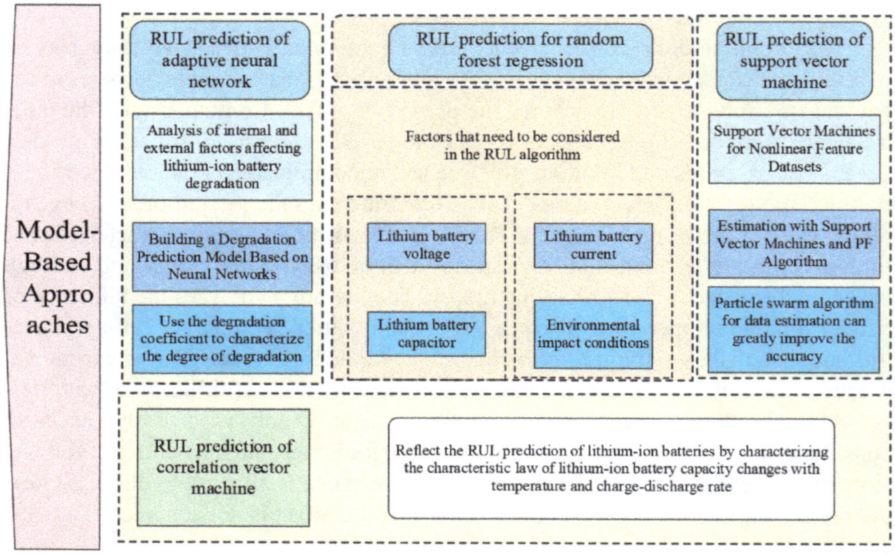

Figure 6.22 Schematic diagram of SOH estimation algorithm based on RUL algorithm

it does not need to explore the internal mechanism of the battery, which is shown in Figure 6.22.

The main prediction process of this method is as follows: collecting data, extracting characteristic parameters, predicting the degradation trend, identifying and analyzing the failure. The data-driven prediction method is a common method for predicting the RUL of lithium-ion batteries at this stage. This method analyzes the failure threshold and degradation data to obtain the changing trend and law of capacity degradation data, to complete the prediction of lithium-ion battery RUL. Common prediction methods based on data-driven methods mainly include the following: adaptive neural network, random forest regression, support vector machine (SVM), correlation vector machine and PF.

1. RUL prediction of adaptive neural network

A NN has good generalization ability and builds nonlinear complex relational models, so it has been widely used in image processing, text recognition, medical treatment, banking, finance, government, agriculture and national defense, and other fields.

Many experiments have tried to determine the degradation characterization parameters by analyzing the internal and external factors that affect the degradation of lithium-ion batteries, establishing a prediction model for the degree of degradation of lithium-ion batteries based on adaptive neural networks and using the degradation coefficient to characterize the degree of lithium-ion battery degradation.

The model is trained, verified and simulated using data at different discharge depths to verify the reliability and applicability of the model in predicting the deterioration degree of lithium-ion batteries. There are also related literature that use recursive NN to predict temporal data streams with abnormal change points and have achieved good results.

2. RUL prediction for random forest regression

   The literature proposes a SOC estimation method for lithium-ion batteries based on a random forest regression algorithm. First, a random forest regression model is constructed. The battery current, voltage and temperature are used as the input of the model, and the SOC value is used as the output of the model. According to the random forest regression algorithm training, prediction model, and finally, uses this model to estimate the SOC of LIB.

3. RUL prediction of support vector machine

   SVMs are often used to deal with datasets with nonlinear characteristics. The literature predicts the life of LIBs according to the life prediction method of the SVM and the life prediction method of PF, and finally evaluates the effect of various prediction methods. The data-driven method can be effectively applied to the life prediction of LIBs and has high engineering application value. A prediction algorithm based on improved support vector regression has been proposed in the relevant literature. The PSO algorithm is used to optimize the penalty coefficient and hyperparameters of support vector regression to improve its prediction ability. Some literatures use the SVM method to classify the extracted optimal features of LIBs train a classification model, and then analyze the results of the model classified by the optimal features to obtain a data set, and finally according to the SVM method. The data set for training prediction ensures that a more suitable regression model can be obtained for the prediction of RUL of lithium-ion batteries.

4. RUL prediction of correlation vector machine

   The correlation vector machine is a sparse probability model similar to the support vector machine. It is a new supervised learning method, which is different from the support vector machine learning machine based on structural risk minimization. The vector machine is trained under the Bayesian framework, and when the prior parameters are known, the irrelevant points are removed based on the active correlation decision theory, and then the sparse model is obtained.

Relevant literature put forward the RUL prediction method of correlation vector machine, based on historical experimental data, establishes the battery failure.

Reduce the model, evaluate the working state of the entire battery, and maintain and replace the equipment in time to ensure the stability of the battery. By combining and improving the kernel function of the correlation vector machine, the performance of the correlation vector machine is optimized, the deviation of the LIB life prediction is reduced and the prediction accuracy is improved. At the same time, there are also literature that use the least squares support vector machine to predict the RUL of LIBs. This method mainly reflects the RUL prediction of LIBs by characterizing the characteristic law of LIB capacity changing with temperature and charge–discharge rate.

5. RUL prediction of PF

Aiming at the inherent defects of traditional filtering technology in dealing with the strong nonlinear characteristics of LIB life, a relatively simple prediction model is proposed, and the PF technology is used to identify the model parameters online in real time.

After comparing the predicted value with the experimental value, it is used to verify the feasibility of the PF-based prediction model. The improved algorithm proposes an ensemble model to characterize the capacity decline trend of the battery and uses the PF algorithm to adjust the model parameters online. There is also a new idea to propose a prediction model based on the combination of PF and radial basis function NN. The model first uses PF to identify the model parameters online and then uses the radial basis function to train the model online.

6. Fusion-based methods

In recent years, fusion-based methods have become a research hotspot in the field of LIB RUL prediction. This method combines a variety of different methods for hybrid prediction and draws on the advantages of different methods to achieve better prediction results, usually including empirical models. The fusion of method and data-driven method, and the method of weighted fusion combining different data-driven methods.

1. Fusion of empirical model method and data-driven method

Considering the online identification of model parameters and state parameters, the fusion of various models and algorithms to predict the RUL of lithium-ion batteries will achieve ideal results [5]. Reference [59] proposed a simple and effective RUL prediction method based on the combination of the model method and the data-driven method. Through the fusion of the model method and data-driven method, the PF algorithm is used to track the process of battery capacity decline, and the double exponential. The empirical degradation model is improved to reduce the difficulty of model parameters and parameter training. Reference [61] proposed a life prediction method based on ensemble NN and PF for the problem of predicting the remaining life of PF when the measurement noise is unknown under the condition of partially observable information. The related new algorithm proposes a lithium-ion battery RUL prediction method that combines PF and an AR model. This method introduces the normalized PF method to solve the uncertainty representation caused by insufficient particle diversity in the PF algorithm. The accuracy is not high, which improves the accuracy of uncertainty expression.

2. The weighted fusion method combined with different data-driven methods

In the traditional method, the method is based on the data-driven least squares support vector machine algorithm and the PF algorithm is introduced. The least-squares support vector machine algorithm is used to train the historical measurement data, and then the future measurement data are predicted, and the PF algorithm is integrated for RUL prediction. The new algorithm idea aims at the problems of complex RUL prediction mechanism model, difficult numerical solution and uncommon use of pure data-driven methods for lithium-ion batteries; a semi-empirical fusion prediction method based on the Gaussian fitting model and PF algorithm is proposed. First, multiple sets of data sets are collected to design the degradation experiment of lithium-ion battery RUL, then the data are analyzed and processed based on MATLAB, and the life decay law, state-space model and observation model are established by the Gaussian fitting method; data for tracking and forecasting.

Chapter 7

Examples and case studies for utility-scale battery operation

7.1 Battery system location and characteristics

7.1.1 Battery management system

The battery management system is closely combined with the battery to detect the battery's voltage, current, and temperature in real time [286]. Besides, it also carries out leakage detection, thermal management, battery balance management, alarm reminder, calculation of the residual capacity and discharge power. According to the voltage, current, and temperature of the battery, the maximum output power is controlled by the algorithm to obtain the maximum driving mileage, the optimal current is controlled by the charging algorithm, and real-time communication is carried out with an onboard general controller, motor controller, energy control system and onboard display system through the CAN bus interface. The basic functional block diagram of the battery management system is shown in Figure 7.1.

As far as the scope of battery management is concerned, it can be divided into two aspects: broad sense and narrow sense. Battery management in a broad sense can involve battery charging and load discharge. To improve the performance of the battery, it can coordinate charging and discharging. When appropriate, the charging and discharging speeds can be adjusted, which is equivalent to the power management and energy management parts of a system [287]. A typical application is the battery management of hybrid electric vehicles. Since the energy supply of a hybrid electric vehicle is not only the battery but also other energy sources, the charging and discharging speed and energy supply proportion of the battery can be arranged according to the actual situation of the battery. In a narrow sense, battery management simply manages various battery states and parameters and provides necessary data communication and other functions [288]. It does not involve charge management and discharge management.

As the battery charging is limited by the power supply capacity and voltage of the charger, it belongs to the extension of battery management and is not controlled by the battery management system. Similarly, the discharge of the battery is mainly affected by the size of the load [289]. The adjustment of

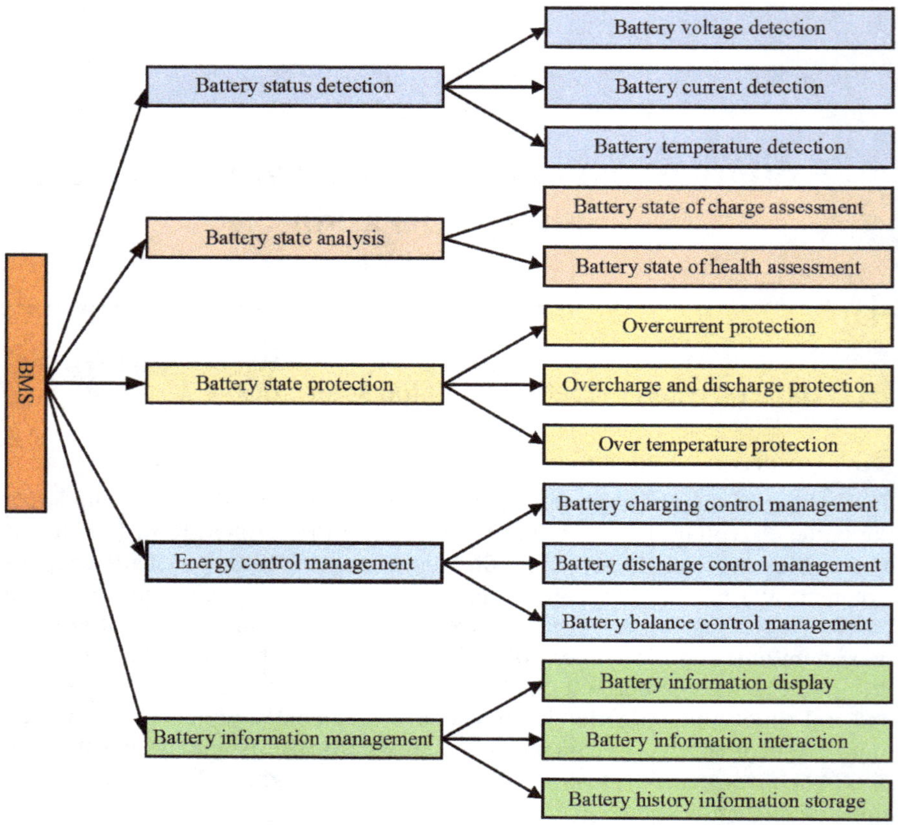

Figure 7.1 The basic function block diagram of BMS

load size is generally adjusted at a higher level of management, rather than by the battery management according to the situation of the battery itself [290]. Battery management only monitors battery status and the status of each cell in the battery and makes appropriate control and adjustment of the battery and its internal components according to its status.

Battery management does not control other components outside the battery but only provides the function of communication data status notification. However, the battery management system can receive the control information of the upper main control module and make the necessary control response. At present, BMS in the general sense refers to the concept of battery management in a narrow sense. Charging management is realized by the charger, and its interface is generally realized by the battery voltage [291]. The charger judges whether to use constant current charging or voltage charging, trickle charging, floating charging, etc. through the battery voltage. According to the information provided by the battery management system, discharge management and load management are adjusted by the upper main control module of battery management.

7.1.2 Battery pack

The capacity of lithium-ion batteries is too large, which is easy to produce high temperatures and induce unsafe factors [292]. Therefore, high-capacity batteries must form battery packs in series and parallel.

After the single batteries are grouped, there will be problems such as overcharge or discharge, excessive current, high temperature, even short circuit, and leakage. For example, when charging or discharging the battery pack in series, some batteries may be fully charged or discharged before other batteries. At this time, continuing to charge or discharge will lead to overcharge or discharge, and the internal side reactions of lithium-ion batteries will lead to problems such as battery capacity reduction, thermal runaway, or internal short circuits [293]. Parallel connection, aging, low temperature, and other conditions will cause the current of some batteries to exceed their carrying capacity and reduce the service life of batteries. The excessive local temperature will reduce the performance of the battery, eventually leading to internal short circuits and thermal runaways, resulting in safety problems. Due to vibration, damp heat, dust, and other factors, the battery will be short-circuited or leaked, endangering the personal safety of drivers and passengers.

One of the functions of BMS is to avoid the above problems of the battery pack. It is necessary to dynamically monitor the working state of the power battery pack, collect the terminal voltage and temperature of each battery, charge and discharge current and the total voltage of the battery pack in real time, and estimate the charging state, safety state, and energy state of each battery [294]. Then, by controlling other devices, the battery can be prevented from overcharging or over-discharging. At the same time, the battery status can be given in time, and the problematic batteries can be selected to maintain the reliability and efficiency of the whole set of batteries.

The main working principle of BMS is that the data acquisition circuit first collects the battery state information data, and then the electronic control unit processes and analyzes the data, sends control instructions to the relevant functional modules in the system according to the analysis results, and transmits the information to the external world.

7.2 SOC estimation based on genetic algorithm and BP neural network

SOC estimation based on a genetic algorithm and BP neural network is to use the genetic algorithm to optimize the initial weight and threshold of the BP neural network so that the optimized neural network can better predict the function output. The flow chart of the GA-BP algorithm is shown in Figure 7.2.

The optimization process can be divided into three parts. First, determine the structure of the network according to the functional requirements. Second, the

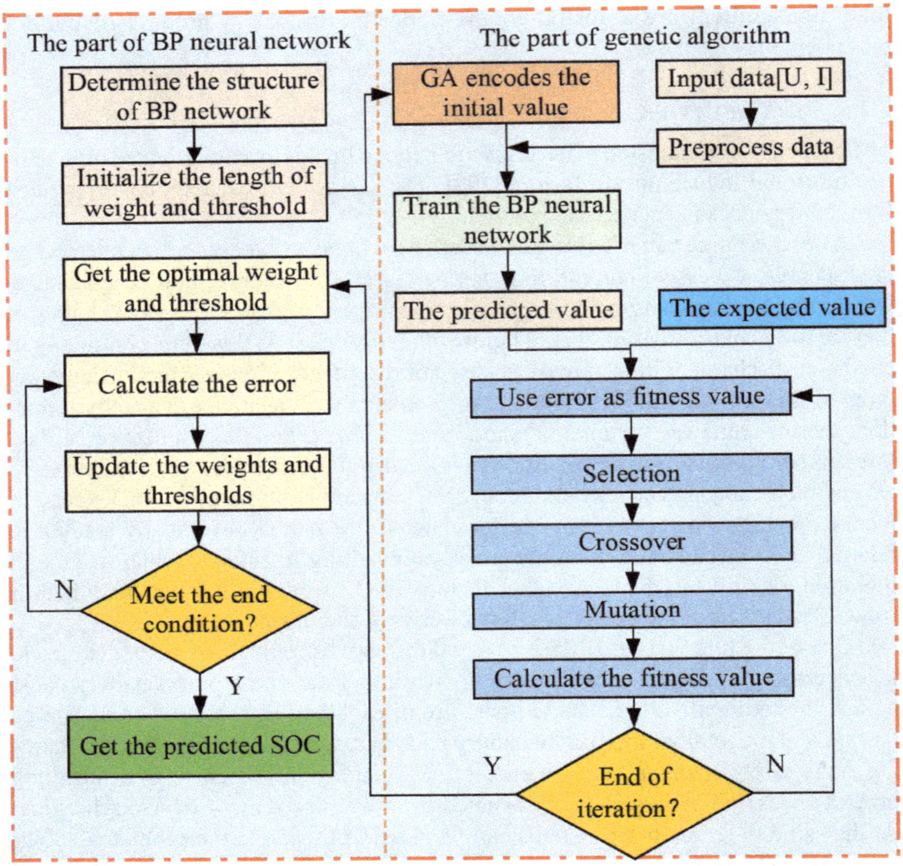

Figure 7.2 The flowchart of the GA-BP algorithm

parameter values of the BP neural network structure are initialized by the genetic algorithm [295]. All initial weights and thresholds in the network are genetically operated to obtain the initial optimal weights and thresholds. Third, maintain the topological structure of the BP neural network, assign the initial optimal weight and threshold to the network, and then use the BP neural network for training, testing, and verification.

7.2.1 BP neural network

BP neural network is a multilayer feedforward neural network. It extends the W-H learning rule and weights the nonlinear differentiable function. BP network is mainly used for function approximation, pattern recognition, classification, and data compression. The flow chart of the BP algorithm is shown in Figure 7.3.

As shown in Figure 7.2, the main characteristics of the network are signal forward transmission and error backpropagation. In the forward transmission, the input signal is

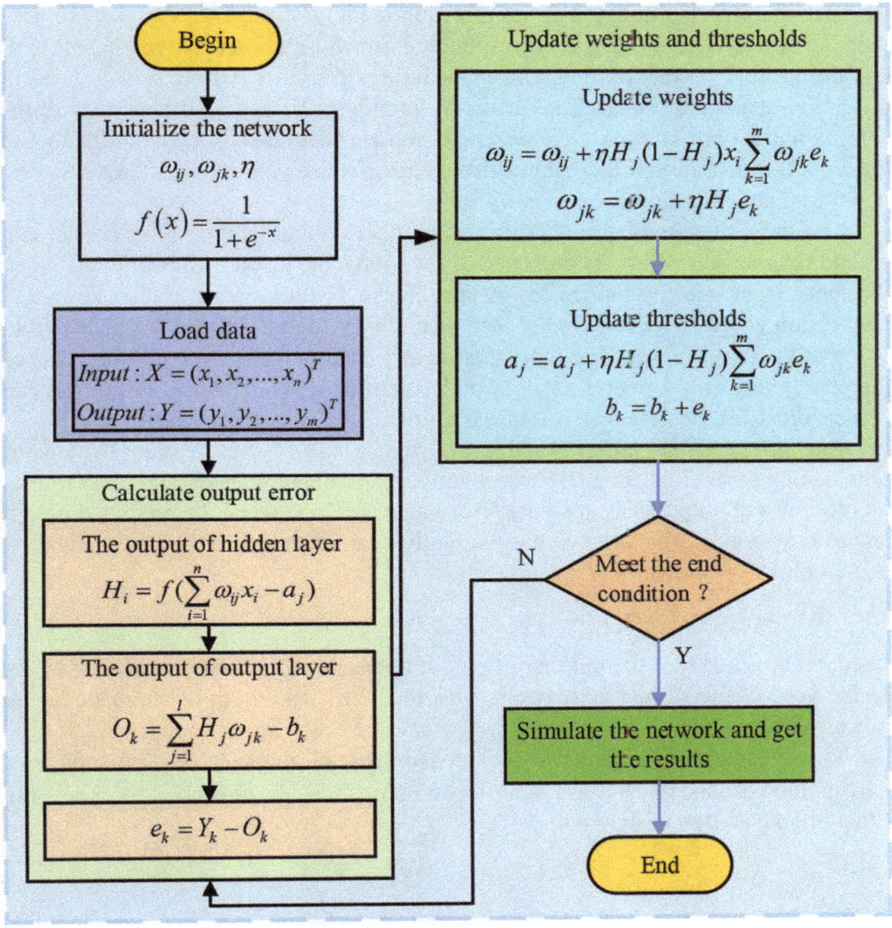

Figure 7.3 *The flowchart of the BP algorithm*

processed layer by layer from the input layer through the hidden layer to the output layer [296]. The state of neurons in each layer only affects the state of neurons in the next layer. If the expected weight and the expected output are not propagated to the neural network, the prediction output is continuously adjusted according to the expected threshold value, and then the BP output is transferred back to the neural network.

7.2.2 Genetic algorithm

A (genetic algorithm) is a parallel random search optimization method that simulates natural genetic mechanisms and biological evolution [297]. It can automatically obtain and accumulate information related to the search space while searching globally, and can adaptively control the search process according to the relevant information of the search space, to obtain the optimal solution to the problem [298].

Because the search strategy and optimization method of the GA only need to provide the objective function and corresponding fitness function to solve the problem, and do not rely on the gradient and other information, it has been widely used in the fields of pattern recognition, artificial intelligence, deep learning, image processing and so on. The basic elements of the BP neural network optimized by the GA include population initialization, fitness function, selection operation, crossover operation, and mutation operation.

The individual coding method is real number coding, and each individual is a real number string, which is composed of four parts: the connection weight between the input layer and the hidden layer, the threshold value of the hidden layer, the connection weight between the hidden layer, and the output layer and the threshold value of the output layer. The individual contains all the weights and thresholds of the neural network. A neural network with its structure, weights and thresholds can be constituted if the network structure is known.

According to the initial weights and thresholds of the BP neural network obtained by individuals, the BP neural network is trained with training data to predict the system output. The absolute error value and E between the predicted output and the expected output are taken as the individual fitness value F. The calculation equation of F is shown in (7.1).

$$F = k\left(\sum_{i=1}^{n} |y_i - o_i|\right) \tag{7.1}$$

In equation (7.1), F is the number of output nodes, y_i is the expected output of the i_{th} the node of the BP neural network, o_i is the predicted output of the i_{th} node and k is a coefficient.

The selection operation of GA includes roulette, tournament, etc. When the roulette method is used to calculate the selection probability p_i of each individual i, the calculation equation is shown in (7.2).

$$\begin{cases} f_i = \frac{k}{F_i} \\ p_i = \frac{f_i}{\sum_{j=1}^{N} f_i} \end{cases} \tag{7.2}$$

In equation (7.2), F_i represents the fitness value of individual i. Since the smaller, the fitness value, the better, the inverse of the fitness value is calculated before individual selection. N represents the number of individuals in the population, and k is a coefficient.

The crossing operation of the k_{th} gene a_k and the l_{th} gene a_l at the j position is realized by using (7.3).

$$\begin{cases} a_{kj} = a_{kj}(1-b) + a_{lj}b \\ a_{lj} = a_{lj}(1-b) + a_{kj}b \end{cases} \tag{7.3}$$

In equation (7.3), b represents a random number in [0,1].

The j_{th} gene of the i_{th} the individual is selected for mutation operation, as shown in (7.4).

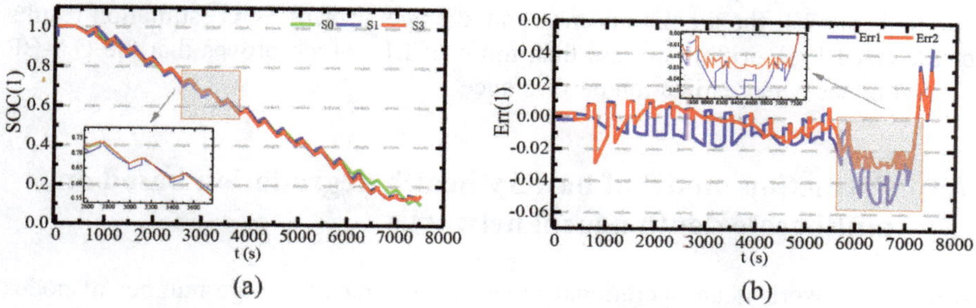

Figure 7.4 The experimental verification results under the DST condition: (a) the SOC estimation results and (b) the SOC estimation error

$$\begin{cases} f(g) = r_1\left(1 - \frac{g}{G_{max}}\right)^2 \\ a_{ij} = \begin{cases} a_{ij} + (a_{ij} - a_{max}) * f(g), r > 0.5 \\ a_{ij} + (a_{min} - a_{ij}) * f(g), r \le 0.5 \end{cases} \end{cases} \tag{7.4}$$

In equation (7.4), a_{max} and a_{min} represent the upper and lower bounds of a_{ij}, r represents the random number, g represents the current iteration, and G_{max} represents the maximum number of evolutions.

7.2.3 Experimental analysis

To verify the accuracy of the BP neural network optimized by GA in SOC prediction is better than the BP neural network, experimental verification is carried out under DST working conditions. The results are shown in Figure 7.4.

In Figure 7.4(a), S0~S2 represents the prediction curves of ampere-hour integral, BP, and GA-BP on SOC, respectively. In Figure 7.4(b), err1~err2, respectively, represents the estimation errors of BP and GA-BP with ampere-hour integral as reference. It can be seen that the output of the GA-BP network is closer to the expected output than the BP network, and the tracking ability of the real curve is stronger. To further analyze the estimation performance of GA-BP, the maximum error, average error, and BP estimation error of GA-BP are compared, as shown in Table 7.1.

Table 7.1 Comparison of estimation results between GA-BP and BP

SOC estimation method	BP	GA-BP
Maximum error	5.2843%	3.1125%
Average error	1.3824	1.0811%

In Table 7.1, the maximum error and average error of SOC estimation results of the GA-BP algorithm are less than those of BP, which proves that the GA-BP algorithm has better estimation performance.

7.3 Prediction model of battery health degradation based on multi-scale depth neural networks

A neural network is an operational model consisting of a large number of nodes (or "neurons") and connections between them [299]. Each node represents a specific output function, called activation function, and activation function [300]. Each connection between two nodes represents a weighted value for the signal passing through the connection, called the weight, which is equivalent to the memory of the artificial neural network. The output of the network depends on the connection method of the network, the weight value and the excitation function are different. The network itself is usually an approximation of a certain algorithm or function in nature, and it may also be an expression of a logic strategy.

Factors affecting the state of health (SOH) of the battery include temperature, depth of discharge and charging rate, etc. However, these indicators cannot directly characterize the performance degradation degree of lithium-ion batteries, and there are certain difficulties in online detection [301]. The actual battery capacity refers to the electric energy stored when the battery is fully charged. It is an important parameter that directly characterizes the deterioration of battery life. Therefore, it is often used as a health indicator to describe the degree of battery health deterioration in the prediction of battery remaining useful life (RUL).

The existing lithium battery RUL prediction methods can be divided into failure physical models and data-driven models. The failure physical model represents the process of lithium battery performance degradation by establishing a mathematical model [302]. However, this method is susceptible to noise and environmental interference, and it is difficult to predict the battery performance. The robustness and adaptability are poor [303]. The data-driven method is currently the mainstream method for battery life prediction due to its advantages of high flexibility and easy implementation. This type of algorithm is extracted from the performance degradation data of the battery. Life characteristic parameters, through modeling output prediction results, provide decision-making information for system maintenance [304]. Local regeneration phenomenon will occur in the process of battery capacity degradation, and the volatility generated by this phenomenon will seriously affect the performance of the prediction algorithm. In practical situations, it is difficult to make effective predictions.

Since the battery capacity data is easy to collect and can directly reflect the health status of the battery, the battery capacity is used as the SOH index to describe the degradation trend of the battery, which is shown in Figure 7.5.

The overall framework of the algorithm is shown in Figure 7.1. First, through the ensemble empirical mode decomposition (EEMD) and correlation analysis (CA) decompose the data into main trend and fluctuation data and analyze the two types

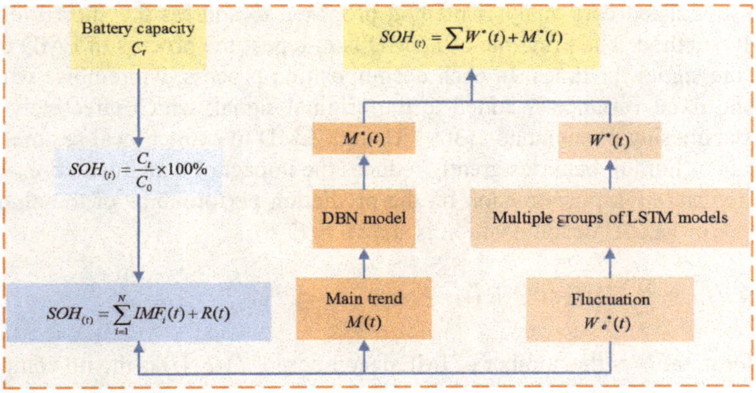

Figure 7.5 Life prediction of lithium batteries based on multiscale decomposition

of data separately to avoid the adverse effect of the regeneration phenomenon during the battery degradation process on the prediction [305]. Based on the excellent performance of the deep neural networks in feature extraction and time series analysis, deep belief network (DBN) and LSTM are used to model two types of data, respectively, and the outputs of each group of models are effectively integrated to obtain prediction results.

7.3.1 Definition of SOH

During the use of lithium batteries, with the increase in charge and discharge cycles, the SOH generally shows a downward trend. Battery capacity is an important indicator of SOH [306]. When the capacity declines to a certain extent, the battery will not continue to work normally and reach the end of its life. The capacity of the lithium battery is used to characterize the SOH, and the definition formula in (7.5).

$$SOH_{(t)} = \frac{C_t}{C_0} \times 100\% \tag{7.5}$$

In equation (7.5), C_t is the capacity of the lithium-ion battery in the charge-discharge cycle, and C_0 is the initial capacity of the lithium battery. When the capacity of the lithium battery drops to the failure threshold, it can be considered that the lithium battery has reached the end of life, and the interval corresponding to the current capacity to the failure threshold is the RUL of the battery.

7.3.2 Multiscale decomposition based on EEMD and CA

Empirical mode decomposition (EMD) is a method for analyzing non-stationary signals [307]. It decomposes the signal into intrinsic mode function (IMF) and margin and adaptively represents the signal in the signal [308]. When EMD is decomposed, it is easy to produce a modal aliasing phenomenon, which leads to the appearance of wrong IMF components, so that IMF loses its specific physical meaning. EEMD is a noise-assisted data analysis method proposed to address the shortcomings of the EMD method. The principle of EEMD is to repeat the process of EMD decomposing the signal for times. In each decomposition process, white noise with zero mean and fixed variance is added to the original signal, which effectively solves the modal aliasing phenomenon [309]. Using EEMD to extract local features in the SOH data of lithium batteries greatly reduces the impact of data fluctuations caused by the regeneration phenomenon on the prediction performance of the algorithm. The result of EEMD decomposition is shown in (7.6).

$$SOH_{(t)} = \sum_{i=1}^{N} IMF_i\left(t\right) + R\left(t\right) \tag{7.6}$$

In the formula, N is the number of IMF decomposed, $IMF_i\left(t\right)$ is the ith component, and $R\left(t\right)$ is the remainder.

Considering that more data groups may lead to a larger final cumulative error, the accuracy of the prediction model will decrease. By calculating the correlation coefficient between $SOH_{(t)}$ and $IMF_i\left(t\right)$, several groups of IMF components with a strong correlation with the original signal are found. To achieve good generalization for different data, a correlation threshold $CT = max\left(IMF_i\right)/N$ is set to select the IMF of the correlation coefficient $r_k > CT$, and the K groups of $IMF_i\left(t\right)$ with strong correlation and the residual $R\left(t\right)$ are combined into $M\left(t\right)$ for prediction, as shown in (7.7).

$$M\left(t\right) = \sum_{k=1}^{K} IMF_k\left(t\right) + R\left(t\right) \tag{7.7}$$

By using EEMD and CA, the global degradation trend and local features in the SOH time series are effectively separated, so that the local rebirth phenomenon can be improved and the prediction performance of the SOH prediction model is improved. In addition, the robustness of the prediction method can be further improved by adaptively setting the correlation threshold [310]. Through the screening method, the IMF data with strong correlation are combined with the margin, which not only retains the local features but also avoids the problem of excessive grouping and excessive cumulative error.

7.3.3 A prediction model based on DBN

DBN is a probabilistic generation model, a deep network composed of multiple restricted Boltzmann machines (RBM) stacked and a regression layer (or classification layer) [311]. In the RBM system, the total energy function of the visible layer node and the hidden layer node is defined in (7.8).

$$E\left(v,h\right) = -\sum_{i=1}^{m} v_i c_i - \sum_{j=1}^{n} h_j b_j - \sum_{i=1}^{m}\sum_{j=1}^{n} v_i h_j w_{ij} \tag{7.8}$$

In the formula, m and n are the numbers of nodes in the visible layer and the hidden layer, respectively; w_{ij} is the connection weight, c is the bias term of the visible layer, and b is the paranoid term of the hidden layer; b, c, w constitute the parameter set θ of the model. The probability distribution obtained based on the probability model is as in (7.9) and (7.10).

$$P(v, h; \theta) = \frac{1}{Z(\theta)} exp(-E(v, h; \theta)) \tag{7.9}$$

$$Z(\theta) = \sum_{v,h} exp(-E(v, h; \theta)) \tag{7.10}$$

In equation (7.10), $Z(\theta)$ is the partition function. Since there is no connection in the DBN layer, the probability of activation of hidden nodes and explicit nodes can be calculated by the following (7.11) and (7.12).

$$P(h_j|v) = \sigma\left(b_j + \sum_{i=1}^{m} W_{i,j} v_i\right) \tag{7.11}$$

$$P(v_j|h) = \sigma\left(c_i + \sum_{j=1}^{n} W_{i,j} h_j\right) \tag{7.12}$$

To obtain the optimal solution of the model, the maximum likelihood method is used to determine the parameter set θ in the RBM, and the partial derivative of the parameter set θ is updated by the loss function. After the DBN training is completed, the model parameters are used to map the display nodes of the bottom input layer to the top layer, and the output results are obtained y, as shown in (7.13).

$$y = f(v; \theta) \tag{7.13}$$

Taking the battery capacity as the health index, the global degradation process separated by EEMD decomposition will show a gradual downward trend with time [312]. Because DBN can build a deep model through time series, find the hidden rules in the process of battery degradation through historical data, and adaptively fit the downward trend, to achieve efficient prediction. Use DBN to predict the main degradation trend of lithium batteries, in (7.14) take time $t - d + 1, \ldots, t$ as input, where d is the size of the input time window, and use the following formula to predict the trained model in h steps in advance.

$$M^*(t + h) = DBN(t - d + 1, \ldots, t) \tag{7.14}$$

The whole training process of DBN consists of two stages: in the pretraining stage, the unsupervised greedy mechanism is used to train layer by layer from the bottom up, transforming the low-level features to the high-level features, and finally sending them to the top regression layer; in the reverse fine-tuning stage, the network weights are fine-tuned from top to bottom to achieve the purpose of reducing errors.

7.3.4 A prediction model based on LSTM

LSTM is a special cyclic neural network (recurrent neural network, RNN). Compared with the traditional RNN, LSTM adds a cell state c in the hidden layer, improves the hidden layer of RNN, and solves the problem that RNN is easy to fall into gradient disappearance or gradient explosion when dealing with long series data, and has more effective long-term prediction ability.

The core of LSTM consists of three "gates" of input, forget and output [313]. The network structure can effectively forget the useless information before, save the effective input information, and determine the information that needs to be output, to process long-term sequences more efficiently. The operation is shown in (7.15–7.20).

$$i_t = \sigma\left(W_i * \left[y_{t-1}, x_t\right] + b_i\right) \tag{7.15}$$

$$f_t = \sigma\left(W_f * \left[y_{t-1}, x_t\right] + b_f\right) \tag{7.16}$$

$$o_t = \sigma\left(W_o * \left[y_{t-1}, x_t\right] + b_o\right) \tag{7.17}$$

$$\tilde{C}_t = tanh\left(W_C * \left[y_{t-1}, x_t\right] + b_C\right) \tag{7.18}$$

$$C_t = f_t \times C_{t-1} + i_t \times \tilde{C}_t \tag{7.19}$$

$$y_t = o_t tanh\left(C_t\right) \tag{7.20}$$

In the formula, x represents the input vector, y represents the output vector, and i, f, o, C represent the input gate, the forgetting gate, the output gate, and the cell state, respectively, and the matrix W, b is the weight parameter and bias term to be trained, respectively, $\sigma(*)$ is the sigmoid nonlinear function.

The output value of the model is calculated according to (7.11) ~ (7.16), and the error term and weight gradient of each LSTM cell are inversely calculated according to the defined error (7.21) and (7.22), to update the network parameters:

$$S = \left(y_t - y_t^*\right)^2 \tag{7.21}$$

$$loss = \frac{-1}{N} \sum_{t=1}^{N} \ln S \tag{7.22}$$

In (7.22), N is the number of samples, y_t^* is the predicted value, and y_t is the actual value.

By introducing a memory gate unit and threshold limitation, LSTM realizes the effective use of long-distance information and responds in time to data with large changes [314]. Through multidoor cooperation, LSTM has good robustness, so it can effectively deal with the problems of gradient disappearance and gradient explosion. LSTM has a significant advantage in performing long-term time series prediction, and (7.23) is used for h-step advance prediction:

$$W^*\left(t+h\right) = LSTM\left(W_{t-d+1}^*, \ldots, W_t\right) \tag{7.23}$$

The LSTM model is used to predict the fluctuations caused by the regeneration phenomenon in the process of battery capacity degradation. The actual data $W_{t-d+1}^*, \ldots, W_t$ corresponding to time $t-d+1, \ldots, t$ is used as the input, which is shown in Figure 7.6.

The training process of the LSTM model is shown in Figure 7.2. In the training process, according to the size of the time window, the required data is taken from the training set and calculated by LSTM, and the output value is obtained. The loss value and weight gradient are calculated by the loss function, and the network parameters are optimized. Repeat the above calculation and update steps until all training data are completed.

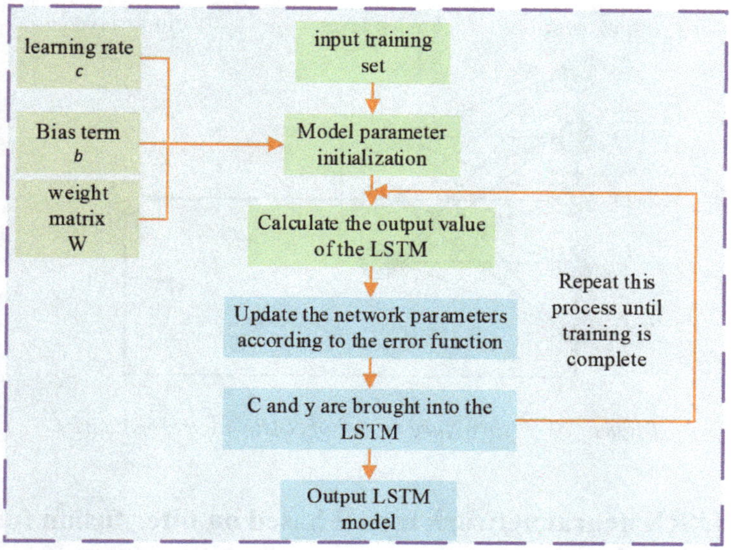

Figure 7.6 *Flow chart of long short-term memory network training*

7.3.5 Combined model prediction framework

The process of the proposed prediction model includes three modules: data pre-processing, network training, and model prediction [315]. The specific steps of the integrated model prediction are as follows:

Take the collected battery capacity time series data as the health indicator SOH, and use (7.5) to obtain.

According to equation (7.6), EEMD is decomposed to obtain several components and the remainder.

Carry out correlation analysis on the components, use (7.7) to accumulate the components with strong correlation and the residual, as the main trend $M(t)$ of the index value, and the residual component IMF is regarded as the data fluctuation $W_e(t)$.

The DBN model and the LSTM model are trained by (7.8)~(7.13), (7.15)~(7.22), and the trained models are trained according to (7.14), (7.23). Obtain the output results $M(t)$ of the DBN model and the output results $W_e(t)$ of multiple groups of LSTM models, and integrate them using (7.24):

$$SOH^*(t) = \sum_{e=1}^{E} W_e^*(t) + M^*(t) \tag{7.24}$$

SOH^*t is the prediction result of the proposed method. $M^*(t)$ and several groups of $W_e^*(t)$ are uniformly divided into a training set and data set, and are substituted into DBN and LSTM models for training and prediction, respectively.

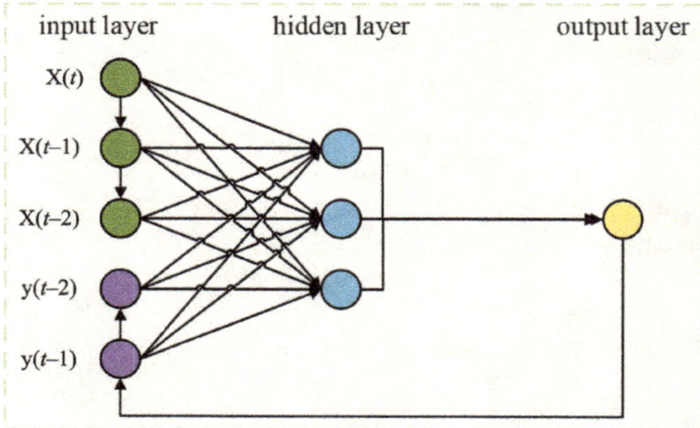

Figure 7.7 Structure model of NARX neural network

7.4 NARX neural network model based on filter fusion for SOC estimation

7.4.1 NARX neural network

NARX neural network (based on the nonlinear autoregressive with exogenous inputs neural network) [316]. NARX is a model for describing nonlinear discrete systems. Expressed as:

$$y(t) = f\{u(t - D_u), \ldots, u(t-1), u(t), y(t - D_y), \ldots, y(t-1)\} \tag{7.25}$$

In equation (7.25), $u(t), y(t)$ are the input and output of the network at time t, respectively; Du is the maximum order of the input delay; Dy is the maximum order of the output delay; and $u(t - Du), \ldots, u(t-1)$ is the historical input relative to time t; $y(t - Dy), \ldots, y(t-1)$ is the historical output relative to time t; f is the nonlinear function obtained by the network fitting, which is shown in Figure 7.7.

The structure of the NARX neural network includes an input layer, a hidden layer, and an output layer [317]. The number of input layer nodes is set according to the number of input values, and the number of output layer nodes is set according to the number of predicted values. By reasonably setting the number of hidden layers and nodes, the NARX neural network is constructed. NARX neural network is similar to the training method of the backpropagation neural network.

The NARX neural network adds a delay and feedback mechanism, so it enhances the memory ability of historical data and is a dynamic neural network [318]. The NARX network is suitable for time series forecasting and has been applied to solve nonlinear series forecasting problems in various fields.

NARX recurrent neural network is used to establish the battery model [319]. NARX (nonlinear autoregressive with external input) recurrent neural network with the delay and feedback functions can keep the input and output of a previous moment and add it to the calculation of the next moment. Therefore, better estimation results are achieved using a small amount of data.

The open-loop NARX neural network feeds back the real values to the input, and the structural expression is defined in (7.26).

$$\hat{y}(t) = \hat{f}(y_o(t-1); u(t-1)) = \hat{f}(y(t-1), y(t-2)) \ldots, u(t-1), u(t-2) \ldots, u(-n_u)) \tag{7.26}$$

$\hat{y}$ is estimation SOC, y_o is the output, u_n is the input, n_u are input memory. To obtain a more accurate estimation of SOC, the open-loop NARX neural network is used. However, the phenomenon of the exploding gradient and vanishing gradient has become an important factor restricting the NARX neural network, and therefore, the NARX dynamic neural network combined with LSTM is proposed to estimate the SOC of the battery.

7.4.2 LSTM neural network

As a new machine learning technology, the LSTM neural network uses the traditional recurrent neural network to address the problem of the exploding gradient and vanishing gradient [320]. The LSTM cell nucleus is used to replace the cell nucleus of the traditional dynamic neural network; therefore, it has long-term memory capabilities. The information forgotten by the LSTM cell nucleus feeds important information to the hidden layer, including the forgetting gate, input gate, and output gate. The memory cell of an LSTM recurrent neural network is shown in Figure 7.8.

The LSTM recurrent neural network is an architecture that maps the output series (SOC) through the input series (voltage, current, temperature, etc.) [321]. The input gate is used as an evaluation index for updating current information. Before generating new memories, the gate network needs to determine whether the current information is important and decide whether to remember. The forgetting gate only evaluates the information of the memory cell and evaluates whether to forget information from the past. The self-recurrent connection generates new memories after merging the past and current memory information. The purpose of the output gate is to separate the final memory. The mathematical expression is shown in (7.27) ~ (7.31).

$$i_t = \sigma\left(w_{xi}x_t + w_{hi}x_{t-1} + b_i\right) \tag{7.27}$$

$$f_t = \sigma\left(w_{xf}x_t + w_{hf}x_{t-1} + b_f\right) \tag{7.28}$$

$$o_t = \sigma\left(w_{xo}x_t + w_{ho}x_{t-1} + b_o\right) \tag{7.29}$$

$$c_t = f_t \times C_{t-1} + tanh\left(w_{xc}x_t + w_{hc}h_{t-1} + b_c\right) \tag{7.30}$$

$$h_t = o_t tanhc_t \tag{7.31}$$

Among them i_t, f_t, o_t, and c_t represent the input, forget, output, and cell state at time t, respectively. σ and *tanh* are the sigmoid activation function and the hyperbolic tangent activation function, respectively.

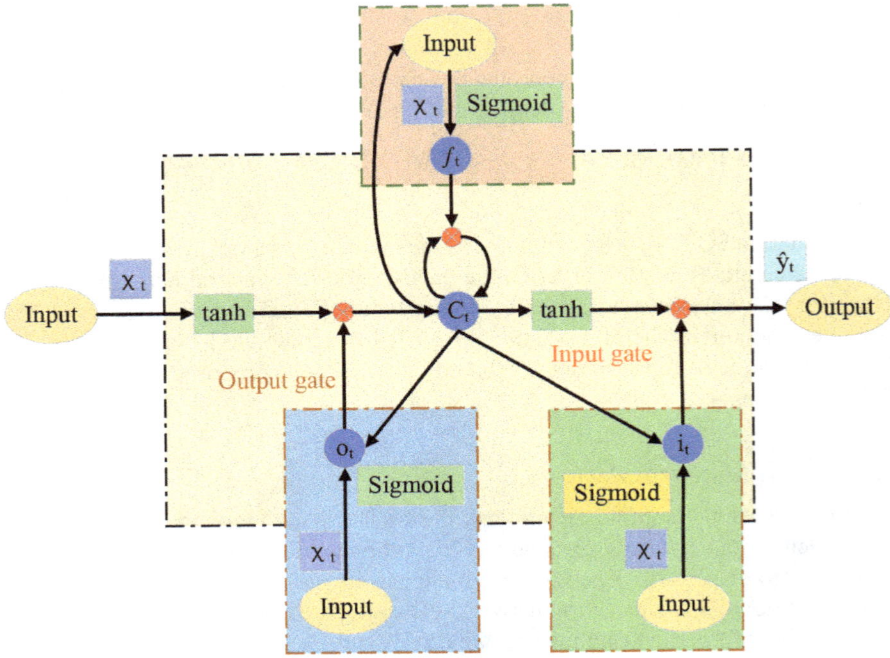

Figure 7.8 Structure of the LSTM recurrent neural network

7.4.3 Hybrid NARX and LSTM model

The model combining NARX and LSTM is proposed to integrate the advantages of each model and obtain stronger estimation performance [322]. The most important feature of the hybrid NARX dynamic neural network is the embedding of LSTM memories, which provide jump-ahead connections in the time-unfolding network. These jumps ahead connections provide a shorter path for the propagation of gradient information, reducing the long-term dependence on the recurrent neural network [323]. Due to addressing the nonlinear system identification and latching problem, the NARX dynamic neural network is chosen for the estimation of SOC. The proposed model not only improves the estimation accuracy but also prevents the overfitting caused by long-term dependence on the structure of the recurrent neural network.

In addition, our embedded LSTM memory is not sophisticated embedded memory cells and simply consists of tapped delayed information to other neurons [324]. The hybrid NARX and LSTM model is shown in Figure 7.5. The hybrid NARX and LSTM model is mainly divided into two sections. First, the SOC of the battery is preliminarily predicted by the NARX recurrent neural network using the input current and voltage [325]. Second, the initial predicted SOC and measured current and voltage are transferred to the LSTM recurrent neural network to predict the SOC, which is shown in Figure 7.9.

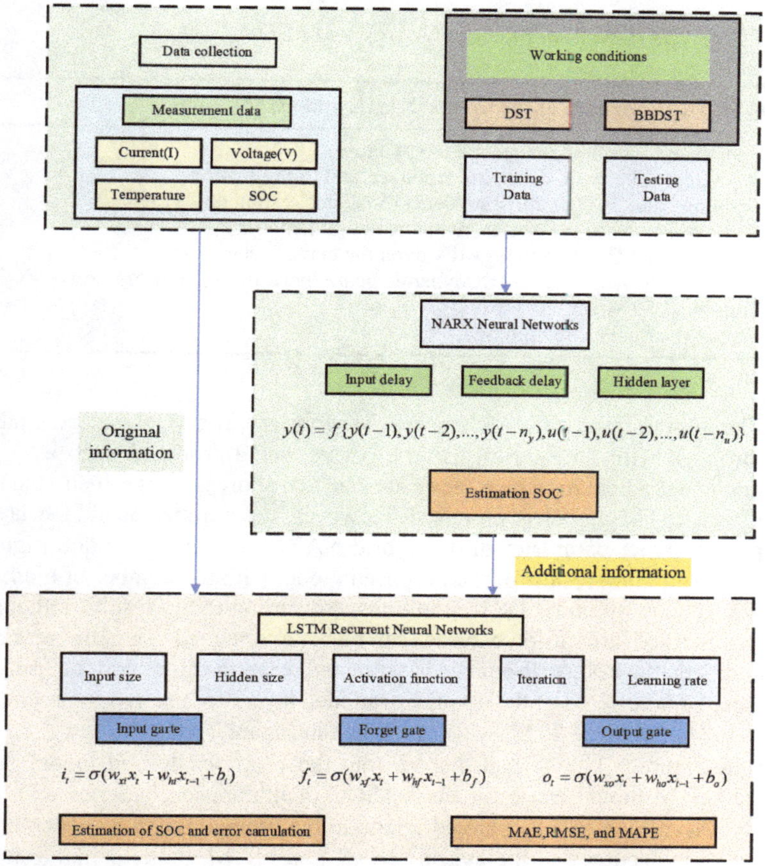

Figure 7.9 Schematic diagram of the hybrid NARX and LSTM model for SOC estimation

To better explain the LSTM-NARX model, Table 7.2 shows the brief running steps of the mixed model algorithm. Real-time data for lithium-ion batteries (current, voltage, and SOC) are used as training data and testing data for the hybrid NARX and LSTM models under DST conditions.

The training data includes current and voltage, which is divided into three parts (70% training data, 15% verification data, and 15% test data). Input delay, feedback delay, and hidden layers are set to 1:2, 1:2, and 20, respectively. The primary estimation of SOC for lithium-ion batteries is obtained through the NARX dynamic neural network, and then 3,000 samples including primary SOC estimations as well as initial current and voltage are input into the LSTM model as training data. The LSTM number is 70, the number of iterations is 100, and the learning rate is 0.01. After training, the trained hybrid NARX and LSTM model is used and the test samples are input for the final SOC estimation, which is shown in Figure 7.10.

Table 7.2 The algorithm of hybrid NARX and LSTM

Algorithm: The algorithm of the hybrid NARX and LSTM model
Step1. Input data (current, voltage, and temperature)
Step2. Input delay, feedback delay, hidden layer, and train function
Step3. Training date (70%), validation data (15%), and testing data (15%)
Step4. Train the model of the NARX dynamic neural network
Step5. Estimate the SOC with the NARX dynamic neural network
Step6. Add the SOC from the NARX dynamic neural network to the database
Step7. Train the LSTM neural network
Step8. Output the final estimation SOC

Although the standard NARX neural network can effectively address the non-linear problem with time series, as time passes, some problems such as gradient explosion and gradient disappearance can occur. To address this problem, the hybrid NARX and LSTM model is proposed. However, the number of hidden layers is an important hyper-parameter of the hybrid NARX and LSTM model. Figure 7.6 shows the estimation results of the proposed model with the number of hidden layers at 60, 70, and 80 under DST conditions. According to the results, although the estimated SOC of lithium-ion batteries can be better obtained using different hidden layer numbers, the best performance is estimated when the number of neurons is 70. Therefore, 70 is selected as the number of hidden layers for the hyper-parameters of the hybrid NARX and LSTM, which is shown in Figure 7.11.

From Figure 7.11, we can discern that the proposed hybrid model and the standard LSTM model can estimate the SOC of lithium-ion batteries with higher accuracy. According to the enlarged image in the figures, it can be concluded that the estimation accuracy of the hybrid NARX and LSTM model is higher than the standard LSTM. The accuracy improves because the proposed model has two-stage feature preprocessing and combines the estimation results of the standard NARX

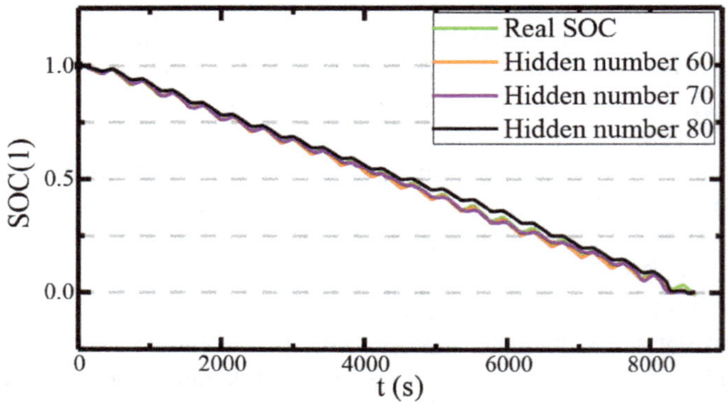

Figure 7.10 Results of SOC estimation

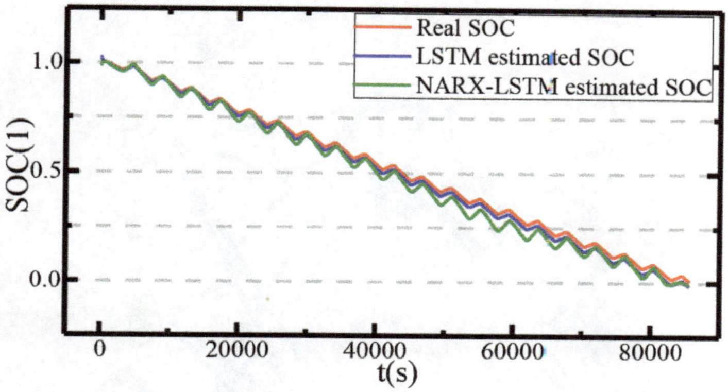

Figure 7.11 *Results of SOC estimation under DST condition*

dynamic neural network with the original input data as the input data for the secondary training, which has high estimation accuracy.

7.5 SOC estimation based on drosophila algorithm and BP neural network

7.5.1 BP neural network

At present, in addition to traditional filtering algorithms, neural network algorithms have also been widely used in the study of SOC estimation of lithium-ion batteries [326]. BP neural network has been widely used in different states estimation of lithium-ion batteries. Its basic topology is shown in Figure 7.12.

According to the topology diagram of BP neural network, its structure can be roughly divided into three layers, input layer, hidden layer, and output layer, respectively [327]. The input layer and output layer at the front end and the last segment of the model have one and only layer, and the hidden layer does not communicate directly with the outside world. The number of layers can be one or more layers. When the structure is used, if the number of hidden layers is set too much, the computational load and complexity of the model will be greatly increased, and the convergence speed will also be slowed down [328]. When dealing with nonlinear mapping problems, the three-layer neural network structure with one hidden layer can solve most problems.

The advantage of BP neural network is that there is not only the forward transfer of data in the whole algorithm operation but also the reverse transfer of the error between the output and the expectation, so as to continuously correct the weights of each connection and reduce the error [329]. The excellent generalization and fault-tolerance ability of BP neural network, as well as its strong processing ability for nonlinear mapping problems, make it widely used in the field of forecasting and have a very wide range of application value.

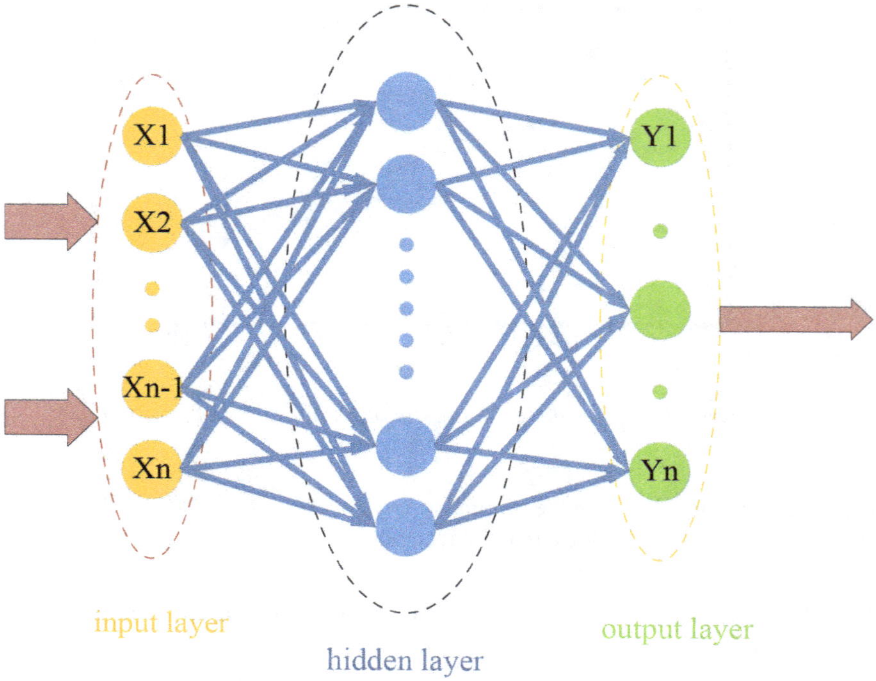

Figure 7.12 Basic Architecture of BP Neural Network

7.5.2 Drosophila optimization algorithm

The evolutionary calculation is a shared name, which refers to the "survival, unsatisfactory elimination" of Darwin evolution, and this concept is actually simulating the calculation method established by the natural circulation process, such as the early GAs [330]. However, the backpower began to move the gravity of gravity to the animal's foraging behavior and group behavior, the earliest ant colony algorithm, the algorithm is to release the body's secretion, so-called Felimon can find the shortest path to achieve an optimization. The calculations subsequently proposed, that is, Particle Swarm Optimization [331]. This algorithm is the foraging behavior of the Simulation Birdroup. By iteration, you can search for the surrounding area of the nearest bird, and finally close to the food, the best solution to the research.

These two algorithms are developed by the animal population, so some scholars are called group intelligence or group intelligence algorithms [332]. They also have to search for the best solution through iteration, and therefore also belong to the field of evolutionary calculations. In any case, the organisms must have their own physical energy and methods of living, these methods are also the direction in which many scholars work. In 2011, the foraging behavior from the fruit flies was inspired, based on the biological behavior simulation in Figure 7.13, the related particle swarm algorithm is proposed.

Figure 7.13 *Swarm intelligence algorithm idea source*

The Drosophila optimization algorithm is a new method to seek global optimization based on the foraging behavior of Drosophila [333]. Drosophila is superior to other species in sensory perception, especially smell and vision. The olfactory organs of fruit flies can collect various odors floating in the air very well, and can even smell food sources 40 kilometers away. Then, after flying near the food location, you can also use your keen vision to find where the food and companions gather, and fly in that direction. The principle of fruit fly optimization algorithm (FOA) is shown in Figure 7.14.

Drosophila algorithm is an algorithm that seeks the global optimal solution based on the foraging behavior of Drosophila. Drosophila, an organism that exists anywhere in the world, has extremely powerful olfactory and visual abilities, and based on this, its foraging properties provide a good source of algorithms. When it looks for food, its roadmap is the basic idea of we sort out the optimization algorithm of fruit flies.

The steps to search for food in Drosophila groups are as follows:

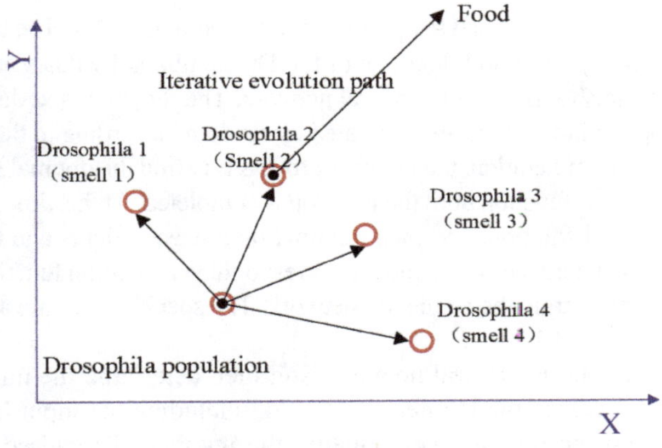

Figure 7.14 *The principle of FOA*

1. Random initialize the position of Drosophila.
   INIT X_AXIS
   INIT Y_AXIS
2. Give the fruit flies autonomous use of the random distance and direction of food.
   Xi = x_axis + random value
   Yi = y_axis + random value
3. Since the position of the food cannot be known, the distance (disc) of the origin is estimated, and the taste determination value (S) is calculated, and this value is the diagram of the distance.
   Disti = SQRT (xi ^ 2 + yi ^ 2)
   Si = 1 / DISTI
4. The taste concentration determination value (S) enters the taste concentration determination function (or a fitness function) to find the taste concentration of the individual position of the Drosity.
   Smelli = function (Si)
5. Find the fruit flies in the drosive concentration in the Drosophila group (see a large value).
   [Bestsmell BestIndex] = Max (SMell)
6. Keep the optimal taste concentration value and the coordinates of x, y, and the Drosophila group flew into this position.
   Smellbest = bestsmell
   X_AXIS = X (BestIndex)
   Y_axis = y (BestIndex)
7. Enter the iterative optimization, repeat steps 2 to 5, and determine whether the taste concentration is better than the proderation of the proderation, if so, step 6 is implemented.

7.5.3 Drosophila algorithm to optimize BP neural network

The Drosophila algorithm is used to optimize the BP neural network. The basic idea is as follows [334]. The distance and direction of the Drosophila individuals are regarded as the weights and thresholds in the BP neural network. The direction and distance of each group of Drosophila individuals are generated by random, according to the basic idea of optimization algorithm iteration, the iterative process is to find the optimal solution of the neural network fitness function until the iteration is completed. At this time, it is indicated to obtain the optimal solution, and then the final optimized weights and thresholds are brought into the BP neural network, and the corresponding activation function is selected. Then bring the data to train the BP neural network. The specific steps are as follows.

1. First initialize the BP neural network, and then determine the number of neurons in each layer in the BP neural network, including the input layer, the hidden layer and the output layer. Finally, the number of required weights and thresholds can be calculated according to the number of neurons in each layer.

2. Take each weight and threshold as a Drosophila individual and initialize their positions. The coordinate is shown in (7.32). Using a random function, assign a random direction and distance to each weight and threshold to search for the global optimal value, The result is shown in (7.33).

$$\begin{cases} x_i = x_0 \\ y_i = y_0 \end{cases} \tag{7.32}$$

$$\begin{cases} x_i = x_d + S \\ y_i = y_d + S \end{cases} \tag{7.33}$$

In equation (7.33), x_i, y_i are their respective horizontal and vertical coordinates, x_d, y_d, S represents random direction and distance.

3. Using the method of calculating distance in mathematics, the result is shown in (7.33). Bring the S into the adaptation function, the adaptation function is shown in (7.34).

$$\begin{cases} D_i = \sqrt{x_i^2 + y_i^2} \\ S_i = 1/D_i \end{cases} \tag{7.34}$$

$$T_i = F\left(S_i\right) \tag{7.35}$$

In (7.34), the distance D between the individual and the origin, the taste concentration judgment value S, where S is the reciprocal of the distance, where T is the adaptation function.

4. Find the minimum value, and write down the corresponding weights and thresholds are obtained.

$$G = min\left(T\right) \tag{7.36}$$

where G is the corresponding weight.

5. Repeat steps 2 to 4, and judge whether the value of the fitness function is better than the weights and thresholds of the previous generation, and if so, execute step 6.

$$\begin{cases} x_i = x_b \\ y_i = y_b \end{cases} \tag{7.37}$$

The coordinates here are the updated values of the coordinates.

6. The optimal weights and thresholds obtained in the iterative process are substituted into the BP neural network. Substitute the prepared test data into the trained BP neural network for verification.

7.5.4 Conclusion

This chapter combines the BP neural network with the Drosophila algorithm to give full play to the advantages of the two and overcome the local minimum defect of the BP neural network. The empirical analysis of the data through MATLAB shows that the optimized algorithm has less error and more accurate evaluation, which has certain research and application value.

7.6 SOH estimation based on sparrow search optimization-extreme learning machine network

7.6.1 Extreme learning machine network

Extreme learning machine (ELM) is a type of machine learning system or method based on feedforward neuron network (FNN), which is suitable for supervised learning and unsupervised learning problems.

ELM is regarded as a special kind of FNN in research, or an improvement of FNN and its backpropagation algorithm [335]. Calculate the output weight.

Traditional ELMs have a single hidden layer, which is considered to have low learning rate and generalization when compared with other shallow learning systems such as single layer perceptron and support vector machine. Ability may have advantages. Some improved versions of ELM gain deep structure by introducing autoencoders to build or stack hidden layers, enabling representation learning.

ELM can be discussed as a learning strategy (such as an improvement to the BP framework) or as a class of neural network construction. For the latter, the standard ELM uses the structure of the single layer feedforward neuron network (SLFN) [336]. Specifically, the composition of SLFN includes an input layer, a hidden layer and an output layer, where the output function of the hidden layer, as shown in (7.38).

$$f_L = \sum_{i=1}^{l} \beta_i h_i(x) = h(x)\beta \tag{7.38}$$

In equation (7.38), x is the input of the neural network and β is the output weight, $h(x)$ is called a feature map or activation function, and its function is to map the data of the input layer from its original space to the feature space of the ELM:

$$h(x) = G(\alpha_i, b_i, x) \tag{7.39}$$

In equation (7.39), α_i and b_i are the parameters of the feature map, α_i also known as node parameters in ELM research, where it is the input weights.

Since the feature map from the input layer to the hidden layer in ELM is random or artificially given and not adjusted, the feature map of ELM is random. According to the general approximation theorem, the feature map can be any nonlinear piecewise continuous function (piecewise continuous function).

Different hidden layer nodes can have different mapping functions, and the nodes of the neural network are also named by their feature maps, such as sigmoid nodes, radial basis function nodes, etc. In addition to the above mapping functions, the nodes of SLFN can also be other encapsulated computing units, such as fuzzy inference systems and other secondary neural networks. The basic block diagram of ELM is shown in Figure 7.15.

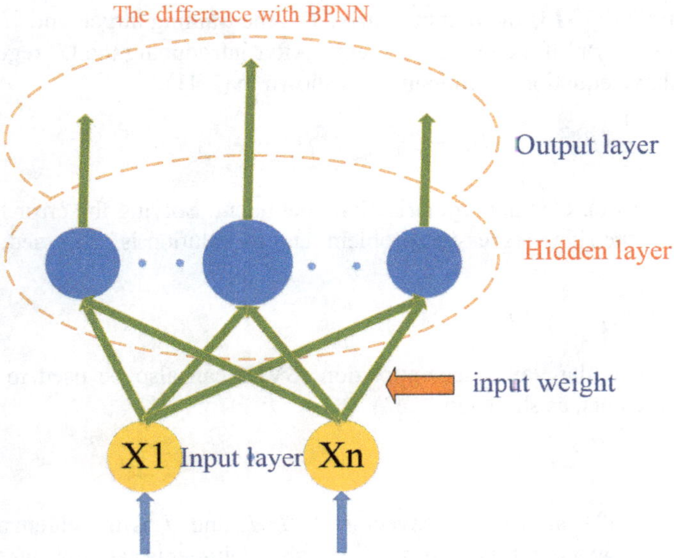

Figure 7.15 *Basic block diagram of extreme learning machine*

ELM only needs to solve the output weights, so it is a linear-in-the-parameter model, and its learning process is easy to converge at the global minimum. Knowing N groups of learning data, the following steps are taken to learn the ELM containing L hidden layer nodes and M output layer nodes:

1. Randomly assign node parameters: at the beginning of the calculation, the node parameters of SLFN are randomly generated, that is, the node parameters are independent of the input data. The random generation here can obey any continuous probability distribution.
2. Calculate the output matrix of the hidden layer: the size of the output matrix of the hidden layer is N rows and L columns, that is, the number of rows is the number of input training data, and the number of columns is the number of nodes in the hidden layer. The output matrix is essentially the result of mapping N input data to L nodes.
3. Solving the output weight: the size of the output weight matrix of the hidden layer is L rows and M columns, that is, the number of rows is the number of nodes in the hidden layer, and the number of columns is the number of nodes in the output layer. Different from other algorithms, in the ELM algorithm, the output layer can have no error nodes. The core of the ELM algorithm is to solve the output weight to minimize the error function.

The L2 loss function used by ELM is expressed, as shown in (7.40).

$$\min_{\beta \in R_{L\times m}} \|H\beta - T\|^2 \tag{7.40}$$

In equation (7.40), $\boldsymbol{H}$ is the output matrix, $\boldsymbol{T}$ is the training target, and $\|H\beta - T\|$ is the Frobenius norm of the matrix elements. After introducing the L2 regularization term, the above equation is reobtained, as shown in (7.41).

$$\min_{\beta \in R_{L \times m}} \frac{1}{2} \|\beta\|^2 + \frac{C}{2} \|H\beta - T\|^2 \tag{7.41}$$

In equation (7.41), $\boldsymbol{C}$ is the regularization coefficient. Solving the error function is equivalent to the ridge regression problem, and its solution is expressed, as shown in (7.42).

$$\beta* = \left(H^T H + \tfrac{1}{C}\right)^{-1} H^T H \tag{7.42}$$

In addition, singular value decomposition (SVD) can also be used to solve the weight coefficients, as shown in (7.43).

$$H\beta = \sum_{i=1}^{N} u_i \frac{d_i^2}{d_i^2 + C} u_i^T X \tag{7.43}$$

In equation (7.43), u_i is the eigenvector of $H^T H$, and d_i is the eigenvalue of $\boldsymbol{H}$. Studies have shown that relatively small weight coefficients can improve the stability and generalization ability of SLFN.

7.6.2 Sparrow search optimization

The sparrow search algorithm (SSA) was proposed in 2020. SSA is mainly inspired by the foraging behavior and antipredation behavior of sparrows [337]. The algorithm is relatively novel, and has the advantages of strong optimization ability and fast convergence speed.

To establish the mathematical model of the SSA, the main rules are as follows:

1. The finder usually has a high energy reserve and is responsible for searching for areas with abundant food in the entire population, providing all joiners with a foraging area and direction. In the establishment of the model, the level of energy reserve depends on the fitness value of the sparrow individual.
2. Once the sparrow detects a predator, the individual begins to chirp as an alarm signal. When the alarm value is greater than the safe value, the finder will take the joiner to other safe areas for foraging.

   The identities of discoverers and joiners are dynamic. As long as the better food source can be found, each sparrow can become a discoverer, but the proportion of discoverers and joiners in the entire population remains the same. In other words, if one sparrow becomes the discoverer, another sparrow becomes the joiner at the same time.
3. The lower the energy of the joiners, the worse their foraging position in the whole population. Some hungry joiners are more likely to fly elsewhere for more energy.
4. During the foraging process, the joiner can always search for the finder that provides the best food, and then obtain food from the best food or forage around

the finder. At the same time, some joiners may constantly monitor the finders and compete for food resources in order to increase their predation rate.

5. When aware of danger, the sparrows at the edge of the group will quickly move to the safe area to get a better position, and the sparrows in the middle of the group will move randomly to get close to other sparrows.

The algorithm flow can be divided into the following seven steps:

Step 1: Initialize the population, the number of iterations and initialize the ratio of predators and joiners.

Step 2: Calculate the fitness value and sort.

Step 3: Update the predator position.

Step 4: Update the position of the joiner.

Step 5: Update the position of the alerter.

Step 6: Calculate the fitness value and update the sparrow position.

Step 7: Whether the stop condition is met, exit if so, and output the result, otherwise, repeat Step 2 to 6.

7.6.3 Sparrow search optimization-extreme learning machine network

As can be seen from the foregoing, the initial weights and thresholds of ELM are randomly generated. However, the initial weights and thresholds generated each time are satisfactory [338]. In this chapter, the SSA is used to optimize the initial weights and thresholds. The fitness function is designed as the MSE of the error on the training set:

fitness = argmin (MSE pridect).

After training, the fitness function selects the MSE error, the smaller the MSE error, the higher the degree of coincidence between the predicted data and the original data. The output of the final optimization is the best initial weight and threshold. The network trained with the best initial weight threshold is then tested on the test dataset.

Chapter 8
Summary and prospect for AI-based battery status monitoring

8.1 Lessons learned

As the most important material found in the industrial age, energy plays an irreplaceable supporting role in the progress and development of human society. Party and state leaders also deeply recognized the importance of energy and made important instructions: 'Throughout the historical context of social progress and development, every major progress of human civilization is also accompanied by the continuous improvement and development of energy' [339]. As the evolution of the international energy crisis and the pollution caused by fossil energy has drawn international public attention, the international energy landscape is undergoing profound changes [340]. The biggest change is that renewable energy, mainly clean energy, has an increasing proportion in the country's total energy.

Energy storage technology includes many technical routes. Several common schemes include pumped water energy storage, electrochemical energy storage, flywheel energy storage, compressed air energy storage, etc. However, among them, electrochemical energy storage has many advantages [341]. The construction period is short, the effect is quick, and the configuration is flexible, so it is recognized as the most potential and fastest-growing direction in the field of energy storage [342]. In electrochemical energy storage, battery energy storage systems account for the largest installed capacity and market share.

In the battery energy storage system, the battery used as energy storage is not used as a single unit, but the battery cells are connected in series and parallel through a circuit to form a battery pack, to achieve the required voltage and electrical body differences required by the system, resulting in the same. The increased inconsistency of the different cells in a battery pack leads to internal discharges, resulting in energy loss and reduced lifespan, as well as increased probability of failure and increased operating and maintenance costs.

For the battery energy storage system, the accuracy and reliability of the battery state of charge (SOC) and the state of health (SOH) built into the battery management system (BMS) are very important [343]. By reviewing the relevant research work in this field at home and abroad, it can be found that most of the existing prediction research on the SOC and SOH is still based on the battery equivalent model or from the perspective of the internal electrochemical mechanism of the battery. The vast amount of battery operation data in practical engineering applications has not been effectively utilized, and it is difficult to substantially improve the prediction accuracy and generalization ability.

This chapter mainly studies the application of artificial intelligence (AI) to battery condition monitoring [344]. Chapter 2 introduces the characteristics of practical-scale lithium-ion battery systems. Chapter 3 introduces AI-based technologies: equivalent modeling and parameter identification. Chapter 4 describes the application of AI in utility-scale battery systems. The main content of Chapter 5 is the AI-based SOC estimation. The sixth chapter is AI-based parameter estimation. The seventh chapter is examples and case studies of utility-scale battery operation. And this chapter is the summary and prospect of battery state monitoring based on AI.

Regarding the state monitoring of batteries, the more important and commonly used are SOC monitoring, SOH monitoring, state of power monitoring and state of energy monitoring [345]. Among them, SOC is usually used to describe the remaining power of the battery and is the most critical part of the battery system. However, the technical problems such as the difficulty of modeling the SOC estimation system, the large estimation error, and the inability of dynamic online prediction have greatly restricted the development of the battery [346]. Although a lot of research has been done on SOC at home and abroad, its estimation accuracy still cannot meet the requirements of modern people; in addition, most of the current research on batteries at home and abroad is still at the level of static research, and it is difficult to achieve dynamic online continuous monitoring. And with the increase of its output, the same amount of waste batteries will be produced, but the cyclicity of the battery makes it still usable in low-power electrical appliances after being eliminated from high-power products, so the detection of battery residual energy is one of today's new opportunity [347]. However, how to establish a high-precision lithium battery charging and discharging model, dynamically and accurately estimate the SOC of the battery online, and detect the residual energy of the used battery to improve its utilization rate is an unavoidable problem in the application of lithium batteries [348]. Therefore, it is of great significance to carry out high-precision modeling of its SOC, realize dynamic online estimation and predict the residual energy of used batteries [349]. With the development of battery technology, SOC estimation methods have been quite rich, including some traditional methods, such as discharge method, open-circuit voltage (OCV) method, ampere-hour method, load method, etc., as well as some novel methods such as Kalman method, fuzzy logic method, neural network method, etc. However, using a certain estimation method alone has certain defects. The following are several commonly used battery SOC estimation methods, which are shown in Figure 8.1.

From Figure 8.1, we can see that the current commonly used SOC estimation methods mainly include: the OCV method, discharge experiment method, ampere-hour integration method, Kalman filter (KF) method, and neural network method.

Among them, the OCV method has shown that there is a certain correlation between the terminal voltage of the battery and the SOC of the battery through a large number of experimental studies [350]. By drawing the terminal voltage-SOC curve, the SOC at each terminal voltage can be obtained [351]. However, this

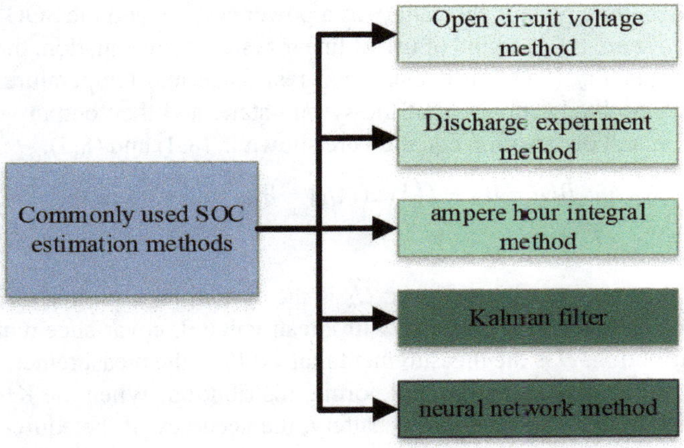

Figure 8.1 Several commonly used battery SOC estimation methods

method takes a lot of time to obtain a stable terminal voltage value to obtain an accurate initial SOC value, and the efficiency is also low, which is not conducive to the online estimation of SOC.

The discharge experiment method is the easiest and most accurate among the traditional SOC prediction and estimation methods, but its efficiency is not high [352]. This method continuously discharges with a small rate of constant current and stops when the discharge reaches the cut-off voltage. The product of the discharge current and time is the remaining capacity of the battery, as shown in (8.1) and (8.2).

$$SOC(t) = \frac{Q_{remaining}(t)}{Q_{rated}} \tag{8.1}$$

$$Q_{remaining}(t) = \int_{t}^{t_1} Idt \tag{8.2}$$

Among them: t_1 is the discharge cut-off time, and I is the constant discharge current. The main principle of the ampere-hour integration method is: on the premise that the initial remaining power value is known, use the integral to calculate the amount of charge that changes during the charging and discharging process of the battery, and then subtract the amount of charge that changes from the known initial value SOC_0 of the battery. In the current SOC value map, the specific calculation formula is as shown in (8.3).

$$SOC = SOC_0 - \frac{1}{C_N} \int \eta I dt \tag{8.3}$$

Among them, SOC_0 is the initial value; C_N is the rated capacity of the battery; η is the coulombic efficiency; I is the current in the loop. The ampere-hour integration method is simple to calculate and can be estimated online in real time, but it cannot accurately estimate the initial value. As the measurement progresses, the error will continue to accumulate, and it cannot deal with the self-discharge problem of the power battery, resulting in insufficient accuracy.

The KF method regards the battery as a power system, and the SOC is only one state of the system. KF is a kind of use of linear system state equation, through input observation quantity (U_k), current, internal resistance and temperature, and other variables, the optimal estimation of the system state, and then output voltage. The state equation and observation equation are shown in (8.4) and (8.5).

$$X_k = AX_{k-1} + BU_k + W_k = f(X_{k-1}, U_k) + W_k \tag{8.4}$$

$$Z_k = HX_k + V_k = g(X_k, U_k) \tag{8.5}$$

where A is the state transition matrix; U_k is the system input vector; B is the input gain matrix; W_k is the process noise with mean value 0, covariance matrix Q, and normal distribution. H is the measurement matrix; V_k is the measurement noise with zero mean, R covariance matrix, and normal distribution. When the KF method is used to estimate the SOC of a power battery, the accuracy of the equivalent circuit model (ECM) of a power battery is very important, which greatly affects the accuracy of algorithm estimation.

The main idea of the neural network algorithm is to continuously adjust the weight and bias of the model in the learning process to reduce the error of the model, thereby improving the accuracy of the model [353]. The algorithm can run in a nonlinear system, and the battery is just such a system. It is difficult to establish a relatively accurate mathematical model for the whole process. Under certain conditions, the artificial neural network (ANN) with nonlinear and strong learning ability can estimate the remaining capacity of the battery quickly and accurately.

BP (Back Propagation) neural network is a network with multiple hidden layers and multiple outputs [354]. It modifies the weights of the network layer by layer through the backpropagation of errors and continuously learns the sample data to calculate the output value. When there is a gap between the actual output and the set target output, the error is back-propagated to the input layer of the network to correct the weights of each layer until the output result matches the target value set by the network [355]. The BP algorithm generally adopts a three-layer network structure, as shown in Figure 8.2.

In Figure 8.2, $x = (x_1, x_2, \cdots\cdots, x_n)^T$ is the input vector, $Y = (Y_1, \cdots\cdots, Y_m)^T$ is the output vector, v_{ij}, ω_{jk} are the distance between the input layer and the hidden layer, respectively, the connection weights and the connection weights between the hidden layer and the output layer. The output layer is shown in (8.6) and (8.7).

$$o_k = f(net_k) \quad k = 1, 2, \cdots\cdots, l \tag{8.6}$$

$$net_k = \sum_{j=0}^{m} \omega_{jk} y_i - \alpha_k \quad k = 1, 2, \cdots\cdots l \tag{8.7}$$

where $o(k)$ is the output of k nodes in the output layer, w_{jk} is the weight coefficient from the input layer to the hidden layer, y_i is the input, and α_k is the threshold. For the hidden layer, there are (8.8) and (8.9).

$$y_i = f(net_j) \quad j = 1, 2, \cdots\cdots m \tag{8.8}$$

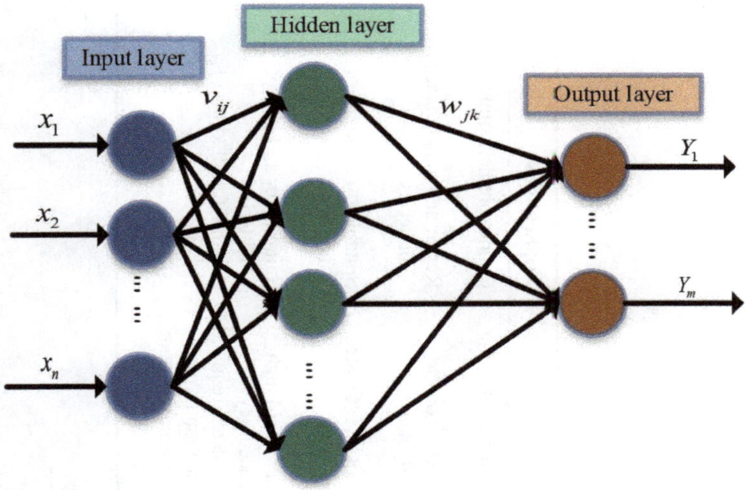

Figure 8.2 BP network structure

$$net_j = \sum_{i=0}^{n} v_{ij}x_i - \alpha_j \quad j = 1, 2, \cdots\cdots, m \tag{8.9}$$

where v_{ij} is the weight coefficient of the hidden layer and the output layer, and α_j is the threshold. In the above formula, the transformation function $f(x)$ adopts the unipolar Sigmoid function, as shown in (8.10).

$$f(x) = \frac{1}{1 + e^{-x}} \tag{8.10}$$

The neural network method does not need an accurate mathematical model but can learn the internal law of a nonlinear system by learning sample data, and a good neural network model can approximate nonlinear mapping with arbitrary precision. However, the standard BP neural network algorithm also has some shortcomings in the application.

1. The network is easy to forget the old samples in the training process.
2. Easy to fall into a local minimum.
3. The network convergence speed is slow.
4. The number of hidden layer nodes mostly relies on empirical formulas, lacking professional theoretical guidance.

Through the analysis of the above-mentioned methods, each method has certain limitations and there is still a large room for improvement. Choosing an appropriate SOC estimation method is the key to the development of battery technology. Comparison of several commonly used SOC prediction methods.

Table 8.1 Advantages and disadvantages of several common SOC estimation methods

SOC estimation method	Advantage	Disadvantages	Applications
Open-circuit voltage method	Simple and easy, high precision	The battery needs to be left for a long time	Cannot predict online
Discharge experiment method	Simple and easy to implement	The prediction accuracy depends on the measuring instrument and takes a long time	Repair and maintenance of batteries
Ampere hour integral method	A simple calculation, with high precision, can be measured online [356]	The initial SOC value cannot be estimated, and there is a cumulative error	Suitable for all kinds of batteries
Kalman filter	Both the SOC estimate and its estimate error can be obtained	A large amount of computation and high requirements for model accuracy	Suitable for all kinds of batteries
Neural network method	No precise mathematical model is required, fast, convenient, high-precision and dynamic prediction can be performed	A large number of sample data are required for training, and the estimation error is greatly affected by test data and training method [357]	Suitable for all kinds of batteries

SOH estimation is the basis of RUL prediction and will directly affect the accuracy of RUL prediction. However, accurately estimating SOH under real operating conditions remains a challenge. This is because it is often impossible to measure SOH directly, and due to the complexity of design and fabrication, as well as environmental and operating conditions, it is difficult to estimate all factors that affect cell performance degradation. Nonetheless, there are still three main parameters that can help estimate SOH: number of cycles, capacity and internal resistance. Since capacity is an important measure of battery capacity, SOH. There are currently no model-based methods available to accurately describe the dynamic behavior of all lithium-ion batteries aging under various operating states [358]. Because the simplified model of the physicochemical reactions occurring in the battery cannot fully reflect the complex working mode faced by the battery, the SOH estimation accuracy based on the model method is seriously affected. By monitoring the value of SOH in real time, it is possible to understand the health of each cell in the battery pack. Replacing aging batteries early can have far-reaching implications for extending battery life, reducing accidents caused by battery aging, improving battery pack performance, and accurately estimating EV mileage. The condition of battery failure is that the SOH reaches 80%, and the accurate estimation of the SOH of the battery can provide a basis for the prediction of RUL. At present, the most accurate method for estimating the SOH of a battery is the full discharge method, also known as the capacity method. The idea of the traditional capacity method is that the ratio of the current maximum capacity of the battery to the maximum actual capacity of the new battery is the current SOH of the battery. From the perspective of battery capacity, it is necessary to measure and calculate the discharge amount of battery SOC from 100 to 0%. Its mathematical expression is shown in (8.11).

$$SOH = \frac{C_M}{C_N} \tag{8.11}$$

Among them, C_M is the discharge amount when the battery SOC value is discharged from 100% to 0%, and C_N is the discharge amount when the battery's initial SOC value is discharged from 100% to 0%. However, this method cannot estimate the SOH value of the battery online, but can only be estimated offline, and it takes a long time to estimate the SOH, and the battery must be completely discharged. For the complete discharge process of the battery, based on the SOC of the battery, it can be considered that the fully discharged state of the battery corresponds to the battery SOC of 100%, and the fully discharged state of the battery corresponds to the battery SOC of 0%, therefore, the complete discharge process of the battery is the process in which the SOC drops from 100% to 0%. This deep discharge is an overdischarge, which reduces the life of the battery.

As the SOH of the battery changes, the internal resistance will also change. The increase of the internal resistance makes the battery age, so SOH can be defined from the perspective of the internal resistance of the battery, as shown in (8.12).

$$SOH_R = \frac{R_{EOL} - R_{NOW}}{R_{EOL} - R_{NEW}} \tag{8.12}$$

where R_{EOL} represents the internal resistance of the battery when the battery SOH is 80%, R_{NEW} represents the internal resistance of the battery SOH is 100%, and R_{NOW} represents the internal resistance of the battery in its current state. This SOH estimation scheme requires an accurate value of internal resistance R.

Cycle number method: determine the stage of battery life by counting the number of cycles of the battery. The battery life is terminated when the cycle number reaches a certain level. This method needs to consider the influence of battery working environment, battery charge and discharge, self-discharge, and other factors on battery life. Different batteries have different working conditions, so this method is not high precision and has low efficiency.

Impedance measurement: ohmic resistance, Solid Electrolyte Interphase (SEI) film resistance, charge transfer impedance, double-layer capacitance, and other parameters can be obtained by electrochemical impedance spectroscopy (EIS). Galeotti *et al.* found that the ohmic resistance changed exponentially with battery life, and this relationship became more and more obvious with the continuous charging and discharging cycle of the battery. However, EIS was greatly affected by different types of batteries, which may lead to large errors in actual use.

Destructive methods: to explain the cell aging mechanism from the perspective of the cell microscopic mechanism, the commonly used methods are Raman spectroscopy, X-ray diffraction, scanning electron microscopy, etc., these methods need to strip the battery, irreversible damage to the battery, so it can only be studied in the laboratory. Li *et al*. found that the structure of the positive electrode changed, including decomposition of active substances and changes in stress, and led to the gradual aging of batteries under storage conditions.

The method is based on an ECM; using the ECM composed of basic circuit components such as resistance, inductance, and capacitance, the corresponding model parameters are obtained through a large number of long-term experimental data, and then the SOH of the battery is predicted.

The method based on the electrochemical model: this method contains many nonlinear and coupled partial differential equations. It can describe the electrochemical dynamics inside the battery in detail, but it has not been well promoted and applied because of its huge amount of calculation, difficulty in solving, and low efficiency, which is not conducive to the online real-time estimation of the battery state.

SOH represents the degree of battery performance aging. At present, most researchers use two parameters of battery ohmic resistance R and capacity to predict battery SOH [359]. From the perspective of practical engineering applications, the mainstream method for estimating SOH is based on the establishment of battery ECM.

8.2 Open questions

This book studies battery state monitoring based on AI and introduces the principles and applications of batteries based on AI technology from different aspects [360]. By introducing utility-scale lithium-ion battery system characteristics, AI-based techniques: equivalent modeling and parameter identification, AI applications in utility-scale battery systems, AI-based state-of-charge estimation, parameter estimation, utility-scale batteries operation examples and case studies, etc., comprehensively expound the research process of battery state monitoring based on AI.

Commonly used AI algorithms mainly include decision trees, random forest, logistic regression, linear regression, naive Bayes, neural network, support vector machine (SVM), K-nearest neighbor algorithm, etc. This book mainly introduces and discusses some AI algorithms and other algorithms that can be further studied in the future. Among them, neural networks are suitable for an input that may fall into at least two categories: Neural Networks (NN) consists of several layers of neurons and the connections between them. The first layer is the input layer, and the last layer is the output layer [361]. Each layer has its classifier in the hidden layer and output layer. Input is input into the network, activated, and the calculated scores are transferred to the next layer, which activates the neural layer behind. Finally, the scores on the nodes of the output layer represent the scores belonging to various types. The classification result of the following example is Class 1. The same input is transmitted to different nodes, so different results can be obtained because each node has different weights and biases, which is also forward propagation. The artificial neuron is the basic unit of the Artificial Neural Networks (ANN), which is a multi-input, single-output nonlinear component, as shown in Figure 8.3.

In Figure 8.3, an artificial neuron receives one or more inputs (representing excitatory and inhibitory postsynaptic potentials at neural dendrites) and sums them to produce an output (or activation, representing the neuron passing along its axon, action potential). Usually, each input is weighted individually and the sum is passed through a nonlinear function called the activation function or transfer function.

Transfer functions usually have a sigmoid shape, but they may also take the form of other nonlinear functions, piecewise linear functions or step functions. They are also typically monotonically increasing, continuous, differentiable, and bounded [362]. The sigmoid neuron network is shown in Figure 8.4.

As can be seen in Figure 8.4, sigmoid neurons are very close to perceptron, but modifying their weights and offsets very little can cause only small changes in the output [363]. This is the decisive fact that enables sigmoid neural networks to learn.

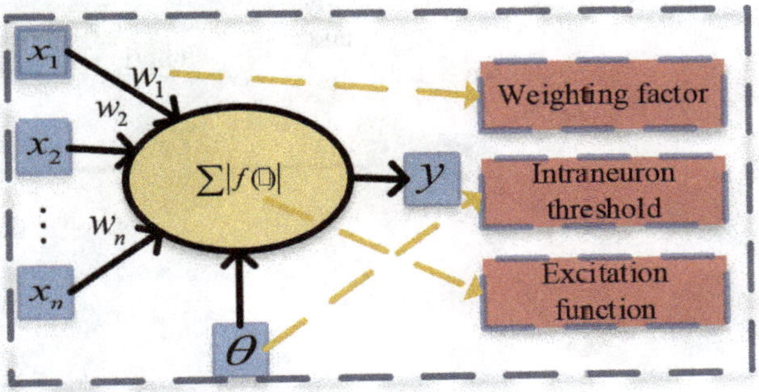

Figure 8.3 Schematic diagram of artificial neuron

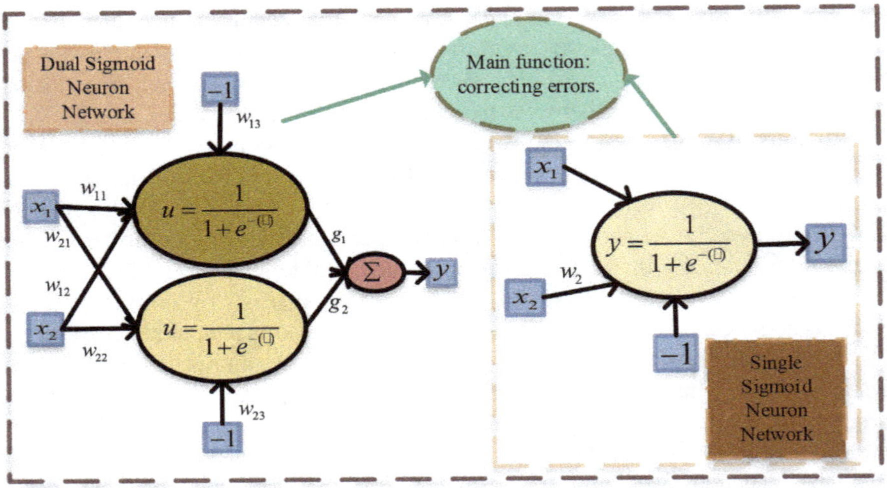

Figure 8.4 The sigmoid neuron network

The ANN is essentially multiple neurons connected according to certain rules, M–P neuron is one of them, because it has only one layer, the learning ability is very limited, until the BP algorithm appeared, the basic structure is shown in Figure 8.5.

From Figure 8.5, an M–P neuron model consists of input weights, activation functions, and outputs. Due to its limited learning and fitting ability, it is gradually rarely used, and an in-depth study of the BP algorithm is called the general trend. This book covers some of the specifics of neural networks in detail, and there are deeper layers to be explored. From this book, we find that

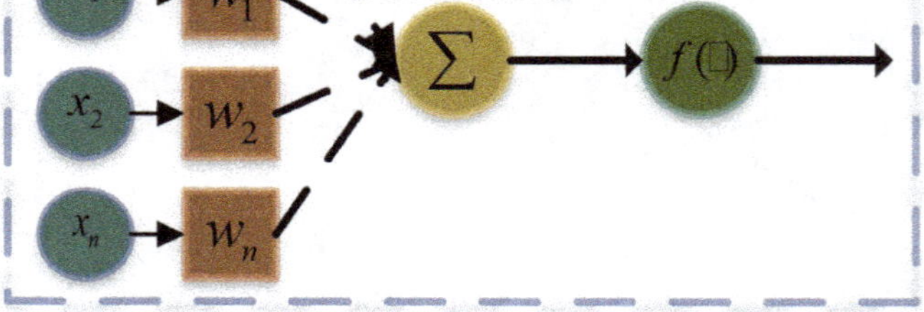

Figure 8.5 The basic structure of the M–P neuron model

there are some problems to be solved in the current research and problems that fail to propose more effective solutions.

1. There are few kinds of literature on the relationship between internal resistance and SOC, partly because the equipment for measuring internal resistance is applied to electric vehicles, which will increase a lot – part of the cost. The online internal resistance detection method with high reliability and easy implementation can be studied.
2. At the expense of estimation time, the estimation accuracy of SOH and RUL is improved. In the future, research can also be carried out to shorten the prediction time while improving the prediction accuracy.
3. At this stage, energy problems have gradually made finding and using clean and efficient renewable new energy an inevitable choice for the development of human society, and rationally improving the utilization rate of existing resources is also a top priority. Among them, it is still relatively difficult to accurately estimate the remaining power of the battery pack, and the estimation error will lead to a low power utilization rate of the battery.
4. Although the current research at home and abroad uses different algorithm models to estimate the SOC of lithium batteries, there is a problem with the estimation error of the SOC. For example, although the ampere-hour integration method and the OCV method are combined, there is still the problem of SOC estimation caused by the current accumulation error. Another example is the systematic error problem introduced by converting the nonlinear algorithm model into a linear model through the extended KF algorithm.
5. Based on a data-driven way of accurate estimate on the battery charged state, which adopts the characteristics of the training sample, mainly from BMS collected features such as voltage, current, temperature, if the future can be collected by other sensors characteristics of some electrochemical changes inside the battery, it can better represent the battery charged shape change, it will improve the accurate estimation of battery SOC in charge state.
6. Although the lithium-ion battery SOC estimation algorithm based on short-term and long-term memory network storage has been optimized by batchnormalization (BN) algorithm and adaptive denoising encoder, the model is larger and requires greater training resources and time. Model the pruning and calculating the force of embedded systems, or with the aid of edge-calculating, 5G and other technologies to guide the computing process to the cloud, and finally apply the algorithm to the BMS.
7. Due to the limitation of the conditions, the charge–discharge experiments of the battery in this book are carried out at room temperature, and the temperature range is relatively small. Therefore, the next step will be to conduct experiments on SOC and secondary batteries over a wider temperature range to more accurately predict the residual energy of SOC and old batteries.
8. When the BP network prediction model is established, only the voltage U, current I, and temperature T are selected as three main influencing factors for input parameters, the prediction model is established without considering the

impact of recycling and aging on the estimation of battery SOC in the future, its impact factor are willfully considered, which makes more accurate estimation precision.

9. The neural network method is suitable for all kinds of batteries, has a strong learning ability for nonlinear systems, and can simulate the dynamic characteristics of batteries; the disadvantage is that it requires a large amount of sample data for training, the calculation amount is large, the computer requirements are high, and the SOC estimation is difficult. The error is also affected by the training data and training method. In addition, the selection and determination of the number of input variables will also directly affect the accuracy and computational complexity of the neural model.

In addition, through this research and study, through the exploration of AI-based battery state monitoring, we not only found some problems to be solved but also found some problems that need to be expanded and deepened.

1. The data used in the experiments in this book are relatively single, and this single battery data may have an impact on the generalization ability of the model. In the future, we will continue to complete the collection of charge and discharge data for other batteries, and then retrain the model.
2. To facilitate the neural network to receive lithium battery data normally, this chapter processes all input data into vectors of the same length in the data preprocessing process. The long-term memory neural network can receive variable-length data. In the future, we will try to input the variable-length data into the network for training, which will reduce the time for data preprocessing and improve the real-time performance of prediction.
3. This book optimizes the hyperparameters of neural networks through Bayesian optimization, but hyperparameter optimization is only a branch of automatic machine learning (ML). In the follow-up work, I will continue to try to introduce architecture search technology to achieve fully automatic training of neural networks.
4. In the real operating environment, there are still factors such as the internal resistance value of the battery, the number of cycles, and the degree of aging, which will all have an impact on the SOC of the battery. At this stage, there is no clear conclusion to study the influence of these factors on lithium batteries, which can be used as the direction of subsequent in-depth research, so that the SOC of lithium batteries can be more accurately estimated.
5. In terms of algorithms, it can be considered to further add a variety of algorithms for combined comparison, such as unscented Kalman filter, particle filter algorithm, etc.
6. This book does not consider the impact of lithium-ion battery overcharge and overdischarge and the inconsistency of the SOC and SOH collaborative estimation algorithm. In the future, we should consider adding battery balancing technology to link SOC and SOH with battery balancing.

8.3 Opportunities and challenges

With the rapid development of big data and deep neural network, AI has made great breakthroughs in machine vision [364], speech analysis, and natural language processing, and has been applied to smart cities, smart homes, industrial manufacturing, and other fields, gradually changing the operation mode of society and people's way of life [365]. In recent years, with the rapid development of communication technology, high-end electronic mobile devices, electric vehicles, and space technology, people rely on mobile power more and more urgently, so they also put forward higher requirements for the performance of rechargeable power [366]. Therefore, the introduction of AI into the status monitoring of utility-scale batteries is an inevitable trend of science and industry development, it will inject new vitality into the battery industry and come to an unprecedented breakthrough [367]. Considering the characteristics of AI and the actual engineering requirements of battery state estimation, this section summarizes the opportunities and challenges of AI in the status monitoring of utility-scale batteries.

Opportunities

1. New method

The neural network has a strong nonlinear processing ability and self-learning ability, especially suitable for dealing with the nonlinear internal structure of lithium batteries [368]. Because the information is distributed and stored in the neurons of the network, the fault tolerance and robustness are high. As shown in Figure 8.6, there is a schematic diagram of a typical neural network model.

As can be seen from Figure 8.6, the signals are summed after being weighted with different weights, and then processed by the activation function to output the results. The corresponding calculation principle of the neural network is shown in (8.13).

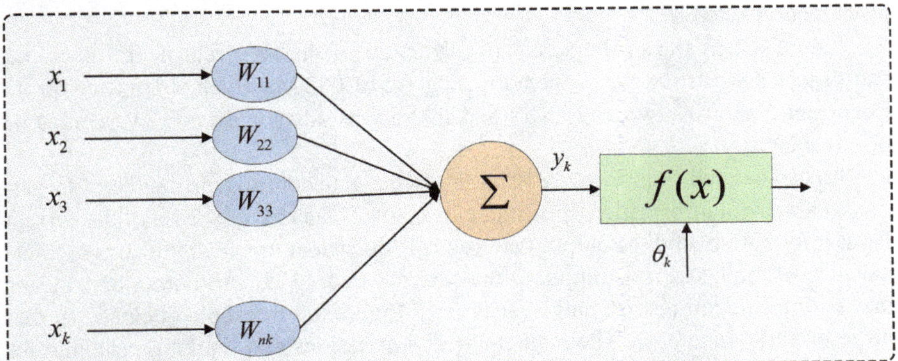

Figure 8.6 Schematic diagram of a typical neural network model

$$\begin{cases} y = \sigma(wx + b) \\ b_k \theta_j j W_{ik} ik V_{kj} \\ h_k = \sigma(\sum_{i=1}^{m} W_{ik} x_i + b_k) \\ y_j = \sigma(\sum_{k=1}^{l} V_{kj} h_k + \theta_j) \\ (x, y)(E) \\ \bar{y} = (\bar{y}_1, \bar{y}_2, \ldots, \bar{y}_n) \\ E = \frac{1}{2} \sum_{j=1}^{n} (\bar{y}_j - y_j)^2 \end{cases} \tag{8.13}$$

where the input quantity corresponding to x is expressed as $x = [x_0, x_1, x_2, \cdots, x_n]$, the weight value connected to it is $w = [w_0, w_1, w_2, \cdots, w_n]$, the corresponding threshold is b, and the final activation function is $\sigma(x)$. The output is y.

Moreover, the neural network has a strong ability for information synthesis, processing quantitative and qualitative information at the same time, and coordinating various input information relations well [369]. Therefore, the neural network does not need to build the battery model [370], but only needs to build the network structure and pay attention to the input and output of the neural network. Some researchers have used BP neural network, Convolutional Neural Network (CNN), and Long Short-Term Memory (LSTM) to achieve the prediction of battery state [371]. The BP neural network model for SOC and SOH estimation is shown in Figure 8.7.

A logical relationship among the input layer, hidden layer and an output layer of the BP neural network can be seen in Figure 8.7 [372]. The input layer parameters are voltage, current, temperature, internal resistance, and other parameters related to SOC and SOH estimation accuracy [373]. The number of nodes in the hidden layer can be set according to the complexity of the processing problem and the number of input parameters [374]. The output of the output layer is the prediction target SOC of the neural network.

Figure 8.8(a) shows the schematic structure of the recurrent neural network, and Figure 8.8(b) shows the schematic diagram of LSTM, which is improved from Recurrent Neural Networks (RNN) and inherits its advantages and overcomes its disadvantages.

As can be seen from Figure 8.8(b), the forget gate controls the degree of forgetting of historical information, the input gate controls the memory level of the current input information, and the output gate controls the extent to which the current internal state of the network is output to the external state [375]. An output of 1 means that all information passes, and 0 means that Figure 8.8 threshold is closed so that no information can pass. The mathematical expressions and working principle for the gates are, respectively, shown as follows (8.14)–(8.17).

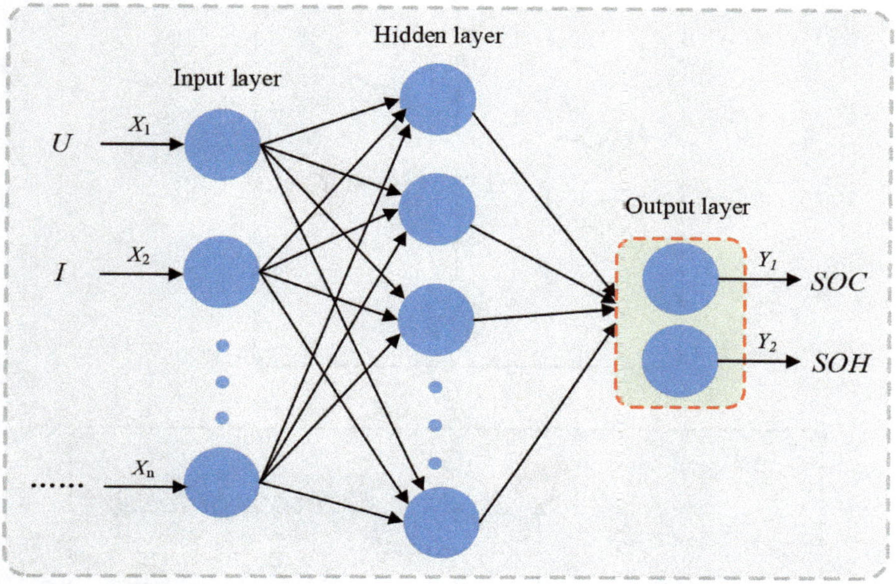

Figure 8.7 BP neural network model for SOC and SOH estimation

- Forget gate:

$$f_t = \sigma * \left(W_f * \left[x_t, h_{t-1}\right] + b_f\right) \tag{8.14}$$

- Input gate:

$$\begin{cases} i_t = \sigma * \left(W_i * \left[x_t, h_{t-1}\right] + b_i\right) \\ \widetilde{C}_t = tanh * \left(W_c * \left[x_t, h_{t-1}\right] + b_c\right) \end{cases} \tag{8.15}$$

- Cell state update:

$$C_t = f_t * C_{t-1} + i_t * \widetilde{C}_t \tag{8.16}$$

- Output gate:

$$\begin{cases} o_t = \sigma * \left(W_o * \left[x_t, h_{t-1}\right] + b_o\right) \\ h_t = o_t * tanh * \left(C_t\right) \end{cases} \tag{8.17}$$

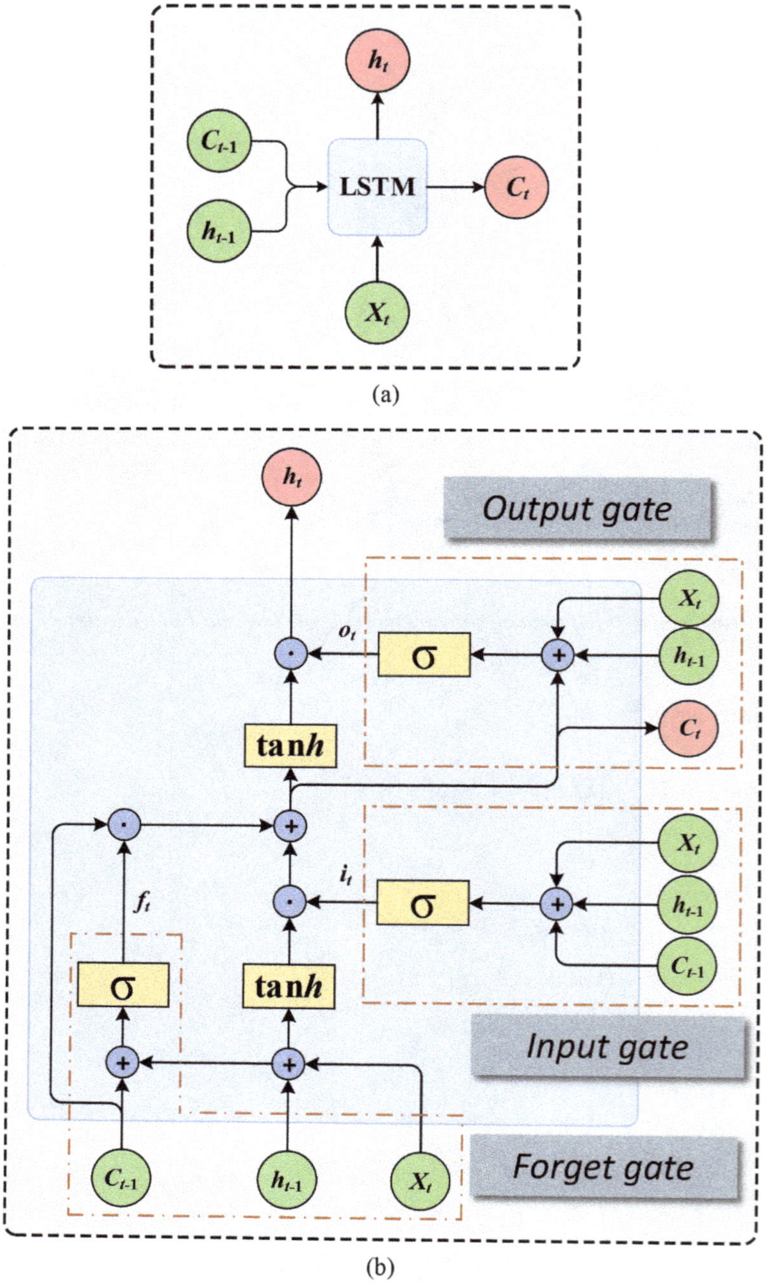

Figure 8.8 *Schematic diagram of LSTM. (a) Schematic diagram of recurrent neural network structure. (b) Schematic diagram of LSTM.*

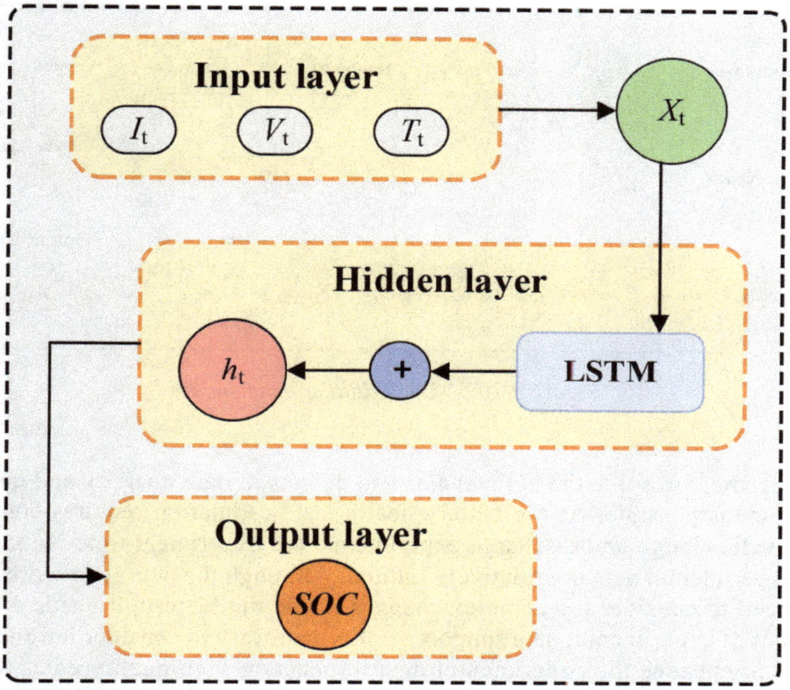

Figure 8.9 The model of SOC estimation based on LSTM

Each gate in the network has a weight W_f, W_i, W_c, and W_o associated with the forget gate, input gate, memory cell, and output gate, respectively. Also, each gate of the network possesses a bias b_f, b_i, b_c, and b_o vector by the forget gate, input gate, memory cell, and output gate, respectively, to enhance the flexibility of the network to adapt to the training data for accurate SOC. The model of SOC estimation based on LSTM is shown in Figure 8.9.

But considering that the development of the neural network is still not mature, there is still a large space for improvement and development potential, so there is still a lot of work to be done in the study of neural networks.

2. New idea

Generative adversarial networks (GAN) were proposed by Goodfellow in 2014 and hailed as the most imaginative deep learning network model in recent years due to his pioneering ideas. Its variants, such as Conditional Generative Adversarial Networks (CGAN), Deep Convolution Generative Adversarial Networks (DCGAN), Wasserstein Generative Adversarial Networks (WGAN), and Wasserstein Generative Adversarial Networks with gradient penalty (WGAN-GP), can generate the same distributed data through training. In actual personal research, it is more practical to make full use of existing data for data

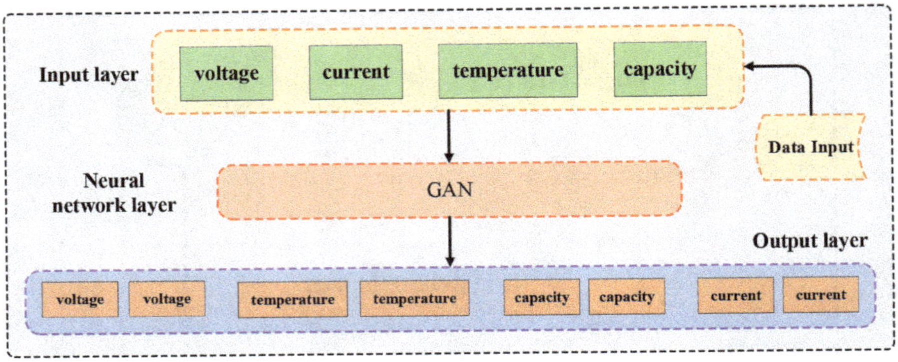

Figure 8.10 GAN data enhancement

enhancement to solve the deficiencies and defects of data quantity and quality as much as possible. As the battery health state estimation requires continuous cyclic charge and discharge experiments, the experiment time is long, so the experimental data are relatively limited. Although the neural network does not need to consider the complex changes inside the battery, it needs a large amount of experimental data support, so it is necessary to consider introducing GAN to enhance the experimental data to generate high-quality data, which will help speed up the process of research. GAN has great potential. It is a new idea to design a model that can generate high-quality time series and finally apply it to data enhancement. The process of GAN data enhancement is shown in Figure 8.10.

As can be seen from Figure 8.10, the source data are processed by the input layer and then enter the GAN network for data generation. This model consists of two basic neural networks, namely, generator neural network and discriminator neural network. One is used to generate the content, and the other is used to evaluate the generated content. The former tries to produce data that are closer to the real, and the latter, in turn, tries to distinguish real data from generated data more perfectly. Thus, the two networks make progress in the confrontation and continue to fight after progress. The data obtained by the generative network will be more and more perfect and close to the real data so that the desired data can be generated.

3. Big data analysis

With the rapid development of information technology, cloud computing, and cloud storage, the field of big data analytics is now rapidly expanding into all science and engineering domains. Due to the limited storage and computing capacity of common BMS, traditional battery health estimation methods are limited. Therefore, big data analysis based on a cloud computing platform can be used to monitor the battery health status in real time. In the monitoring data of new energy vehicles, the data fields related to charging include charging state, SOC, current, total voltage, single voltage and temperature, data collection time, etc. SOH evaluation is an important step in the life prediction of a power battery, which can avoid the system unexpected interruption and other dangerous events caused by battery life problems. In addition, it can also make use of the advantages of big data analysis to carry out a comprehensive evaluation of battery status over a long period and multidimension, to further improve the safety of the battery and optimize the performance of the battery in all aspects. At present, the main application fields of monitoring big data of new energy vehicles are fault warning of vehicle and power battery, health evaluation of power battery, analysis of user travel and charging behavior characteristics, emission reduction evaluation of electric vehicles, and performance analysis of key components. It can be seen that big data analysis can fully explore the value of data, what is more, it can be combined with AI, which will be the development trend of battery status monitoring in the future.

With the increasing integrity and accuracy of battery system big-data, integrated state estimation and prediction, combined with intelligent optimization algorithms, such as deep learning and reinforcement learning, will become growingly popular in the community of battery management and integration. But AI brings challenges as well as opportunities.

Challenges

1. Algorithms issues

Factorisation Neural Network (FNN)-based approaches can achieve reasonable results; however, performance is constrained due to storage capacity and execution time [376]. The regression and probabilistic approaches are robust against noise, uncertainty, and data-overfitting issues; however, they fail to provide accurate solutions in high-dimensional and nonlinear systems. The time series-based methods can predict future predictions accurately; nevertheless, the selection of past information and feedback steps is difficult to determine.

2. Implementation issues

Deep learning algorithms have provided excellent results in SOC, SOH, and RUL prediction, they need good quality and a vast amount of data as well as fast and expensive computing processors [377]. What is more, AI-based methods often need a large amount of data for training, and the training process is time-consuming. Therefore, how to improve computational efficiency and reducing cost is the key to the better application of AI.

3. Data issues

Accurate data are the premise of correct estimation and analysis. Therefore, high-precision sensors and data acquisition equipment are needed to obtain accurate data. Only when the collected data reach a certain accuracy can high-quality data be obtained, thus serving for further big data analysis and AI algorithms.

To sum up, the opportunities and challenges for AI in the status monitoring of utility-scale batteries are as shown in Figure 8.11. The challenges and opportunities coexist, and it certainly has great development potential in the future.

The opportunities and challenges for AI in the status monitoring of utility-scale batteries are discussed in detail in this chapter [378]. The opportunities it faces are mainly the new methods and ideas brought by AI, while the challenges it faces are

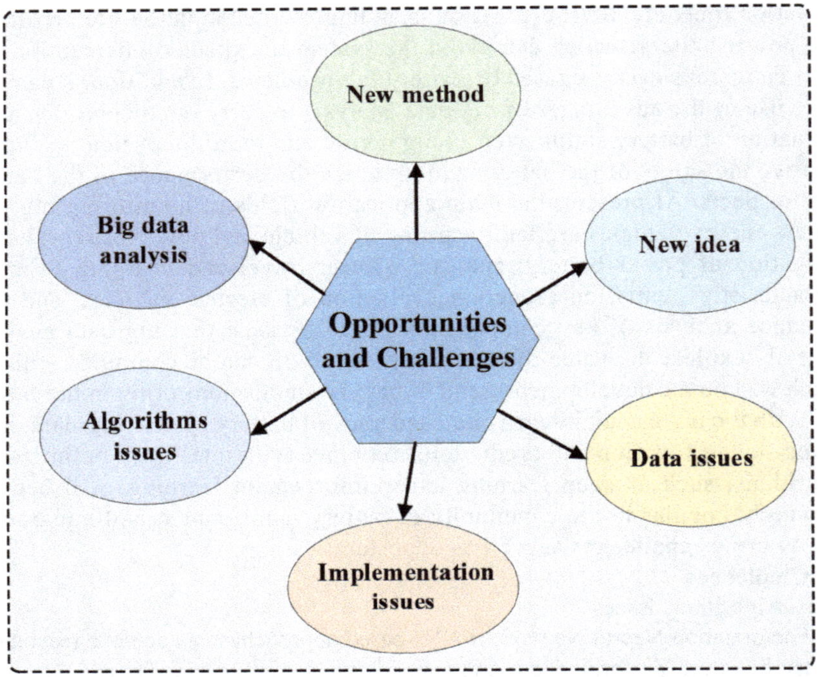

Figure 8.11 *Opportunities and challenges for AI in the status monitoring of utility-scale batteries*

also the limitations of AI at present. However, it is worth being optimistic that the development of AI is now extremely rapid, so opportunities are faced with challenges, and challenges also bring opportunities.

8.4 Development trend of AI in BMS

Over the years, the development of lithium-ion batteries has made remarkable progress, various innovative technologies have emerged, the battery performance has been improved, the charging is faster and safer, and the battery is longer and more durable [379]. Among these technologies, using AI to optimize BMS is one of the hot topics recently.

At present, the method based on AI has been more and more widely used in battery state estimation. For the health state estimation of the battery, in addition to evaluating the current health state, the remaining useful life (RUL) estimation is also essential for practical application. AI is a branch of computer science and is considered to be one of the three cutting-edge technologies (genetic engineering, nanoscience, and AI) in the twenty-first century. This is because it has achieved rapid development in recent 30 years, has been widely used in many subject fields, and has achieved fruitful results [380]. AI has gradually become an independent branch and has become a system in theory and practice. At present, AI has been more and more widely used in the field of computers, and has been applied for a robot, new energy measurement and control, control systems and simulation systems [381]. An AI algorithm is a supplement to the traditional mathematical algorithm, which has a strong classification ability and linear/nonlinear regression ability [382]. With the improvement in the integrity and accuracy of big data onto battery systems, the integrated state estimation and prediction combined with intelligent optimization algorithms such as deep learning and reinforcement learning will attract more and more attention in the field of battery management and integration.

Most AI systems currently used have the ability to learn from experience. The most widely used way to make machines do this is through an algorithm architecture called ML, which has powerful functions to evaluate cubes (i.e., data sets containing multiple variables), find patterns of data, and unlock applications that are difficult to use other methods [383]. The discovery function of modern ML algorithms depends on the quantity, quality, and accuracy of data. Therefore, the first step of any ML-based approach is to train the ML model and evaluate it where possible after building a suitable and sufficiently complete data set. In the most common cause (supervised model), this is achieved by using a part of the data set to train the algorithm (training step) [384]. The prediction ability of the algorithm is evaluated by comparing the value predicted by the model with the data not used for the training step.

There is great interest in developing a highly accurate aging model that allows earlier failure prediction, greater interpretability, and wider application to various cycle conditions [385]. Trained ML technology can potentially combine high accuracy and low computational cost, which makes it very interesting to accurately predict aging models and battery life.

To ensure the reliability of lithium batteries of their service life, real-time and accurate diagnosis based on their electrochemical characterization is very important [386]. Therefore, different battery state parameters need to be considered [387]. Among them, SOC, SOH, and RUL are the main efforts of industry and academia, especially for automotive applications.

The SOC of a lithium battery is defined as the percentage of the remaining charge relative to the fully charged condition [388]. Accurate SOC estimation is very important to optimizing operation strategy and balancing battery cells in the battery pack [389]. SOC estimation based on electrochemical technology has been put into extensive research work. Coulomb counting is one of the most effective methods to obtain battery SOC, or battery OCV is the basis of many proposed SOC estimators. In addition, EIS can be used to determine it by the electrochemical impedance model.

SOH reflects the current capacity of the battery to store and provide energy relative to the battery of the beginning of its service life, which makes it suitable to assess its degree of degradation [390]. It can quantitatively estimate the ratio of the actual battery capacity of its initial capacity [391]. Together with the battery resistance, these are the two parameters usually used to calculate the SOH of lithium-ion batteries (LIB).

RUL, together with SOH, is one of the main LIB parameters to evaluate its current health status [392]. It can be defined as the remaining load cycle (or time) before the battery reaches its service life (end of life (EOL)), or until its SOH reaches 0%.

The determination of the above parameters (SOC, SOH, and RUL) is based on electrochemical characterization technology [393]. Therefore, it is very important to record the charging and discharging voltage versus a specific capacity curve. In academia, cycling is usually (but not always) (291) carried out at a constant current, although the test protocol, i.e., more similar to real-life applications (as constant power), is preferred. Another widely used diagnostic tool is EIS, which records the response (usually current) to small sinusoidal disturbances (usually voltage) at different frequencies [394]. This technology provides information about the impedance of the battery of different time scales, and can also provide information about the possible degradation mechanism of the battery. Alternatively, the current pulse is also used to determine the battery resistance and its change in the use of the battery.

The electrochemical characterization of lithium batteries is the basis to ensure the reliability of its service life [395]. It provides an accurate diagnosis of its state through different parameters (such as SOC, SOH, and RUL). In this line, the battery life prediction obtained through offline experimental analysis has been solved by physical and semi-empirical models [396]. Considering various degradation modes and their coupling with thermal and mechanical phenomena inside the battery, the success is limited, which makes it challenging to develop a model with reasonable calculation cost.

To overcome the above shortcomings, the ML method has been implemented in the electrochemical characterization of LIB to evaluate online estimation, life prediction, performance, and safety [397]. Various customized models have been developed to address these topics, and this section thoroughly analyzes their advantages and disadvantages in terms of battery physical characteristics, parameterization, training, and uncertainty.

In terms of performance and safety prediction, an ELM method is proposed to model the battery and predict the evolution of temperature, voltage, and power [398]. In addition, the thermal behavior of the battery is analyzed by the ANN model, and the safety problem is solved.

About aging prediction, the rise of available computing resources has unlocked the use of ML and deep learning, using a large amount of data onto RUL prediction [399]. At present, the data-driven methods in battery aging prediction include supervised and unsupervised algorithms, such as time series analysis, ANN, SVM, Relevance Vector Machine (RVM), and probability estimation, such as Gaussian Process Regression (GPR). The most commonly used are those based on probability estimation, SVM and NN series, in this order. All this has brought (and will bring in the future) new insights beyond the current level of experts to better understand battery performance and the failure process [400]. In particular, we believe that regression algorithms based on time series will leave a mark shortly because they deal with a large amount of data [401]. However, in the long run, we believe that algorithms that can make accurate and realistic predictions with a more limited amount of data will eventually prevail because they require fewer resources to build data sets and training steps.

Online estimation, i.e., SOC monitoring and prediction of lithium battery through the data-driven model, is usually based on the analysis of data generated by a specific battery system [402]. The scientific articles revised for this chapter can conclude that the most commonly used ML methods of lithium-ion battery online estimation are SVM, NN family, probability estimation, and integrated learning technology [403]. In particular, SVM is the most popular one of SOH and RUL online estimation.

Other advanced data-driven models applied to electrochemical characterization, aging, and temperature evolution are being developed for lithium batteries. It is expected that the key feature set that minimizes the input parameters will improve the accuracy and reduce the overall simulation time [404]. However, these models need a large amount of data onto proper training, resulting in a relatively high computational cost of training steps [405]. Nevertheless, after the training process, a highly accurate model can provide information about LIB, SOX, and RUL at a very low computational cost. Another aspect to consider is the lack of any physical insight into data-driven models, which partially hinders their application [406]. These shortcomings can be overcome by combining the advantages of different model methods of the so-called hybrid methods, such as physical and data-driven models.

Due to the battery capacity/power characteristics, it will gradually decline from actual use until its EOL [407]. Therefore, how to use the early aging stage of a small sample to build a reliable AI method to accurately estimate the RUL or aging path of the battery is extremely important to advanced battery management [408]. In addition, external characteristic parameters such as current, voltage, and temperature are usually selected as training input, so the AI method based on training data cannot describe the internal electrochemical behavior of the battery, which greatly limits their application for different working conditions [409]. Moreover, considering the high requirements of AI-based methods of the quantity and quality of training data and the high time-consuming of the training process, accurate estimation of state variables [410], considering the electrochemical parameters inside the battery of algorithm training, and improving calculation efficiency will be the main challenges faced by AI-based methods.

Using AI to evaluate the health of the battery packs [411]. When evaluating the health status of the battery pack, AI uses particle swarm optimization genetic algorithm to construct a neural network and carries out sharing and active screening of effective information for various information of battery, such as the correlation between battery capacity, SOC, OCV, and inconsistency of battery [412]. Finally, it estimates the SOC and OCV of the battery in the form of the particle filter. The advantage is that it can accurately detect the critical points of various performance indicators of the battery, and use the actual service conditions of the battery as the algorithm standard to predict the health status, such as the change of the maximum energy storage of the battery pack [413]. At the same time, it can avoid the influence of noise and other factors on the test results [414], and greatly improve the efficiency and accuracy.

Therefore, in the future, to solve the problem that the amount of data collected for battery operation data in SOC estimation is small and cannot be collected and transmitted in real time [415], the Internet of Things technology will be used to collect and transmit a large number of process data including measurement, operation and so on, which have been generated in the practical application of lithium-ion battery engineering, to facilitate the subsequent estimation algorithm to process and predict it in real time [416]. The structured data obtained by data analysis and feature extraction can be input into the ML model, which can effectively to improve the accuracy and efficiency of state estimation [417]. The development of the Internet of Things and other technologies makes it easier to obtain high-quality massive data, and also makes the prospect of AI methods more and more clear. In addition to the common algorithms such as ANN and SVM [418], the generation of countermeasure neural network has also been applied. This method can quickly expand the data set, infer the experimental data onto unknown battery systems, and speed up the research and development of new batteries. At the same time [419], to solve the problem that the traditional SOC estimation algorithm model has poor learning ability and can not effectively reflect the time-varying nonlinear characteristics of lithium-ion battery SOC, the long-term and short-term memory network with good prediction ability for time series data and sufficient model complexity in AI algorithm is selected [420], and it is further optimized and improved to achieve a certain error accuracy.

In the future, with the development of AI technology, the BMS will focus on the in-depth development of the following technologies: the accuracy of the basic data acquisition of the BMS will be further improved [421], and the estimation depth of the internal state parameters of BMS depends on the basic data acquisition of the battery. The basic data onto the battery have the characteristics of a large amount of collection, high accuracy requirements, and complex internal circuits [422]. The collection of basic data will be the key research direction.

The safety performance and fault diagnosis technology of BMS have been further improved [423]. The safety performance includes the safety of thermal management and the safety of battery high voltage [424]. How to more stably control the internal thermal balance of battery and the monitoring of high voltage insulation, avoid the leakage of high voltage and high current, and prevent injury to vehicles and drivers [425]. Especially when the vehicle is impacted by external forces, a lot of experimental research is needed on the safety performance of battery management, when the above measures fail [426], the BMS can make a timely and effective judgment and early warning to ensure the safety of personnel and vehicles [427], which will be the key research direction.

Data-driven algorithms and intelligent algorithms promote the development of intelligent BMS [428]. In recent years, the development of data-driven and intelligent networking technology further provides a broad space for the development of BMS technology [429]. With wisdom! For the application of driving technology, many companies actively explore the application of wireless BMS technology in new energy battery vehicles, which is shown in Figure 8.12.

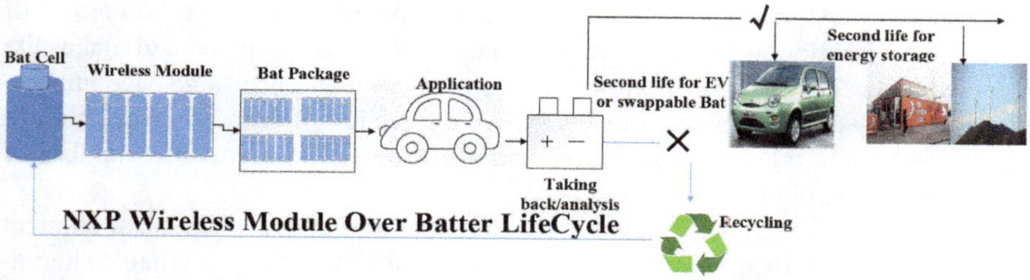

Figure 8.12 . Application of wireless BMS technology in new energy battery vehicles

The wireless BMS technical scheme designed by a company will realize computing including battery management algorithm in the cloud, reduce the load of local onboard computing [430], further release the storage space in the vehicle, and provide a new direction for reducing the cost of electric vehicles.

To improve the efficiency of battery research, many new tools have been developed to meet this requirement. Among them, AI, especially its fruitful branch of ML, is a promising method, which can help people overcome the method of dealing with a large number of variables and data.

ML can help researchers effectively address both the parameters and data challenges of LIB, and research and development of battery technologies beyond the lithium-ion battery, such as sodium-ion batteries, all-solid-state batteries, lithium batteries, and supercapacitors. The method of ML is shown in Figure 8.13. To achieve this goal, several challenges need to be addressed, such as defining widely accepted standards for battery research and development (R&D), integrating system data disclosure, identifying the most appropriate descriptor for an ML model or identifying associated errors, which is shown in Figure 8.13.

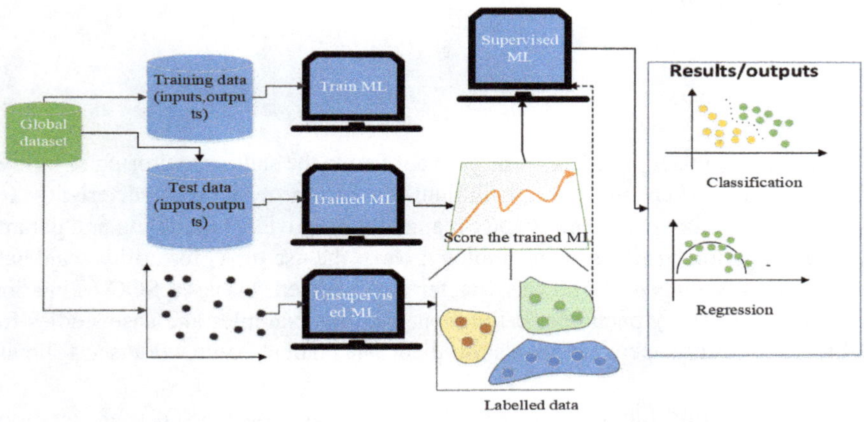

Figure 8.13 ML method: overall principle of work for supervised, unsupervised and classification, and regression methods

Generally, AI, especially ML, is expected to overcome the main limitations of battery optimization, which often leads to a combinatorial explosion and makes the detailed rendering of chemistry and electrode/battery manufacturing space impractical. ML-based methods can allow selective navigation of such chemical, formulation, and operating condition spaces and are expected to reduce the number of experiments and/or calculations required.

From a theoretical point of view, ML can support the development of efficient force fields and create new opportunities for material simulation and reliable alternative models. This can also promote the development of a multiscale modeling framework with a reasonable computational cost. In addition, ML may become a powerful experimental enhancement tool to identify the reaction mechanism directly from electrochemical results, such as cyclic voltammetry. On the one hand, some ML applications in the field of the battery have been widely studied in the scientific literature, such as online and offline estimation of battery SOH, SOC, and RUL. On the other hand, several promising applications of AI/ML are surprisingly poorly studied in the field of batteries. Among them, battery manufacturing and battery material characterization are obvious examples. The use of data-driven methods will have a profound impact on the industrial facilities of modern society and lead to the industrial 4.0 revolution. Battery manufacturing is no exception. A dedicated data warehouse will be needed shortly. Despite this obvious trend, academic research on this subject is still rare in the literature, which requires greater efforts in this regard.

Today, it is clear that the capabilities and potential of AI, ML, in particular, are attracting increasing interest to gain new insights into batteries of all sizes: from material design and synthesis to manufacturing, from materials to electrochemical characterization. Many hopes for data-driven approaches applied to batteries, but significant progress needs to be made in the field before it leads to a revolution in AI commitment.

8.5 Conclusion

1. Work summary

First of all, this topic of AI research is applied to the status monitoring of utility-scale batteries. Then, utility-scale lithium-ion battery system characteristics are proposed in this book, and the chapter realized the equivalent modeling and parameter identification based on AI technology. Next, the use of AI for utility-scale battery systems is analyzed. And this chapter also realized AI-based SOC estimation and AI-based battery parameter estimation. Finally, examples and case studies for utility-scale battery operation are also studied. The main research work is as follows:

1. The estimation methods of SOC, SOH, state of power (SOP), and state of energy (SOE) of lithium-ion batteries and several standard batteries based on AI are briefly described. The main influencing factors are analyzed. By comparing

several estimation methods of SOC, according to the different working states of the battery, the prediction method is determined.

2. After analyzing the factors affecting the estimation of battery SOC, it is found that the charging and discharging voltage, current, and ambient temperature are the main factors. Therefore, the charging and discharging voltage U, current I, and ambient temperature T of the battery are taken as input parameters of the neural network. SOC is taken as output parameters to establish the neural network intelligent prediction model. The experiment selects appropriate batteries to carry out constant discharge current tests, part of the obtained data as training samples, and imports the established network prediction model for verification and analysis. The experiment shows that the intelligent prediction model improves the prediction accuracy of the SOC of the battery and has strong practicability.
3. Accurate estimation of the SOC of the battery pack is mainly reflected in the following three aspects:
    1. First, the experiments characterize the battery discharge depth and study the life of the battery cell.
    2. Second, SOC estimation as the starting condition of battery equalization problem can study the balance management state of the battery pack.
    3. Finally, SOC estimation can be used to formulate braking energy recovery strategies to study energy recovery.
4. The chapter analyzes the equivalent modeling methods based on AI technology, including second-order equivalent circuit modeling, compound equivalent circuit modeling, voltage matching equivalent circuit modeling, and improved second-order equivalent circuit modeling. Complex test conditions are introduced, and the chapter also provides methods for offline and online parameter identification. Offline parameter identification is divided into point calculation strategies, curve fitting methods and model parameter validation and evaluation, and double exponential fitting strategies recursive least squares. Bias compensation recursive least square method, forgetting factor recursive least square method, and improved multi-innovation least square method are used for AI-based online parameter identification.
5. Through the equivalent modeling and parameter identification based on AI analysis, the chapter introduced the use of AI for utility-scale battery systems, analyzed the selection criteria for the choice of AI techniques, this chapter expounds on the development status and development prospect of monitoring of utility-scale batteries development with AI. Basic parameters for condition monitoring based on AI are discussed, including input and correction voltage, the capacity of internal state parameters, the internal resistance of state parameters, the polarization resistance and capacitance of internal resistance state parameters, and calibration energy density.
6. The experiments make full use of battery energy and reduce the loss of residual energy detection under the national standard. This chapter did some old and new levels of different battery charging and discharging experiments and chose

representative pieces of batteries to obtain their voltage characteristic curve using the MATLAB platform. The discharge characteristic curve of a parameter (voltage, current, and voltage change rate) is selected in the neural network model for the above test experiment. Results show that the sample test error is small. It has a good application prospect.
7. The lithium battery SOC monitoring system is established, which communicates with EBC-A10 through the USB port to realize real-time display and storage of battery parameters, and MATLAB Script realizes dynamic prediction of lithium battery SOC.
8. According to the obtained peak current and cut-off voltage, the SOP of straight L sampling points can be obtained, and the SOP calculation is shown in (8.18).

$$\begin{cases} SOP_{max}^{dis,V} = I_{max}^{dis,V} \cdot U_{L,min} \\ SOP_{min}^{chg,V} = I_{min}^{chg,V} \cdot U_{L,max} \end{cases} \tag{8.18}$$

2. Job outlook

Through the above work, this chapter has made some progress in estimating the SOC of the battery. However, due to the limitations of capacity and experimental conditions, there are still some shortcomings in the study of lithium batteries, which need to be discussed and studied in the following aspects:

1. This chapter uses a tester to do battery charging and discharging experiments. The tester can only do the monomer battery charging and discharging experiments each time. However, we mainly used the battery pack in real life without considering the difference between single cells. We should improve the tester even to complete the experiment on the battery pack.
2. Due to the limitation of the conditions, the charge–discharge experiments of the battery in this chapter are carried out at room temperature, and the temperature range is relatively small. Therefore, the next step will be to conduct experiments on SOC and secondary batteries over a more comprehensive temperature range to more accurately predict SOC and the residual energy of old batteries.

3. Opportunities and challenges

In recent years, the battery industry has developed rapidly. Still, at the same time, it is also faced with challenges such as security risks that cannot be effectively predicted, long-testing cycles, high costs, and difficulty in guaranteeing product consistency.

Traditional battery prediction techniques, such as physical modeling and active signal injection, are based on experience to monitor key indicators such as voltage, current, and temperature and to ensure they are within safe ranges. However, the influencing factors in the use of batteries are complex and diverse. Different users, locations and environmental factors may interfere with the results, and the reliability of the safe operating range obtained based on engineering experience is unknown. At the same time, these two technologies cannot track battery pack abnormalities that are within the safe zone.

The utility-scale battery status monitoring system based on AI extracts the hidden information behind the surface based on macro indicators. Compared with the traditional model, it has a significant improvement in intelligence and depth of data analysis, and it can gain insight into battery anomalies more quickly and accurately.

Another advantage of battery condition monitoring based on AI is its versatility. Due to its solid empirical attribute, traditional technology needs to be customized according to specific scenarios and application conditions and has poor portability. However, data-driven intelligent prediction only needs fine-tuning in details and can be widely applied to different customers and systems.

What is more, the status of battery health prediction in complex industrial scenarios is faced with many difficulties. On the one hand, there are many uncontrollable interference factors in real scenarios. On the other hand, SOH is limited by the requirements of practical application, and the information dimension that can be obtained in industrial scenarios is insufficient. AI is different from traditional solutions based on the first principles of physics. AI-based battery status monitoring provides AI solutions organically combined with a highly coupled and highly complex environment from the perspective of data, realizing deep mining and efficient and accurate extraction of abnormal battery information.

AI-based status monitoring of utility-scale batteries does not require analysis of aging mechanisms but requires a large amount of training and test data. The effectiveness of AI in industrial scenarios is highly dependent on the amount of accumulated data.

Batteries have a wide range of applications, from consumer electronics, electric cars, railways, and oceans to power grid storage systems. But one of the key requirements for consumer acceptance of electric vehicles is a smaller battery pack for a longer range and lower cost. All of these goals depend on accurate estimates of SOH and life expectancy predictions under a variety of operating conditions. More accurate life prediction will improve battery life at all stages of battery technology.

The deterioration of lithium-ion batteries also begins after the production of the battery due to the electronic chemical reactions within the battery. This process leads to the deterioration of the active material inside the battery, and as a result, the internal resistance of the battery increases, which means more loss and capacity decay inside the battery. Estimating battery capacity can give us some useful information. In this way, we can compare the capacity of the battery (C_{batt}) with its initial capacity value (C_{init}) to obtain the current attenuation degree of the battery. In general, we consider the end life of a battery to be 80% of its initial capacity, and the SOH calculation is shown in (8.19).

$$\text{SOH} = 1 - \frac{C_{init} - C_{batt}}{0.2C_{init}} \quad (8.19)$$

Here, SOH can vary between 0 and 1, with 0 indicating the end of battery life (C_{batt} = 80% C_{init}). The coefficient 0.2 in the denominator of the formula comes from (8.20).

$$0.2C_{init} = C_{init} - 0.8C_{init} \quad (8.20)$$

First, it can shorten product development cycles by illustrating failure mechanisms, especially when models can be combined with experiments in a closed loop.

Second, it can be used to optimize manufacturing solutions.

Third, improved life prediction guarantees lower warranty and insurance costs, that can do timely preventive maintenance, lower upfront capital costs by reducing over-engineering, and longer life by better control of charge and discharge.

Finally, it holds out the prospect of better battery second-life applications that support a circular economy around battery manufacturing, reusing and recycling. Demand for battery second-life applications is expected to outstrip raw material supply and refining over the next decade.

Unfortunately, neither data-driven nor physics-based approaches alone can solve the battery life prediction challenge from field data. Physics-based modeling is challenged by a large number of coupling and nonlinear degradation mechanisms that are almost impossible to observe from electrical measurements and difficult to parameterize. Therefore, it is difficult to find a degradation model that will satisfy everyone, as many different mechanisms and formulas can explain similar observed degradation behaviors. In addition, mechanical or acoustic measurements have been proposed as a means of addressing the lack of observation, but they are not yet widely used.

At the same time, the data-driven approach suffers from a 'dimensional disaster' in which the amount of data required to capture all combinations of operational conditions grows rapidly with the number of investigated conditions. Further complicating the thing is that the picture is the relatively slow pace of obtaining battery life data, which requires months or years of experimentation for each change in chemical composition, shape coefficient or manufacturing process. In addition, trade secrets limit the amount of data available to individual researchers. The inherent nonlinear, path-dependent nature of battery degradation further exacerbates this problem.

The criteria for determining the EOL may vary from application to application but usually occur when the battery can no longer meet the requirements for typical service conditions, such as range, running time or maximum power capacity. The key parameters that affect the EOL are capacity (available energy) and internal resistance (available power). Battery aging depends on internal factors, such as manufacturing variability and packaging design, and external factors, such as temperature and intensity of use, so it is difficult to predict, especially in using scenarios outside the laboratory. Existing literature reviews and views consider SOH prediction, lifetime prediction, and fusion approaches based on physical and data-driven models, but historically these methods have been limited to fairly small data sets under controlled conditions.

In the laboratory, battery cycles can be repeated consistently and conditions such as current and temperature can be tightly controlled. Laboratory tests can be used to build models to explore how different operating conditions affect cycle life, thereby determining warranties, maintenance schedules, and system size. Operating conditions used in laboratory tests include driving patterns and schedules tailored to very specific applications, such as specific electric vehicle models. Laboratory testing can also be used to guide the development of new battery chemicals, optimize

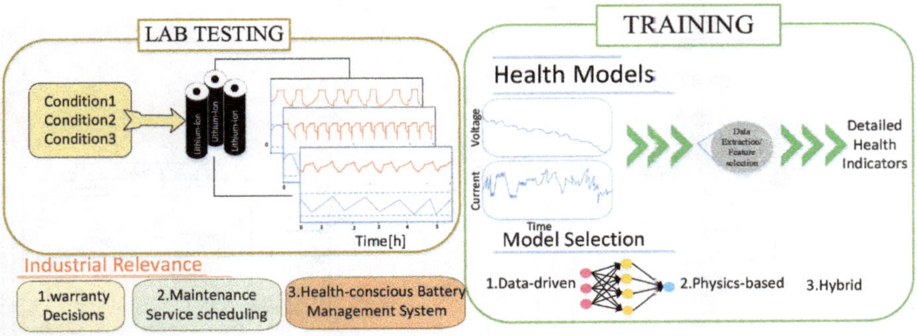

Figure 8.14 Predictions using laboratory data

battery design and improve manufacturing processes, such as studying the impact of parameters such as humidity or formation cycles on performance that is shown in Figure 8.14.

Based on extensive studies of battery test data in modeling laboratories, methods for estimating battery life can be broadly divided into four categories: empirical aging model, pure data-driven approach, feature-based data-driven approach, and physics-based approach, as shown in Figure 8.15.

Laboratory battery testing is limited by the number of test channels available and the time available for testing. Ultimately, what matters is battery performance in real-world applications, as shown in Figure 8.16.

When a battery can no longer meet the capacity or power requirements of its primary application, it can sometimes be reused for different 'second-life' applications. This is good for the environment and economically viable, especially with

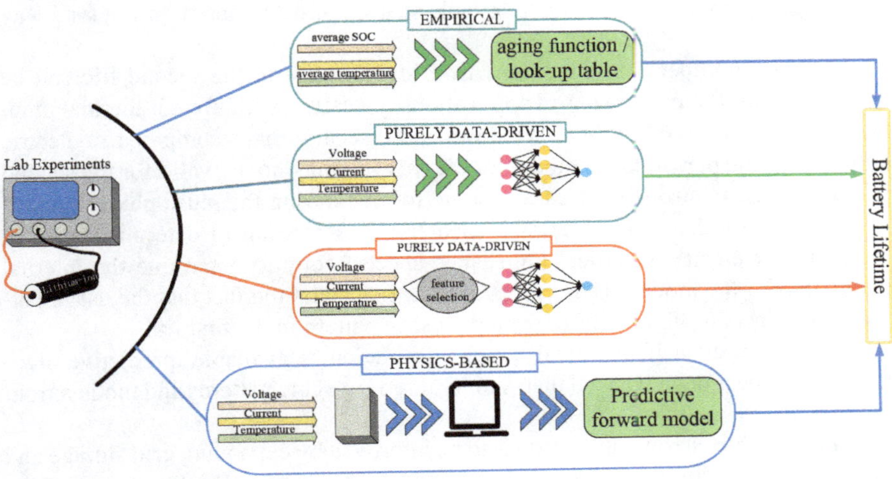

Figure 8.15 Different methods of battery life estimation

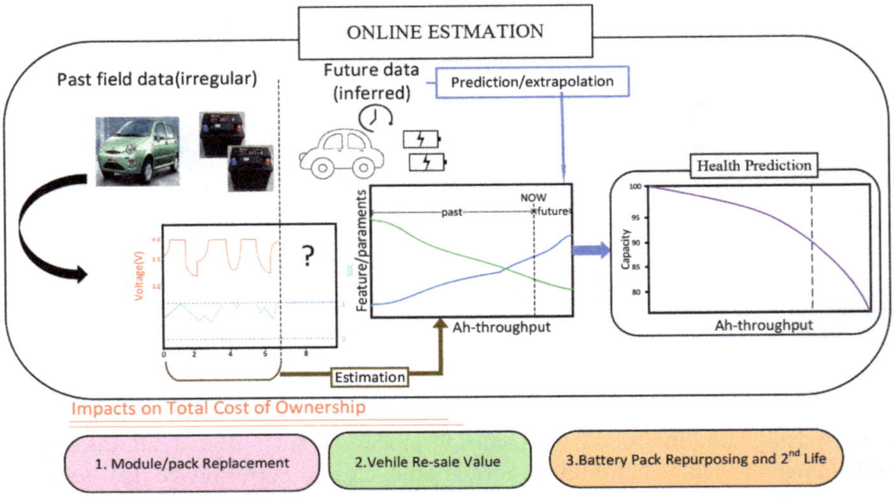

Figure 8.16 Use field data for prediction

the growing market for lithium-ion batteries for electric vehicles, which can then be reused for the grid or fixed storage. The challenge here is to predict the RUL of the battery under its intended new application, which may be quite different from its previous use, to determine whether this new use is feasible, and to provide information for ensuring decisions.

There are two particular problems:

First of all, the historical data that may be used for the first time may not be available due to trade secrets or other reasons, and even if it is available, it may not have the required reliability or fidelity. While initiatives such as battery passports are being implemented to facilitate reliable sharing of data, there is still a long way to go.

Second, the operating conditions from the first life to the second life can be quite different. For example, grid applications typically have different thermal management systems and low SOC windows and discharge rates compared to electric vehicles, so the performance model for the first life may not be valid for the second life. However, controlled cycles can be performed during the reuse phase, such as a slow deep discharge to determine capacity, pulse testing to determine internal resistance or electric chemical impedance spectroscopy to determine the internal parameters of the model. RUL estimation must consider the fact that the usage pattern in the second life will differ from the usage pattern in the first life.

Therefore, even if historical usage information is available, predictive algorithms must be more advanced than simply inferring aging patterns and models from the first life.

If renewable energy integration is to be improved by expanding grid storage and rapidly replacing internal combustion engines with electric vehicles as part of the solution to keep climate warming below 2°C of previous industrial levels, the cost of batteries must be reduced. To solve this problem, there is a great need to accurately estimate battery health, diagnose degradation, predict life in different usage scenarios, and detect faults. To reliably infer future behavior from available data, a combination of data-driven and physics-based models would be superior to purely data-driven or purely physics-based approaches. Different methods apply to different cases. Alternatively, companies can use large-scale data to fine-tune models they develop in laboratory data. Developing and training a battery life prediction model requires a large amount of data covering the entire range of operating conditions. In addition, more in-depth characterization of batteries that have reached their useful life can be performed to determine their true SOH and cycle life, thus building a database for supervised learning. Alternatively, individual batteries that exhibit particular behavior can be selected for disassembly experiments to reveal the aging process of these batteries and thus verify the predictions of physics-based models. To achieve this, further research is needed in multiple areas and collaboration between disciplines is strengthened:

1. Further research should develop a consistent large-scale open-source database for battery performance and life data and should train and validate life prediction algorithms. Ideally, these should include laboratory and field data and share common standards for data and metadata.
2. Further research should develop fast and accurate physics-based models, with a particular focus on understanding which key degradation sub-models are important and how to estimate and track key parameters. Such complex models have traditionally required a very high level of expertise to implement, but it may be made easier with open-source modeling packages.
3. Further research should develop extensible algorithms for combining data-driven approaches with physics-based models. Examples include field inversion and ML paradigms, neural differential equations, state-space model of the Gaussian process, and general differential equations.
4. Further research should explore methods to diagnose without historical data whether batteries have aged beyond the inflection point and to evaluate whether batteries can be used for second-life applications.
5. Further research should quantify the value generated by improved SOH estimation and life predictions.
6. Further research should address data privacy issues to ensure the anonymity and intelligent aggregation of individual user data without losing insight into diagnosis and prediction.

4. Development trend of AI in BMS

BMS are critical to the cost, power, and safety of electric vehicles. The BMS can monitor the current and voltage of a single battery, control the balance of charge and discharge of a single battery, observe the charge and discharge current and temperature of the battery, estimate the remaining power of the battery, communicate with the vehicle monitoring system and the vehicle charger in real-time bus, and coordinate the control and optimization of electric vehicle power distribution. As one of the key technologies of electric vehicles, the BMS plays an indispensable role, but the balance problem and the reliability of collected data need to be improved when estimating SOC accurately.

With the strong support of the state, the development of electric vehicles is very rapid, but it is still in the early stage. As one of the key battery technologies, BMS has made great breakthroughs and research achievements at the present stage, but there are still many areas to be improved and corrected.

1. Accurate estimation of battery SOC will continue to be the focus of future research.
2. The balance of the battery pack is still a difficult problem, and it is urgent to put forward a reasonable plan to make an efficient and comprehensive balance of the battery pack.
3. The security performance of BMS is also very important. Only with good security, BMS will be favored by consumers. Among them, the thermal management system is the key to ensuring the safety performance of BMS and it will be the research direction in the future.
4. At present, most mature BMS abroad are only developed for a single specific battery or battery pack with poor versatility, so the research and development of more general BMS will become the future development direction.

5. Conclusion

AI for status monitoring of utility-scale batteries is the inevitable trend of battery condition monitoring industry development. The BMS is the guarantee of stable and efficient operation of electric vehicles, it plays a key role in the development of electric vehicles, and it is the inevitable power to promote the development of electric vehicles. The BMS has always been a hot research direction and will make more progress in future research.

References

[1] Abd-Elmagid M.A., Ferdowsi A., Dhillon H.S., Saad W. 'Deep reinforcement learning for minimizing age-of-information in UAV-assisted networks'. *IEEE Global Communications Conference*; Waikoloa, HI, 2019. Available from https://ieeexplore.ieee.org/xpl/mostRecentIssue.jsp?punumber=8968653

[2] Abdelmoula R., Khamis A., Karray F. 'A deep learning-based noise-resilient keyword spotting engine for embedded platforms'. *Image Analysis and Recognition*. 2019;**11663**:134–46.

[3] Abdelshafy A.M., Jurasz J., Hassan H., Mohamed A.M. 'Optimized energy management strategy for grid connected double storage (pumped storage-battery) system powered by renewable energy resources'. *Energy*. 2020;**192**:116615.

[4] Abdullah S., Asghar M.N., Ashraf M., Abbas N. 'An energy-efficient message scheduling algorithm with joint routing mechanism at network layer in internet of things environment'. *Wireless Personal Communications*. 2020;**111**(3):1821–35.

[5] Abe K., Amey J., Andreopoulos C., *et al*. 'Characterization of nuclear effects in muon-neutrino scattering on hydrocarbon with a measurement of final-state kinematics and correlations in charged-current pionless interactions at T2K'. *Physical Review D*. 2018;**98**(3).

[6] Adams R.A., Varma A., Pol V.G. 'Temperature dependent electrochemical performance of graphite anodes for K-ion and li-ion batteries'. *Journal of Power Sources*. 2019;**410–411**:124–31.

[7] Adcock M., Sonder F., Schättin A., Gennaro F., de Bruin E.D. 'A usability study of a multicomponent video game-based training for older adults'. *European Review of Aging and Physical Activity*. 2020;**17**(1):3.

[8] Adekanye T., Saleh A., Okunola A., Osavbie I., Jegede K. 'Energy efficiency and use of a parametric method for poultry production'. *IOP Conference Series*; Kwara State, Nigeria, 2020.

[9] Adekoya O.B. 'Long memory in the energy consumption by source of the united states: fractional integration, seasonality effect and structural breaks'. *Estudios de Economía*. 2020;**47**(1):31–48.

[10] Adeogun A.G., Ganiyu H.O., Ladokun L.L., Ibitoye B.A. 'Evaluation of hydrokinetic energy potentials of selected rivers in Kwara State, Nigeria'. *Environmental Engineering Research*. 2020;**25**(3):267–73.

[11] Adepoju A.A., Doumbia M., Williams Q.L. 'Graphene nanoplatelet additives for high C-rate lifepo4 battery cathodes'. *JOM*. 2020;**72**(9):3170–75.

[12] Afrasiabi M., Mohammadi M., Rastegar M., Kargarian A. 'Multi-agent microgrid energy management based on deep learning forecaster'. *Energy*. 2019;**186**(115873):115873.

[13] Afshari M., Abbasi R., Sovizi M.R. 'Evaluation of nanometer-sized zirconium oxide incorporated al—mg—ga—sn alloy as anode for alkaline aluminum batteries'. *Transactions of Nonferrous Metals Society of China*. 2020;**30**(1):90–98.

[14] Afzal A., Mohammed Samee A.D., Abdul Razak R.K., Ramis M.K. 'Thermal management of modern electric vehicle battery systems (MEVBS)'. *Journal of Thermal Analysis and Calorimetry*. 2021;**144**(4):1271–85.

[15] Agarwal S., Brahmadathan S.S., Hariharan K.S., *et al*. 'Communication—precursors to abrupt capacity loss in li-ion batteries'. *Journal of The Electrochemical Society*. 2020;**167**(14):140502.

[16] Ah-Lung G., Flamme B., Ghamouss F., Maréchal M., Jacquemin J. 'Guidelines for designing highly concentrated electrolytes for low temperature applications'. *Chemical Communications*. 2020;**56**(68):9830–33.

[17] Ahammer A., Schober T. 'Exploring variations in health-care expenditures-what is the role of practice styles?'. *Health Economics*. 2020;**29**(6):683–99.

[18] Ahmad A.B., Ooi C.A., Ishak D. 'State-of-charge balancing control for optimal cell utilisation of a grid-scale three-phase battery energy storage system using hybrid modular multilevel converter topology without redundant cells'. *IEEE Access: Practical Innovations, Open Solutions*. 2021;**9**:53920–35.

[19] Ahwiadi M., Wang W. 'An adaptive particle filter technique for system state estimation and prognosis'. *IEEE Transactions on Instrumentation and Measurement*. 2020;**69**(9):6756–65.

[20] Akedo J. 'Room temperature impact consolidation and application to ceramic coatings: aerosol deposition method'. *Journal of the Ceramic Society of Japan*. 2020;**128**(3):101–16.

[21] Akeyo O.M., Rallabandi V., Jewell N., Ionel D.M. 'The design and analysis of large solar pv farm configurations with DC-connected battery systems'. *IEEE Transactions on Industry Applications*. 2020;**56**(3):2903–12.

[22] Akhavan-Hejazi H., Taylor Z., Mohsenian-Rad H. 'Optimal cell removal to enhance operation of aged grid-tied battery storage systems'. *IEEE Transactions on Sustainable Energy*. 2021;**12**(1):739–42.

[23] Aktas A. 'Design and implementation of adaptive battery charging method considering the battery temperature'. *IET Circuits, Devices & Systems*. 2020;**14**(1):72–79.

[24] Al-Gabalawy M., Mahmoud K., Darwish M.M.F., Dawson J.A., Lehtonen M., Hosny N.S. 'Reliable and robust observer for simultaneously estimating state-of-charge and state-of-health of lifepo4 batteries'. *Applied Sciences*. 2021;**11**(8):3609.

[25] Al-Ghussain L., Ahmad A.D., Abubaker A.M., Abujubbeh M., Almalaq A., Mohamed M.A. 'A demand-supply matching-based approach for mapping renewable resources towards 100% renewable grids in 2050'. *IEEE Access: Practical Innovations, Open Solutions*. 2021;**9**:58634–51.

[26] Al-Ghussain L., Darwish Ahmad A., Abubaker A.M., Mohamed M.A. 'An integrated photovoltaic/wind/biomass and hybrid energy storage systems towards 100% renewable energy microgrids in university campuses'. *Sustainable Energy Technologies and Assessments*. 2021;**46**(101273):101273.

[27] Al-Saffar M., Musilek P. 'Reinforcement learning-based distributed BESS management for mitigating overvoltage issues in systems with high pv penetration'. *IEEE Transactions on Smart Grid*. 2020;**11**(4):2980–94.

[28] Al-Shalabi M., Anbar M., Wan T.-C., Alqattan Z. 'Energy efficient multi-hop path in wireless sensor networks using an enhanced genetic algorithm'. *Information Sciences*. 2019;**500**:259–73.

[29] Mauger A., Julien C.M., Goodenough J.B., Zaghib K. 'Tribute to Michel armand: from rocking chair – li-ion to solid-state lithium batteries'. *Journal of The Electrochemical Society*. 2019;**167**(7):070507.

[30] Alasmary W. 'An innovative smartphone-based solution for traffic rule violation detection'. *International Journal of Advanced Computer Science and Applications*. 2020;**11**(1):625–36.

[31] Albertengo G., Debele F.G., Hassan W., Stramandino D. 'On the performance of web services, google cloud messaging and firebase cloud messaging'. *Digital Communications and Networks*. 2020;**6**(1):31–37.

[32] Albertus P., Manser J.S., Litzelman S. 'Long-duration electricity storage applications, economics, and technologies'. *Joule*. 2020;**4**(1):21–32.

[33] Aleshin A.N., Bugaev A.S., Ruban O.A., Saraikin V.V., Tabachkova N.Yu., Shchetinin I.V. 'The energy state of epitaxial layers in a multilayer heterostructure, grown on a (001) Gas substrate'. *Crystallography Reports*. 2020;**65**(1):138–46.

[34] Alessi B., Macias-Montero M., Maddi C., Maguire P., Svrcek V., Mariotti D. 'Bridging energy bands to the crystalline and amorphous states of Si QDS'. *Faraday Discussions*. 2020;**222**:390–404.

[35] Alexanderson S., Henter G.E., Kucherenko T., Beskow J. 'Style-Controllable speech-driven gesture synthesis using normalising flows'. *Computer Graphics Forum*. 2020;**39**(2):487–96.

[36] Alexoudi A., Patrikelis P., Fasilis T., Deftereos S., Sakas D., Gatzonis S. 'Effects of anodal tdcs on motor and cognitive function in a patient with multiple system atrophy'. *Disability and Rehabilitation*. 2020;**42**(6):887–91.

[37] Ali A. 'Performance evaluation of ZVS/ZCS high efficiency AC/DC converter for high power applications'. *Bulletin of the Polish Academy of Sciences-Technical Sciences*. 2020;**68**(4):793–807.

[38] Ali A., Raisz D., Mahmoud K., Lehtonen M. 'Optimal placement and sizing of uncertain PVS considering stochastic nature of PEVS'. *IEEE Transactions on Sustainable Energy*. 2020;**11**(3):1647–56.

[39] Álvarez-Caballero A., Blanco C., Couso I., Sánchez L. 'Identification of li-ion battery models through monotonic echo serial networks for coarse data'. *Logic Journal of the IGPL*. 2020;**28**(1):109–20.

[40] Ambalkar A.A., Panmand R.P., Kawade U.V, *et al*. 'Facile synthesis of SNO_2 @ carbon nanocomposites for lithium-ion batteries'. *New Journal of Chemistry*. 2020;**44**(8):3366–74.

[41] Amin H.M.A., Apfel U.P. 'Metal-Rich chalcogenides as sustainable electrocatalysts for oxygen evolution and reduction: state of the art and future perspectives'. *European Journal of Inorganic Chemistry*. 2020;**2020**(28): 2679–90.

[42] Amin K., Zhang J., Zhou H.-Y., *et al*. 'Surface controlled pseudo-capacitive reactions enabling ultra-fast charging and long-life organic lithium ion batteries'. *Sustainable Energy & Fuels*. 2020;**4**(8):4179–85.

[43] Amini M.R., Wang H., Gong X., Liao-McPherson D., Kolmanovsky I., Sun J. 'Cabin and battery thermal management of connected and automated hevs for improved energy efficiency using hierarchical model predictive control'. *IEEE Transactions on Control Systems Technology*. 2020;**28**(5):1711–26.

[44] Amores M., Wada K., Sakaushi K., Nishihara H. 'Reversible energy storage in layered copper-based coordination polymers: unveiling the influence of the ligand's functional group on their electrochemical properties'. *The Journal of Physical Chemistry C*. 2020;**124**(17):9215–24.

[45] Ando K., Matsuda T., Imamura D. 'Degradation diagnosis of lithium-ion batteries using AC impedance technique in fixing the state of charge of an electrode'. *Journal of Energy Chemistry*. 2021;**53**:285–89.

[46] Ando T., Sato Y., Matsuyama T., Sakuda A., Tatsumisago M., Hayashi A. 'High-rate operation of sulfur/mesoporous activated carbon composite electrode for all-solid-state lithium-sulfur batteries'. *Journal of the Ceramic Society of Japan*. 2020;**128**(5):233–37.

[47] Anjana P.K., Babu B., Shaijumon M.M., Thirumurugan A. 'Lithium-ion-based electrochemical energy storage in a layered vanadium formate coordination polymer'. *ChemPlusChem*. 2020;**85**(6):1137–44.

[48] Anjum S.S. 'An optimal management modelling of energy harvesting and transfer for iot-based RF-enabled sensor networks'. *Ad Hoc & Sensor Wireless Networks*. 2020;**46**(1–2):83–112.

[49] Ansean D., Garcia V.M., Gonzalez M., *et al*. 'Lithium-ion battery degradation indicators via incremental capacity analysis'. *IEEE Transactions on Industry Applications*. 2019;**55**(3):2992–3002.

[50] Apostolova R.D., Shembel E.M. '$LiMn_2O_4$–norit at a low temperature in comparison with $LiMn_2O_4$—MWCNT and $LiMn_2O_4$–EUZ–м graphite in the prototype li-battery'. *Surface Engineering and Applied Electrochemistry*. 2020;**56**(4):533–40.

[51] Araujo D., Batista N., Carvalho P., *et al*. 'Renewable hybrid systems: characterization and tendencies'. *IEEE Latin America Transactions*. 2020;**18**(1):102–12.

[52] Araujo N.B., Nielsen T.R., Barca M.L, *et al*. 'Brazilian version of the European cross-cultural neuropsychological test battery (CNTB-BR): diagnostic accuracy across schooling levels'. *Revista Brasileira de Psiquiatria (Sao Paulo, Brazil)*. 2020;**42**(3):286–94.

[53] Archana S., Athika M., Elumalai P. 'Supercapattery and full-cell lithium-ion battery performances of a [Ni (Schiff base)] -derived ni/nio/nitrogen-doped carbon heterostructure'. *New Journal of Chemistry*. 2020;**44**(29):12452–64.

[54] Arefi A., Ledwich G., Nourbakhsh G., Behi B. 'A fast adequacy analysis for radial distribution networks considering reconfiguration and DGS'. *IEEE Transactions on Smart Grid*. 2020;**11**(5):3896–909.

[55] Arévalo P., Cano A., Jurado F. 'Comparative study of two new energy control systems based on PEMFC for a hybrid tramway in Ecuador'. *International Journal of Hydrogen Energy*. 2020;**45**(46):25357–77.

[56] Ariyoshi K., Tanimoto M., Yamada Y. 'Impact of particle size of lithium manganese oxide on charge transfer resistance and contact resistance evaluated by electrochemical impedance analysis'. *Electrochimica Acta*. 2020;**364**:137292.

[57] Ariyoshi K., Yuzawa K., Yamada Y. 'Reaction mechanism and kinetic analysis of the solid-state reaction to synthesize single-phase $Li_2Co_2O_4$ spinel'. *Journal of Physical Chemistry C*. 2020;**124**(15):8170–77.

[58] Arnaiz M., Shanmukaraj D., Carriazo D., *et al*. 'A transversal low-cost pre-metallation strategy enabling ultrafast and stable metal ion capacitor technologies'. *Energy & Environmental Science*. 2020;**13**(8):2441–49.

[59] Arroyo-de Dompablo M.E., Ponrouch A., Johansson P., Palacín M.R. 'Achievements, challenges, and prospects of calcium batteries'. *Chemical Reviews*. 2020;**120**(14):6331–57.

[60] Artnaseaw A. 'Synthesis and characterization of multi-walled carbon nanotubes from silkworm cocoons to use as the cathode material for al-air battery'. *International Journal of Green Energy*. 2020;**17**(13):846–52.

[61] ArunKumar S., Kumar B.V., Pandi M. 'Artificial bee colony optimization based energy-efficient wireless network interface selection for industrial mobile devices'. *Computer Communications*. 2020;**154**:1–10.

[62] Atalay S., Sheikh M., Mariani A., Merla Y., Bower E., Widanage W.D. 'Theory of battery ageing in a lithium-ion battery: capacity fade, nonlinear ageing and lifetime prediction'. *Journal of Power Sources*. 2020;**478**(229026):229026.

[63] Atamuradov V., Medjaher K., Camci F., Zerhouni N., Dersin P., Lamoureux B. 'Machine health indicator construction framework for failure diagnostics and prognostics'. *Journal of Signal Processing Systems*. 2020;**92**(6):591–609.

[64] Auguscik A., Durajski A., Szczęśniak R. 'First-principles study on electronic and magnetic properties of sc-, ti- and zn-doped graphene'. *Acta Physica Polonica A*. 2020;**137**(5):958–60.

[65] Augutis J., Krikštolaitis R., Martišauskas L., Urbonienė S., Urbonas R., Ušpurienė A.B. 'Analysis of energy security level in the Baltic states based on indicator approach'. *Energy*. 2020;**199**:117427.

[66] Aung H., Soon J.J., Goh S.T., Lew J.M., Low K.-S. 'Battery management system with state-of-charge and opportunistic state-of-health for a miniaturized satellite'. *IEEE Transactions on Aerospace and Electronic Systems*. 2020;**56**(4):2978–89.

[67] Ayama T. 'Microstructure and charge-discharge mechanism of a Li_3CuS_2 positive electrode material for all-solid-state lithium-ion batteries'. *Acs Applied Energy Materials*. 2021;**4**(6):6290–95.

[68] Baba T., Takao N., Honda Y., *et al*. 'A spatially-resolved *operando* high-energy confocal X-ray diffraction method for observing non-uniform degradation phenomena in A practical cylindrical lithium-ion battery'. *Electrochemistry*. 2020;**88**(2):63–68.

[69] Badmos O., Kopp A., Bernthaler T., Schneider G. 'Image-based defect detection in lithium-ion battery electrode using convolutional neural networks'. *Journal of Intelligent Manufacturing*. 2020;**31**(4):885–97.

[70] Bai Y., Zhao K., Liu Y., Stein P., Xu B.-X. 'A chemo-mechanical grain boundary model and its application to understand the damage of li-ion battery materials'. *Scripta Materialia*. 2020;**183**:45–49.

[71] Baik J.-H., Lee S.Y., Kim K, *et al*. 'Enhanced cycle stability of rechargeable Li-O_2 batteries using immobilized redox mediator on air cathode'. *Journal of Industrial and Engineering Chemistry*. 2020;**83**:14–19.

[72] Bajpai S., Upadhyay A., Sati H., Pandey R.M., Chaterjee P., Dey A.B. 'Hindi version of Addenbrooke ' S cognitive examination III: distinguishing cognitive impairment among older Indians at the lower cut-offs'. *Clinical Interventions in Aging*. 2020;**15**:329–39.

[73] Balakrishnan S., Sofianos M.V., Paskevicius M., Rowles M.R., Buckley C.E. 'Destabilized calcium hydride as a promising high-temperature thermal battery'. *The Journal of Physical Chemistry C*. 2020;**124**(32):17512–19.

[74] Ball S., Clark J., Cookson J. 'Battery materials technology trends and market drivers for automotive applications challenges for science and industry in electric vehicles growth'. *Johnson Matthey Technology Review*. 2020;**64**(3):287–97.

[75] Ballestas-Barrientos A.R., Xia Q., Masters A.F., Ling C.D., Maschmeyer T. 'Interfacial reactions between lithium and grain boundaries from anatase tio_2-TUD-1 electrodes in lithium-ion batteries with enhanced capacity retention'. *ACS Omega*. 2020;**5**(13):7584–92.

[76] Bambini V., Bischetti L., Bonomi C.G., Arcara G., Lecce S., Ceroni M. 'Beyond the motor account of amyotrophic lateral sclerosis: verbal humour and its relationship with the cognitive and pragmatic profile'. *International Journal of Language & Communication Disorders*. 2020;**55**(5):751–64.

[77] Banday A., Shahid R., Meena S.S., Yusuf S.M., Murugavel S. 'Effect of crystallite size on the phase transition behavior of heterosite $FePo_4$'. *Physical Chemistry Chemical Physics*. 2020;**22**(27):15478–87.

[78] Banerjee A., Wang X., Fang C., Wu E.A., Meng Y.S. 'Interfaces and interphases in all-solid-state batteries with inorganic solid electrolytes'. *Chemical Reviews*. 2020;**120**(14):6878–933.

[79] Banguero E., Correcher A., Pérez-Navarro Á., García E., Aristizabal A. 'Diagnosis of a battery energy storage system based on principal component analysis'. *Renewable Energy*. 2020;**146**:2438–49.

[80] Barbosa L., Luna-Lama F., González Peña Y., Caballero A. 'Simple and eco-friendly fabrication of electrode materials and their performance in high-voltage lithium-ion batteries'. *ChemSusChem*. 2020;**13**(4):838–49.

[81] Barcellona S., Piegari L. 'Integrated electro-thermal model for pouch lithium ion batteries'. *Mathematics and Computers in Simulation*. 2021;**183**:5–19.

[82] Barrett D.H., Haruna A. 'Artificial intelligence and machine learning for targeted energy storage solutions'. *Current Opinion in Electrochemistry*. 2020;**21**:160–66.

[83] Basia A., Simeu-Abazi Z., Gascard E., Zwolinski P. 'Review on state of health estimation methodologies for lithium-ion batteries in the context of circular economy'. *CIRP Journal of Manufacturing Science and Technology*. 2021;**32**:517–28.

[84] Bayat P., Baghramian A. 'A novel self-tuning type-2 fuzzy maximum power point tracking technique for efficiency enhancement of fuel cell based battery chargers'. *International Journal of Hydrogen Energy*. 2020;**45**(43):23275–93.

[85] Bayati M., Abedi M., Gharehpetian G.B., Farahmandrad M. 'Sinusoidal-ripple current control in battery charger of electric vehicles'. *IEEE Transactions on Vehicular Technology*. 2020;**69**(7):7201–10.

[86] Bednar D.J., Reames T.G. 'Recognition of and response to energy poverty in the United States'. *Nature Energy*. 2020;**5**(6):432–39.

[87] Beelen H., Bergveld H.J., Donkers M.C.F. 'Joint estimation of battery parameters and state of charge using an extended Kalman filter: a single-parameter tuning approach'. *IEEE Transactions on Control Systems Technology*. 2021;**29**(3):1087–101.

[88] Beelen H., Mundaragi Shivakumar K., Raijmakers L., Donkers M.C.F., Bergveld H.J. 'Towards impedance-based temperature estimation for Li-ION battery packs'. *International Journal of Energy Research*. 2020;**44**(4):2889–908. Available from https://onlinelibrary.wiley.com/toc/1099114x/44/4

[89] Ben Ali J., Azizi C., Saidi L., Bechhoefer E., Benbouzid M. 'Reliable state of health condition monitoring of li-ion batteries based on incremental support vector regression with parameters optimization'. *Proceedings of the Institution of Mechanical Engineers Part I-Journal of Systems and Control Engineering*; 2020. pp. 1–14.

[90] Benkhelifa F., ElSawy H., Mccann J.A., Alouini M.-S. 'Recycling cellular energy for self-sustainable iot networks: a spatiotemporal study'. *IEEE Transactions on Wireless Communications*. 2020;**19**(4):2699–712.

[91] Benmouna A., Becherif M., Depernet D., Dépature C., Boulon L. ‘Nonlinear control and optimization of hybrid electrical vehicle under sources limitation constraints’. *International Journal of Hydrogen Energy*. 2020;**45**(19):11255–66.

[92] Bera A.K., Yusuf S.M. ‘Temperature-dependent na-ion conduction and its pathways in the crystal structure of the layered battery material $Na_2Ni_2TeO_6$’. *Journal of Physical Chemistry C*. 2020;**124**(8):4421–29.

[93] Bhayo B.A., Al-Kayiem H.H., Gilani S.I.U., Ismail F.B. ‘Power management optimization of hybrid solar photovoltaic-battery integrated with pumped-hydro-storage system for standalone electricity generation’. *Energy Conversion and Management*. 2020;**215**:112942.

[94] Bi K., Liu Y., Tian B., Wang L., Yan W. ‘Distributed state-of-charge balancing control strategy for modular super capacitor energy storage system’. *IEEE International Conference on Industrial Technology (ICIT)*; Buenos Aires, Argentina, 2020. pp. 953–58. Available from https://ieeexplore.ieee.org/xpl/mostRecentIssue.jsp?punumber=9047988

[95] Bi Y., Yin Y., Choe S.-Y. ‘Online state of health and aging parameter estimation using a physics-based life model with a particle filter’. *Journal of Power Sources*. 2020;**476**:228655.

[96] Bi Z., Reiner M.A., Keoleian G.A., Zhou Y., Wang M., Lin Z. ‘Wireless charging and shared autonomous battery electric vehicles (W+SABEV): synergies that accelerate sustainable mobility and greenhouse gas emission reduction’. *Mitigation and Adaptation Strategies for Global Change*. 2020;**25**(3):397–411.

[97] Bian X., Liu L., Yan J., Zou Z., Zhao R. ‘An open circuit voltage-based model for state-of-health estimation of lithium-ion batteries: model development and validation’. *Journal of Power Sources*. 2020;**448**:227401.

[98] Bian X., Wei Z., He J., Yan F., Liu L. ‘A two-step parameter optimization method for low-order model-based state-of-charge estimation’. *IEEE Transactions on Transportation Electrification*. 2021;7(2):399–409.

[99] Bian X., Wei Z., He J., Yan F., Liu L. ‘A novel model-based voltage construction method for robust state-of-health estimation of lithium-ion batteries’. *IEEE Transactions on Industrial Electronics*. 2021;**68**(12):12173–84.

[100] Billot N., Beyer M., Koch N., Ihle C., Reinhart G. ‘Development of an adhesion model for graphite-based lithium-ion battery anodes’. *Journal of Manufacturing Systems*. 2021;**58**:131–42.

[101] Boaretto N., Meabe L., Martinez-Ibañez M., Armand M., Zhang H. ‘Review—polymer electrolytes for rechargeable batteries: from nanocomposite to nanohybrid’. *Journal of The Electrochemical Society*. 2020;**167**(7):070524.

[102] Bölle F.T., Mathiesen N.R., Nielsen A.J., Vegge T., Garcia-Lastra J.M., Castelli I.E. ‘Autonomous discovery of materials for intercalation electrodes’. *Batteries & Supercaps*. 2020;**3**(6):488–98.

[103] Borlaug B., Salisbury S., Gerdes M., Muratori M. ‘Levelized cost of charging electric vehicles in the united states’. *Joule*. 2020;**4**(7):1470–85.

[104] Boyle J., Littler T., Foley A. 'Battery energy storage system state-of-charge management to ensure availability of frequency regulating services from wind farms'. *Renewable Energy*. 2020;**160**:1119–35.

[105] Brietzke B., Solovej J.P. 'The second-order correction to the ground state energy of the dilute Bose gas'. *Annales Henri Poincaré*. 2020;**21**(2):571–626.

[106] Bruch M., Millet L., Kowal J., Vetter M. 'Novel method for the parameterization of a reliable equivalent circuit model for the precise simulation of a battery cell's electric behavior'. *Journal of Power Sources*. 2021;**490**(1):229513.

[107] Bugryniec P.J., Davidson J.N., Brown S.F. 'Computational modelling of thermal runaway propagation potential in lithium iron phosphate battery packs'. *Energy Reports*. 2020;**6**:189–97.

[108] Bui Y.-H., Hussain A., Kim H.-M. ' Double deep Q-learning-based distributed operation of battery energy storage system considering uncertainties '. *IEEE Transactions on Smart Grid*. 2020;**11**(1):457–69.

[109] Bukhalkin D.D., Semenov A.P., Novikov A.A., *et al*. 'Phase change materials in energy: current state of research and potential applications'. *Chemistry and Technology of Fuels and Oils*. 2020;**55**(6):733–41.

[110] Busquets-Monge S., Filba-Martinez A., Alepuz S., *et al*. 'Multibattery-fed neutral-point-clamped DC–AC converter with soc balancing control to maximize capacity utilization'. *IEEE Transactions on Industrial Electronics*. 2020;**67**(1):16–27.

[111] Cai G., Holoubek J., Xia D, *et al*. 'An ester electrolyte for lithium-sulfur batteries capable of ultra-low temperature cycling'. *Chemical Communications (Cambridge, England)*. 2020;**56**(64):9114–17.

[112] Cai Z., Ma Y., Huang X., *et al*. 'High electrochemical stability al-doped spinel limn2o4 cathode material for li-ion batteries'. *Journal of Energy Storage*. 2020;**27**:101036.

[113] Cao B., Li X.F. Recent progress on carbon-based anode materials for Na-ion batteries. *Acta Physico-Chimica Sinica*. 2020;**36**(5).

[114] Cao J., Harrold D., Fan Z., Morstyn T., Healey D., Li K. 'Deep reinforcement learning-based energy storage arbitrage with accurate lithium-ion battery degradation model'. *IEEE Transactions on Smart Grid*. 2020;**11**(5):4513–21.

[115] Cao X., Xu W., Liu X., Peng J., Liu T. 'A deep reinforcement learning-based on-demand charging algorithm for wireless rechargeable sensor networks'. *Ad Hoc Networks*. 2021;**110**(485):102278.

[116] Cao Y., Chen J.J.J., Barteau M.A. 'Systematic approaches to improving the performance of polyoxometalates in non-aqueous redox flow batteries'. *Journal of Energy Chemistry*. 2020;**50**:115–24.

[117] Carter R., Parker J.F., Sassin M.B, *et al*. 'Initiated chemical vapor deposition of ultrathin polymer coatings at graphite electrodes for enhanced performance in Li-ion batteries'. *Journal of The Electrochemical Society*. 2020;**167**(6):060510.

[118] Casino S., Niehoff P., Börner M., Winter M. ‘Protective coatings on silicon particles and their effect on energy density and specific energy in lithium ion battery cells: a model study’. *Journal of Energy Storage*. 2020;**29**: 101376.

[119] Cavalheiro G.M., Iriyama T., Nelson G.J., Huang S., Zhang G. ‘Effects of nonuniform temperature distribution on degradation of lithium-ion batteries’. *Journal of Electrochemical Energy Conversion and Storage*. 2020; **17**(2).

[120] Cen J.W., Jiang F.M. ‘Li-ion power battery temperature control by a battery thermal management and vehicle cabin air conditioning integrated system’. *Energy for Sustainable Development*. 2020;**57**:141–48.

[121] Cen Z.H., Kubiak P. ‘Lithium-ion battery SOC/SOH adaptive estimation via simplified single particle model’. *International Journal of Energy Research*. 2020;**44**(15):12444–59.

[122] Chakir A., Tabaa M., Moutaouakkil F., *et al*. ‘Optimal energy management for a grid connected PV-battery system’. *Energy Reports*. 2020;**6**:218–31.

[123] Chang G., Cui X., Li Y., Ji Y. ‘Effects of reciprocating liquid flow battery thermal management system on thermal characteristics and uniformity of large lithium-ion battery pack’. *International Journal of Energy Research*. 2020;**44**(8):6383–95.

[124] Chang Q., Jin Y., Jia M., Yuan Q., Zhao C., Jia M. ‘Sulfur-doped cop@ nitrogen-doped porous carbon hollow tube as an advanced anode with excellent cycling stability for sodium-ion batteries’. *Journal of Colloid and Interface Science*. 2020;**575**:61–68.

[125] Che Y., Deng Z., Lin X., Hu L., Hu X. ‘Predictive battery health management with transfer learning and online model correction’. *IEEE Transactions on Vehicular Technology*. 2021;**70**(2):1269–77.

[126] Chehade A.A., Hussein A.A. ‘A collaborative gaussian process regression model for transfer learning of capacity trends between li-ion battery cells’. *IEEE Transactions on Vehicular Technology*. 2020;**69**(9):9542–52.

[127] Chen A., Zhang W., Zhang C., Huang W., Liu S. ‘A temperature and current rate adaptive model for high-power lithium-titanate batteries used in electric vehicles’. *IEEE Transactions on Industrial Electronics*. 2020;**67**(11):9492–502.

[128] Chen C.-H., Brosa Planella F., O’Regan K., Gastol D., Widanage W.D., Kendrick E. ‘Development of experimental techniques for parameterization of multi-scale lithium-ion battery models’. *Journal of The Electrochemical Society*. 2020;**167**(8):080534.

[129] Chen F.X., Deng H.J., Shao Z.G. ‘Decentralised control method of battery energy storage systems for soc balancing and reactive power sharing’. *IET Generation, Transmission & Distribution*. 2020;**14**(18):3702–09.

[130] Chen G.-J., Liu Y.-H., Wang S.-C., Luo Y.-F., Yang Z.-Z. ‘Searching for the optimal current pattern based on grey wolf optimizer and equivalent circuit model of li-ion batteries’. *Journal of Energy Storage*. 2021;**33**(3):101933.

[131] Chen G., Liu Z., Su H., Zhuang W. 'Electrochemical-distributed thermal coupled model-based state of charge estimation for cylindrical lithium-ion batteries'. *Control Engineering Practice*. 2021;**109**:104734.

[132] Chen H., Chen J., Lu H., Yan C., Liu Z. 'A modified MPC-based optimal strategy of power management for fuel cell hybrid vehicles'. *IEEE/ASME Transactions on Mechatronics*. 2020;**25**(4):2009–18.

[133] Chen H., Xu B.-B., Ping Q.-S, *et al.* '$Co_2B_2O_5$ as an anode material with high capacity for sodium ion batteries'. *Rare Metals*. 2020;**39**(9):1045–52.

[134] Chen H.J., Huang H.Y.S. 'Modeling and simulation of the non-equilibrium process for a continuous solid solution system in lithium-ion batteries'. *International Journal of Solids and Structures*. 2021;**212**:124–42.

[135] Chen J., He Z., Sun S., Guo S., Chen Z. 'Efficient linear schemes with unconditional energy stability for the phase field model of solid-state dewetting problems'. *Journal of Computational Mathematics*. 2020;**38**(3):452–68.

[136] Chen J.C., Chen T.-L., Liu W.-J., Cheng C.C., Li M.-G. 'Combining empirical mode decomposition and deep recurrent neural networks for predictive maintenance of lithium-ion battery'. *Advanced Engineering Informatics*. 2021;**50**:101405.

[137] Chen J.M. 'All solid polymer electrolytes for lithium batteries'. *Progress in Chemistry*. 2020;**32**(4):481–96.

[138] Chen J., Feng X., Jiang L., Zhu Q. 'State of charge estimation of lithium-ion battery using denoising autoencoder and gated recurrent unit recurrent neural network'. *Energy*. 2021;**227**(9):120451.

[139] Chen K., Chen Y., Song M., Wang S. 'Multi-parameter structure design of parallel mini-channel cold plate for battery thermal management'. *International Journal of Energy Research*. 2020;**44**(6):4321–34.

[140] Chen L., An J., Wang H., Zhang M., Pan H. 'Remaining useful life prediction for lithium-ion battery by combining an improved particle filter with sliding-window gray model'. *Energy Reports*. 2020;**6**:2086–93.

[141] Chen L., An J., Wang H., Zhang M., Pan H. 'Remaining useful life prediction for lithium-ion battery by combining an improved particle filter with sliding-window gray model'. *Energy Reports*. 2020;**6**:2086–93.

[142] Chen L., Chen J., Wang H., *et al.* 'Remaining useful life prediction of battery using a novel indicator and framework with fractional grey model and unscented particle filter'. *IEEE Transactions on Power Electronics*. 2020;**35**(6):5850–59.

[143] Chen L., Chen Y., Lopes A.M., Kong H., Wu R. 'State of charge estimation of lithium-ion batteries based on fuzzy fractional-order unscented Kalman filter'. *Fractal and Fractional*. 2021;**5**(3):91.

[144] Chen L., Huang L., Chen G., Zhang X., Chen Y. 'Highly graphitized porous carbon-feni3 fabricated from oleic acid for advanced lithium-sulfur batteries'. *Chemistry (Weinheim an Der Bergstrasse, Germany)*. 2020;**26**(41):8926–34.

[145] Chen L., Wang H., Liu B., Wang Y., Ding Y., Pan H. 'Battery state-of-health estimation based on a metabolic extreme learning machine

combining degradation state model and error compensation'. *Energy*. 2021;**215**(119078):119078.

[146] Chen L., Wang S., Jiang H., Fernandez C., Zou C. 'Decreasing weight particle Swarm optimization combined with unscented particle filter for the non-linear model for lithium battery state of charge estimation'. *International Journal of Electrochemical Science*. 2020;**15**(10):10104–16.

[147] Chen L., Zhang M., Ding Y., *et al*. 'Estimation the internal resistance of lithium-ion-battery using a multi-factor dynamic internal resistance model with an error compensation strategy'. *Energy Reports*. 2021;**7**:3050–59.

[148] Chen L., Hu M., Cao K., *et al*. 'Core temperature estimation based on electro-thermal model of lithium-ion batteries'. *International Journal of Energy Research*. 2020;**44**(7):5320–33. Available from https://onlinelibrary.wiley.com/toc/1099114x/44/7

[149] Chen M., Lin S., Song W., Lv J., Feng Z. 'Electrical and thermal interplay in lithium-ion battery internal short circuit and safety protection'. *International Journal of Energy Research*. 2020;**44**(8):6745–57.

[150] Chen N., Zhang P., Dai J., Gui W. 'Estimating the state-of-charge of lithium-ion battery using an H-infinity observer based on electrochemical impedance model'. *IEEE Access: Practical Innovations, Open Solutions*. 2020;**8**:26872–84.

[151] Chen R., Xu X., Peng S, *et al*. 'A flexible and safe aqueous zinc–air battery with a wide operating temperature range from −20 to 70 °C'. *ACS Sustainable Chemistry & Engineering*. 2020;**8**(31):11501–11.

[152] Chen R., Li Q., Yu X., Chen L., Li H. 'Approaching practically accessible solid-state batteries: stability issues related to solid electrolytes and interfaces'. *Chemical Reviews*. 2020;**120**(14):6820–77.

[153] Chen S., Qiu L., Cheng H.-M. 'Carbon-based fibers for advanced electrochemical energy storage devices'. *Chemical Reviews*. 2020;**120**(5):2811–78.

[154] Chen S., Zhang J., Wang Z.. 'Electrocatalytic $NiCo_2O_4$ nanofiber arrays on carbon cloth for flexible and high-loading lithium-sulfur batteries'. *Nano Letters*. 2021;**21**(12):5285–92.

[155] Chen T., Liu W., Zhuo Y., *et al*. 'Single-phase P2-type layered oxide with cu-substitution for sodium ion batteries'. *Journal of Energy Chemistry*. 2020;**43**:148–54.

[156] Chen W.-K., Zhang Y., Jiang B., Fang W.-H., Cui G. 'Efficient construction of excited-state hessian matrices with machine learning accelerated multi-layer energy-based fragment method'. *The Journal of Physical Chemistry. A*. 2020;**124**(27):5684–95.

[157] Chen X., Liu Z., Wang J., Yang C., Long B., Zhou X. 'An adaptive prediction model for the remaining life of an li-ion battery based on the fusion of the two-phase wiener process and an extreme learning machine'. *Electronics*. 2021;**10**(5):540.

[158] Chen X., Shi M., Zhou J., *et al*. 'Distributed cooperative control of multiple hybrid energy storage systems in a DC microgrid using consensus protocol'. *IEEE Transactions on Industrial Electronics*. 2020;**67**(3):1968–79.

[159] Chen X.-D., Yang H.-Y., Wun J.-S., Wang C.-H., Li L.-L. 'Life prediction of lithium-ion battery based on a hybrid model'. *Energy Exploration and Exploitation*. 2020;**38**(5):1854–78.

[160] Chen X.B., Chen X.P., Chen X.W. 'A novel framework for lithium-ion battery state of charge estimation based on Kalman filter Gaussian process regression'. *International Journal of Energy Research*. 2021;**45**(9):13238–49. Available from https://onlinelibrary.wiley.com/toc/1099114x/45/9

[161] Chen X.-D., Yang H.-Y., Wun J.-S., Wang C.-H., Li L.-L. 'Life prediction of lithium-ion battery based on a hybrid model'. *Energy Exploration & Exploitation*. 2020;**38**(5):1854–78.

[162] Chen X., Gao Z., Ma R., Huang X. 'Hybrid extended-unscented Kalman filters for continuous-time nonlinear fractional-order systems involving process and measurement noises'. *Transactions of the Institute of Measurement and Control*. 2020;**42**(9):1618–31.

[163] Chen X., Tang J., Li W., Xie L. 'Operational reliability and economy evaluation of reusing retired batteries in composite power systems'. *International Journal of Energy Research*. 2020;**44**(5):3657–73.

[164] Chen X., Song S., Ji J., Tang Z., Todo Y. 'Incorporating a multiobjective knowledge-based energy function into differential evolution for protein structure prediction'. *Information Sciences*. 2020;**540**:69–88.

[165] Chen X.-S., Gao Y., Zhu G.-R., *et al*. 'Multifunctional interlayer with simultaneously capturing and catalytically converting polysulfides for boosting safety and performance of lithium-sulfur batteries at high-low temperatures'. *Journal of Energy Chemistry*. 2020;**50**:248–59.

[166] Chen Y., He Y., Li Z., Chen L., Zhang C.. 'Remaining useful life prediction and state of health diagnosis of Lithium-ion battery based on second-order central difference particle filter'. *IEEE Access: Practical Innovations, Open Solutions*. 2020;**8**:37305–13.

[167] Chen Y. 'A combined robust approach based on auto-regressive long short-term memory network and moving horizon estimation for state-of-charge estimation of lithium-ion batteries'. *International Journal of Energy Research*. 2021:1–16.

[168] Chen Y., Li M., Yang B., *et al*. 'Variable-parameter T-circuit-based ipt system charging battery with constant current or constant voltage output'. *IEEE Transactions on Power Electronics*. 2020;**35**(2):1672–84.

[169] Chen Y., Liu X., Fathy H.K., Zou J., Yang S. 'A graph-theoretic framework for analyzing the speeds and efficiencies of battery pack equalization circuits'. *International Journal of Electrical Power & Energy Systems*. 2018;**98**:85–99.

[170] Chen Y., Lu Y., Liu Z., *et al*. 'Efficient dissolution of lithium-ion batteries cathode LiCoO2 by polyethylene glycol-based deep eutectic solvents at mild temperature'. *ACS Sustainable Chemistry & Engineering*. 2020;**8**(31):11713–20.

[171] Chen Y.-C., Hsu Y.-K. 'Ultrafast carrier transport through an advanced thick electrode with a high areal capacity for aqueous lithium-ion batteries'. *ChemSusChem*. 2020;**13**(13):3479–87.

[172] Chen Y.-C., Lin K.-C., Wu C.-Y., Chen C.-J., Hsieh Y.-W. 'Determinants of quality of life in the older residents of long-term care facilities using the world Health organization International classification of functioning, disability and health framework in Taiwan'. *Disability and Rehabilitation.* 2020;**42**(16):2325–33.

[173] Chen Z., Chen L., Shen W., Xu K. 'Remaining useful life prediction of lithium-ion battery via a sequence decomposition and deep learning integrated approach'. *IEEE Transactions on Vehicular Technology.* 2022;**71**(2):1466–79.

[174] Chen Z., Xiao J., Shu X., Shen S., Shen J., Liu Y. 'Model-based adaptive joint estimation of the state of charge and capacity for lithium–ion batteries in their entire lifespan'. *Energies*. 2020;**13**(6):1410.

[175] Chen Z., Zhao H., Shu X., Zhang Y., Shen J., Liu Y. 'Synthetic state of charge estimation for lithium-ion batteries based on long short-term memory network modeling and adaptive H-infinity filter'. *Energy*. 2021;**228**(1): 120630.

[176] Chen Z., Zhou J., Zhou F., Xu S. 'State-of-charge estimation of lithium-ion batteries based on improved H infinity filter algorithm and its novel equalization method'. *Journal of Cleaner Production*. 2021;**290**:125180.

[177] Cheng F., Hu Y., Zhao L.X. 'Analysis of weak solutions for the phase-field model for lithium-ion batteries'. *Applied Mathematical Modelling.* 2020;**78**:185–99.

[178] Cheng G., Wang X., He Y. 'Remaining useful life and state of health prediction for lithium batteries based on empirical mode decomposition and a long and short memory neural network'. *Energy*. 2021;**232**:121022.

[179] Cheng H., Wang L., Xu L., Ge X., Yang S. 'An integrated electrified powertrain topology with SRG and SRM for plug-in hybrid electrical vehicle'. *IEEE Transactions on Industrial Electronics*. 2020;**67**(10):8231–41.

[180] Cheng X.M., Tang Y., Wang Z.P. 'Thermal property measurements of a large prismatic lithium-ion battery for electric vehicles'. *Journal of Thermal Science*. 2021;**30**(2):477–92.

[181] Cheng Y., Song D., Wang Z., Lu C., Zerhouni N. 'An ensemble prognostic method for lithium-ion battery capacity estimation based on time-varying weight allocation'. *Applied Energy*. 2020;**266**:114817.

[182] Cheng Z., Liu M., Ganapathy S., *et al.* 'Revealing the impact of space-charge layers on the li-ion transport in all-solid-state batteries'. *Joule.* 2020;**4**(6):1311–23.

[183] Chien Y.-C., Menon A.S., Brant W.R., Brandell D., Lacey M.J. 'Simultaneous monitoring of crystalline active materials and resistance evolution in lithium-sulfur batteries'. *Journal of the American Chemical Society.* 2020;**142**(3):1449–56.

[184] Cho S.J., Choi H., Youk J.H. 'Evaluation of PBI nanofiber membranes as a high-temperature resistance separator for lithium-ion batteries'. *Fibers and Polymers*. 2020;**21**(5):993–98.

[185] Choi E., Chang S. 'A temperature-dependent state of charge estimation method including hysteresis for lithium-ion batteries in hybrid electric vehicles'. *IEEE Access: Practical Innovations, Open Solutions*. 2020;**8**:129857–68.

[186] Choi J., Yoon S.U., Lee M.E., *et al*. 'High-performance nanohybrid anode based on FeS2 nanocubes and nitrogen-rich graphene oxide nanoribbons for sodium ion batteries'. *Journal of Industrial and Engineering Chemistry*. 2020;**81**:61–66.

[187] Choi S.-W., Oh S.-T., Kim M.-W., Lee I.-O., Lee J.-Y. 'Interleaved isolated single-phase pFc converter module for three-phase eV charger'. *IEEE Transactions on Vehicular Technology*. 2020;**69**(5):4957–67.

[188] Choi Y.-H., Park H., Lee S., Jeong H.-D. 'Synthesis and electrochemical performance of π-conjugated molecule bridged silicon quantum dot cluster as anode material for lithium-ion batteries'. *ACS Omega*. 2020;**5**(15):8629–37.

[189] Chu A., Allam A., Cordoba Arenas A., Rizzoni G., Onori S. 'Stochastic capacity loss and remaining useful life models for lithium-ion batteries in plug-in hybrid electric vehicles'. *Journal of Power Sources*. 2020;**478**(228991):228991.

[190] Chu M., Li H., Liao X., Cui S. ' Reinforcement learning-based multiaccess control and battery prediction with energy harvesting in IOT systems '. *IEEE Internet of Things Journal*. 2019;**6**(2):2009–20.

[191] Chub A., Vinnikov D., Kosenko R., Liivik E., Galkin I. 'Bidirectional DC–DC converter for modular residential battery energy storage systems'. *IEEE Transactions on Industrial Electronics*. 2020;**67**(3):1944–55.

[192] Chun H., Kim J., Yu J., Han S. 'Real-time parameter estimation of an electrochemical lithium-ion battery model using a long short-term memory network'. *IEEE Access: Practical Innovations, Open Solutions*. 2020;**8**:81789–99.

[193] Clemente A., Montiel M., Barreras F., Lozano A., Costa-Castello R. 'Vanadium redox flow battery state of charge estimation using a concentration model and a sliding mode observer'. *IEEE Access: Practical Innovations, Open Solutions*. 2021;**9**:72368–76.

[194] Cong X., Zhang C., Jiang J., Zhang W., Jiang Y. 'A hybrid method for the prediction of the remaining useful life of lithium-ion batteries with accelerated capacity degradation'. *IEEE Transactions on Vehicular Technology*. 2020;**69**(11):12775–85.

[195] Contestabile P., Crispino G., Di Lauro E., Ferrante V., Gisonni C., Vicinanza D. 'Overtopping breakwater for wave energy conversion: review of state of art, recent advancements and what lies ahead'. *Renewable Energy*. 2020;**147**:705–18.

[196] Conway B., Jewell K., Tanner C. 'Flexible lithium-ion conducting composite electrolyte'. *Batteries & Supercaps*. 2020;**3**(7):653–57.

[197] Copley R.J., Cumming D., Wu Y., Dwyer-Joyce R.S. 'Measurements and modelling of the response of an ultrasonic pulse to a lithium-ion battery

as a precursor for state of charge estimation'. *Journal of Energy Storage*. 2021;**36**:102406.

[198] Corno M., Pozzato G. 'Active adaptive battery aging management for electric vehicles'. *IEEE Transactions on Vehicular Technology*. 2020;**69**(1):258–69.

[199] Couto L.D., Schorsch J., Job N., Léonard A., Kinnaert M. 'State of health estimation for lithium ion batteries based on an equivalent-hydraulic model: an iron phosphate application'. *Journal of Energy Storage*. 2019;**21**:259–71.

[200] Cui G.L. 'Reasonable design of high-energy-density solid-state lithium-metal batteries'. *Matter*. 2020;**2**(4):805–15.

[201] Cui S., Joe I. 'A dynamic spatial-temporal attention-based GRU model with healthy features for state-of-health estimation of lithium-ion batteries'. *IEEE Access: Practical Innovations, Open Solutions*. 2021;**9**:27374–88.

[202] Cui X.B., Hu T.T. 'State of health diagnosis and remaining useful life prediction for lithium-ion battery based on data model fusion method'. *IEEE Access: Practical Innovations, Open Solutions*. 2020;**8**:207298–307.

[203] Dey S., Shi Y., Smith K., Colclasure A., Li X. 'From battery cell to electrodes: real-time estimation of charge and health of individual battery electrodes'. *IEEE Transactions on Industrial Electronics*. 2020;**67**(3): 2167–75.

[204] Diaz C., Quintero V., Perez A., *et al*. 'Particle-filtering-based prognostics for the state of maximum power available in lithium-ion batteries at electromobility applications'. *IEEE Transactions on Vehicular Technology*. 2020;**69**(7):7187–200.

[205] DingG.-C.,ZhuL.-M.,YangQ,*etal*.'NaV_3O_8/poly(3,4-ethylenedioxythiophene) composites as high-rate and long-lifespan cathode materials for reversible sodium storage'. *Rare Metals*. 2020;**39**(8):865–73.

[206] Cui X., Zeng J., Zhang H., *et al*. 'Simplification strategy research on hard-cased li-ion battery for thermal modeling'. *International Journal of Energy Research*. 2020;**44**(5):3640–56. Available from https://onlinelibrary.wiley.com/toc/1099114x/44/5

[207] Cui X., Garg A., Trang Thao N., Trung N.T. 'Machine learning approach for solving inconsistency problems of li-ion batteries during the manufacturing stage'. *International Journal of Energy Research*. 2020;**44**(11):9194–204.

[208] Ding L., Han Q.-L., Ning B., Yue D. 'Distributed resilient finite-time secondary control for heterogeneous battery energy storage systems under denial-of-service attacks'. *IEEE Transactions on Industrial Informatics*. 2020;**16**(7):4909–19.

[209] Doenges K., Egido I., Sigrist L., Lobato Miguelez E., Rouco L. 'Improving AGC performance in power systems with regulation response accuracy margins using battery energy storage system (BESS)'. *IEEE Transactions on Power Systems*. 2020;**35**(4):2816–25.

[210] Dong G., Yang F., Wei Z., Wei J., Tsui K.-L. 'Data-driven battery health prognosis using adaptive Brownian motion model'. *IEEE Transactions on Industrial Informatics*. 2020;**16**(7):4736–46.

[211] Dong Y., Li Y., Shi H., *et al*. 'Graphene encapsulated iron nitrides confined in 3D carbon nanosheet frameworks for high-rate lithium ion batteries'. *Carbon*. 2020;**159**:213–20.

[212] Dongmo S., Stock D., Alexander Kreissl J.J., *et al*. 'Implications of testing a zinc-oxygen battery with zinc foil anode revealed by operando gas analysis'. *ACS Omega*. 2020;**5**(1):626–33.

[213] Dörfler S., Althues H., Härtel P., Abendroth T., Schumm B., Kaskel S. 'Challenges and key parameters of lithium-sulfur batteries on pouch cell level'. *Joule*. 2020;**4**(3):539–54.

[214] Drummond R., Bizeray A.M., Howey D.A., Duncan S.R. 'A feedback interpretation of the doyle–fuller–newman lithium-ion battery model'. *IEEE Transactions on Control Systems Technology*. 2020;**28**(4):1284–95.

[215] Du D., Hu X., Zeng D., *et al*. 'Water-insoluble side-chain-grafted single ion conducting polymer electrolyte for long-term stable lithium metal secondary batteries'. *ACS Applied Energy Materials*. 2020;**3**(1):1128–38.

[216] Du R., Hu X., Xie S., Hu L., Zhang Z., Lin X. 'Battery aging- and temperature-aware predictive energy management for hybrid electric vehicles'. *Journal of Power Sources*. 2020;**473**:228568.

[217] Du W., Gao W., Yang T., *et al*. 'Cobalt nanoparticles embedded into free-standing carbon nanofibers as catalyst for room-temperature sodium-sulfur batteries'. *Journal of Colloid and Interface Science*. 2020;**565**:63–69.

[218] Duan B., Zhang Q., Geng F., Zhang C. 'Remaining useful life prediction of lithium-ion battery based on extended Kalman particle filter'. *International Journal of Energy Research*. 2020;**44**(3):1724–34. Available from https://onlinelibrary.wiley.com/toc/1099114x/44/3

[219] Cui Z., He S.-A., Liu Q., Zou R. 'Multifunctional $NiCO_2O_4$ nanosheet-assembled hollow nanoflowers as a highly efficient sulfur host for lithium-sulfur batteries '. *Dalton Transactions*. 2020;**49**(20):6876–83.

[220] Duan W., Song C., Peng S., Xiao F., Shao Y., Song S. 'An improved gated recurrent unit network model for state-of-charge estimation of lithium-ion battery'. *Energies*. 2020;**13**(23):6366.

[221] Cunha R.P., Lombardo T., Primo E.N., Franco A.A. 'Artificial intelligence investigation of NMC cathode manufacturing parameters interdependencies'. *Batteries & Supercaps*. 2020;**3**(1):60–67.

[222] Dahbi M., Doubabi S., Rachid A., Oulad-Abbou D. 'Performance evaluation of electric vehicle brushless direct current motor with a novel high-performance control strategy with experimental implementation'. *Proceedings of the Institution of Mechanical Engineers Part I-Journal of Systems and Control Engineering*. 2020;**234**(3):358–69.

[223] Danilov D.L., Chen C., Jiang M., Eichel R.-A., Notten P.H.L. 'On the conversion of NDP energy spectra into depth concentration profiles for thin-films all-solid-state batteries'. *Radiation Effects and Defects in Solids*. 2020;**175**(3–4):367–82.

[224] Duan X., Liu Y., Zhou X., Zou P., Liu J. 'Experimental investigation energy balance and distribution of a turbocharged GDI engine fuelled with ethanol

and gasoline blend under transient and steady-state operating conditions'. *Thermal Science*. 2020;**24**(1 Part A):243–57.

[225] Das B., Pandey B. 'Can we constrain the dark energy equation of state parameters using configuration entropy?'. *Monthly Notices of the Royal Astronomical Society*. 2020;**492**(3):3928–39.

[226] Dubarry M., Baure G., Pastor-Fernández C., Yu T.F., Widanage W.D., Marco J. 'Battery energy storage system modeling: a combined comprehensive approach'. *Journal of Energy Storage*. 2019;**21**:172–85.

[227] Duraisamy T., Kaliyaperumal D. 'Adaptive passive balancing in battery management system for e-mobility'. *International Journal of Energy Research*. 2021;**45**(7):10752–64.

[228] Das R., Dash D., Sarkar M.K. 'HTMS: fuzzy based hierarchical trust management scheme in WSN'. *Wireless Personal Communications*. 2020;**112**(2):1079–112.

[229] Datta U., Kalam A., Shi J. 'Battery energy storage system control for mitigating pv penetration impact on primary frequency control and state-of-charge recovery'. *IEEE Transactions on Sustainable Energy*. 2020;**11**(2):746–57.

[230] Day C., Greig K., Massey A., Peake J., Crossley D., Dryfe R.A.W. 'Utilizing cyclic voltammetry to understand the energy storage mechanisms for copper oxide and its graphene oxide hybrids as lithium-ion battery anodes'. *ChemSusChem*. 2020;**13**(6):1504–16.

[231] El Kharbachi A. 'Pseudo-ternary LiBh4 center dot LiCl center dot P2S5 system as structurally disordered bulk electrolyte for all-solid-state lithium batteries'. *Physical Chemistry Chemical Physics*. 2020;**22**(25):13872–79.

[232] El-Sehiemy R.A., Hamida M.A., Mesbahi T. 'Parameter identification and state-of-charge estimation for lithium-polymer battery cells using enhanced sunflower optimization algorithm'. *International Journal of Hydrogen Energy*. 2020;**45**(15):8833–42.

[233] De Cauwer C., Verbeke W., Van Mierlo J., Coosemans T. 'A model for range estimation and energy-efficient routing of electric vehicles in real-world conditions'. *IEEE Transactions on Intelligent Transportation Systems*. 2020;**21**(7):2787–800.

[234] De Gennaro M., Paffumi E., Martini G., Giallonardo A., Pedroso S., Loiselle-Lapointe A. 'A case study to predict the capacity fade of the battery of electrified vehicles in real-world use conditions'. *Case Studies on Transport Policy*. 2020;**8**(2):517–34.

[235] Delaporte N., Ossonon D.B., Zaghib K., Bélanger D. 'Fabrication of current collectors and binder-free electrodes on separators used in lithium-ion batteries'. *Batteries & Supercaps*. 2020;**3**(7):638–46.

[236] Deleebeeck L., Veltzé S. 'Electrochemical impedance spectroscopy study of commercial li-ion phosphate batteries: a metrology perspective'. *International Journal of Energy Research*. 2020;**44**(9):7158–82.

[237] Demircali A., Koroglu S. 'Jaya algorithm-based energy management system for battery- and ultracapacitor-powered ultralight electric vehicle'. *International Journal of Energy Research*. 2020;**44**(6):4977–85.

[238] Deng J., Bae C., Denlinger A., Miller T. 'Electric vehicles batteries: requirements and challenges'. *Joule*. 2020;**4**(3):511–15.

[239] Deng L., Shen W., Wang H., Wang S. 'A rest-time-based prognostic model for remaining useful life prediction of lithium-ion battery'. *Neural Computing and Applications*. 2021;**33**(6):2035–46.

[240] Deng Y., Zhang Y., Luo F., Mu Y. 'Operational planning of centralized charging stations utilizing second-life battery energy storage systems'. *IEEE Transactions on Sustainable Energy*. 2021;**12**(1):387–99.

[241] Deng Y.-P., Liang R., Jiang G., Jiang Y., Yu A., Chen Z. ' The current state of aqueous Zn-based rechargeable batteries '. *ACS Energy Letters*. 2020;**5**(5):1665–75.

[242] Deng Z., Hu X., Lin X., Kim Y., Li J. 'Sensitivity analysis and joint estimation of parameters and states for all-solid-state batteries'. *IEEE Transactions on Transportation Electrification*. 2021;7(3):1314–23.

[243] Deng Z., Hu X., Lin X., Xu L., Che Y., Hu L. 'General discharge voltage information enabled health evaluation for lithium-ion batteries'. *IEEE/ASME Transactions on Mechatronics*. 2021;**26**(3):1295–306.

[244] Deng Z., Hu X., Lin X., Xu L., Li J., Guo W. 'A reduced-order electrochemical model for all-solid-state batteries'. *IEEE Transactions on Transportation Electrification*. 2021;7(2):464–73.

[245] Eleftheroglou N., Mansouri S.S., Loutas T., *et al*. 'Intelligent data-driven prognostic methodologies for the real-time remaining useful life until the end-of-discharge estimation of the lithium-polymer batteries of unmanned aerial vehicles with uncertainty quantification'. *Applied Energy*. 2019;**254**(113677):113677.

[246] Eroğlu F., Kurtoğlu M., Vural A.M. 'Bidirectional DC–DC converter based multilevel battery storage systems for electric vehicle and large-scale grid applications: a critical review considering different topologies, state-of-charge balancing and future trends'. *IET Renewable Power Generation*. 2021;**15**(5):915–38.

[247] Li J., Ye M., Meng W., Xu X., Jiao S. 'A novel state of charge approach of lithium ion battery using least squares support vector machine'. *IEEE Access: Practical Innovations, Open Solutions*. 2020;**8**:195398–410.

[248] Li J., Wang D., Deng L., *et al*. 'Aging modes analysis and physical parameter identification based on a simplified electrochemical model for lithium-ion batteries'. *Journal of Energy Storage*. 2020;**31**(6):101538.

[249] Li K., Xu Z., Liu Q., *et al*. 'Space-confined construction of nitrogen-rich cobalt porphyrin-derived nanoparticulates anchored on activated carbon for high-current lithium thionyl chloride battery'. *Electrochimica Acta*. 2020;**353**:136543.

[250] Li K., Zhou P., Lu Y., Han X., Li X., Zheng Y. 'Battery life estimation based on cloud data for electric vehicles'. *Journal of Power Sources*. 2020;**468**(228192):228192.

[251] Li K., Xue W., Tan G., Denzer A.S. 'A state of the art review on the prediction of building energy consumption using data-driven technique and evolutionary algorithms'. *Building Services Engineering Research and Technology*. 2020;**41**(1):108–27.

[252] Aaboud M. 'Search for charged Higgs bosons decaying via H-+/ -- > tau (+/-) nu (tau) in the tau plus jets and tau plus lepton final states with 36 FB (-1) of PP collision data recorded at root s=13 tev with the ATLAS experiment'. *Journal of High Energy Physics*. 2018;**9**.

[253] Wang F., Amogne Z.E., Tseng C., Chou J. 'A hybrid method for online cycle life prediction of lithium-ion batteries'. *International Journal of Energy Research*. 2022;**46**(7):9080–96.

[254] Wang J., Huang S., Wu D., Lu N. 'Operating a commercial building HVAC load as a virtual battery through airflow control'. *IEEE Transactions on Sustainable Energy*. 2021;**12**(1):158–68.

[255] Wang K., Liu C., Sun J., *et al*. 'State of charge estimation of composite energy storage systems with supercapacitors and lithium batteries'. *Complexity*. 2021;**2021**:1–15.

[256] Wang L.-S., Fang Y.-Y., Zhao T, *et al*. 'Lithium-ion cell inconsistency analysis based on three-parameter weibull probability model'. *Rare Metals*. 2020;**39**(4):392–401.

[257] Faddel S., Saad A.A., Youssef T., Mohammed O. 'Decentralized control algorithm for the hybrid energy storage of shipboard power system'. *IEEE Journal of Emerging and Selected Topics in Power Electronics*. 2020;**8**(1):720–31.

[258] Wang L., Lu D., Wang X., Pan R., Wang Z. 'Ensemble learning for predicting degradation under time-varying environment'. *Quality and Reliability Engineering International*. 2020;**36**(4):1205–23.

[259] Fan D., Lu D., Wu S., *et al*. 'Mechanism of a lithiated interlayer for improving the cycle life of high voltage li-ion batteries using a commercial carbonate electrolyte'. *The Journal of Physical Chemistry C*. 2020;**124**(15):8057–66.

[260] Wang P., Yang L., Wang H., Tartakovsky D.M., Onori S. 'Temperature estimation from current and voltage measurements in lithium-ion battery systems'. *Journal of Energy Storage*. 2021;**34**:102133.

[261] Wang Q., Gu H., Ye M., Wei M., Xu X. 'State of charge estimation for Lithium-ion battery based on NarX recurrent neural network and moving window method'. *IEEE Access: Practical Innovations, Open Solutions*. 2021;**9**:83364–75.

[262] Fan X., Zhang W., Wang Z., An F., Li H., Jiang J. 'Simplified battery pack modeling considering inconsistency and evolution of current distribution'. *IEEE Transactions on Intelligent Transportation Systems*. 2021;**22**(1):630–39.

[263] Wang Q., Wang Z., Zhang L., Liu P., Zhang Z. 'A novel consistency evaluation method for series-connected battery systems based on real-world operation data'. *IEEE Transactions on Transportation Electrification*. 2021;**7**(2):437–51.

[264] Wang S., Jin S., Bai D., Fan Y., Shi H., Fernandez C. 'A critical review of improved deep learning methods for the remaining useful life prediction of lithium-ion batteries'. *Energy Reports*. 2021;**7**:5562–74.

[265] Wang S., Fernandez C., Yu C., Fan Y., Cao W., Stroe D.-I. 'A novel charged state prediction method of the lithium ion battery packs based on the composite equivalent modeling and improved splice Kalman filtering algorithm'. *Journal of Power Sources*. 2020;**471**(228450):228450.

[266] Fang Y., Zhang Q., Zhang H, *et al.* 'State-of-charge estimation technique for lithium-ion batteries by means of second-order extended Kalman filter and equivalent circuit model: great temperature robustness state-of-charge estimation'. *IET Power Electronics*. 2021;**14**(8):1515–28. Available from https://onlinelibrary.wiley.com/toc/17554543/14/8

[267] Wang W.Z., Preindl M. 'Dual cell links for battery-balancing auxiliary power modules: a cost-effective increase of accessible pack capacity'. *IEEE Transactions on Industry Applications*. 2020;**56**(2):1752–65.

[268] Farjah A., Ghanbari T. 'Early ageing detection of battery cells in battery management system'. *Electronics Letters*. 2020;**56**(12):616–19.

[269] Wang X., Li R., Dai H., Zhang N., Chen Q., Wei X. 'A novel dual time scale life prediction method for lithium-ion batteries considering effects of temperature and state of charge'. *International Journal of Energy Research*. 2021;**45**(10):14692–709.

[270] Wang X., Wang S., Junhan H., Carlos F., Yicong Z., Lei C. 'A novel Gaussian particle swarms optimized particle filter algorithm for the state of charge estimation of lithium-ion batteries'. *International Journal of Electrochemical Science*. 2020;**15**(10):10632–48.

[271] Wang Y., Yang Z., Lin F., An X., Zhou H., Fang X. 'A hybrid energy management strategy based on line prediction and condition analysis for the hybrid energy storage system of tram'. *IEEE Transactions on Industry Applications*. 2020;**56**(2):1793–803.

[272] Wang Y., Xu Y., Ma S., *et al.* 'Low temperature performance enhancement of high-safety lithium–sulfur battery enabled by synergetic adsorption and catalysis'. *Electrochimica Acta*. 2020;**353**:136470.

[273] Wang Z., Liu N., Guo Y. 'Adaptive sliding window LSTM NN based RUL prediction for lithium-ion batteries integrating LTSA feature reconstruction'. *Neurocomputing*. 2021;**466**:178–89.

[274] Watari D., Taniguchi I., Onoye T. 'SOH aware system-level battery management methodology for decentralized energy network'. *IEICE Transactions on Fundamentals of Electronics, Communications and Computer Sciences*. 2020;**E103.A**(3):596–604.

[275] Wei L., Jia L., An Z., Dang C. 'Experimental study on thermal management of cylindrical li-ion battery with flexible microchannel plates'. *Journal of Thermal Science*. 2020;**29**(4):1001–09.

[276] Wei M., Gu H., Ye M., Wang Q., Xu X., Wu C. 'Remaining useful life prediction of lithium-ion batteries based on Monte Carlo dropout and gated recurrent unit'. *Energy Reports*. 2021;**7**:2862–71.

[277] Wei Z., Dong G., Zhang X., Pou J., Quan Z., He H. 'Noise-immune model identification and state-of-charge estimation for lithium-ion battery using bilinear parameterization'. *IEEE Transactions on Industrial Electronics*. 2021;**68**(1):312–23.

[278] Werner M., Pandolfi A., Weinberg K. 'A multi-field model for charging and discharging of lithium-ion battery electrodes'. *Continuum Mechanics and Thermodynamics*. 2021;**33**(3):661–85.

[279] Wu L., Liu K., Pang H., Jin J. 'Online soc estimation based on simplified electrochemical model for lithium-ion batteries considering current bias'. *Energies*. 2021;**14**(17):5265.

[280] Wu Y., Xue Q., Shen J., Lei Z., Chen Z., Liu Y. 'State of health estimation for lithium-ion batteries based on healthy features and long short-term memory'. *IEEE Access: Practical Innovations, Open Solutions*. 2020;**8**:28533–47.

[281] Xia X., Tang L. 'A fast-validated computational model for cylindrical lithium-ion batteries under multidirectional mechanical loading'. *International Journal of Energy Research*. 2021;**45**(3):4410–28.

[282] Xia Z.Y., Abu Qahouq J.A. 'State-of-charge balancing of lithium-ion batteries with state-of-health awareness capability'. *IEEE Transactions on Industry Applications*. 2021;**57**(1):673–84.

[283] Xiao B., Xiao B., Liu L.S. 'Rapid measurement method for lithium-ion battery state of health estimation based on least squares support vector regression'. *International Journal of Energy Research*. 2021;**45**(4):5695–709.

[284] Xie S.B., Qi S.W., Lang K. 'A data-driven power management strategy for plug-in hybrid electric vehicles including optimal battery depth of discharging'. *IEEE Transactions on Industrial Informatics*. 2020;**16**(5):3387–96.

[285] Xie Y., Wang S., Fernandez C., *et al*. 'Improved gray wolf particle filtering and high-fidelity second-order autoregressive equivalent modeling for intelligent state of charge prediction of lithium-ion batteries'. *International Journal of Energy Research*. 2021;**45**(13):19203–14.

[286] Feng F., Hu X., Hu L., Hu F., Li Y., Zhang L. 'Propagation mechanisms and diagnosis of parameter inconsistency within li-ion battery packs'. *Renewable and Sustainable Energy Reviews*. 2019;**112**:102–13.

[287] Feng F., Hu X., Liu K., *et al*. 'A practical and comprehensive evaluation method for series-connected battery pack models'. *IEEE Transactions on Transportation Electrification*. 2020;**6**(2):391–416.

[288] Feng F., Teng S., Liu K., *et al.* 'Co-estimation of lithium-ion battery state of charge and state of temperature based on a hybrid electrochemical-thermal-neural-network model'. *Journal of Power Sources*. 2020;**455**:227935.

[289] Feng L., Ding J., Han Y.Y. 'Improved sliding mode based EKF for the SOC estimation of lithium-ion batteries'. *Ionics*. 2020;**26**(6):2875–82.

[290] Feng X., Ren D., He X., Ouyang M. 'Mitigating thermal runaway of lithium-ion batteries'. *Joule*. 2020;**4**(4):743–70.

[291] Feng X., Weng C., He X., *et al.* 'Online state-of-health estimation for li-ion battery using partial charging segment based on support vector machine'. *IEEE Transactions on Vehicular Technology*. 2019;**68**(9):8583–92.

[292] Feng X., Xu C., He X., Wang L., Gao S., Ouyang M. 'A graphical model for evaluating the status of series-connected lithium-ion battery pack'. *International Journal of Energy Research*. 2019;**43**(2):749–66.

[293] Feng Y., Dong Z. 'Optimal energy management with balanced fuel economy and battery life for large hybrid electric mining truck'. *Journal of Power Sources*. 2020;**454**(227948):227948.

[294] Feng Y., Xue C., Han Q.-L., Han F., Du J. 'Robust estimation for state-of-charge and state-of-health of lithium-ion batteries using integral-type terminal sliding-mode observers'. *IEEE Transactions on Industrial Electronics*. 2020;**67**(5):4013–23.

[295] Macias Fernandez A., Kandidayeni M., Boulon L., Chaoui H. 'An adaptive state machine based energy management strategy for a multi-stack fuel cell hybrid electric vehicle'. *IEEE Transactions on Vehicular Technology*. 2020;**69**(1):220–34.

[296] Finegan D.P., Quinn A., Wragg D.S., *et al.* 'Spatial dynamics of lithiation and lithium plating during high-rate operation of graphite electrodes'. *Energy & Environmental Science*. 2020;**13**(8):2570–84.

[297] Finegan D.P., Zhu J., Feng X., *et al.* 'The application of data-driven methods and physics-based learning for improving battery safety'. *Joule*. 2021;**5**(2):316–29.

[298] Fu Z., McOwen D., Zhang L., *et al.* 'Predicting the flexural strength of li-ion-conducting garnet type oxide for solid-state-batteries'. *Journal of the American Ceramic Society*. 2020;**103**(9):5186–95.

[299] Galazutdinova Y., Al-Hallaj S., Grágeda M., Ushak S. 'Development of the inorganic composite phase change materials for passive thermal management of li-ion batteries: material characterization'. *International Journal of Energy Research*. 2020;**44**(3):2011–22.

[300] Gan C., Yan W., Zhang Y., Jiang Q., Tang J. 'Research progress of two-dimensional layered and related derived materials for nitrogen reduction reaction'. *Sustainable Energy & Fuels*. 2021;**5**(13):3260–77.

[301] Gao H.D., Cui L.R., Dong Q.L. 'Reliability modeling for a two-phase degradation system with a change point based on a wiener process'. *Reliability Engineering & System Safety*. 2020;**193**:106601.

[302] Gao T., Wang Z., Chen S., Guo L. 'Hazardous characteristics of charge and discharge of lithium-ion batteries under adiabatic environment and hot environment'. *International Journal of Heat and Mass Transfer*. 2019;**141**:419–31.

[303] Gao Y., Rojas T., Wang K., *et al*. 'Low-temperature and high-rate-charging lithium metal batteries enabled by an electrochemically active monolayer-regulated interface'. *Nature Energy*. 2020;**5**(7):534–42.

[304] Gao Y., Wang C., Zhang J., *et al*. 'Graphite recycling from the spent lithium-ion batteries by sulfuric acid curing–leaching combined with high-temperature calcination'. *ACS Sustainable Chemistry & Engineering*. 2020;**8**(25):9447–55.

[305] Garg A., Shaosen S., Gao L., Peng X., Baredar P. 'Aging model development based on multidisciplinary parameters for lithium-ion batteries'. *International Journal of Energy Research*. 2020;**44**(4):2801–18.

[306] Garg A., Singh S., Li W., *et al*. 'Illustration of experimental, machine learning, and characterization methods for study of performance of li-ion batteries'. *International Journal of Energy Research*. 2020;**44**(12):9513–26.

[307] Ge Y., Zhang F., Ren Y. 'Lithium ion battery health prediction via variable mode decomposition and deep learning network with self-attention mechanism'. *Frontiers in Energy Research*. 2022;**10**:1–15.

[308] Geng P., Xu X.Y., Tarasiuk T. 'State of charge estimation method for lithium-ion batteries in all-electric ships based on lstm neural network'. *Polish Maritime Research*. 2020;**27**(3):100–08.

[309] Geng Z., Wang S., Lacey M.J., Brandell D., Thiringer T. 'Bridging physics-based and equivalent circuit models for lithium-ion batteries'. *Electrochimica Acta*. 2021;**372**(9):137829.

[310] Ghaeminezhad N., Ouyang Q., Hu X., Xu G., Wang Z. 'Active cell equalization topologies analysis for battery packs: a systematic review'. *IEEE Transactions on Power Electronics*. 2021;**36**(8):9119–35.

[311] Ghannoum A., Nieva P. 'Graphite lithiation and capacity fade monitoring of lithium ion batteries using optical fibers'. *Journal of Energy Storage*. 2020;**28**:101233.

[312] Gholizadeh M., Yazdizadeh A. 'Systematic mixed adaptive observer and EKF approach to estimate SOC and SOH of lithium–ion battery '. *IET Electrical Systems in Transportation*. 2020;**10**(2):135–43.

[313] Ghosal A., Halder S., Das S.K. 'Distributed on-demand clustering algorithm for lifetime optimization in wireless sensor networks'. *Journal of Parallel and Distributed Computing*. 2020;**141**:129–42.

[314] Sangrós Giménez C., Helmers L., Schilde C., Diener A., Kwade A. 'Modeling the electrical conductive paths within all-solid-state battery electrodes'. *Chemical Engineering & Technology*. 2020;**43**(5):819–29.

[315] Gomes R., Bhattacharyya A.J. 'Carbon nanotube-templated covalent organic framework nanosheets as an efficient sulfur host for room-temperature

metal–sulfur batteries'. *ACS Sustainable Chemistry & Engineering*. 2020;**8**(15):5946–53.

[316] Gong C., Hu Y., Gan C., Chen G., Alkahtani M. 'Modeling, analysis, and attenuation of uncontrolled generation for IPMSM-based electric vehicles in emergency'. *IEEE Transactions on Industrial Electronics*. 2020;**67**(6):4453–62.

[317] Gong Z., van de Ven B.A.C., Gupta K.M., *et al*. 'Distributed control of active cell balancing and low-voltage bus regulation in electric vehicles using hierarchical model-predictive control'. *IEEE Transactions on Industrial Electronics*. 2020;**67**(12):10464–73.

[318] Gonzalez-Castellanos A.J., Pozo D., Bischi A. 'Non-ideal linear operation model for li-ion batteries'. *IEEE Transactions on Power Systems*. 2020;**35**(1):672–82.

[319] Gou B., Xu Y., Feng X. 'State-of-health estimation and remaining-useful-life prediction for lithium-ion battery using a hybrid data-driven method'. *IEEE Transactions on Vehicular Technology*. 2020;**69**(10):10854–67.

[320] Gozdur R., Przerywacz T., Bogdański D. 'Low power modular battery management system with a wireless communication interface'. *Energies*. 2021;**14**(19):6320.

[321] Greenbank S., Howey D. 'Automated feature extraction and selection for data-driven models of rapid battery capacity fade and end of life'. *IEEE Transactions on Industrial Informatics*. 2022;**18**(5):2965–73.

[322] Guo D., Yang G., Feng X., Han X., Lu L., Ouyang M. 'Physics-based fractional-order model with simplified solid phase diffusion of lithium-ion battery'. *Journal of Energy Storage*. 2020;**30**:101404.

[323] Guo F., Hu G., Xiang S., Zhou P., Hong R., Xiong N. 'A multi-scale parameter adaptive method for state of charge and parameter estimation of lithium-ion batteries using dual Kalman filters'. *Energy*. 2019;**178**:79–88.

[324] Guo F., Hu G., Zhou P., Hu J., Sai Y. 'State of charge estimation in electric vehicles at various ambient temperatures'. *International Journal of Energy Research*. 2020;**44**(9):7357–70.

[325] Guo P.Y., Cheng Z., Yang L. 'A data-driven remaining capacity estimation approach for lithium-ion batteries based on charging health feature extraction'. *Journal of Power Sources*. 2019;**412**:442–50.

[326] Hafiz F., Awal M.A., Queiroz A.R. de, Husain I. Real-time stochastic optimization of energy storage management using deep learning-based forecasts for residential PV applications. *IEEE Transactions on Industry Applications*. 2020;**56**(3):2216–26.

[327] Han W., Zou C., Zhou C., Zhang L. 'Estimation of cell soc evolution and system performance in module-based battery charge equalization systems'. *IEEE Transactions on Smart Grid*. 2019;**10**(5):4717–28.

[328] Li J.B. 'Joint estimation of state of charge and state of health for lithium-ion battery based on dual adaptive extended Kalman filter'. *International Journal of Energy Research*. 2021:1–16.

[329] Han Z., Ai Y., Jiang X., *et al.* 'Pre-polymerization enables controllable synthesis of nanosheet-based porphyrin polymers towards high-performance li-ion batteries'. *Chemistry*. 2020;**26**(46):10433–38.

[330] Hao X., Wang H.W., Ouyang M.G. 'A novel state-of-charge-based method for plug-in hybrid vehicle electric distance analysis validated with actual driving data'. *Mitigation and Adaptation Strategies for Global Change*. 2020;**25**(3):459–75.

[331] Hasan R., Scott J. 'Comments on "state of charge-dependent polynomial equivalent circuit modeling for electrochemical impedance spectroscopy of lithium-ion batteries"'. *IEEE Transactions on Power Electronics*. 2020;**35**(4):1.

[332] Hashemi S.R., Bahadoran Baghbadorani A., Esmaeeli R., Mahajan A., Farhad S. 'Machine learning-based model for lithium-ion batteries in BMS of electric/hybrid electric aircraft'. *International Journal of Energy Research*. 2021;**45**(4):5747–65.

[333] Haus B., Mercorelli P. 'Polynomial augmented extended Kalman filter to estimate the state of charge of lithium-ion batteries'. *IEEE Transactions on Vehicular Technology*. 2020;**69**(2):1452–63.

[334] He J., Wei Z., Bian X., Yan F. 'State-of-health estimation of lithium-ion batteries using incremental capacity analysis based on voltage–capacity model'. *IEEE Transactions on Transportation Electrification*. 2020;**6**(2): 417–26.

[335] He L., Hu M., Wei Y., Liu B., Shi Q. 'State of charge estimation by finite difference extended Kalman filter with HPPC parameters identification'. *Science China Technological Sciences*. 2020;**63**(3):410–21.

[336] He Y., Li Q., Zheng X., Liu X. 'Equivalent hysteresis model based soc estimation with variable parameters considering temperature'. *Journal of Power Electronics*. 2021;**21**(3):590–602.

[337] He Z., Li Y., Sun Y., *et al.* ' State-of-charge estimation of lithium ion batteries based on adaptive iterative extended Kalman filter '. *Journal of Energy Storage*. 2021;**39**:102593.

[338] Hein S., Danner T., Latz A. 'An electrochemical model of lithium plating and stripping in lithium ion batteries'. *ACS Applied Energy Materials*. 2020;**3**(9):8519–31.

[339] Hennessy M.G., Moyles I.R. 'Asymptotic reduction and homogenization of a thermo-electrochemical model for a lithium-ion battery'. *Applied Mathematical Modelling*. 2020;**80**:724–54.

[340] Heo K., Im J., Lee J.-S., *et al.* 'High-rate blended cathode with mixed morphology for all-solid-state li-ion batteries'. *Journal of Electrochemical Science and Technology*. 2020;**11**(3):282–90.

[341] Hidalgo-Reyes J.I., Gómez-Aguilar J.F., Alvarado-Martínez V.M., López-López M.G., Escobar-Jiménez R.F. 'Battery state-of-charge estimation using fractional extended Kalman filter with mittag-leffler memory'. *Alexandria Engineering Journal*. 2020;**59**(4):1919–29.

[342] Holoubek J., Yu M., Yu S., *et al.* 'An all-fluorinated ester electrolyte for stable high-voltage li metal batteries capable of ultra-low-temperature operation'. *ACS Energy Letters*. 2020;**5**(5):1438–47.

[343] Hong J., Lee D., Jeong E.-R., Yi Y. 'Towards the swift prediction of the remaining useful life of lithium-ion batteries with end-to-end deep learning'. *Applied Energy*. 2020;**278**(115646):115646.

[344] Hoque M.A., Nurmi P., Kumar A., *et al*. 'Data driven analysis of lithium-ion battery internal resistance towards reliable state of health prediction'. *Journal of Power Sources*. 2021;**513**:230519.

[345] Hossain Lipu M.S., Hannan M.A., Karim T.F., *et al*. 'Intelligent algorithms and control strategies for battery management system in electric vehicles: progress, challenges and future outlook'. *Journal of Cleaner Production*. 2021;**292**(126044):126044.

[346] Hou J., Yang Y., Gao T. 'A variational Bayes based state-of-charge estimation for lithium-ion batteries without sensing current'. *IEEE Access: Practical Innovations, Open Solutions*. 2021;**9**:84651–65.

[347] Hsieh Y.Z. 'Prediction of battery discharge states based on the recurrent neural network'. *Journal of Internet Technology*. 2020;**21**(1):113–20.

[348] Hu J., He H., Wei Z., Li Y. 'Disturbance-immune and aging-robust internal short circuit diagnostic for lithium-ion battery'. *IEEE Transactions on Industrial Electronics*. 2022;**69**(2):1988–99.

[349] Hu X., Deng Z., Lin X., Xie Y., Teodorescu R. 'Research directions for next-generation battery management solutions in automotive applications'. *Renewable and Sustainable Energy Reviews*. 2021;**152**:111695.

[350] Hu X., Yang X., Feng F., Liu K., Lin X. 'A particle filter and long short-term memory fusion technique for lithium-ion battery remaining useful life prediction'. *Journal of Dynamic Systems, Measurement, and Control*. 2021;**143**(6):1–13.

[351] Hu X., Boulon L., Sharkh S., *et al*. 'IEEE access special section editorial: battery energy storage and management systems'. *IEEE Access: Practical Innovations, Open Solutions*. 2020;**8**:123098–103.

[352] Hu X., Che Y., Lin X., Deng Z. 'Health prognosis for electric vehicle battery packs: a data-driven approach'. *IEEE/ASME Transactions on Mechatronics*. 2020;**25**(6):2622–32.

[353] Hu X., Che Y., Lin X., Onori S. 'Battery health prediction using fusion-based feature selection and machine learning'. *IEEE Transactions on Transportation Electrification*. 2021;**7**(2):382–98.

[354] Hu X., Jiang H., Feng F., Liu B. 'An enhanced multi-state estimation hierarchy for advanced lithium-ion battery management'. *Applied Energy*. 2020;**257**:114019.

[355] Hu X., Liu W., Lin X., Xie Y., Foley A.M., Hu L. 'A control-oriented electro-thermal model for pouch-type electric vehicle batteries'. *IEEE Transactions on Power Electronics*. 2021;**36**(5):5530–44.

[356] Hu X., Xu L., Lin X., Pecht M. 'Battery lifetime prognostics'. *Joule*. 2020;**4**(2):310–46.

[357] Hu X., Zhang K., Liu K., Lin X., Dey S., Onori S. 'Advanced fault diagnosis for lithium-ion battery systems: a review of fault mechanisms, fault features, and diagnosis procedures'. *IEEE Industrial Electronics Magazine*. 2020;**14**(3):65–91.

[358] Hua X., Zhang C., Offer G. 'Finding a better fit for lithium ion batteries: a simple, novel, load dependent, modified equivalent circuit model and parameterization method'. *Journal of Power Sources*. 2021;**484**:229117.

[359] Huang D., Chen Z., Zhou S. 'Model prediction-based battery-powered heating method for series-connected lithium-ion battery pack working at extremely cold temperatures'. *Energy*. 2021;**216**:119236.

[360] Huang H., Wang H., Gu J., Wu Y. 'High-dimensional model representation-based global sensitivity analysis and the design of a novel thermal management system for lithium-ion batteries'. *Energy Conversion and Management*. 2019;**190**:54–72.

[361] Huang J. (Allen), Shoai Naini S., Miller R., *et al*. 'Development of a heat pipe–based battery thermal management system for hybrid electric vehicles'. *Proceedings of the Institution of Mechanical Engineers, Part D-Journal of Automobile Engineering*. 2020;**234**(6):1532–43.

[362] Huang X., State Key Laboratory of Automotive Safety and Energy, Tsinghua University, Beijing 100084, P R China 'Study on modeling, experimentation and state of charge estimation of parallel connected lithium-ion batteries'. *International Journal of Electrochemical Science*. 2020;**15**(2):1264–86.

[363] Huo Y., Hu W., Li Z., Rao Z. 'Research on parameter identification and state of charge estimation of improved equivalent circuit model of li-ion battery based on temperature effects for battery thermal management'. *International Journal of Energy Research*. 2020;**44**(14):11583–96.

[364] Li C., Cui N., Wang C., Zhang C. 'Simplified electrochemical lithium-ion battery model with variable solid-phase diffusion and parameter identification over wide temperature range'. *Journal of Power Sources*. 2021;**497**(15):229900.

[365] Huo Y.T., Pang X.W., Rao Z.H. 'Investigation on the effects of temperature equilibrium strategy in battery thermal management using phase change material'. *International Journal of Energy Research*. 2020;**44**(9): 7660–73.

[366] Li D., Yang L. 'Remaining useful life prediction of lithium battery based on sequential CNN–LSTM method'. *Journal of Electrochemical Energy Conversion and Storage*. 2021;**18**(4):1–13.

[367] Ibrahim A., Guo J., Wang Y., Zheng Y., Lei B., Jiang F. 'Performance of serpentine channel based li-ion battery thermal management system: an experimental investigation'. *International Journal of Energy Research*. 2020;**44**(13):10023–43.

[368] Li D., Zhang Z., Liu P., Wang Z., Zhang L. 'Battery fault diagnosis for electric vehicles based on voltage abnormality by combining the long short-term memory neural network and the equivalent circuit model'. *IEEE Transactions on Power Electronics*. 2021;**36**(2):1303–15.

[369] Imran R.M., Li Q., Flaih F.M.F. 'An enhanced lithium-ion battery model for estimating the state of charge and degraded capacity using an optimized extended Kalman filter'. *IEEE Access: Practical Innovations, Open Solutions*. 2020;**8**:208322–36.

[370] Li D.D., Yang L. 'Identification of spatial temperature gradient in large format lithium battery using a multilayer thermal model'. *International Journal of Energy Research*. 2020;**44**(1):282–97.

[371] Li H., Wang Y., He X., Li Q., Lian C., Wang Z. 'Effects of structure parameters on the thermal performance of a ternary lithium-ion battery under fast charging conditions'. *Energy & Fuels*. 2020;**34**(7):8891–904.

[372] Islam S.M.R., Park S.-Y., Balasingam B. 'Unification of internal resistance estimation methods for li-ion batteries using hysteresis-free equivalent circuit models'. *Batteries*. 2020;**6**(2):32.

[373] Li H., Gu S., Tao B, *et al*. 'Highly wrinkled nio nanosheet-based hierarchical structure/reduced fluorographene composite for enhanced performance of lithium-sulfur battery'. *Journal of the Taiwan Institute of Chemical Engineers*. 2020;**111**:205–11.

[374] Li H., 'State of charge estimation for lithium-ion battery models based on a thermoelectric coupling model'. *International Journal of Electrochemical Science*. 2020;**15**(5):3807–24.

[375] Jafari M., Malekjamshidi Z., Zhu J., Khooban M.-H. 'A novel predictive fuzzy logic-based energy management system for grid-connected and off-grid operation of residential smart microgrids'. *IEEE Journal of Emerging and Selected Topics in Power Electronics*. 2020;**8**(2):1391–404.

[376] Jaumaux P., Liu Q., Zhou D., *et al*. 'Deep-eutectic-solvent-based self-healing polymer electrolyte for safe and long-life lithium-metal batteries'. *Angewandte Chemie*. 2020;**59**(23):9134–42.

[377] Jena R. 'An empirical case study on Indian consumers' sentiment towards electric vehicles: a big data analytics approach'. *Industrial Marketing Management*. 2020;**90**:605–16.

[378] Jenkins B., Krupadanam A., Annaswamy A.M. 'Fast adaptive observers for battery management systems'. *IEEE Transactions on Control Systems Technology*. 2020;**28**(3):776–89.

[379] Jeon J.-W., Biswas M.C., Patton C.L., Wujcik E.K. 'Water-processable, sprayable $LiFePO_4$/graphene hybrid cathodes for high-power lithium ion batteries'. *Journal of Industrial and Engineering Chemistry*. 2020;**84**:72–81.

[380] Ji H., Zhang W., Pan X., *et al*. 'State of health prediction model based on internal resistance'. *International Journal of Energy Research*. 2020;**44**(8):6502–10.

[381] Jianfang J., Keke W., Xiaoqiong P., Yuanhao S., Jie W., Jianchao Z. 'Multi-Scale prediction of RUL and SOH for lithium-ion batteries based on WNN-UPF combined model'. *Chinese Journal of Electronics*. 2021;**30**(1):26–35.

[382] Jia L., Tsui C.-Y., Au O.C., Jia K. 'A low-power motion estimation architecture for HEVC based on A new sum of absolute difference computation'. *IEEE Transactions on Circuits and Systems for Video Technology*. 2020;**30**(1):243–55.

[383] Jia L.L. 'Interface reconstruction study by functional scanning probe microscope in lithium-ion battery research'. *Chinese Journal of Structural Chemistry*. 2020;**39**(2):200–05.

[384] Jia N., Liu J., Liu Y., *et al.* 'In situ conversion of iron sulfide (FeS) to iron oxyhydroxide (γ-feooh) on N, S co-doped porous carbon nanosheets: an efficient electrocatalyst for the oxygen reduction reaction and zinc-air batteries'. *J Colloid Interface Sci.* 2020;**558**:323–33.

[385] Jia Y., Dong Z.Y., Sun C., Chen G. 'Distributed economic model predictive control for a wind–photovoltaic–battery microgrid power system'. *IEEE Transactions on Sustainable Energy*. 2020;**11**(2):1089–99.

[386] Jiang C., Wang S., Wu B., Fernandez C., Xiong X., Coffie-Ken J. 'A state-of-charge estimation method of the power lithium-ion battery in complex conditions based on adaptive square root extended Kalman filter'. *Energy*. 2021;**219**:119603.

[387] Jiang H., Chen B., Li H., *et al.* 'State-of-charge estimation of lithium-ion batteries from a gas-liquid dynamics model including the direct temperature input'. *Journal of Energy Storage*. 2021;**39**:102622.

[388] Jiang J.B., Li D.G., Dou R.Z .. 'A lattice Boltzmann modeling and analysis of the thermal convection in a lithium-ion battery'. *Computers & Mathematics with Applications*. 2019;**77**(10):2695–706.

[389] Jiang L., Deng Z., Tang X., Hu L., Lin X., Hu X. 'Data-driven fault diagnosis and thermal runaway warning for battery packs using real-world vehicle data'. *Energy*. 2021;**234**:121266.

[390] Jin B. 'Lithium battery soc estimation based on internal resistance power consumption'. *Automotive Engineering*. 2020;**42**(8):1008–15.

[391] Joshi U., Kumar R. 'A novel deep neural networks based path prediction'. *Cluster Computing*. 2020;**23**(4):2915–24.

[392] Juston M., Damay N., Forgez C., *et al.* 'Direct determination of a single battery internal resistances distribution using a heterogeneous model'. *Mathematics and Computers in Simulation*. 2021;**183**:20–33.

[393] Kaabinejadian A., Ahmadi H.A., Moghimi M. 'Investigation of porous media effects on lithium-ion battery thermal management'. *Journal of Thermal Analysis and Calorimetry*. 2020;**141**(5):1619–33.

[394] Kamath D., Shukla S., Arsenault R., Kim H.C., Anctil A. 'Evaluating the cost and carbon footprint of second-life electric vehicle batteries in residential and utility-level applications'. *Waste Management*. 2020;**113**:497–507.

[395] Kara A. 'A data-driven approach based on deep neural networks for lithium-ion battery prognostics'. *Neural Computing and Applications*. 2021;**33**(20):13525–38.

[396] Kasai N., Kato M., Ishida N., Kitamura N., Idemoto Y. 'Safety test methods simulating internal short circuit and the mechanism for safety improvement of li-ion batteries by heat resistant separators'. *Electrochemistry*. 2020;**88**(6):555–59.

[397] Kato M., Sano H., Kiyobayashi T., Yao M. 'Viologen derivatives extended with aromatic rings acting as negative electrode materials for use in rechargeable molecular ion batteries'. *ChemSusChem*. 2020;**13**(9):2379–85.

[398] Ke C., Fan D., Chen C., Li X., Jiang M., Hu X. 'Two-dimensional tetragonal transition-metal carbide anodes for non-lithium-ion batteries'. *Physical Chemistry Chemical Physics*. 2020;**22**(24):13680–88.

[399] Kelly J.C., Dai Q., Wang M. 'Globally regional life cycle analysis of automotive lithium-ion nickel manganese cobalt batteries'. *Mitigation and Adaptation Strategies for Global Change*. 2020;**25**(3):371–96.

[400] Khaleghi Rahimian S., Forouzan M.M., Han S., Tang Y. 'A generalized physics-based calendar life model for li-ion cells'. *Electrochimica Acta*. 2020;**348**(136343):136343.

[401] Khalfi J., Boumaaz N., Soulmani A., Laadissi E.M. 'Nonlinear modeling of lithium-ion battery cells for electric vehicles using a hammerstein–wiener model'. *Journal of Electrical Engineering & Technology*. 2021;**16**(2):659–69.

[402] Khalid A., Sarwat A.I. 'Unified univariate-neural network models for lithium-ion battery state-of-charge forecasting using minimized akaike information criterion algorithm'. *IEEE Access: Practical Innovations, Open Solutions*. 2021;**9**:39154–70.

[403] Khan H.F., Hanif A., Ali M.U., Zafar A. 'A lagrange multiplier and sigma point Kalman filter based fused methodology for online state of charge estimation of lithium-ion batteries'. *Journal of Energy Storage*. 2021;**41**(24): 102843.

[404] Khan N., Ullah F.U.M., Afnan, Ullah A., Lee M.Y., Baik S.W. 'Batteries state of health estimation via efficient neural networks with multiple channel charging profiles'. *IEEE Access: Practical Innovations, Open Solutions*. 2021;**9**:7797–813.

[405] Khayamy M., Nasiri A., Okoye O. 'Development of an equivalent circuit for batteries based on a distributed impedance network'. *IEEE Transactions on Vehicular Technology*. 2020;**69**(6):6119–28.

[406] Kim G.Y., Park S.R., Yu J.S. 'Design and characteristics of low-resistance lithium-ion battery pack and its fast charging method for smart phones'. *International Journal of Energy Research*. 2021;**45**(12):17631–46.

[407] Kim J.K. 'Faraway small drone detection based on deep learning'. *International Journal of Computer Science and Network Security*. 2020;**20**(1):149–54.

[408] Kim J.Y., Kim J., Kang S.H., *et al*. 'Efficient cell design and fabrication of concentration-gradient composite electrodes for high-power and high-energy-density all-solid-state batteries'. *ETRI Journal*. 2020;**42**(1):129–37.

[409] Kim M.-S., Lee B.-H., Park J.-H., *et al*. 'Operando identification of the chemical and structural origin of li-ion battery aging at near-ambient temperature'. *Journal of the American Chemical Society*. 2020;**142**(31):13406–14.

[410] Kim S., Choi Y.Y., Kim K.J., Choi J.-I. 'Forecasting state-of-health of lithium-ion batteries using variational long short-term memory with transfer learning'. *Journal of Energy Storage*. 2021;**41**(6058):102893.

[411] Kim T., Adhikaree A., Pandey R., *et al.* 'An on-board model-based condition monitoring for lithium-ion batteries'. *IEEE Transactions on Industry Applications*. 2019;**55**(2):1835–43.

[412] Kim W., Lee P.-Y., Kim J., Kim K.-S. 'A robust state of charge estimation approach based on nonlinear battery cell model for lithium-ion batteries in electric vehicles'. *IEEE Transactions on Vehicular Technology*. 2021;**70**(6):5638–47.

[413] Kim Y., Lin X., Abbasalinejad A., Kim S.U., Chung S.H. 'On state estimation of all solid-state batteries'. *Electrochimica Acta*. 2019;**317**(3):663–72.

[414] Ko M., Chae S., Ma J, *et al.* 'Author correction: scalable synthesis of silicon-nanolayer-embedded graphite for high-energy lithium-ion batteries'. *Nature Energy*. 2020;**5**(4):344.

[415] Kong D., Wang S., Ping P. 'State-of-health estimation and remaining useful life for lithium-ion battery based on deep learning with Bayesian hyperparameter optimization'. *International Journal of Energy Research*. 2022;**46**(5):6081–98.

[416] Kong H., Yan J., Wang H., Fan L. 'Energy management strategy for electric vehicles based on deep Q-learning using Bayesian optimization'. *Neural Computing and Applications*. 2020;**32**(18):14431–45.

[417] Kong L., Hu X., Gui G., Su Y., Pecht M. 'Computed tomography analysis of li-ion battery case ruptures'. *Fire Technology*. 2020;**56**(6):2565–78.

[418] Koseoglou M., Tsioumas E., Jabbour N., Mademlis C. 'Highly effective cell equalization in a lithium-ion battery management system'. *IEEE Transactions on Power Electronics*. 2020;**35**(2):2088–99.

[419] Krauskopf T., Richter F.H., Zeier W.G., Janek J. 'Physicochemical concepts of the lithium metal anode in solid-state batteries'. *Chemical Reviews*. 2020;**120**(15):7745–94.

[420] Kuipers M., Schröer P., Nemeth T., Zappen H., Blömeke A., Sauer D.U. 'An algorithm for an online electrochemical impedance spectroscopy and battery parameter estimation: development, verification and validation'. *Journal of Energy Storage*. 2020;**30**:101517.

[421] L. Kulova T. 'A brief review of post-lithium-ion batteries'. *International Journal of Electrochemical Science*. 2020;**15**(8):7242–59.

[422] Satheesh Kumar S., Immanuel Selvakumar A., Kumar S.S., Selvakumar A.I. 'Maximum power point tracking and power flow management of hybrid renewable energy system with partial shading capability: a hybrid technique'. *Transactions of the Institute of Measurement and Control*. 2020;**42**(12):2276–96.

[423] Kwabi D.G., Ji Y., Aziz M.J. 'Electrolyte lifetime in aqueous organic redox flow batteries: a critical review'. *Chemical Reviews*. 2020;**120**(14):6467–89.

[424] Kwak B., Kim M., Kim J. 'Inrush current reduction technology of DAB converter for low-voltage battery systems and DC bus connections in DC microgrids'. *IET Power Electronics*. 2020;**13**(8):1528–36.

[425] Kwak E., Jeong S., Kim J.H., Oh K.Y. 'Prediction of compression force evolution over degradation for a lithium-ion battery'. *Journal of Power Sources*. 2021;**483**(229079):229079.

[426] Laforgue A., Yuan X.-Z., Platt A, *et al*. 'Effects of fast charging at low temperature on a high energy Li-ion battery'. *Journal of the Electrochemical Society*. 2020;**167**(14):140521.

[427] Lai J., Xing Y., Chen N., Li L., Wu F., Chen R. 'Electrolytes for rechargeable lithium-air batteries'. *Angewandte Chemie (International Ed. in English)*. 2020;**59**(8):2974–97.

[428] Lai X., He L., Wang S., *et al*. 'Co-estimation of state of charge and state of power for lithium-ion batteries based on fractional variable-order model'. *Journal of Cleaner Production*. 2020;**255**(3):120203.

[429] Lai X., Yi W., Cui Y., *et al*. 'Capacity estimation of lithium-ion cells by combining model-based and data-driven methods based on a sequential extended Kalman filter'. *Energy*. 2021;**216**(2):119233.

[430] Lan T., Jermsittiparsert K., T. Alrashood S., Rezaei M., Al-Ghussain L., A. Mohamed M. 'An advanced machine learning based energy management of renewable microgrids considering hybrid electric vehicles' charging demand'. *Energies*. 2021;**14**(3):569.

Index